Capybara

José Roberto Moreira • Katia Maria P.M.B. Ferraz
Emilio A. Herrera • David W. Macdonald
Editors

Capybara

Biology, Use and Conservation of an Exceptional Neotropical Species

Springer

Editors
José Roberto Moreira
Embrapa Recursos Genéticos e Biotecnologia
Brasília, Brazil

Emilio A. Herrera
Departamento de Estudios Ambientales
Universidad Simón Bolívar
Caracas, Venezuela

Katia Maria P.M.B. Ferraz
Departamento de Ciências Florestais
Escola Superior de Agricultura
"Luiz de Queiroz"
Universidade de São Paulo
São Paulo, Brazil

David W. Macdonald
Wildlife Conservation Research Unit
Zoology Department, University of Oxford
The Recanati-Kaplan Centre
Oxford, United Kingdom

ISBN 978-1-4614-3999-8 ISBN 978-1-4614-4000-0 (eBook)
DOI 10.1007/978-1-4614-4000-0
Springer NewYork Heidelberg Dordrecht London

Library of Congress Control Number: 2012941611

Printed on acid-free paper

Springer is part of Springer Science+Business Media (www.springer.com)

Foreword

Large neotropical herbivore mammals are characterized by the wide variety of hystricognath rodents among them. These include pacas, agoutis, vizcachas, coypus, maras, and pacaranas, which in this region carry out ecological functions similar to those of ungulates on other continents. Among them, the capybara stands out because of its large size, gregarious and semi-aquatic habits, wide distribution in South America, and local abundance in flooded savannas and wetlands and on the margins of bodies of water. It is the largest living rodent in the world and the last remnant of a long line of gigantic grass-eating rodents that evolved in South America over millions of years.

The capybara attracted the attention of explorers and writers in the New World from the sixteenth century onward. They commented on its large size and gregarious habits, the use of its meat by local communities, and its incursions into crops near water, often linking it with pigs, as its Linnean name of 1766, *Sus hydrochaeris,* indicates. The pioneering naturalists in South America, such as Georg Marcgraf (1610–1644), Felix de Azara (1746–1811), and Johann Rengger (1795–1832) provided more detailed descriptions of the animal and its natural history. Over a century ago, meticulous studies of the paleontology of the Hydrochoeridae family began, especially in Argentina. There have also been parasitological studies, management experiments in zoos, and records kept of capybara hide exports and of the trade in its meat.

It is therefore surprising that ecological studies of the capybara, such a peculiar and widely distributed rodent, so important as a resource for many communities, and also commanding high market prices, should have only started about 40 years ago, and mainly in Venezuela and Colombia. Next came many important projects in Brazil and Argentina, covering the use of its habitats, diet, reproduction, behavior and social structure, estimates of abundance, population dynamics and productivity, as well as techniques for keeping capybaras in captivity. There have also been experiments in management and sustainable harvest of natural populations, and legal instruments to control these areas. As a result, the capybara is today one of the most studied and best known native mammals of South America.

However, a large number of the most original investigations were often unpublished theses or university monographs, or institutional projects and databases that are difficult to access. In consequence, the results of the studies are dispersed among numerous national and international journals and in the minutes of conferences and are often beyond the reach of the people and institutions interested in them. Almost all of the publications on this rodent that exist in book form are manuals for rearing it in captivity.

For this reason, we consider that it is extremely useful and appropriate to publish this work, which provides a global overview of up-to-date knowledge on the capybara's biology, ecology, and management, in 24 chapters by notable specialists on each topic. This book, arising from a coordinated effort by a multidisciplinary and preeminently South American team, thus brings together the recent results of many studies and offers an excellent starting point for better plans that cover conservation, sustainable use, production, and successful commerce of this valuable giant rodent.

Juhani Ojasti

Preface

For the aboriginal peoples of South America, wildlife was an important source of protein and clothing. Throughout the period of colonization by Europeans, and during the growth of nations, wildlife has always contributed to the wealth of the continent. Early settlers described wild animals as plentiful, but since the 1970s the populations of many have been rapidly depleted. Historically, the importance of this once vital indigenous resource was neglected. Furthermore, until the mid-twentieth century very little was known by biologists about South America's fauna. However, the last three decades of that century, and the first of the twenty-first century, have seen knowledge about neotropical wildlife grow rapidly.

The implications, scientific, political, economic, and even ethical, of using wildlife present topical, intricate, and challenging questions to wildlife specialists and wider society. In this context, the capybara has interest beyond its own specific case because, of neotropical wild mammals, it has the greatest potential for production. Furthermore, as the following chapters reveal, the consumptive use of wild capybaras raises a different set of issues to those associated with farming this species. We present the capybara, therefore, not merely as rivetingly interesting as a species in its own right, but as an exceptional model to inform thinking about wider, interdisciplinary wildlife issues. Among the features that give the capybara this special potential for use are its high prolificacy, rapid growth rate, herbivorous diet, resistance to disease, social behavior, relative tameness, and susceptibility to captive rearing.

Historically, the species has been eaten all over its range, especially by poor, rural, and traditional communities engaged in subsistence hunting. More recently, in large urban settlements, some city dwellers consider it a delicacy. The scene is set, therefore, with several South American countries committed to the sustainable use of capybaras, while others have encouraged capybara rearing in captivity. Each offers different insights and opportunities, and each poses its particular set of challenges and problems, and our purpose in creating this book is to identify, disentangle, and learn from these.

Since the mid-twentieth century, governments and institutions in various countries have fostered studies of the biology, use, and conservation of the capybara, accumulating copious technical information. These data have been scattered across different research centers and universities and, sometimes, recorded in variously inaccessible publications. It was more than time to gather, evaluate, synthesize, and make available all of the diverse knowledge about this species. As the chapters that follow will reveal, there are many and diverse topics relevant to understanding the capybara's biology, and weaving that understanding through considerations of their conservation and management. However, at the foundation of this invigorating interdisciplinarity lie fundamental questions about the species' habitat use, social system, and behavioral ecology.

This book also represents a personal journey for us, the editors, because one of the first steps in answering these fundamental questions was taken by one of us (DWM) in the early 1980s with a field study of capybaras in Venezuela. This research set a trajectory joined by others (EAH and JRM, also editors) who were doctoral students at Oxford in Macdonald's WildCRU – in turn, we have fostered a second generation of students, several of whom are authors of chapters in this book. This tradition, over three generations of researchers, inspired us to compile this book and to assemble the foremost students of capybara biology to synthesize the vast amount of knowledge now presented herein. We hope that this compendious account will inspire future neotropical mammalogists and conservationists. The insights are relevant to a spectrum of species, whether endangered or used.

The book is organized into three parts, each addressing a particular area in capybara biology, use, and conservation. The first part covers the evolution, biology, and ecology of capybaras. In this part, the opening chapter describes the species and its life history, scientific nomenclature, and distribution, providing a general overview of capybara biology and the basic framework for the rest of the book. The next two chapters cover the paleontological and genetic perspectives of capybara evolution, with leading researchers in each sphere presenting their different opinions on capybara systematics. Two chapters then describe capybara feeding habits and digestive physiology, and three cover capybara reproduction. Chapter 9 describes capybara infectious diseases and offers recommendations for their treatment and control. The next two chapters present a comprehensive overview of social behavior and chemical communication in capybaras, providing a foundation for understanding the species' management in the wild and rearing in captivity.

The chapters in the second part include broad and up-to-date reviews on capybara production. The first summarizes the diverse uses of the species throughout its range, and these are elaborated in four chapters devoted to different aspects of captive rearing, such as management, behavior, feeding, and reproduction, providing compendious information from practitioners on how to rear capybaras safely, healthily, and economically. Information on the capybara's behavior in captivity is deployed to inform the highest standards of husbandry. The part concludes with a simulation of the effects of various factors on the sustainability of harvesting capybaras, illustrating the role of scientific evidence in informing management decisions.

The third and last part deals with current conservation issues, describing the invaluable experiences of capybara use and conservation in Venezuela, Colombia, Brazil, and Argentina. These country-based chapters describe the history of capybara exploitation in each, presenting the different legislation, forms of management, human-capybara conflicts, and conservation status.

Across its vast range, some capybara populations appear to be used sustainably, while others are perceived as pests and judged by some to need control. In both situations, reliable estimates of population densities are necessary, and methodologies for estimating these are explained in Chap. 22. Addressing an emerging topical issue, Chap. 23 describes the capybara's role in the epidemiology of Brazilian Spotted Fever.

In assembling this material, we have attempted to take an even-handed and impartial editorial role; we hope our contributors present the information necessary to inform critical evaluation of science and policy. Personally, we as editors do not advocate here any particular position on trade in capybaras, but we do advocate exposing the issues and facilitating discussion of them. With this in mind, we present an overview in the concluding chapter so that the reader can assess the conditions for a successful capybara management program under varying circumstances and local legal systems and cultures. Since each chapter is self-contained, they can be read in any order, and readers can concentrate on the topic of their interest.

The capybara is the largest living rodent. Its exceptional features (see Chap. 24) have made it an emblematic component of a range of ecosystems in much of South America. It is arguably the most important native mammalian herbivore in the ecology of the wetlands and savannas of this subcontinent. The adaptations that make the capybara a fascinating member of neotropical faunas, and an exceptional model for the study of behavioral ecology and social systems, also make it a candidate for sustainable harvest over much of its range, since its meat and hides are highly valuable. With this book, we hope to make biologists, conservationists, wildlife managers, and policymakers more conscious of the great value and ecological importance of this species and of the diverse neotropical fauna as a whole. We also aim to convince the reader of the beauty and value of wildlife, in general, and capybaras, in particular, to contribute to their sustainable management and conservation.

Brasília, Distrito Federal, Brazil — José Roberto Moreira
Piracicaba, São Paulo, Brazil — Katia Maria P.M.B. Ferraz
Caracas, Venezuela — Emilio A. Herrera
Oxford, UK — David W. Macdonald

Contents

Part I Biology, Ecology and Evolution

1 **Taxonomy, Natural History and Distribution of the Capybara** 3
José Roberto Moreira, Martin R. Alvarez, Teresa Tarifa, Víctor Pacheco, Andrew Taber, Diego G. Tirira, Emilio A. Herrera, Katia Maria P.M.B. Ferraz, Juanita Aldana-Domínguez, and David W. Macdonald

2 **Paleontology, Evolution and Systematics of Capybara** 39
María Guiomar Vucetich, Cecilia M. Deschamps, and María E. Pérez

3 **Phylogenetics of Caviomorph Rodents and Genetic Perspectives on the Evolution of Sociality and Mating Systems in the Caviidae** 61
Rodney L. Honeycutt

4 **Foraging Strategies and Feeding Habits of Capybaras** 83
Guillermo R. Barreto and Rúben D. Quintana

5 **Capybara Digestive Adaptations** 97
Emilio A. Herrera

6 **Reproductive Morphology and Physiology of the Male Capybaras** 107
Tarcízio A.R. Paula and Nicholas J. Walker

7 **Morphology and Reproductive Physiology of Female Capybaras** 131
Maria Angélica Miglino, Tatiana Carlesso dos Santos, Claudia Kanashiro, and Rosa Helena dos Santos Ferraz

8 Capybara Demographic Traits 147
José Roberto Moreira, Helga Wiederhecker, Katia Maria P.M.B. Ferraz, Juanita Aldana-Domínguez, Luciano M. Verdade, and David W. Macdonald

9 Diseases of Capybara 169
Gerardo Rubén Cueto

10 Capybara Scent Glands and Scent-Marking Behavior 185
David W. Macdonald and Emilio A. Herrera

11 Capybara Social Behavior and Use of Space: Patterns and Processes 195
Emilio A. Herrera

Part II Production

12 Products and Uses of Capybaras 211
Max S. Pinheiro and José Roberto Moreira

13 Confined and Semi-confined Production Systems for Capybaras 229
Sérgio L.G. Nogueira-Filho, Max S. Pinheiro, Selene S.C. Nogueira

14 Social and Spatial Relationships of Capybaras in a Semi-confined Production System 243
Katia Maria P.M.B. Ferraz, Patrícia Izar, Takechi Sato, and Sílvia M. Nishida

15 Feeds and Nutrition of Farmed Capybaras 261
Alcester Mendes and Sérgio L.G. Nogueira-Filho

16 The Impact of Management Practices on Female Capybara Reproductive Parameters in Captivity 275
Sérgio L.G. Nogueira-Filho and Selene S.C. Nogueira

17 The Sustainable Management of Capybaras 283
José Roberto Moreira, Luciano M. Verdade, Katia Maria P.M.B. Ferraz, and David W. Macdonald

Part III Conservation

18 Capybaras as a Source of Protein: Utilization and Management in Venezuela 305
Emilio A. Herrera and Guillermo R. Barreto

19 Conservation and Use of the Capybara and the Lesser Capybara in Colombia 321
Juanita Aldana-Domínguez, Maria Isabel Vieira-Muñoz, and Patricia Bejarano

20 **Capybara Production in Brazil: Captive Breeding or Sustainable Management?** 333
José Roberto Moreira and Max S. Pinheiro

21 **Use of Capybaras in Argentina** 345
Rubén D. Quintana and María Luisa Bolkovic

22 **Counting Capybaras** 357
Luciano M. Verdade, José Roberto Moreira, and Katia Maria P.M.B. Ferraz

23 **Brazilian Spotted Fever: The Role of Capybaras** 371
Marcelo B. Labruna

24 **The Capybara Paradigm: From Sociality to Sustainability** 385
David W. Macdonald, Emilio A. Herrera, Katia Maria P.M.B. Ferraz, and José Roberto Moreira

Appendix: List of Reviewers 409

Index 413

Contributors

Juanita Aldana-Domínguez Independent consultant on biodiversity and landscape, Puerto Colombia, Atlantico, Colombia

Martin R. Alvarez Departamento de Ciências Biológicas, Universidade Estadual de Santa Cruz, Ilhéus, Bahia, Brazil

Guillermo R. Barreto Departamento de Biología de Organismos, Universidad Simón Bolívar, Caracas, Venezuela

Patricia Bejarano Conservación Internacional Colombia, Bogotá, DC, Colombia

María Luisa Bolkovic Dirección de Fauna Silvestre, Secretaría de Ambiente y Desarrollo Sustentable de la Nación, Buenos Aires, Argentina

Gerardo Rubén Cueto Instituto de Ecología Genética y Evolución de Buenos Aires (IEGEBA), UBA-CONICET, Facultad de Ciencias Exactas y Naturales, Universidad de Buenos Aires, Ciudad Universitaria, Buenos Aires, Argentina

Cecilia M. Deschamps Departamento Paleontología Vertebrados, Museo de La Plata, CIC, Comisión de Investigaciones Científicas de la Provincia de Buenos Aires, La Plata, Argentina

Katia Maria P.M.B. Ferraz Departamento de Ciências Florestais, Escola Superior de Agricultura "Luiz de Queiroz", Universidade de São Paulo, Piracicaba, São Paulo, Brazil

Rosa Helena dos Santos Ferraz Universidade Federal de Mato Grosso, Cuiabá, Brazil

Emilio A. Herrera Departamento de Estudios Ambientales, Universidad Simón Bolívar, Caracas, Venezuela

Rodney L. Honeycutt Natural Science Division, Pepperdine University, Malibu, CA, USA

Patrícia Izar Departamento de Psicologia Experimental, Instituto de Psicologia, Universidade de São Paulo, São Paulo, Brazil

Claudia Kanashiro Universidade de São Paulo, São Paulo, Brazil

Marcelo B. Labruna Departamento de Medicina Veterinária Preventiva e Saúde Animal, Faculdade de Medicina Veterinária e Zootecnia, Universidade de São Paulo, São Paulo, Brazil

David W. Macdonald Wildlife Conservation Research Unit Zoology Department, University of Oxford, The Recanati-Kaplan Centre Oxford, Tubney, Abingdon, United Kingdom

Alcester Mendes Departamento de Ciências Agrárias e Ambientais, Universidade Estadual de Santa Cruz, Ilhéus, Bahia, Brazil

Maria Angélica Miglino Universidade de São Paulo, São Paulo, Brazil

José Roberto Moreira Embrapa Recursos Genéticos e Biotecnologia, Parque Estação Biológica, Brasília, Distrito Federal, Brazil

Silvia M. Nishida Departamento de Fisiologia, Instituto de Biociências, Universidade Estadual Paulista "Júlio de Mesquita Filho", Botucatu, São Paulo, Brazil

Selene S.C. Nogueira Departamento de Ciências Biológicas, Laboratório de Etologia Aplicada, Universidade Estadual de Santa Cruz, Ilhéus, Bahia, Brazil

Sérgio L.G. Nogueira-Filho Departamento de Ciências Agrárias e Ambientais, Universidade Estadual de Santa Cruz, Ilhéus, Bahia, Brazil

Víctor Pacheco Museo de Historia Natural, Universidad Nacional Mayor de San Marcos, Peru

Tarcízio A.R. Paula Departamento de Veterinária, Universidade Federal de Viçosa, Viçosa, Brazil

María E. Pérez Museo Paleontológico Egidio Feruglio, CONICET, Consejo Nacional de Investigaciones Científicas y Técnicas, Trelew, Argentina

Max S. Pinheiro Embrapa Clima Temperado, Pelotas, Rio Grande do Sul, Brazil

Takechi Sato Departamento de Psicologia Experimental, Instituto de Psicologia, Universidade de São Paulo, São Paulo, Brazil

Tatiana Carlesso dos Santos Departamento de Zootecnia, Universidade Estadual de Maringá, Paraná, Brazil

Rúben D. Quintana Instituto de Investigación e Ingeniería Ambiental (3iA), Universidad Nacional de San Martín, San Martín, Pcia. de Buenos Aires, Argentina

Andrew Taber Arlington, VA, USA

Teresa Tarifa Coleccion Boliviana de Fauna, La Paz, Bolivia
Caldwell, ID, USA

Diego G. Tirira Fundación Mamíferos y Conservación, Quito, Ecuador

Luciano M. Verdade Laboratório de Ecologia Isotópica, CENA, Universidade de São Paulo, Piracicaba, São Paulo, Brazil

María Isabel Vieira-Muñoz Instituto de Investigación de Recursos Biologicos Alexander von Humboldt, Bogotá, DC, Colombia

María Guiomar Vucetich Departamento Paleontología Vertebrados, Museo de La Plata, CONICET, Consejo Nacional de Investigaciones Científicas y Técnicas, La Plata, Argentina

Nicholas J. Walker Laboratório de Sistemática Molecular, Departamento de Biologia Animal, Universidade Federal de Viçosa, Viçosa, Brazil

Helga Wiederhecker SHIS QI 23 CH 3, Brasília, Distrito Federal, Brazil

Part I
Biology, Ecology and Evolution

Chapter 1
Taxonomy, Natural History and Distribution of the Capybara

José Roberto Moreira, Martin R. Alvarez, Teresa Tarifa, Víctor Pacheco, Andrew Taber, Diego G. Tirira, Emilio A. Herrera, Katia Maria P.M.B. Ferraz, Juanita Aldana-Domínguez, and David W. Macdonald

1.1 Introduction

When the Iberian colonists arrived in South America in the late fifteenth century, they encountered a diverse and previously unimagined fauna. The unusual anatomy and behavior of these species intrigued the early explorers. In their reports they named the new-found endemic animals after the most analogous European species.

J.R. Moreira (✉)
Embrapa Recursos Genéticos e Biotecnologia, Parque Estação Biológica, Final W5 Norte, Brasília, Distrito Federal 70770-900, Brazil
e-mail: alencar_moreira@yahoo.com.br

M.R. Alvarez
Departamento de Ciências Biológicas, Universidade Estadual de Santa Cruz, Rodovia Ilhéus Itabuna km 16, Ilhéus, Bahia 45662-900, Brazil
e-mail: malva@uesc.br

T. Tarifa
Coleccion Boliviana de Fauna, La Paz, Bolivia
3407 Fairoaks Circle, Caldwell, ID 83605, USA
e-mail: teresa_tarifa@yahoo.com

V. Pacheco
Museo de Historia Natural, Universidad Nacional Mayor de San Marcos, Av. Arenales 1256, Lima 14, Peru
e-mail: vpachecot@unmsm.edu.pe

A. Taber
2319N. Glebe Road, Arlington, VA 22207, USA
e-mail: chacotaber@hotmail.com

D.G. Tirira
Fundación Mamíferos y Conservación, Dionisio Bras N35-290 y Manosca, Quito, Ecuador
e-mail: diego_tirira@yahoo.com

J.R. Moreira et al. (eds.), *Capybara: Biology, Use and Conservation of an Exceptional Neotropical Species*, DOI 10.1007/978-1-4614-4000-0_1,

In 1576, for example, Pero de Gândavo (2004) described the capybara (*Hydrochoerus hydrochaeris*) as "a type of pig." However, capybaras were sufficiently unlike any known European species for most explorers to simply adopt a phonetic representation of the local name. Therefore, in 1557, the capybara was called *catiuare* by the German Hans Staden (1557), *capiyûára* in 1560 by the Spaniard José de Anchieta (1997), and *capijuara* in 1625 by the Portuguese Fernão Cardim (1980). The name capybara actually originates from a word in the indigenous Tupi, which in the sixteenth century was the most widely spread language in South America: *kapii'gwara* meaning grass eater (*ka'pii* = "grass" + *gwara* = "eater"; Houaiss et al. 2004).

In the narratives of these early explorers, the capybara was described in terms of its exotic appearance, unusual habits, and usefulness. Staden (1557:174) noted:

> There is an animal named Catíuare; abides on land and in the water. The reeds that grow by the banks of the fresh water, that they eat. When anything alarms them they flee into the water to the bottom. Are larger than a sheep, have a head in the manner of a hare, but larger, and short ears; have a stumpy tail, fairly long legs and run fast on land from one body of water to another. Its hair is dark grey; has three lumps on each foot; tastes like pork.

Some explorers described capybaras from the narrative of others. These second-hand descriptions, along with the then common view of nature in Christian cultures as being above all the fruit of the Creator's fertile and eternal power, led to a number of imaginative classifications. The Anglo-Irish Oliver Goldsmith (1870), in 1774, listed the species as being among the "quadrupeds of the hog kind" (Fig. 1.1). Goldsmith (1870:351) wrote that "It seizes the fish, upon which it preys, with its hoofs and feet, and carries them to the edge of the lake to devour them at its ease." This is almost certainly a confusion with the giant otter found in the same general habitat.

The first detailed description of the capybara by a naturalist came in the mid-seventeenth century from observations in Pernambuco State, Brazil. At that time, northeastern Brazil was occupied by the Dutch, and among their scientific

E.A. Herrera
Departamento de Estudios Ambientales, Universidad Simón Bolívar, Apdo. 89000, Caracas, 1080-A, Venezuela
e-mail: eherre@usb.ve

K.M.P.M.B. Ferraz
Departamento de Ciências Florestais, Escola Superior de Agricultura "Luiz de Queiroz", Universidade de São Paulo, Av. Pádua Dias 11, Piracicaba, São Paulo 13418-900, Brazil
e-mail: katia.ferraz@usp.br

J. Aldana-Domínguez
Independent consultant on biodiversity and landscape, Calle 16, No, 1 F-12, Puerto Colombia, Atlantico, Colombia
e-mail: juanitaldana@gmail.com

D.W. Macdonald
Wildlife Conservation Research Unit Zoology Department, University of Oxford
The Recanati-Kaplan Centre Oxford, Tubney House, Abingdon Road, Tubney, Abingdon OX13 5QL, United Kingdom
e-mail: david.macdonald@zoo.ox.ac.uk

Wildlife Conservation Research Unit
Zoology Department, University of Oxford
The Recanati-Kaplan Centre Oxford,

Fig. 1.1 Capybara classified by Oliver Goldsmith, in 1774, as one of the "quadrupeds of the hog kind" (From Goldsmith (1870))

commission was the German naturalist Georg Marcgrave (Vanzolini 1996). Marcgrave, writing in 1648, predated Linneaus and his work would probably have remained unrecognized if it were not for the inclusion of 39 of his species descriptions by Linnaeus, in the twelfth edition of his *Sistema Naturae* in 1766, among these the capybara.

Marcgrave's (1648) description of the capybara in the *Historia Naturalis Brasiliae* is the definitive reference used by all subsequent naturalists up to Linnaeus (Jonston 1650; Piso 1658; Ray 1693; Barrère 1741; Brisson 1756, 1762; Linnaeus 1758, 1766; Buffon 1764). Thus, it seems strange that Linnaeus (1766) defined the type locality of the capybara as Suriname. Since both Brisson (1762) and Buffon (1764) cite works from French Guiana and indicate the habitat as "Guyana and Brazil," it is possible that Linnaeus confused the Guianas. Marcgrave worked in a Dutch colony in America and it is plausible to imagine that Linnaeus assumed that he was referring to Dutch Guiana –Suriname. The type locality for the capybara should more properly be considered the São Francisco River, Brazil, which is the location mentioned by Marcgrave (Mones 1975).

In this introductory chapter, we seek to characterize the subject of this book – the capybara. First we detail the taxonomic history of the capybara and clarify its scientific name – an issue of recent controversy. We then describe the anatomy, physiology, and ecology of the capybara, concluding with its distribution (which remains uncertain in many areas) to delineate the scope of the following chapters in this volume.

1.2 Taxonomic History of the Capybara

The history of the scientific nomenclature for the capybara is long and turbulent. Recently, there has been great inconsistency in the genus name adopted and in its spelling. Much of the debate is due to uncertainty over which name takes historical precedence, but some variants seem to be the result of simple spelling errors. It is important to clarify which name is now valid, the first name available, and the evidence available for the choice.

Marcgrave (1648:230) provided the first name for the capybara based on a naturalistic description: "Capy-bara Brasiliensibus: Porcus est fluviatilis." The first mention of the capybara in the binomial system (the presently accepted scientific naming system) was made by Linnaeus 1758, in describing the guinea pig, which he called *Mus porcellus* (Linnaeus 1758:59). Earlier, in 1756, and later in 1762, Brisson classified the capybara in a separate genus he named *Hydrochoerus* (Brisson 1762:80). Linnaeus, in 1766, sought to correct his earlier mistake and classified the capybara among pigs, as *Sus hydrochaeris* (Linnaeus 1766:103). Since then the capybara has received several generic names, none regularly used (Mones 1984), and the vast majority being a variation of *Hydrochoerus* Brisson, 1762, among them *Hydrochaeris* Brünnich, 1772, *Hydrochaerus* Erxleben, 1777, and *Hydrochoeris* Allen, 1916.

As the first two editions of the work of Brisson, from 1756 and 1762, entitled *Regnum Animale*, did not consistently use binomial names for species, they should not be considered for taxonomic classification (Gentry 1994). However, in 1911 the International Commission on Zoological Nomenclature (ICZN) decided to consider available (Opinion 37) some generic names of birds presented in the work of Brisson, 1760, entitled *Ornithologia*, since they were on the Official Lists and Indexes of Names and Works in Zoology. With the acceptance of some names for birds, Tate (1935 cited by Gentry 1994) suggested the possibility of using the genus names for mammals proposed by Brisson (1762) since some (among them *Hydrochoerus*) were already in use at the time. Meanwhile, Hopwood (1947) proposed the opposite. Since *Regnum Animale* of Brisson (1762) was not Linnean, Hopwood proposed that *Hydrochaeris* from the work of Brünnich (1772), entitled *Zoologiæ fundamenta*, should be the generic name of the capybara (Brünnich 1772:36), despite not being regularly used. The generic name *Hydrochaeris* Brünnich, 1772 was included on the Official List in 1954, in Opinion 236 (Gentry 1994).

This inclusion, however, had no effect on general use, and until the 1980s *Hydrochoerus* Brisson, 1762 was still the most widely used genus for the capybara, including in the now classic book by Ojasti (1973, 2011). The adoption of *Hydrochaeris* Brünnich, 1772 only became commonplace following publication by the influential reference guides *Walker's Mammals of the World* (Nowak and Paradiso 1983) and *Mammal Species of the World* (Wilson and Reeder 1993).

However, *Hydrochaeris* Brünnich, 1772 was already a *nomen oblitum,* which had not been used for two centuries (Mones 1973). In 1994, the ICZN (Case 2928) proposed that Brisson's *Regnum Animale*, 1762, should be rejected for nomenclatural purposes, but supported the retention of 11 of the genus names for mammals (Gentry 1994). Among the retained genera was that of the capybara – *Hydrochoerus* Brisson, 1762 – because of its established use in the zoological literature for over 230 years and the importance of stability in the use of names. The publication of Opinion 1894 on March 31, 1998, in the Bulletin of Zoological Nomenclature officially recognized the genus *Hydrochoerus* Brisson, 1762, for the capybara (ICZN 1998). It also recognized the species *hydrochaeris*, *Sus* Linnaeus, 1766, from the twelfth edition of *Systema Naturae*, as the specific name of the type species of *Hydrochoerus* Brisson, 1762.

The lesser capybara (*Hydrochoerus isthmius*) of eastern Panama, northwestern Venezuela, and northern and western Colombia is not on the ICZN Official List (Melville and Smith 1987; Smith 2001). Nevertheless, Mones (1984) and Mones and Ojasti (1986) recognize *H. isthmius* as a distinct species from *H. hydrochaeris* based on anatomical differences, particularly its smaller size, and a genetic study (Peceño 1983), and the species is included in the latest edition of *Mammal Species of the World* (Wilson and Reeder 2005). The first listing available for the species is Goldman, 1912, from the sixtieth volume of Smithsonian Miscellaneous Collection.

In sum, the generic and specific names to be used for the capybara and the lesser capybara are *H. hydrochaeris* and *H. isthmius*, respectively. The name that should be adopted for the capybara family is Hydrochoeridae (Vucetich et al. 2012), not Hydrochaeridae. However, it is worth noting that some authors consider that the capybara is a member of the subfamily Hydrochoerinae within the family Caviidae (Rowe and Honeycutt 2002; Honeycutt 2012).

The latest edition of *Mammal Species of the World* (Wilson and Reeder 2005) adopted Opinion 1894 and uses the generic name *Hydrochoerus* Brisson, 1762, for the capybara. However, these authors incorrectly presented the name proposed by Brisson, 1762, as *Hydrochoeris*, a misspelling that was no more than a typographic error, but which risks further muddying the waters with a third option. This source of confusion has been perpetuated as the misspelling has been adopted by, for example, Lim and Engstrom (2005) and by Lim et al. (2005a). The error was corrected in later reprints of *Mammal Species of the World*, but not in the publisher's website.

1.3 Natural History of the Capybara

The most striking characteristic of the capybara (*H. hydrochaeris*) is its status as the largest living rodent, with adults weighing 49–50 kg, (range 35–65 kg; Fig. 1.2; Mones and Ojasti 1986). There is no difference in weight between the sexes, but there are differences in body mass across its geographical distribution, with capybaras in Venezuela smaller than those of central and southeastern Brazil and Argentina and those found in northeastern Brazil being smaller still (Mones and Ojasti 1986; Emilio A. Herrera, Martin R. Alvarez and José R. Moreira, personal observations). The other species of the genus *H. isthmius* – the lesser capybara – weighs around

Fig. 1.2 Adult female of *Hydrochoerus hydrochaeris* Linnaeus, 1766 (Photo by J.R. Moreira)

28 kg (Trapido 1949b). *H. isthmius* is smaller in all external and cranial measurements than *H. hydrochaeris* and features thicker and wider frontal bones, shorter and thicker pterygoids, and a longer diastema (Trapido 1949b). The diploid number of *H. hydrochaeris* is 66 (FN = 102). The X chromosome is metacentric and large, while the Y is telocentric and small (Saez et al. 1971). The karyotype of *H. hydrochaeris* is monomorphic (Mones and Ojasti 1986). The *H. isthmius* karyotype is different, possessing 2n = 64, FN = 104 (Peceño 1983).

The capybara has a robust, oblong form, with a total length of 1.2 m and height at the shoulders of 0.6 m, short legs and a body covered with rough, dark brown or reddish fur composed of sparse, long, brittle hairs (Fig. 1.2; Mones and Ojasti 1986). The head is large with an elongated skull, high and thick zygomatic arches (Fig. 1.3), a broad rostrum, truncated snout, and cleft upper lip. The ears are small, short, and hairless, with a mobile fold that closes the ear canal. The nostrils, eyes, and ears are positioned near the top of the head, adaptations to a semiaquatic life style (Ojasti 1973).

Capybara extremities are short relative to the body volume, and the hind legs are longer than the forelegs. The forefeet have four toes and the hind feet three, all partially webbed (also an adaptation to a semi-aquatic life), with strong, thick nails similar to perissodactyl hoofs (Mones and Ojasti 1986). The tail is vestigial. Unlike other rodents, the capybara has subcutaneous sweat glands sparsely distributed throughout the body (Pereira et al. 1980). The epidermis is undulating, with numerous folds and cavities. Each hair emerges at an acute angle. The dermis contains numerous hair follicles, usually in groups of three with associated sebaceous gland and arrector pili muscle (Pereira et al. 1980).

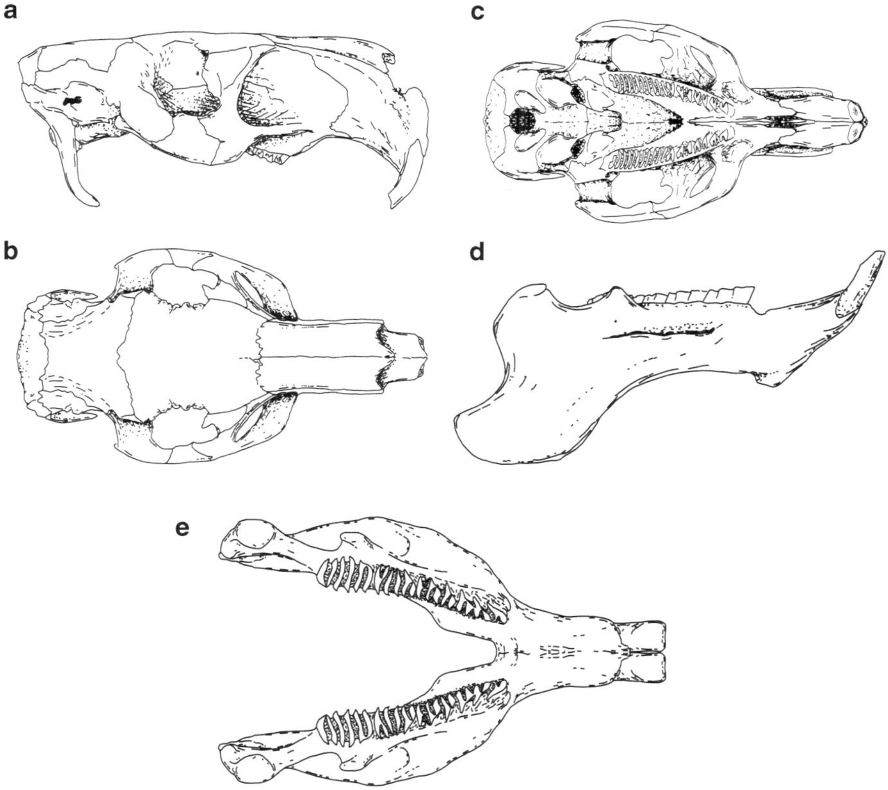

Fig. 1.3 The cranium and lower jaw of *Hydrochoerus hydrochaeris*: (**a**) right view of the cranium; (**b**) dorsal view of the cranium; (**c**) ventral view of the cranium; (**d**) right view of the lower jaw; (**e**) dorsal view of the lower jaw

The dental formula is 2 (i 1/1, c 0/0, p 1/1, m 3/3), with a total of 20 teeth and a large diastema (gap between the incisors and the first cheek tooth; Fig. 1.3). Capybaras' teeth grow continuously. They possess elasmodont molars consisting of a considerable number of transverse dentine laminae, covered with enamel and joined by intermediate layers of cement, almost as thick as the laminae themselves (Mones and Ojasti 1986). The incisors are strong and highly specialized for cutting grass. The last molar of the upper jaw is as long as the previous three molariform teeth together.

The capybara has anatomical and physiological adaptations for its herbivorous diet (Escobar and González-Jiménez 1976; Herrera 2012a) as selective grazer (González-Jiménez and Escobar 1975; Barreto and Quintana 2012). As a monogastric herbivore, it has a simple stomach with a volume of approximately 2 l when fully grown (Parra and Gonzalez-Jiménez 1972). The cecum is well developed, occupying three quarters of the entire volume of the digestive tract (Garrod 1876), where anaerobic microbial fermentation occurs (Baldizán et al. 1983). To increase the efficiency of protein utilization, the capybara engages in cecotrophic behavior (ingestion of the cecal content) during the morning, when feces are protein-rich due to the presence of microbes used in the fermentation of grass consumed during the previous afternoon and night (Herrera 1985, 2012a; Mendes and Nogueira-Filho 2012).

A nasal gland with the appearance of a dark protuberance is more developed in adult males (Macdonald et al. 1984; Macdonald and Herrera 2012), even though it can be visible in some females. Both sexes show a pair of anal glands but with different positions inside the cloaca (Ojasti 1973). These glands are well developed but different in both sexes. The glands of females are in the form of a pocket producing an abundant, pasty secretion. The glands of males, on the other hand, are open, dry and covered with hair modified to loosen easily and coated by a crystalline substance (Macdonald et al. 1984; Macdonald and Herrera 2012).

The male has no externally apparent scrotum and possesses a bacculum (penile bone; Paula and Walker 2012). The female genitalia include a bipartite uterus with a split cervix, characteristic of rodents (Ojasti 1973; Miglino et al. 2012). They have 5–6 pairs of teats (Moreira 1995).

1.3.1 Ecological and Life History Characteristics

Capybaras are semiaquatic and usually most active during the afternoon and at night (Macdonald 1981). As sweat glands are not well developed, they remain in the water or under shade to regulate their body temperature (Herrera 1986, 2012b). Capybaras also use water for mating, to escape from predators, and as a place to eat their preferred aquatic plants. A capybara group rests in the morning, spends most of the early afternoon in the water, and grazes from late afternoon until dawn (Azcárate et al. 1980; Herrera 2012b).

The capybara is a social animal that lives in family groups of 5–14 adult individuals, usually including a dominant male, one or two subordinate males, and several (probably related) females (Herrera and Macdonald 1987; Herrera 2012b). It is also a sedentary species; home ranges vary from 5 to 16 ha and usually include a large area of grassland for foraging, a permanent body of water, and an area of dry land for resting (Herrera and Macdonald 1989; Herrera 2012b). Many males are found as peripheral elements to the group. In forested habitat, they live in pairs or trios along rivers (Soini 1992).

The gestation period of a capybara is on average 150.6 days (Zara 1973; López-Barbella 1987; Miglino et al. 2012), with the females isolating themselves from the group at birth and during the first days post-partum. An average of 4.2 young are born weighing approximately 1.5 kg each (Table 1.1). In the wild on Marajó Island, Brazil, capybaras produce on average one litter per year (Moreira and Macdonald 1996; Moreira et al. 2012). Females have an estrous cycle of 7.5 days on average (López-Barbella 1982). A female is sexually mature at 12 months of age. Females

Table 1.1 Life history traits of the capybara (*Hydrochoerus hydrochaeris*) and the lesser capybara (*Hydrochoerus isthmius*)

Species	Body weight (g)	Gestation length (days)	Age at first reproduction (years)	Litter size	Weight at birth (g)	Births per year
Hydrochoerus hydrochaeris	48,900[a]	150.6[b]	2.00[c]	4.2[a]	1500.0[c]	1.0[a]
Hydrochoerus isthmius[d]	20,000	108.0	–	3.5	1100.0	–

[a]Moreira (1995)
[b]López-Barbella (1987)
[c]Ojasti (1973)
[d]Trapido (1949b)

are able to reproduce throughout the year, but in some regions, there is a distinct seasonal peak in births. For example, the breeding season of capybaras on Marajó Island occurs in December, during the early rainy season (Moreira and Macdonald 1996; Moreira et al. 2012). In Venezuela, females are receptive and a lot of sexual activity is observed 2 weeks after the onset of the rains at the end of April (Ojasti 1973; Herrera 1998). A peak in births then occurs 5 months later at the end of September and October (Ojasti 1973). The estimated fertility observed in Marajó Island was 2.59 females/female/year (Moreira and Macdonald 1996).

The evolution of capybara behavioral patterns is attributed to its role as a large grazing herbivore and prey for large carnivores (Herrera 1986). Historically, big cats such as jaguars (*Panthera onca*) and pumas (*Puma concolor*) have been the main predators of capybaras on land, and caimans (*Melanosuchus niger* and *Caiman* spp.) in water. Young are often attacked by caimans; snakes (*Boa constrictor* and *Eunectes murinus*); crab-eating foxes (*Cerdocyon thous*); small cats (*Leopardus* spp.) and some birds, like the caracara (*Polyborus plancus*); and especially black vultures (*Coragyps atratus*; Ojasti 1973; Azcárate et al. 1980; Jorgenson 1986; Yaber and Herrera 1994). Today, in areas where their natural predators are no longer found, the capybara is prey for humans and packs of feral dogs (*Canis lupus familiaris*; Macdonald 1981; Ojasti 1991).

1.4 Geographic Distribution

The genus *Hydrochoerus* has two species with distinct distributions. The lesser capybara – *H. isthmius* – is distributed to the west of the Andes, in Panama, Colombia and Venezuela (Fig. 1.4). The larger species – *H. hydrochaeris* – is found east of the Andes, from Venezuela to the mouth of the de la Plata River in Argentina (Fig. 1.4). Chile is the only country in South America that has no capybaras, and Panama is the only Central American country where they are found. Both species inhabit a wide variety of lowland habitats near ponds, lakes, rivers, streams, reservoirs, and swamps (Mones and Ojasti 1986). These habitats include gallery forests, seasonally flooded

Fig. 1.4 Distribution of the genus *Hydrochoerus*. Shaded area: *Hydrochoerus hydrochaeris*. Diagonal stripes: *Hydrochoerus isthmius*

savannas, and wetlands (Moreira and Macdonald 1996). The maximum elevation recorded for the capybara is 1,500 m in the Chapada dos Veadeiros National Park, Goiás State, Brazil (Moreira 1995).

In the following, we describe in more detail the areas throughout the range of both species of capybara where their presence has been doubtful in previous works (Ojasti 1973; Mones and Ojasti 1986). The frequency with which records of capybara occurrence are represented on maps (below) can rarely be taken as an indication of their relative abundance, but rather of the variously haphazard activities of recorders.

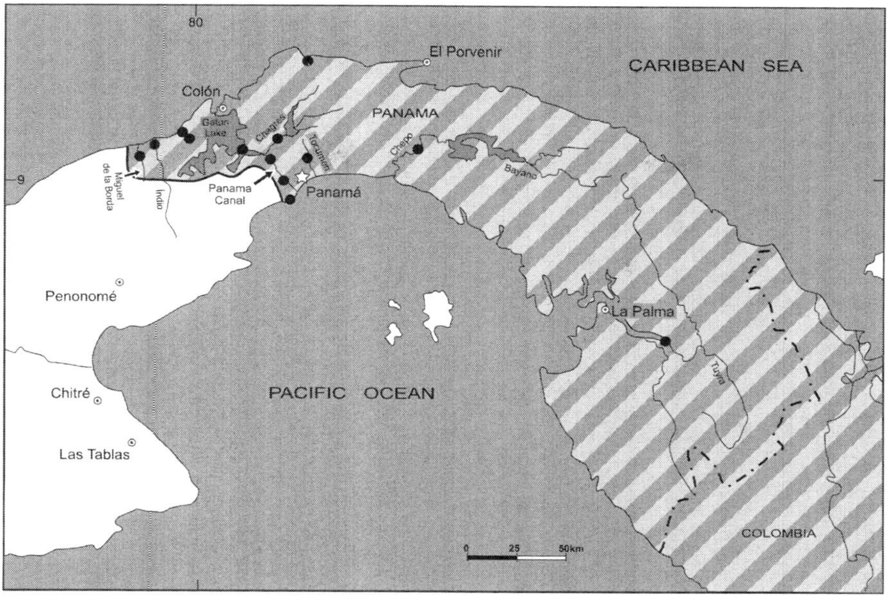

Fig. 1.5 Western limit of *Hydrochoerus isthmius* distribution in Panama (Trapido 1947, 1949a; Voss and Emmons 1996; Eric D. Núñez and Venicio Wilson personal communications). *Black dots* show records of the species' presence

1.4.1 Panama: Northern Limit of Hydrochoerus isthmius

In the 1940s, Trapido (1947, 1949a) reported the presence of the lesser capybara (*H. isthmius*) in Panama, in the Tuyra River valley, and the expansion of its western range (Fig. 1.5). At the time, the species was also found in the Bayano River valley along the Pacific coast to the Tocumen, Cabuya, and Cabra Rivers, near Panama City. There is extensive wetland habitat suitable for the capybara along the Pacific coast, from the Bayano to the Tocumen River. Trapido (1947) subsequently found that the species had colonized the Chagres River and was present at the Panama Canal. It was later seen on Barro Colorado Island and other small islands in Gatun Lake (Voss and Emmons 1996; Emilio A. Herrera personal observation).

In 2002, the species was seen to have occupied an expanded distribution along the Caribbean coast, facilitated by deforestation for cattle pastures (Venicio Wilson, personal communication). In the Province of Colón, the capybara is already found in the Índio and the Miguel de la Borda Rivers, west of the Panama Canal (Fig. 1.5; Eric D. Núñez personal communication). *H. isthmius* is known locally as poncho or capibara, and in other areas as lanco, ponche, or conejo poncho (Rodríguez-Mahecha et al. 1995).

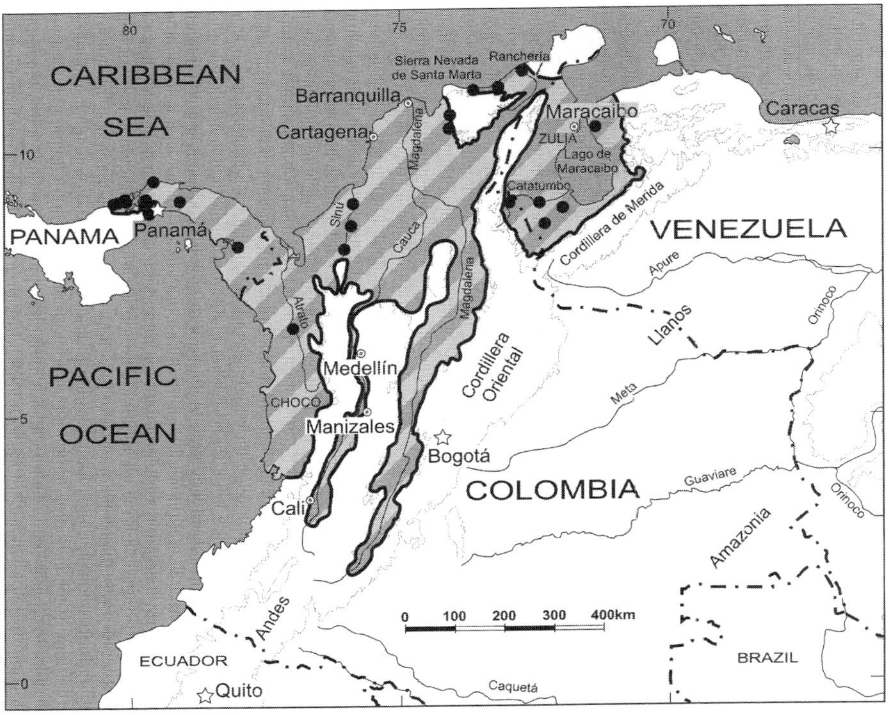

Fig. 1.6 Distribution of *Hydrochoerus isthmius* in Panama (Trapido 1947, 1949a; Voss and Emmons 1996; Eric D. Núñez and Venicio Wilson personal communications), Colombia (León 1974; Torres and Sanabria 1976), and Venezuela (Ojasti 1973). *Black dots* show records of the species' presence

1.4.2 Colombia

Both species of capybara occur in Colombia, separated by the Andes. *H. isthmius* is found in the northern parts of the country along the Caribbean coast, the lowland headwaters of the Catatumbo and Rancheria Rivers and the rivers to the north and west of the Sierra Nevada de Santa Marta (Fig. 1.6; León 1974). They are also found in the valleys of the Magdalena, Cauca, Sinú, and Atrato Rivers and in the Department of Chocó (Torres and Sanabria 1976). The species is known as ponche or cacó culopando, lancha, lanche, lancho, lanco, piropiro, and tinajo-ponche (Rodríguez-Mahecha et al. 1995). Little information is available, but populations of *H. isthmius* are thought to be small.

H. hydrochaeris, known locally as chigüiro, is distributed across the savannas of the Llanos Orientales in the Departments of Arauca, Casanare, Meta, and Vichada (Fig. 1.7), where it reaches the highest recorded densities (Aldana-Domínguez et al. 2002; Rodríguez et al. 2003; Caro et al. 2005; Aldana-Domínguez and Ángel-Escobar 2007). It is also found in the rainforests of the Departments of Caquetá, Putumayo, and Amazonas (Fuerbringer 1974; Concha and Vargas 1980) where their

Fig. 1.7 Distribution of *Hydrochoerus hydrochaeris* in Colombia (Fuerbringer 1974; Concha and Vargas 1980; Usma 1991; Emmons 1997; Aldana-Domínguez et al. 2002; Rodríguez et al. 2003; Caro et al. 2005; Aldana-Domínguez and Ángel-Escobar 2007), Venezuela (Ojasti 1973; Eisenberg 1989; Lim and Engstrom 2005), Guyana (Lim and Engstrom 2005; Lim et al. 2005a), Suriname (Eisenberg 1989; Lim et al. 2005b), and French Guiana (Voss and Emmons 1996; Voss et al. 2001). *Black dots* show records of the species' presence

densities are lower and populations are restricted to river banks (Emmons 1997). In 1986, *H. hydrochaeris* was introduced into the natural range of *H. isthmius* in the Cauca River valley (west of the Andes), and viable populations have subsequently been established in the Laguna de Sonso (Usma 1991) and the La Vieja River (Aldana-Domíngues et al. 2012). Other names by which both species are known in Colombia are: cabiari, capibara, capiguara, capiouara, chigüire, chindó, copiwara, jesus, ronsoco, sancho, and yulo (Mones and Mones 1981; González-Jiménez 1995; Rodríguez-Mahecha et al. 1995; Tirira 2004).

1.4.3 Venezuela

Venezuela is the only other country where both capybara species are present. *H. isthmius* is found in Venezuela only around Lago de Maracaibo (Fig. 1.6) in Zulia State, west of the Andes (Ojasti 1973). It is known locally as piropiro, but is also known as culo-pando and poncho (Mones and Mones 1981). It is separated from *H. hydrochaeris* by the Andes (Cordillera de Mérida). The two species are not sympatric.

H. hydrochaeris is common in the flooded savannas of the States of Apure, Barinas, Guarico, and Cojedes. They are also present in other regions, but not with the same abundance as found in the Llanos (Fig. 1.7). The current distribution of the capybara is a fragmented version of that mapped 40 years ago (Ojasti 1973; Herrera and Barreto 2012). There have been drastic reductions in many regions or even local extinctions. There are also large differences in density among regions, with the highest found in the States of Apure and Barinas (in the southwest). The most common name for *H. hydrochaeris* in Venezuela is chigüire, from the Cumanagotos and Palenques indigenous peoples (Ojasti 1973). In the Llanos they are also known as chindó. Around Venezuela, capybaras are also called capiba, capigua, capybara, cerdo de agua, chancho de agua, chigüiro (Ojasti 1973; Mones and Mones 1981; González-Jiménez 1995; Tirira 2004).

1.4.4 Guyana, Suriname, and French Guiana

H. hydrochaeris is widely distributed in Guyana, Suriname, and French Guiana. They are usually found along major rivers in the savanna and rainforest areas of the sparsely inhabited interior and the coast (Fig. 1.7; Voss and Emmons 1996; Voss et al. 2001; Lim et al. 2005a, b; Lim and Engstrom 2005). As these three countries are in the humid tropics, there is a large amount of suitable wetland habitat for the capybara. The species is not found in areas above 1,000 m in the Guyanan Shield. It is known in Guyana as watras, capybara, laubba, waterhog, waterpig, and thick-nosed tapir (Mones and Mones 1981; Burton Lim personal communication). The common names of the capybara in Suriname are kapoewa, waterzuyn, waterhaas, watra-agoe, and capivard (Mones and Mones 1981). In French Guiana, they are commonly known as cabiai, but are also known as cabionara or cochon d'eau.

Fig. 1.8 Distribution of *Hydrochoerus hydrochaeris* in Brazil (Schaller and Crawshaw 1981; Alho et al. 1987; Moreira 1995; Mourão and Campos 1995; Fonseca et al. 1996; Voss and Emmons 1996; Eisenberg and Redford 1999; Rechenberg et al. 2000; Moreira et al. 2001; Moreira 2004; Oliveira and Bonvicino 2006; Verdade and Ferraz 2006; Ferraz et al. 2007; José R. Moreira personal observation). *Black dots* show records of the species' presence

1.4.5 Brazil

H. hydrochaeris is present in all 26 Brazilian states and the Federal District (Brasilia; Fig. 1.8; Fonseca et al. 1996; Moreira 2004; Oliveira and Bonvicino 2006), usually associated with riparian and lacustrine habitats and wetlands. The highest densities are found in seasonally flooded savanna areas like the Pantanal Matogrossense (Schaller and Crawshaw 1981), the Região dos Lagos of Rio

Grande do Sul State (Oliveira and Bonvicino 2006), and flood plains of major river systems such as the Amazonas (Voss and Emmons 1996), Paraná, and Araguaia. However, it is extinct in most of the Caatinga biome of northeastern Brazil (Moreira 2004).

Capybaras are widely distributed in the agro-ecosystems of São Paulo State, mainly in agricultural habitats in pastoral areas, with a predominance of C4 plants, such as pastures and sugarcane fields that are also associated with strong human presence (Ferraz et al. 2007). In these areas, capybaras can also reach high densities (Verdade and Ferraz 2006). Here, capybaras are considered pests of a variety of crops including sugarcane, corn, rice, banana, and soybeans, and they are alleged to compete for food with livestock.

Apart from agricultural habitats, capybaras also frequent water bodies (rivers, dams and reservoirs) within urban limits, in public parks and residential areas (Moreira and Pinheiro 2012). In several regions of the country, human-capybara conflicts are increasingly common. In these cases, illegal hunting pressure, urbanization, and associated habitat change encourage capybaras to invade urban properties and eat ornamental garden plants; capybaras may drown in swimming pools, cause traffic accidents along streets and roads, and contaminate lawns with ticks (Moreira et al. 2001). In some rare cases, capybaras attack people as well (Rechenberg et al. 2000). Recently, increasing cases of Brazilian spotted fever in southeastern Brazil (Labruna 2012) and the possible association with high capybara densities have triggered controversy over the desirability of the species' spread into urban areas and possible methods of population control.

The species is known as capivara throughout Brazil. In Rio Grande do Sul State it is also known as carpincho or capincho; in the State of Amazonas as cupido; and on Marajó Island (Pará State) as beque. It is possible that the name beque originates from its regional synonym in Portuguese meaning "big nose" due to the nasal gland of the male. The male capybara is called trombudo or caixa (or cachapu) by the inhabitants of Marajó, also because of its large nasal gland. In some places in the interior of Bahia State, capybaras are known as porco-capivara; in southeastern Pará State they are named cunum and in Goiás State cubu (Sálvio Xavier, personal communication).

1.4.5.1 Distribution in Northeastern Brazil

There has been some doubt about the distribution of capybaras in northeastern Brazil because of the region's aridity (Mones and Mones 1981; Mones and Ojasti 1986; Ojasti 1991). They are almost extinct in the Caatinga biome (Fig. 1.9), possibly due to high hunting pressure and habitat alteration (Moreira 2004). However, they are still found along the major rivers such as the São Francisco, Parnaíba, and Paraguaçu and their tributaries such as the Preto (of the São Francisco), Gurguéia (of the Parnaíba), and Bonito Rivers (of the Paraguaçu), as well as along slopes of some ranges such as Chapada Diamantina (Fig. 1.9). A number of isolated populations are also found on the Vaza-Barris River (Bahia State), in Lago Parnaguá (Piaui State),

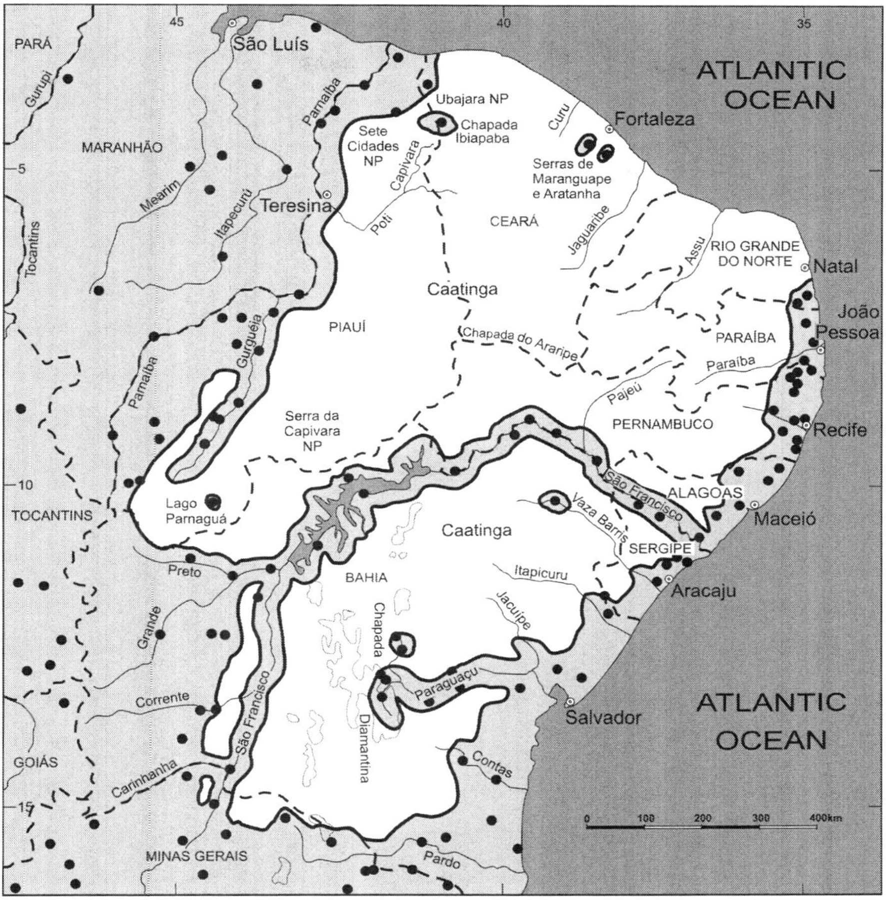

Fig. 1.9 Distribution of *Hydrochoerus hydrochaeris* in the Caatinga of northeastern Brazil (Moreira 2004; José R. Moreira personal observation). *Black dots* show records of the species' presence

and Chapada Ibiapaba (Ceará State), or in areas where they have been reintroduced such as the hills of Maranguape and Aratanha (Ceará State). Along the Atlantic coast, the species is found most commonly in areas of Atlantic Forest biome and the Coastal Zone to the south of Rio Grande do Norte and west of Ceará States. The species is not found along an extensive stretch of coastline between Ceará and Rio Grande do Norte States.

The scarcity and even disappearance of the species in this region have been noted in recent decades (Rocha 1948; Paiva 1973). Capybaras no longer occur in protected areas where once they were found, such as the Ubajara and Sete Cidades National Parks. The capybaras of Chapada do Araripe (Ceará and

Pernambuco States) and the Curu and Jaguaribe Rivers (Ceará State), which still existed in the 1960s (Paiva 1973), are now extinct. The same is the case along the Paraíba River (Paraiba State). Populations in the center of Piaui State, observed by the Scottish botanist George Gardner (1975) in 1839, are now extinct. The capybara no longer occurs in many areas with names that indicate its previous presence, like the Capivara River, a tributary of the Poti River, in Piauí State. In contrast, the National Park of Serra da Capivara (in the south of Piauí State) gets its name from wall paintings in local archeological sites that resemble the capybara (but are perhaps more likely to be the paca – *Cuniculus paca*), not from the presence of the species in the recent past.

1.4.6 Ecuador

H. isthmius is not found in Ecuador, but *H. hydrochaeris* is a common and widely distributed species in the country's Amazon region (Fig. 1.10), where it inhabits tropical rainforests, usually below 400 m altitude (Tirira 2007). It is found in forests in the headwaters of several tributaries of the Amazon and Marañón Rivers, including the Napo, Curaray, Pastaza, and Santiago Rivers, in the Provinces of Sucumbíos, Napo, Orellana, Pastaza, and Morona-Santiago. The highest elevations recorded for capybaras in Ecuador are at the confluence of the Pastaza River and a tributary (1,130 m), in the Province of Pastaza (Rageot and Albuja 1994). Its presence was also documented in several conservation units in the Amazon region (Mena-Valenzuela 1997; Mena-Valenzuela et al. 1997; INEFAN 1998; Tirira 2007). Capybaras are locally known as capibara, capihuara, ronsoco, carpincho, capibaro, chigüiro, chigüire, and yulo (Tirira 2004).

1.4.7 Peru

H. hydrochaeris is widely distributed in eastern Peru, throughout the Amazon River basin, at elevations ranging from 130 to 915 m. Their presence has been recorded in the Region of Amazonas along tributaries of the Marañón River. In the Region of Loreto, capybaras are rarely seen along tributaries of the Amazon River close to Iquitos (Valqui 2001). However, they are abundant on the Yavarí River and its tributaries (Salovaara et al. 2003; Amanzo 2006). They are also found in this Region along the tributaries of the Putumayo (Montenegro and Escobedo 2004), Tigre (Soini et al. 2001), Marañón, and Ucayali rivers (Aquino et al. 2001), and also in the Cordillera Azul National Park (Pacheco and Arias 2001).

In the central region of the country, capybara presence has been recorded in the Region of Ucayali along the Purus River and tributaries (Leite et al. 2003). They are also present in the Huallaga River and tributaries (Hutterer et al. 1995) and in the Regions of Huánuco and Pasco. To the south, there are records of the species in the Region of Cuzco along tributaries of the Urubamba River (Figueroa 2004;

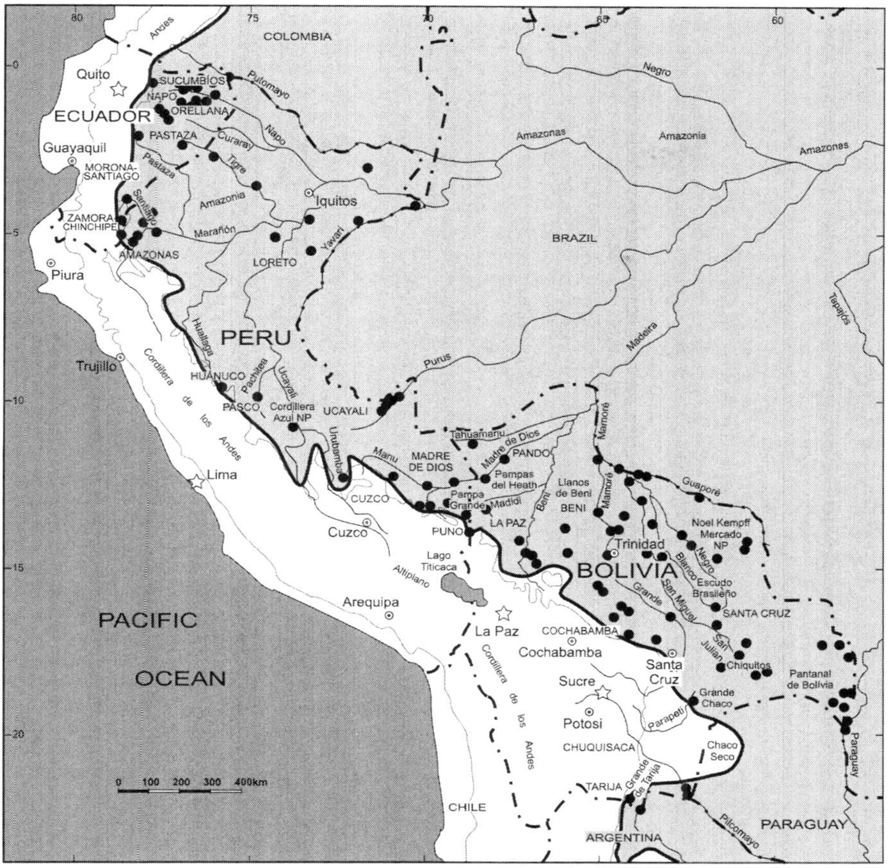

Fig. 1.10 Distribution of *Hydrochoerus hydrochaeris* in Ecuador (Rageot and Albuja 1994; Mena-Valenzuela et al. 1997; INEFAN 1998; Tirira 2007; Diego G. Tirira personal observation), Peru (Grimwood 1969; Patton et al. 1982; Soini and Soini 1992; Emmons et al. 1994; Emmons and Romo 1994; Hutterer et al. 1995; Pacheco and Vivar 1996; Emmons and Feer 1997; Aquino et al. 2001; Pacheco and Arias 2001; Soini et al. 2001; Valqui 2001; Leite et al. 2003; Salovaara et al. 2003; Figueroa 2004; Montenegro and Escobedo 2004; Amanzo 2006; Solari et al. 2006; Victor Pacheco personal observation), and Bolivia (Emmons 1991; Aguirre 1992; Altamirano 1992; Anderson et al. 1993; Arias et al. 1994; Barrera et al. 1994; Taber 1994; Perry et al. 1996; Anderson 1997; Torrico et al. 1997; Emmons 1998; Guinart 1998; Rumiz et al. 1998; Alverson et al. 2000; Tarifa et al. 2001; Maffei et al. 2002; Romo et al. 2002; Rumiz et al. 2002; Cuéllar and Noss 2003; Acosta and Aguanta 2006; Andrew Taber personal observation; N. Bernal personal communication). *Black dots* show records of the species' presence

Victor Pacheco, personal observation). In the Region of Madre de Dios, capybaras have been recorded in tributaries of the Manu (Pacheco and Vivar 1996; Solari et al. 2006) and of the Madre de Dios River (Emmons and Romo 1994; Emmons et al. 1994). In the extreme south of Peru, the capybara is found in Pampa Grande, Region of Puno. The species is known in Peru as ronsoco or capibara.

1.4.8 Bolivia

In Bolivia, the capybara (*H. hydrochaeris*) is commonly known as capibara, capiwara, capiguara, or carpincho. Generally, capybara meat is not appreciated in Bolivia and is usually consumed only by indigenous populations (Andrew Taber personal observation).

It is widely distributed to the east of the Andes, occurring in the phyto-geographic units of the northern wet grasslands (or Llanos de Beni; Fig. 1.10). Capybara presence has been recorded in the northwest of Pando Department, around the Tahuamanu River, where it was considered abundant (Alverson et al. 2000) and in the tributaries of the Madre de Dios River (Aguirre 1992). In the Department of La Paz, they are found in the region of Pampas del Heath (Romo et al. 2002), in the Madidi River (Emmons 1991; Tarifa et al. 2001), and the tributaries of Beni River (Barrera et al. 1994; Perry et al. 1996). The most extensive distribution area for capybaras in Bolivia is the wide savanna floodplains of the Beni and Mamore Rivers, in the Department of Beni (Anderson 1997). They are also found on the tributaries of these two rivers (Altamirano 1992). In the Department of Cochabamba, the capybara has been documented along the tributaries of the Grande River (Fig. 1.10; Rumiz et al. 1998).

Capybaras are found in the Department of Santa Cruz along tributaries of the Mamoré River (Fig. 1.10; Anderson et al. 1993; Acosta and Aguanta 2006). To the north of this department, the species is found in the Blanco and Negro Rivers (Arias et al. 1994) and to the east, in the Noel Kempff Mercado National Park (Emmons 1998). They are also abundant in the Bolivian Pantanal (Torrico et al. 1997; Teresa Tarifa personal observation). In the Chiquitos region of the Department of Santa Cruz they are found in small populations along some rivers such as the San Julian River (Fig. 1.10; Guinart 1998; Rumiz et al. 2002; Cuéllar and Noss 2003).

1.4.9 Distribution in the Gran Chaco Region

The distribution of the capybara in the Gran Chaco region of Bolivia, Paraguay, and Argentina has been questioned by various authors (Mones and Mones 1981; Ojasti 1991). However, although only reported at low densities, capybaras are found in isolated areas of the Gran Chaco in all three countries (Andrew Taber personal observation). In Bolivia, they are present along the San Miguel River to the south of the Sierra de San José de Chiquitos (Cuéllar and Noss 2003). In the north of the Gran Chaco, a few records were obtained in the region of Tucavaca (Maffei et al. 2002), while its presence in the northwest is restricted to the Parapetí River (Taber 1994) and the seasonal wetlands of Izozog. The capybara is also found on the border of Argentina in the floodplains formed by the Grande de Tarija and Bermejo Rivers, and near the common borders of Bolivia, Argentina, and Paraguay along the Pilcomayo River (Fig. 1.11; N. Bernal personal communication).

In the Chaco Seco of Paraguay, capybaras are restricted to permanent streams and swamps (Ziegler et al. 2002). Groups are found along major waterways like the

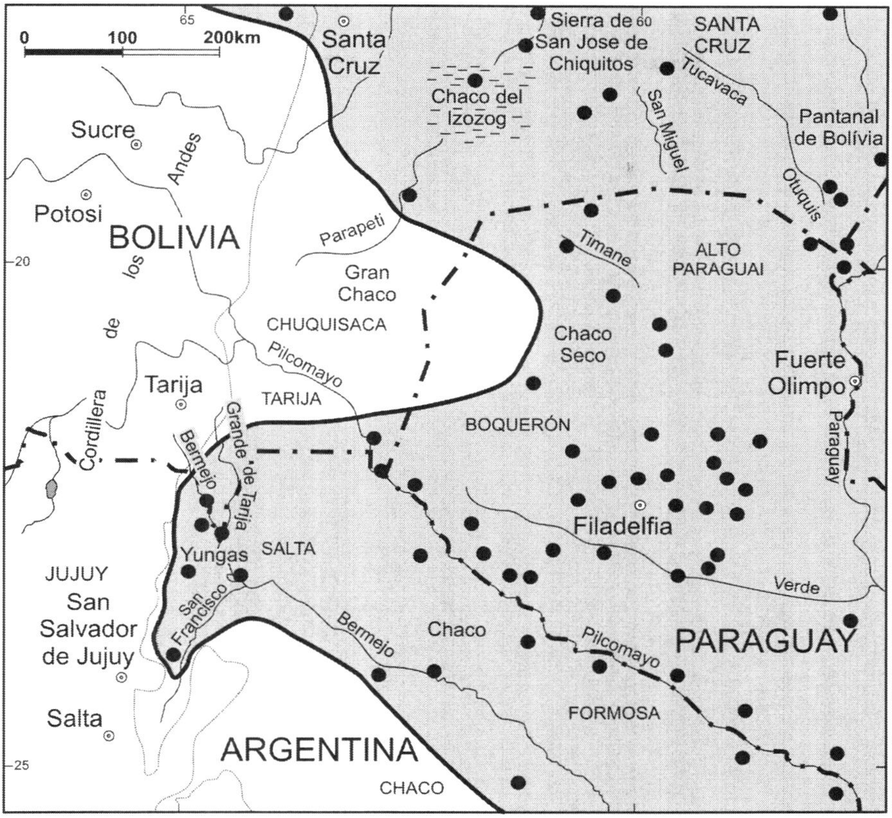

Fig. 1.11 Distribution of *Hydrochoerus hydrochaeris* in the Gran Chaco between Bolivia (Taber 1994; Maffei et al. 2002; Cuéllar and Noss 2003; Andrew Taber personal observation; N. Bernal personal communication), Paraguay (Ziegler et al. 2002; Andrew Taber personal observation), and Argentina (Heinonen and Bosso 1994; Heinonen and Chébez 1997; Andrew Taber and Martin Alvarez personal observations). *Black dots* show records of the species' presence

Pilcomayo River, where individuals and small groups are often sighted. Throughout the remainder of the Chaco Seco, capybaras persist around small, permanent ponds, for example, along the seasonal Timane River basin of the Chaco in northwestern Paraguay (Fig. 1.11; Andrew Taber personal observation), where annual rainfall can be less than 500 mm. Pools and more durable small ponds can be found along old water courses across this vast plain. These water courses no longer drain into the Pilcomayo and Paraguay Rivers basins, except during exceptionally wet years, but still provide semi-permanent habitats for capybaras. Mones and Mones (1981) report that in the Gran Chaco, the capybara is known as yeptahang (in various indigenous languages).

In the Argentine Provinces of Jujuy and Salta, the capybara is found in the Yungas biome (sensu Cabrera 1971). In Jujuy, the species is present in an area that includes the San Francisco River, the largest tributary in the region of the Bermejo River (Fig. 1.11), and in lakes connected by the rivers' meanders (Heinonen and Bosso

1994; Martin Alvarez personal observation). In the Province of Salta, capybaras are seen in the lakes neighboring the upper reaches of the Bermejo River and the lower area of its major tributaries, and on the upper reaches of the San Francisco River (Heinonen and Chébez 1997). All these are areas of tropical mountain forest, with rainfall higher than that of the Chaco Seco. In the Argentine Chaco biome, capybaras are found only along the Bermejo and Pilcomayo Rivers. Unlike the Dry Chaco of Paraguay, the capybara is heavily hunted in the Argentine Chaco.

1.4.10 Paraguay

Capybaras (*H. hydrochaeris*) are widespread in central and eastern parts of Paraguay (Fig. 1.12), but their presence is always associated with the Paraguay and Parana Rivers (Myers 1982; Yahnke et al. 1998) where they are particularly abundant. These regions are dominated by forests and fields. The capybara is among the fauna of seasonally flooded palm savannas of the Paraguayan Humid Chaco. Due to the climate, soil type, and vegetation, the Chaco Seco (in the north) is considered inhospitable and capybaras are found only in wetter areas (Fig. 1.11; Ziegler et al. 2002). The species is known locally as carpincho or capibara.

1.4.11 Uruguay

In Uruguay, *H. hydrochaeris* is found in damp environments, from the banks of the Uruguay River to the Atlantic coast (Fig. 1.12), throughout the country. It is abundant in the Department of Salto, where hunting is allowed, and there were some attempts at captive production. In the interior of the country, capybaras are abundant in the valley of the Negro River along the Atlantic Coast, and on the plains of the Department of Rocha (Bocage 1995; González 2000). Although often hunted, the species is not endangered in Uruguay. However, populations are scarcer in the more populous south. Capybaras in Uruguay are known as carpincho or capincho. Mones and Mones (1981) suggested that the origin of this name was possibly from the indigenous Quechua language in which "rabincho" means "without tail." However, it is possible that the origin of the first part of the word, carpin or capin, is the same as that for the capybara in the Tupi language – ka'pii, which means grass.

1.4.12 Argentina

In Argentina, *H. hydrochaeris* occurs in the Provinces of Jujuy, Salta, Formosa, Chaco, Misiones, Corrientes, Entre Rios, Santa Fe, Santiago del Estero, Cordoba, and Buenos Aires (Fig. 1.12; Alvarez 2002; Alvarez and Martinez 2006; Quintana and Bolkovic 2012). However, the distribution of the species during the eighteenth

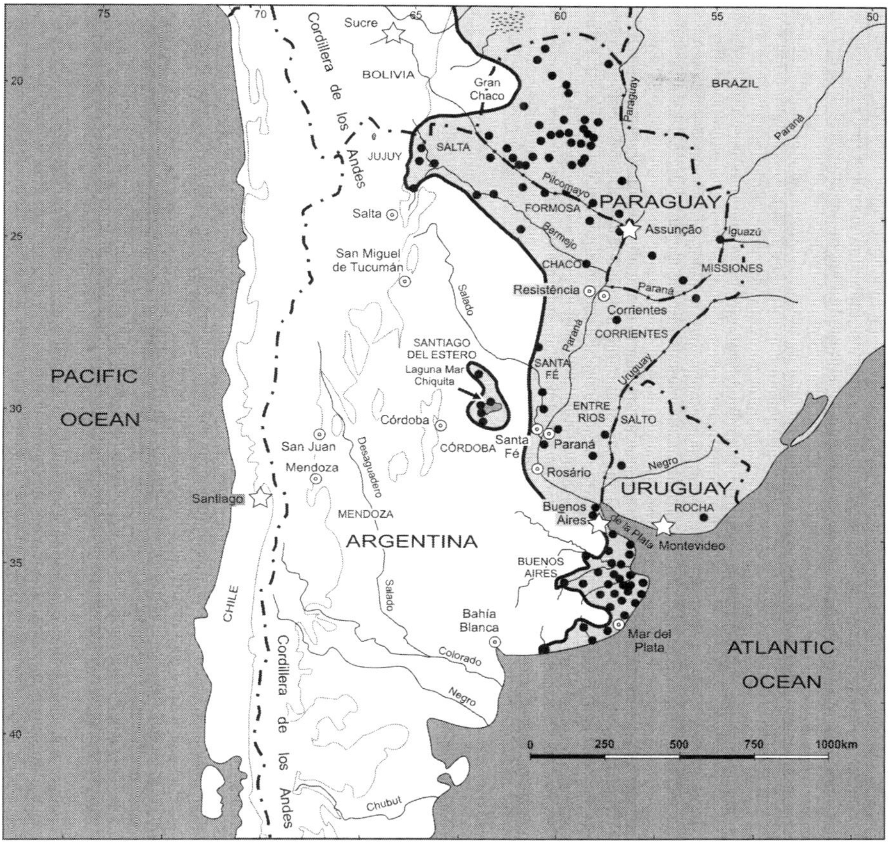

Fig. 1.12 Distribution of *Hydrochoerus hydrochaeris* in Paraguay (Myers 1982; Yahnke et al. 1998; Ziegler et al. 2002; Andrew Taber personal observation), Argentina (Adámoli et al. 1988; Redford and Eisenberg 1992; Goldfeder 1993; Heinonen and Bosso 1994; Miatello 1994; Heinonen and Chébez 1997; Alvarez 2002; Miatello 2003; Alvarez and Martinez 2006; Andrew Taber and Martin Alvarez personal observations), and Uruguay (Bocage 1995; González 2000). *Black dots* show records of the species' presence

century also included habitats to the south of the current distribution and a larger portion of western Argentina, including the Province of Mendoza. For example, the skin of this species was collected in about 1915, from a tributary of the Desaguadero River, in the south of the Province of Mendoza, (Roig 1991). As the drainage systems were changed in the nineteenth century, the species disappeared from the southern portion of its distribution. Subsequently, these areas have experienced desertification and are no longer suitable for capybaras.

Currently, the Bermejo, Pilcomayo, Salado, Paraguay, Iguazu, Paraná, Uruguay, de la Plata Rivers, and other smaller rivers and their tributaries form a prodigious network of water bodies where capybaras remain abundant (Adámoli et al. 1988). However, there are three areas in the country where populations of the species are

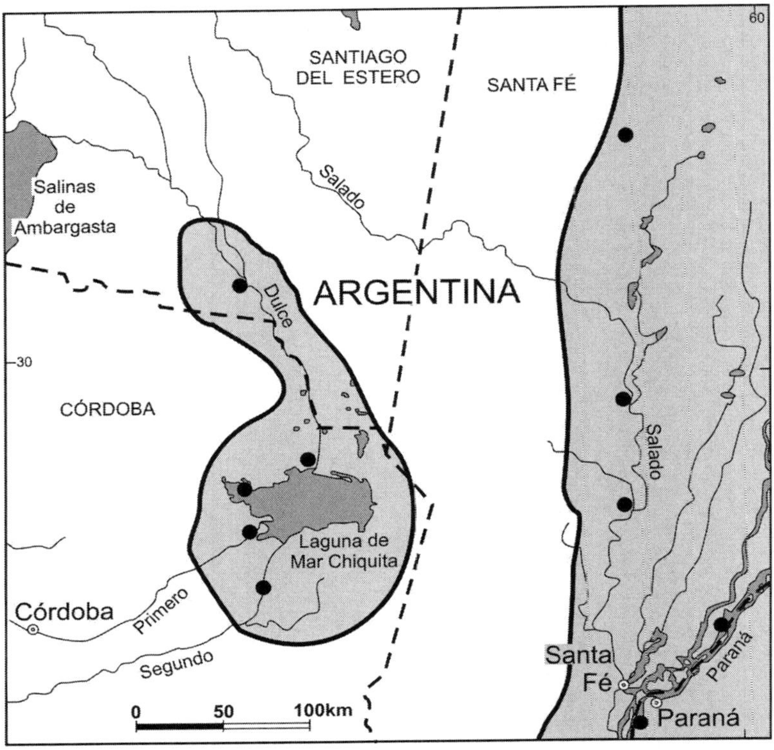

Fig. 1.13 Distribution of *Hydrochoerus hydrochaeris* in the central region of Argentina, in the Provinces of Santiago del Estero and Córdoba (Goldfeder 1993; Miatello 1994, 2003). *Black dots* show records of the species' presence

practically isolated: North, Central, and South (Alvarez and Martinez 2006). The distribution in the northern region was previously described in the section covering the Gran Chaco. The capybara is known in Argentina as carpincho.

1.4.12.1 Central Distribution (Provinces of Santiago del Estero and Córdoba)

Between the Provinces of Santiago del Estero and Cordoba, capybaras can be found in the endorheic basin (a closed drainage basin) of the Laguna Mar Chiquita (Fig. 1.13), where they most likely survive as remnant populations. The Laguna Mar Chiquita is found at the confluence of the Dulce, Primero, and Segundo Rivers. Its waters are brackish, with halophyte vegetation along the water margins (Reati et al. 1997). Although capybaras have low tolerance for brackish water (Ojasti 1973), they seem to select the mouths of the Primero, Segundo, and Dulce Rivers, where salinity is apparently lower than that found in the Laguna Mar Chiquita (Goldfeder 1993).

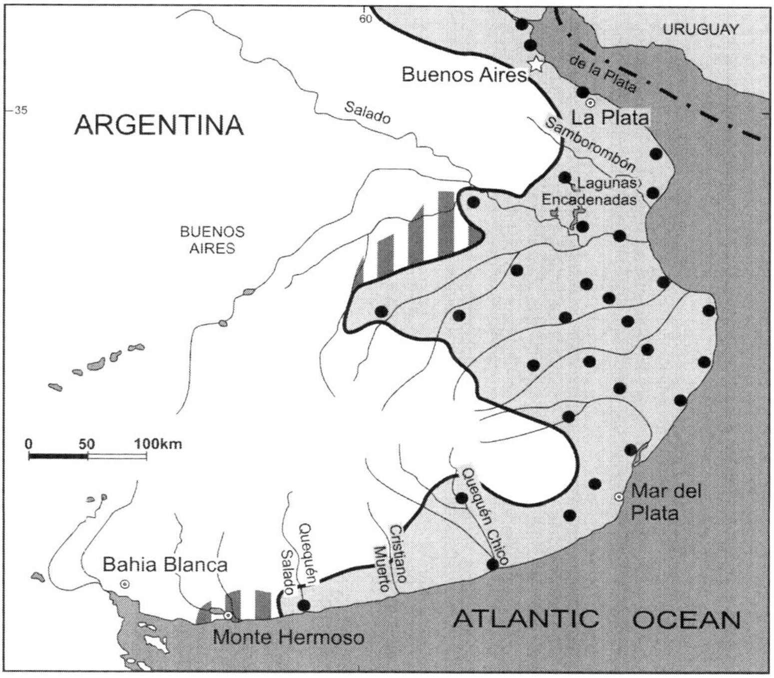

Fig. 1.14 Southern distributional limit of *Hydrochoerus hydrochaeris* in the Province of Buenos Aires, Argentina (Adámoli et al. 1988; Alvarez 2002). *Black dots* show records of the species' presence. *Vertical stripes*: possible presence of *H. hydrochaeris*

In the 1980s, these populations dispersed upstream of the basin along the eastern bank of the Dulce River, occupying the southern tip of the Province of Santiago del Estero (Miatello 1994). The whole region north of the Laguna Mar Chiquita was occupied and, in the 1990s, all the banks of the lagoon and the wetlands of the Dulce River were inhabited by capybaras (Goldfeder 1993; Miatello 2003).

1.4.12.2 Buenos Aires Province: Southernmost Limit of *Hydrochoerus hydrochaeris*

The Province of Buenos Aires is the southern limit of the geographic distribution of the capybara, but the exact location of the limit within the Province is unclear. Alvarez (2002) reported the presence of capybaras to the mid-east of the Province, where there is a profuse network of lakes that form the Encadenadas system of lagoons and the basins of the Samborombón and Salado Rivers. The distribution reaches the south of the Quequén Salado River (Fig. 1.14). Although there are no recent records, capybaras may also inhabit the rivers close to Monte Hermoso. It is possible that the dispersal and establishment of capybara populations in the region is hampered by the salinity of water bodies (Adámoli et al. 1988). However, recurrent floods in this region may alter the distribution of the species. Capybaras are hunted in the eastern portion of this region.

1.5 Final Comments

Inconsistency in the use of the capybara's scientific name is unhelpful, and we emphasize that the species should be referred to by the generic name *Hydrochoerus* Brisson, 1762, and the species name *Hydrochoerus hydrochaeris* Linnaeus, 1766 (for the capybara, living east of the Andes) and *H. isthmius* Goldman, 1912 (for the lesser capybara, found west of the Andes).

Although the genus is not threatened with extinction, some populations deserve particular attention. Of these, we highlight the populations of *H. hydrochaeris* in northeastern Brazil and the Chaco Seco region of Argentina where hunting is intense. A mounting concern is that the populations in the Llanos of Venezuela and Colombia are at increasing risk of overexploitation, and local extinctions cannot be ruled out. Governments, and those responsible for wildlife in the capybara's range states, should take note of the species' value and act to mitigate the risks it faces – the material in this book will, we hope, provide them with the basis for policy and action.

Acknowledgments We thank Hugh Casement for translation of the old German text of Hans Staden and Sérgio Eustáquio de Noronha for geo-referencing capybara occurrence in Brazil. This work would not have been possible without the valuable information provided by Carol Kelloff, Burton Lim, Mark Engstrom, Gerald Urquhart, Venicio Wilson, Eric Yensen, Sálvio do Carmo dos Santos Xavier, Álvaro Mones, and Erick Nuñez. Eric Yensen also helped with the English spelling. JRM would like to thank all the 1,695 city councils, 60 conservation units, and 4 indigenous reserves, which answered his questionnaire about the occurrence of capybaras in Brazil, and acknowledges the financial support of PRODETAB/World Bank. MRA is grateful for the valuable knowledge of Fernando O. Kravetz, Marta Piantanida, and Elio Masoia (*in memoriam*), colleagues at SAREM for information about capybara presence in Argentina, and Universidad de Buenos Aires (UBA, Argentina) for financial support.

References

Acosta SL, Aguanta FA (2006) Lista preliminar de los mamíferos del bosque experimental Elías Meneses, Santa Cruz, Bolivia. Kempffiana 2(1):144–149

Adámoli J, Sennhauser E, Astrada E, Agráz JL (1988) Propuesta para la delimitación del área geográfica potencial de los carpinchos en la Argentina. Informe Exp. 1325. Consejo Federal de Inversiones, Buenos Aires, 133pp

Aguirre LF (1992) Mamíferos. In: Salm H, Marconi M (eds) Reserva Nacional Amazónica Manuripi-Heath: programa de reestructuración (fase II). Pl-480, Lidema Y Cordepando, La Paz, pp 90–97

Aldana-Domínguez J, Forero-M J, Betancur J, Cavelier J (2002) Dinámica y estructura de la población de chigüiros (*Hydrochaeris hydrochaeris*: Rodentia, Hydrochaeridae) de Caño Limón, Arauca, Colombia. Caldasia 24(2):445–458
Aldana-Domínguez J, Ángel-Escobar DC (2007) Evaluación del tamaño y densidad de poblaciones silvestres de chigüiros en el departamento del Casanare. In: Aldana-Domínguez J, Vieira-Muñoz MI, Ángel-Escobar DC (eds) Estudios sobre la ecología del chigüiro (*Hydrochoerus hydrochaeris*), enfocados a su manejo y uso sostenible en Colombia. Instituto Alexander von Humboldt, Bogotá, pp 33–48
Aldana-Domínguez J, Vieira-Muñoz MI, Bejarano P (2012) Conservation and use of the capybara and the lesser capybara in Colombia. In: Moreira JR, Ferraz KMPMB, Herrera EA, Macdonald DW (eds) Capybara: biology, use and conservation of an exceptional neotropical species. Springer, New York, pp 321–332
Alho CJR, Campos MSC, Gonçalves HC (1987) Ecologia de capivara (*Hydrochaeris hydrochaeris*, Rodentia) do Pantanal: II. Atividade, sazonalidade, uso do espaço e manejo. Rev Bras Biol 47:99–110
Altamirano R (ed) (1992) Territorio Indígena Parque Nacional Isidoro-Sécure: inventario de flora y fauna. Ciddebeni, Instituto de Ecología, GTZ, La Paz, p 112
Alvarez MR (2002) Manejo sustentable del carpincho (*Hydrochoerus hydrochaeris*, Linnaeus 1766) en Argentina: un aporte al conocimiento de la biología de la especie desde la cría en cautiverio. Ph.D. thesis, Facultad de Ciencias. Exactas y Naturales, Universidad de Buenos Aires, Buenos Aires, 210pp
Alvarez MR, Martinez RA (2006) Familias Hydrochoeridae, Agoutidae, Dasyproctidae. In: Barquez RM, Díaz MM, Ojeda RA (eds) Mamíferos de Argentina: sistemática y distribución. Sociedad Argentina para el Estudio de los Mamíferos SAREM, Buenos Aires, pp 210–212
Alverson WS, Moskovitz DK, Shopland JM (eds) (2000) Bolivia: Pando, Rio Tahuamanu, vol 1, Rapid biological inventories report. The Field Museum, Chicago, p 79
Amanzo J (2006) Mamíferos medianos y grandes. In: Vriesendorp C et al (eds) Perú: Matsés, vol 16, Rapid biological inventories report. The Field Museum, Chicago, pp 98–106
Anchieta Jde (1997) Carta de São Vicente 1560, vol 7, Caderno. Conselho Nacional da Reserva da Biosfera da Mata Atlântica, São Paulo
Anderson S (1997) Mammals of Bolivia, taxonomy and distribution. Bull Am Mus Nat Hist 231:1–652
Anderson S, Riddle BR, Yates TL, Cook JA (1993) Los mamíferos del Parque Nacional Amboró y la región de Santa Cruz de la Sierra, Bolivia, vol 2, Special publication the Museum of Southwestern Biology. Museum of Southwestern Biology/University of New Mexico, Albuquerque, pp 1–58
Aquino R, Bodmer RE, Gil JG (2001) Mamíferos de la cuenca del río Samiria: ecología poblacional y sustentabilidad de la caza. Impresión Rosegraff, Lima
Arias CS et al (1994) La fauna de La Reserva de Vida Silvestre de los Ríos Blanco y Negro: distribución, diversidad, densidad y pautas para su conservación. In: Plan de manejo de la Reserva de Vida Silvestre de Ríos Blanco y Negro. Fundación Amigos de la Naturaleza and Wildlife Conservation Society, Bolivia, pp 293–413
Azcárate T, Alvarez F, Braza F (1980) Tendencias gregarias del capibara (*Hydrochoerus hydrochaeris*) en los Llanos de Venezuela. Reunión Iberoamericana Zool Vertebrat 1:285–292
Baldizan AR, Dixon E, Parra YR (1983) Digestion in the capybara (*Hydrochoerus hydrochaeris*). S Afr J Anim Sci 13(1):27–28
Barrera S, Guerra JF, Osorio F, Sarmiento J, Villalba L (1994) Territorio Indígena Reserva de la Biosfera Pilón Lajas: reconocimiento preliminar de la fauna. Colección Boliviana de Fauna, Instituto de Ecología y Veterinarios Sin Fronteras. La Paz, p 57
Barrère P (1741) Essai sur l'histoire naturelle de la France equinoxiale ou De'nombrement des plantes, des animaux & des minéraux, qui se trouvent dans l'Isle de Cayenne, les isles de Remire, sur les côtes de la mer, & dans le Continent de la Guyane. Veuve Piget, Paris
Barreto GR, Quintana RD (2012) Foraging strategies and feeding habits of capybaras. In: Moreira JR, Ferraz KMPMB, Herrera EA, Macdonald DW (eds) Capybara: biology, use and conservation of an exceptional neotropical species. Springer, New York, pp 83–96

Bocage A (1995) Mamíferos del Uruguay. Ediciones de la Banda Oriental, Montevideo, 105pp
Brisson AD (1756) Regnum animale in classes IX distributum, sive synopsis methodica sistens generalem *Animalium* distributionem in classes IX, & duarum primarum classium, *Quadrupedum* scilicet & *Cetaceorum*, particularem divisionem in ordines, sectiones, genera & species. Joannem-Baptistam Bauche, Paris
Brisson AD (1762) Regnum animale in classes IX distributum, sive synopsis methodica sistens generalem *Animalium* distributionem in classes IX, & duarum primarum classium, *Quadrupedum* scilicet & *Cetaceorum*, particularem divisionem in ordines, sectiones, genera & species. Theodorum Haak, Leiden
Brünnich MTh (1772) Zoologiæ fundamenta, prælectionibus academicis accommodata. Pelt, Kopenhagen
Buffon G (1764) Histoire naturelle, générale et particuliére, avec la description du Cabinet du roi, vol 12. L'Imprimerie Royale, Paris
Cabrera AL (1971) Fitogeografía de la República Argentina. B Soc Argent Bot XIV(1–2):1–44
Cardim F (1980) Tratados da terra e da gente do Brasil, vol 13, Coleção Reconquista do Brasil. Editora Itatiaia Limitada, São Paulo
Caro C et al (2005) Fortalecimiento al control y seguimiento a las poblaciones naturales de chigüiro para la implementación de proyectos de uso sostenible en Casanare- Municipios de Paz de Ariporo y Hato Corozal. Fase III. Informe final. UNILLANOS-Ministerio de Ambiente, Vivienda y Desarrollo Territorial, Villavicencio
Concha LC, Vargas LF (1980) Estudio del chigüiro amazónico (*Hydrochaeris hydrochaeris*) y algunas observaciones técnicas para su manejo en zoocriaderos. Graduate dissertation, Universidad del Tolima, Ibagué
Cuéllar SE, Noss A (2003) Mamíferos del Chaco y de la Chiquitanía de Santa Cruz. Editorial Fan, Santa Cruz, 202pp
Eisenberg JF (1989) Mammals of the Neotropics: the Northern Neotropics, vol 1, Panama, Colombia, Venezuela, Guyana, Suriname, French Guiana. The University of Chicago, Chicago, 449pp
Eisenberg JF, Redford KH (1999) Mammals of the Neotropics: the Central Neotropics, vol 3, Ecuador, Peru, Bolivia, Brazil. The University of Chicago, Chicago, p 609
Emmons LH (1991) Mammal List. In: Parker TA, Bailey B (eds) A biological assessment of the Alto Madidi Region, vol 1, Rapid Assessment Program Report. Conservation Internacional, Washington, DC, pp 72–73
Emmons LH (1997) Neotropical rainforest mammals. The University of Chicago, Chicago, 307pp
Emmons LH (1998) Mamíferos del Parque Nacional Noel Kempff Mercado. In: Killeen TL, Schulenberg TS (eds) A biological assessment of Parque Nacional Noel Kempff Mercado, Bolivia, vol 10, Rapid assessment program report. Conservation Internacional, Washington, DC, pp 341–347
Emmons LH, Ascorra C, Romo M (1994) Mammals of the Río Heath and Peruvian pampas. In: Foster R, Carr J, Forsyth A (eds) The Tambopata-Candamo Reserved Zone of southeastern Perú: a biological assessment, vol 6, RAP working papers. Conservation International, Washington, DC, pp 146–149
Emmons LH, Feer F (1997) Neotropical rainforest mammals: a field guide, 2nd edn. The University of Chicago Press, Chicago
Emmons LH, Romo M (1994) Mammals of the upper Tambopata//Távara. In: Foster R, Carr J, Forsyth A (eds) The Tambopata-Candamo Reserved Zone of southeastern Perú: a biological assessment, vol 6, RAP working papers. Conservation International, Washington, DC, pp 46–47
Escobar A, González-Jiménez E (1976) Estudio de la competencia alimenticia de los herbívoros mayores del llano inundable con especial al chigüire (*Hydrochoerus hydrochaeris*). Agron Trop 23:215–227
Ferraz KMPMB, Ferraz SFB, Moreira JR, Couto HTZ, Verdade LM (2007) Capybara (*Hydrochoerus hydrochaeris*) distribution in agroecosystems: a cross-scale habitat analysis. J Biogeogr 34:223–230

Figueroa J (2004) Mamíferos grandes. In: Vriesendorp C, Rivera Chávez L, Moskovits D, Shopland J (eds) Perú: Megantoni, vol 15, Rapid biological inventories report. The Field Museum, Chicago, pp 288–293
Fonseca GAB, Hermann G, Leite YLR, Mittermeier RA, Rylands AB, Patton JL (1996) Lista anotada dos mamíferos do Brasil, vol 4, Occasional papers in conservation biology, pp 1–38
Fuerbringer J (1974) El chigüiro, su cría y explotación racional. Temas de orientación agropecuaria 90: 1–59. Bogotá
Gândavo PM (2004) A primeira história do Brasil: história da provincia de Santa Cruz a que vulgarmente chamamos Brasil. Jorge Zahar, Rio de Janeiro
Gardner G (1975) Viagem ao interior do Brasil, principalmente nas províncias do Norte e nos distritos do ouro e do diamante durante os anos de 1836–1841, vol 13, Coleção Reconquista do Brasil. Livraria Itatiaia Editora, São Paulo
Garrod AH (1876) On the caecum coly of the capybara (*Hydrochoerus capybara*). Proc Zoolog Soc Lond 3:20–23
Gentry A (1994) Case 2928, *Regnum Animale* ..., Ed. 2 (M.J. Brisson, 1762): proposed rejection, with the conservation of the mammalian generic names *Philander* (Marsupialia), *Pteropus* (Chiroptera), *Glis*, *Cuniculus* and *Hydrochoerus* (Rodentia), *Meles*, *Lutra* and *Hyaena* (Carnivora), *Tapirus* (Perissodactyla), *Tragulus* and *Giraffa* (Artiodactyla). Bull Zool Nomencl 51:135–146
Goldfeder SD (1993) Potencial del sudoeste de Mar Chiquita (Córdoba, Argentina) para sustentar poblaciones silvestres de carpincho. M.Sc. dissertation. Facultad de Ciencias Exactas, Físicas y Naturales. Universidad Nacional de Córdoba, Argentina
Goldsmith O (1870) A history of the earth and animated nature. Blackie and Son, Glasgow
González E (2000) Lista sistemática, afinidades biogeográficas, hábitos y hábitats de los mamíferos terrestres autóctonos de Uruguay (Mammalia): una introducción. Jornadas Sobre Animales Silvestres, Desarrollo Sustentable y Medio Ambiente, Montevideo, pp 58–73
González-Jiménez E (1995) El Capibara (*Hydrochoerus hydrochaeris*); estado actual de su producción, vol 122, Estudio FAO Producción y Sanidad Animal. FAO, Roma
González-Jiménez E, Escobar A (1975) Digestibilidade comparada entre chigüires (*Hydrochoerus hydrochaeris*), conejos y ovinos, con raciones de diferentes proporciones de forrajes y concentrados. Agron Trop 25:283–290
Grimwood IR (1969) Notes on the distribution and status of some Peruvian mammals 1968, vol 21, Special Publication of the American Committee for International Wild Life Protection. American Committee for International Wild Life Protection, Bronx, pp 1–86
Guinart SD (1998) Los mamíferos de nuestro territorio. Proyecto de Manejo Forestal Sostenible (Bolfor), Santa Cruz, 169pp
Heinonen FS, Bosso A (1994) Nuevos aportes para el conocimiento de la mastofauna del Parque Nacional Calilehua (Provincia de Jujuy, Argentina). Mastozoolog Neotrop 1(1):51–60
Heinonen FS, Chebez JC (1997) Los mamíferos de los parques nacionales de la Argentina. Editorial LOLA, Buenos Aires
Herrera EA (1985) Coprophagy in the capybara, *Hydrochoerus hydrochaeris*. J Zool 207: 616–619
Herrera EA (1986) The behavioural ecology of the Capybara, *Hydrochoerus hydrochaeris*. D.Phil. thesis, University of Oxford, Oxford, UK
Herrera EA (1998) Reproductive strategies of female capybaras: dry-season gestation. In: Dunstone N, Gorman ML (eds) Behaviour and ecology of riparian mammals, Symposia of the Zoological Society of London. Cambridge University Press, London, pp 281–296
Herrera EA (2012a) Capybara digestive adaptations. In: Moreira JR, Ferraz KMPMB, Herrera EA, Macdonald DW (eds) Capybara: biology, use and conservation of an exceptional neotropical species. Springer, New York, pp 97–106
Herrera EA (2012b) Capybara social behavior and use of space: patterns and processes. In: Moreira JR, Ferraz KMPMB, Herrera EA, Macdonald DW (eds) Capybara: biology, use and conservation of an exceptional neotropical species. Springer, New York, pp 185–207

Herrera EA, Barreto GR (2012) Capybaras as a source of protein: utilization and management in Venezuela. In: Moreira JR, Ferraz KMPMB, Herrera EA, Macdonald DW (eds) Capybara: biology, use and conservation of an exceptional neotropical species. Springer, New York, pp 305–319

Herrera EA, Macdonald DW (1987) Group stability and the structure of a capybara population. Symp Zoolog Soc Lond 58:115–130

Herrera EA, Macdonald DW (1989) Resource utilization and territoriality in group-living capybaras (*Hydrochoerus hydrochaeris*). J Anim Ecol 58:667–679

Honeycutt RL (2012) Phylogenetics of Caviomorph Rodents and genetic perspectives on the evolution of sociality and mating systems in the Caviidae. In: Moreira JR, Ferraz KMPMB, Herrera EA, Macdonald DW (eds) Capybara: biology, use and conservation of an exceptional neotropical species. Springer, New York, pp 61–81

Houaiss A, Villar MS, Franco FMM (2004) Dicionário Houaiss da Língua Portuguesa. Ed. Objetiva, Rio de Janeiro

Hopwood AT (1947) The generic names of the mandrill and baboons, with notes on some of the genera of Brisson, 1762. Proc Zoolog Soc Lond 117:533–536

Hutterer R, Verhaagh M, Diller J, Podloucky R (1995) An inventory of mammals observed at Panguana Biological Station, Amazonian Peru. Ecotropica 1:3–20

ICZN (International Commission on Zoological Nomenclature) (1998) Opinion 1894 Regnum animale…, ed. 2 (M.J. Brisson, 1762): rejected for nomenclatural purposes, with the conservation of the mammalian generic names Philander (Marsupialia), Pteropus (Chiroptera), Glis, Cuniculus, and Hydrochoerus (Rodentia), Meles, Lutra and Hyaena (Carnivora), Tapirus (Perissodactyla), Tragulus and Giraffa (Artiodactyla). Bull Zool Nomencl 55:64–71

INEFAN (Instituto Ecuatoriano Forestal y de Áreas Naturales y Vida Silvestre) (1998) Guía de parques nacionales y reservas del Ecuador. INEFAN, Quito

Jonston J (1650) Historiæ natvralis de quadrvpetibvs libri cum æneis figuris. Merianus, Francfurt

Jorgenson JP (1986) Notes on the ecology and behaviour of capybaras in Northeastern Colombia. Vida Silv Neotrop 1:31–40

Labruna MB (2012) Brazilian spotted fever: the role of capybaras. In: Moreira JR, Ferraz KMPMB, Herrera EA, Macdonald DW (eds) Capybara: biology, use and conservation of an exceptional neotropical species. Springer, New York, pp 371–383

Leite PR, Beck H, Velazco P (2003) Mamíferos terrestres y arbóreos de la selva baja de la Amazonía peruana: entre los ríos Manu y Alto Purús. In: Leite Pitman R, Pitman N, Alvarez P (eds) Alto Purús, biodiversidad, conservación y manejo. Duke University, Durham, pp 109–124

León CA (1974) Zoonimia y distribución geográfica del Chigüiro en Colombia. Primer simposio internacional sobre chigüiro y babilla. INDERENA. Santa fe de Bogotá

Lim BK, Engstrom MD, Ochoa JG (2005a) Mammals. In: Hollowel T, Reynolds RP (eds) Checklist of the terrestrial vertebrates of the Guiana Shields, vol 13, Bulletin of the biological society of Washington. National Museum of Natural History, Washington, DC, pp 77–92

Lim BK et al (2005b) Results of the Alcoa Foundation-Suriname expeditions. XIV. Mammals of Brownsberg Nature Park, Suriname. Ann Carnegie Mus 74:225–274

Lim BK, Engstrom MD (2005) Mammals of Iwokrama Forest. Proc Acad Nat Sci Phila 154:71–108

Linnaeus C (1758) Systema Naturae per regna tria naturae secundum classes, ordines, genera, species, cum characteribus, differentiis, synonymis, locis, vol 1, 10th edn, Tomo. Holmiae, Laurentii Salvii, Stockholm

Linnaeus C (1766) Systema Naturae per regna tria naturae secundum classes, ordines, genera, species, cum characteribus, differentiis, synonymis, locis, vol 1, 12th edn, Tomo. Holmiae/ Laurentii Salvii, Stockholm

López-Barbella S (1982) Determinacion del ciclo estral en chigüires (*Hydrochoerus hydrochaeris*). Acta Cient Venez 33:497–501

López-Barbella S (1987) Consideraciones generales sobre la gestacion del chigüire (*Hydrochoerus hydrochaeris*). Acta Cient Venez 38:84–89

Macdonald DW (1981) Dwindling resources and the social behaviour of capybaras, (*Hydrochoerus hydrochaeris*) (Mammalia). J Zool 194:371–391

Macdonald DW, Herrera EA (2012) Capybara scent glands and scent-marking behavior. In: Moreira JR, Ferraz KMPMB, Herrera EA, Macdonald DW (eds) Capybara: biology, use and conservation of an exceptional neotropical species. Springer, New York, pp 185–193

Macdonald DW, Krantz K, Aplin RT (1984) Behavioural, anatomical and chemical aspects of scent marking amongst capybaras (*Hydrochoerus hydrochaeris*) (Rodentia: Caviomorpha). J Zool 202:341–360

Maffei L, Cuéllar E, Noss AJ (2002) Uso de trampas-cámara para la evaluación de mamíferos en el ecotono Chaco-Chiquitanía. Rev Bol Ecol 11:55–65

Marcgrav de Liebstad G (1648) Historiæ rervm natvarlivm Brasiliæ, libri octo: quorum tres priores agvnt de plantis. Quartus de piscibvs. Quintus de avibvs. Sextus de quadrvpedibvs & serpentibvs. Septimus de insectis. Octavus de ipsa regione, & illivs incolis. Franciseum Hackium, Leiden

Melville RV, Smith JDD (1987) Official lists and indexes of names and works in Zoology. International Trust for Zoological Nomenclature, London, UK

Mena-Valenzuela P (1997) Diversidad y abundancia relativa de los mamíferos en Sinangüé, Reserva Ecológica Cayambe-Coca, Sucumbíos, Ecuador. In: Mena PA, Soldi A, Alarcón R, Chiriboga C, Suárez L (eds) Estudios biológicos para la conservación: diversidad, ecología y etnobiología. EcoCiencia, Quito, pp 57–72

Mena-Valenzuela P, Regalado J, Cueva R (1997) Oferta de animales en el bosque y cacería en la comunidad huaorani de Quehueire'ono, zona de amortiguamiento del Parque Nacional Yasuní, Napo, Ecuador. In: Mena PA, Soldi A, Alarcón R, Chiriboga C, Suárez L (eds) Estudios biológicos para la conservación: diversidad, ecología y etnobiología. EcoCiencia, Quito, pp 395–426

Mendes A, Nogueira-Filho SLG (2012) Feeds and nutrition of farmed capybaras. In: Moreira JR, Ferraz KMPMB, Herrera EA, Macdonald DW (eds) Capybara: biology, use and conservation of an exceptional neotropical species. Springer, New York, pp 261–274

Miatello R (1994) Vertebrados en vías de extinción de la provincia de Córdoba. Tras la Huella (3)11, Córdoba

Miatello R (2003) Capítulo de fauna. in Regiones Naturales de la Provincia de Córdoba. Córdoba, Argentina, Serie C. Publicaciones Técnicas. Dirección de Ambiente, Córdoba, pp 1–101

Miglino MA, dos Santos TC, Kanashiro C, Ferraz RHS (2012) Morphology and reproductive physiology of female capybaras. In: Moreira JR, Ferraz KMPMB, Herrera EA, Macdonald DW (eds) Capybara: biology, use and conservation of an exceptional neotropical species. Springer, New York, pp 131–146

Mones A (1973) Estudios sobre la familia *Hydrochoeridae* (Rodentia). I. Introducción e historia taxonomica. Rev Bras Biol 33:277–283

Mones A (1975) Estudios sobre la familia *Hydrochoeridae* (Rodentia), VI. Catálogo anotado de los ejemplares-tipo, vol 1, Comunicaciones Paleontológicas del Museo de Historia Natural de Montevideo. Museo de Historia Natural, Montevideo, pp 99–130

Mones A (1984) Estudios sobre la familia Hydrochoeridae. XIV. Revisión sistemática. Senckenb Biol 65:1–17

Mones A, Mones UK (1981) Estudios sobre la familia Hydrochoeridae (Rodentia), X. Nomenclatura vernácula del genero *Hydrochoerus* Brisson, 1762. Rev Fac Humanidades Cien Cienc Biolog 1:225–260

Mones A, Ojasti J (1986) *Hydrochoerus hydrochaeris*. Mamm Species 264:1–7

Montenegro O, Escobedo M (2004) Mamíferos. In: Pitman N (ed) Perú: Ampiyacu, Apayacu, Yaguas, Medio Putumayo, vol 12, Rapid biological inventories report. The Field Museum, Chicago, pp 254–261

Moreira JR (1995) The reproduction, demography and management of Capybaras (*Hydrochaeris hydrochaeris*) on Marajó Island – Brazil. D.Phil. thesis, University of Oxford, Oxford, UK

Moreira JR (2004) Avaliação do atual estado de uso e da conservação da capivara no Brasil. In: XLI Reunião Anual da Sociedade Brasileira de Zootecnia. Embrapa Gado de Corte, Campo Grande, pp 517–527

Moreira JR, Macdonald DW (1996) Capybara use and conservation in South America. In: Taylor VJ, Dunstone N (eds) The exploitation of mammal populations. Chapman and Hall, London, UK, pp 88–101

Moreira JR, Pinha PRS, Cunha HJ (2001) Capivaras do Lago Paranoá. In: Fonseca FO (ed) Olhares sobre o Lago Paranoá. Secretaria de Meio Ambiente e Recursos Hídricos, Brasília, pp 141–147

Moreira JR, Pinheiro MS (2012) Capybara production in Brazil: captive breeding or sustainable management? In: Moreira JR, Ferraz KMPMB, Herrera EA, Macdonald DW (eds) Capybara: biology, use and conservation of an exceptional neotropical species. Springer, New York, pp 333–344

Moreira JR, Wiederhecker H, Ferraz KMPMB, Aldana-Domínguez J, Verdade LM, Macdonald DW (2012) Capybara demographic traits. In: Moreira JR, Ferraz KMPMB, Herrera EA, Macdonald DW (eds) Capybara: biology, use and conservation of an exceptional neotropical species. Springer, New York, pp 147–167

Mourão G, Campos Z (1995) Survey of broad-snouted caiman *Caiman latirostris*, marsh deer *Blastocerus dichotomus* and capybara *Hydrochaeris hydrochaeris* in the area to be inundated by Porto Primavera dam, Brazil. Biol Conserv 73:27–31

Myers P (1982) Origins and affinities of the mammal fauna of Paraguay. In: Mares MA, Genoways HH (eds) Mammalian biology in South America, vol 6, Special publication series, Pymatuning Laboratory of Ecology. University of Pittsburgh. Pittsburgh, pp 85–93

Nowak RM, Paradiso L (1983) Walker's mammals of the world, 4th edn. Johns Hopkins University Press, Baltimore

Ojasti J (1973) Estudio biologico del chigüire o capibara. FONAIAP, Caracas

Ojasti J (2011) Estudio biologico del chigüire o capibara, 2nd edn. Editorial Equinoccio, Caracas

Ojasti J (1991) Human exploitation of capybara. In: Robinson JG, Redford KH (eds) Neotropical wildlife use and conservation. University of Chicago Press, Chicago, pp 236–252

de Oliveira JA, Bonvicino CR (2006) Ordem Rodentia. In: Reis NRdos, Peracchi AL, Pedro WA, de Lima IP (eds) Mamíferos do Brasil. Nelio R. dos Reis, Londrina, pp 347–406

Pacheco V, Arias L (2001) Mamíferos. In: Alverson WS, Rodriguez LO, Moskovits DK (eds) Perú: Biabo Cordillera Azul, vol 2, Rapid biological inventories report. The Field Museum, Chicago, pp 226–227

Pacheco V, Vivar E (1996) Annotated checklist of the non-flying mammals at Pakitza, Manu Reserve Zone, Manu National Park, Perú. In: Wilson DE, Sandoval A (eds) Manu, the biodiversity of southeastern Peru. Smithsonian Institution, Ed. Horizonte, Lima, pp 577–592

Paiva MP (1973) Distribuição e abundância de alguns mamíferos selvagens no Estado do Ceará. Ciênc Cult 25:442–450

Parra R, González-Jiménez E (1972) Fisiología digestiva del chigüire (*Hydrochoerus hydrochaeris*). 1. Capacidad de los diferentes compartimientos del tracto digestivo. Acta Cient Venez 23(1):30

Patton JL, Berlin B, Berlin EA (1982) Aboriginal perspectives of a mammal community in Amazonian Perú: knowledge and utilization patterns among the Aguaruna Jivaro. In: Mares MA, Genoways HH (eds) Mammalian biology in South America, vol 6, University of Pittsburgh, Pymatuning Laboratory of Ecology. Special Publication Series. University of Pittsburgh, Linesville, pp 111–128

Paula TAR, Walker NJ (2012) Reproductive morphology and physiology of the male capybara. In: Moreira JR, Ferraz KMPMB, Herrera EA, Macdonald DW (eds) Capybara: biology, use and conservation of an exceptional neotropical species. Springer, New York, pp 107–129

Peceño MC (1983) Estudio Citogenética y Genético-evolutivo del "Chigüire", Género Hydrochaeris. Licenciado en Biología. Graduate dissertation, Universidad Simón Bolivar, Sartenejas-Baruta

Pereira JN, McEvan D, Finley E (1980) The structure of the skin of the capybara. Acta Cient Venez 31:361–364

Perry A, Hennessey AB, Ríos BU (1996) Evaluación biológica de la región Beni-Suapi-Beu-Chepite. Reserva de la Biosfera y Territorio Indígena Pilón Lajas, Bolivia. Trex. La Paz, Bolivia. 66pp

Piso W (1658) De Indiae utriusque re naturali et medica libri quatuordecim / / quorum contenta pagina sequens exhibet. Ludovicum et Danielem Elzevirios, Amsterdam, 327pp

Quintana RD, Bolkovic ML (2012) Use of capybaras in Argentina. In: Moreira JR, Ferraz KMPMB, Herrera EA, Macdonald DW (eds) Capybara: biology, use and conservation of an exceptional neotropical species. Springer, New York, pp 345–356

Rageot R, Albuja L (1994) Mamíferos de un sector de la alta Amazonía ecuatoriana: Mera, provincia de Pastaza. Rev Politécnica 19(2):165–208

Ray J (1693) Synopsis methodica animalium quadrupedum et serpentini generis. Vulgarium natas characteristicas, rariorum descriptiones integras exhibens: cum historiis & observationibus anatomicis perquam curiosis. Præmittuntur nonnulla de animalium in genere, sensu, generatione, divisione, & c. S. Smith & B. Walford, London

Reati G, Florín M, Fernández G, Montes C (1997) The Laguna Mar Chiquita (Córdoba, Argentina): a little known, secularly fluctuating, saline lake. Int J Salt Lake Res 5:187–219

Rechenberg E, Bertelli PW, Oechsler A, Debatin T (2000) Ocorrência de três ataques de capivara (*Hydrochaeris hydrochaeris*) à espécie humana no município de Blumenau (SC). In: Resumos do Congresso Brasileiro de Zoologia, vol 23. Universidade Federal de Mato Grosso, Cuiabá, p 639, Resumo MA228

Redford KH, Eisenberg JF (1992) Mammals of the Neotropics: the Southern Cone, vol 2, Chile, Argentina, Uruguay, Paraguay. The University of Chicago, Chicago, p 430

Rocha D (1948) Subsídio para o estudo da fauna cearense. Rev Inst Ceará 62:102–138

Rodríguez M et al (2003) Caracterización de las poblaciones silvestres de chigüiro y sus hábitats en las sabanas anegables del departamento de Casanare con miras a formular una propuesta de conservación y uso sostenible. Informe final. Grupo de Estudios Ecológicos OIKOS – Gobernación de Casanare, Bogotá

Rodríguez-Mahecha JV, Hernández-Camacho JI, Defler TR, Alberico M, Mast RB, Mittermeier RA, Cadena A (1995) Mamíferos colombianos: sus nombres comunes e indígenas, vol 3, Conservation International. Occasional papers in conservation biology., pp 1–56

Roig Vg (1991) Desertification and distribution of mammals in the Southern Cone of South America. In: Mares MA, Schmidly DJ (eds) Latin American mammals: their conservation, ecology, and evolution. University of Oklahoma Press, Norman, pp 239–279, 468pp

Romo M, Luna L, Cornejo AF, Kopper C (2002) Comparación de ocurrencia de especies de mamiferos del Snph [Santuario Nacional Pampas Del Heath] (Rap 1992, e 1996) Con otras localidades de Madre de Dios. In: Montambault JR (ed) Informes de las evaluaciones biologicas Pampas del Heath, Peru Alto Madidi, Bolivia y Pando, Bolivia, vol 24, Rapid Assessment Program bulletin of biological assessment. Conservation Internacional, Washington, DC, pp 105–109, 124pp

Rowe DL, Honeycutt RL (2002) Phylogenetic relationships, ecological correlates, and molecular evolution within the Cavioidea (Mammalian, Rodentia). Mol Biol Evol 19:263–277

Rumiz DI, Eulert CF, Arispe R (1998) Evaluación de la diversidad de mamíferos medianos y grandes en el Parque Nacional Carrasco (Cochabamba – Bolivia). Rev Boliviana Ecol 4:77–90

Rumiz DI et al (2002) La biodiversidad de la estancia San Miguelito, Santa Cruz – Bolivia: una justificación para establecer reservas privadas de conservación, vol 1, Documentos, Serie Biodiversidad. Ecología en Bolivia, La Paz

Saez FA, Drets ME, Brum-Zorilla N (1971) Karyotype of the "Carpincho" *Hydrochaeris hydrochaeris uruguayensis* (Rodentia Hydrochaeridae). Experientia 27:584–585

Salovaara K, Bodmer R, Recharte M, Reyes CF (2003) Diversidad y abundancia de Mamíferos. In: Pitman N, Vriesendorp C, Moskovits D (eds) Perú: Yavarí, vol 11, Rapid biological inventories report. The Field Museum, Chicago, pp 74–82

Schaller GB, Crawshaw PG Jr (1981) Social organization in a capybara population. Säugetierkundlichen Mitteilungen 29:3–16

Smith JDD (2001) Official lists and indexes of names and works in zoology. Supplement 1986–2000. International Trust for Zoological Nomenclature, London

Soini P (1992) Un estudio de la dinámica poblacional del ronsoco o capibara (*Hydrochoerus hydrochaeris*) en el río Pacaya, Perú. Folia Amazonica 5:139–156

Soini P, Dosantos A, Calle A, Arias L (2001) Mamíferos. In: Conservación y Manejo de la Biodiversidad de la Cuenca del Pucacuro. Informe Técnico IIAP, Iquitos, Peru, p 28–35

Soini P, Soini M (1992) Ecologia del ronsoco o capibara (*Hydrochoerus hydrochaeris*) en la reserva nacional Pacaya-Samiria, Peru. Folia Amazonica 4:135–150

Solari S, Pacheco V, Luna L, Velazco PM, Patterson BD (2006) Mammals of the Manu Biosphere Reserve. In: Patterson BD, Stotz DF, Solari S (eds) Mammals and birds of the Manu Biosphere Reserve, Peru, vol 110, Fieldiana: zoology, new series. Field Museum of Natural History, Chicago, pp 13–22

Staden H (1557) Warhaftig Historia vnd beschreibung eyner Landtschafft der Wilden/Nacketen/ Grimmigen Menschfresser Leuthen/in der Newenwelt America gelegen/vor vnd nach Christi geburt im Land zu Hessen vnbekant/biß vff dise ij nechst vergangene jar/Da sie Hans Staden von Homberg auß Hessen durch sein eygne erfarung erkant/vnd yetzo durch den truck an tag gibt. H. Philipsen, Marburg

Taber A (1994) Anexo VI: Lista de mamíferos del área protegida propuesta. In: Parque Nacional y Área Natural de Manejo Integrado Kaa-Iya del Gran Chaco. Propuesta Técnica y Étnica. C.A.B.I. – Capitanía del Alto y Bajo Izozog, Fundación Ïvï-Iyambae and Wildlife Conservation Society, La Paz, pp 39–45

Tarifa T, Aliaga E, Ríos B, Hagaman D (2001) Mamíferos del Parque Nacional Madidi. Hisbol. La Paz, Bolivia. 187pp

Tirira DG (2004) Nombres de los mamíferos del Ecuador. Murciélago Blanco y Museo Ecuatoriano de Ciencias Naturales, vol 5, Publicación especial sobre los mamíferos del Ecuador. Ediciones Murciélago Blanco, Quito

Tirira DG (2007) Guía de campo de los mamíferos del Ecuador, vol 6. Publicación especial sobre los mamíferos del Ecuador. Murciélago Blanco, Quito

Torres E, Sanabria J (1976) Aspectos ecológicos del chigüiro y establecimiento de un zoocriadero. Graduate dissertation, Universidad Nacional de Colombia, Santa Fe de Bogotá

Torrico A, Santivañez JL, Rojas JM, Caballero M (1997) Lista de mamíferos censados en el área de Otuquis. In: Áreas protegidas pantanal de Otuquis y San Matías. Universidad Autónoma Gabriel René Moreno, Museo de Historia Natural Noel Kempff Mercado. Santa Cruz, Bolivia, 60pp. + Anexos

Trapido H (1947) The Isthmian capibara in the Canal Zone. J Mammal 28:408–409

Trapido H (1949a) The Isthmian capibara in the Tocumen Savannas. Panama. J Mammal 30:80

Trapido H (1949b) Gestation period, young, and maximum weight of the isthmian capybara. *Hydrochoerus isthmius* Goldman. J Mammal 30:433

Usma S (1991) Aspectos ecológicos del chigüiro (*Hydrochaeris hydrochaeris*) en la Reserva Natural Laguna de Sonso. Graduate dissertation. Universidad del Valle, Cali

Valqui M (2001) Mammal diversity and ecology of small mammals in Western Amazonia. Ph.D. thesis, University of Florida, Gainesville, xv, p 207

Vanzolini PE (1996) A contribuição zoológica dos primeiros viajantes naturalistas no Brasil. Rev USP 30:190–238

Verdade LM, Ferraz KMPMB (2006) Capybaras on an anthropogenic habitat in Southeastern Brazil. Braz J Biol 66:29–41

Voss RS, Emmons LH (1996) Mammalian diversity in Neotropical Lowland rainforest: preliminary assessment. Bull Am Mus Nat His 230:1–115

Voss RS, Lunde DP, Simmons NB (2001) The mammals of Paracou, French Guiana: a Neotropical lowland rainforest fauna. Part 2. Nonvolant species. Bull Am Mus Nat Hist 263:1–236

Vucetich MG, Deschamps CM, Pérez ME (2012) Paleontology, evolution and systematics of capybara. In: Moreira JR, Ferraz KMPMB, Herrera EA, Macdonald DW (eds) Capybara: biology, use and conservation of an exceptional neotropical species. Springer, New York, pp 39–59

Wilson DE, Reeder DAM (1993) Mammals species of the World, a taxonomic and geographic reference, 2nd edn. Smithsonian Institution Press, Washington, DC

Wilson DE, Reeder DAM (2005) Mammals species of the world, a taxonomic and geographic reference, 3rd edn. Smithsonian Institution Press, Washington, DC

Yáber MC, Herrera EA (1994) Vigilance, group size and social status in capybaras. Anim Behav 48:1301–1307

Yahnke CJ, de Fox IG, Colman F (1998) Mammalian species richness in Paraguay: the effectiveness of National Parks in preserving biodiversity. Biol Conserv 84:263–268

Zara JL (1973) Breeding and husbandry of the capybara *(Hidrochaerus hydrochaeris)* at Evansville Zoo. Int Zoo Yearbook 13:137–139

Ziegler T, Unger J, Feiler A, Kraft R (2002) On the occurrence of the capybara, *Hydrochoerus hydrochaeris* (Linnaeus, 1766), in the Dry Chaco of Paraguay (Mammalia:Rodentia:Hydrocho eridae). Faun Abh Mus Tierkde Dresden 22(27):423–430

Chapter 2
Paleontology, Evolution and Systematics of Capybara

María Guiomar Vucetich, Cecilia M. Deschamps, and María E. Pérez

2.1 Introduction

Caviomorph rodents, the New World Hystricognathi, are one of the most noteworthy groups of mammals in South America. Isolated for more than 30 million years, they have given rise to several extraordinary rodents including the pacas (*Cuniculus* spp.), cavies (*Cavia* spp.), vizcachas (*Lagostomus* spp.), agouties (*Dasyprocta* spp.), as well as two giants: *Phoberomys*, from the late Miocene (6 Ma; mega annum = million years) which probably weighed more than 400 kg (Sánchez Villagra et al. 2003), and *Josephoartigasia monesi* probably from the Pleistocene (2.5 Ma), at about 1,000 kg (Rinderknecht and Blanco 2008). Capybaras (*Hydrochoerus* spp.) are particularly conspicuous because of their size – they are the largest living rodents – and their aggregation in herds (Ojasti 1973; Macdonald 1981; Macdonald et al. 2007). This chapter examines their paleontology and evolution, focusing particularly on another striking feature: their unpaired ever-growing cheek teeth whose very complicated occlusal surface design changes throughout the capybara's life (Vucetich et al. 2005).

M.G. Vucetich (✉)
Departamento Paleontología Vertebrados, Museo de La Plata, CONICET, Consejo Nacional de Investigaciones Científicas y Técnicas, Paseo del Bosque s/n, La Plata B1900FWA, Argentina
e-mail: vucetich@fcnym.unlp.edu.ar

C.M. Deschamps
Departamento Paleontología Vertebrados, Museo de La Plata, CIC, Comisión de Investigaciones Científicas de la Provincia de Buenos Aires, Paseo del Bosque s/n, La Plata B1900FWA, Argentina
e-mail: ceci@fcnym.unlp.edu.ar

M.E. Pérez
Museo Paleontológico Egidio Feruglio, CONICET, Consejo Nacional de Investigaciones Científicas y Técnicas, Av. Fontana 140, Trelew U9100GYO, Argentina
e-mail: mperez@mef.org.ar

J.R. Moreira et al. (eds.), *Capybara: Biology, Use and Conservation of an Exceptional Neotropical Species*, DOI 10.1007/978-1-4614-4000-0_2,

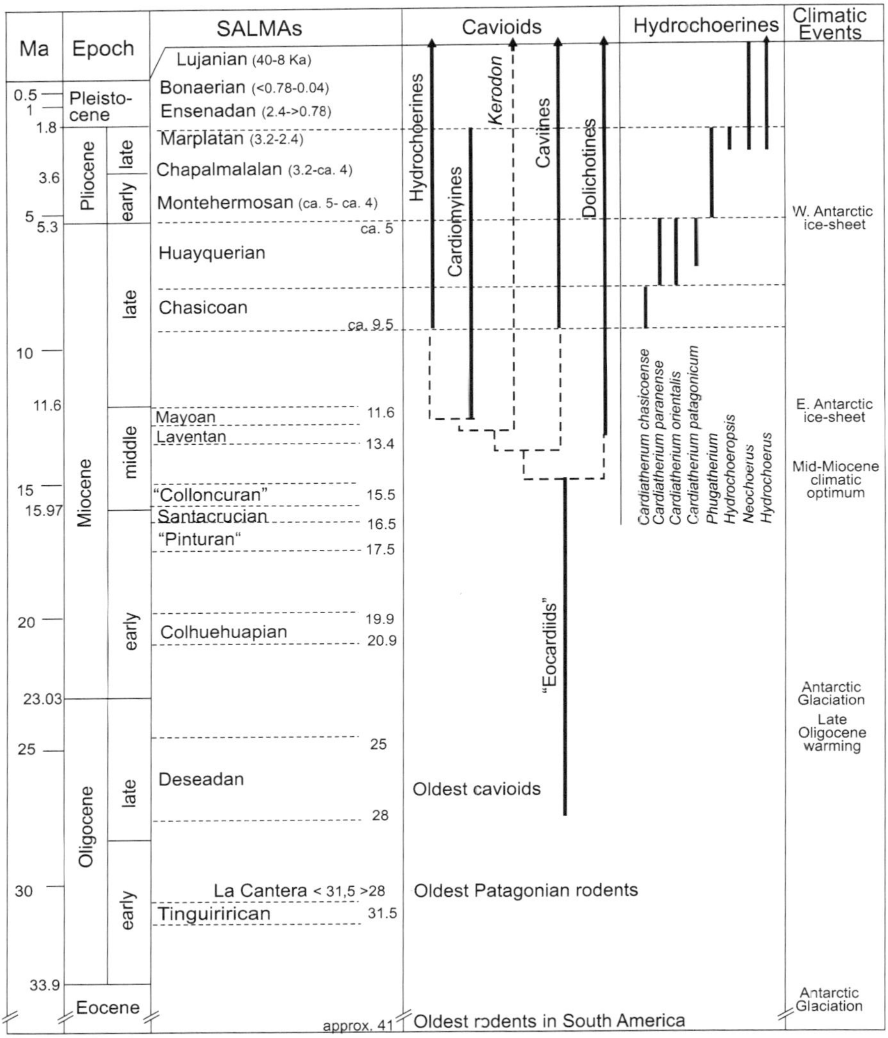

Fig. 2.1 Stratigraphic chart showing the oldest records of caviomorph and cavioid rodents, cavioid phylogenetic relationships, and the stratigraphic record of fossil capybaras against the main global climatic events from Zachos et al. (2001). *Ma* mega annum (million years), *SALMA* South American Land-Mammal Age

Capybaras are undoubtedly allied to the living cavies and Patagonian "hares" (*Dolichotis* spp.) and their extinct relatives. However, the relationships between capybaras and these taxa are not clear. The fossil capybara record begins in the Chasicoan South American Land-Mammal Age (SALMA), during the early-late Miocene (ca. 9–7 Ma), in Central Argentina (Fig. 2.1). Ameghino (1883) was the first to describe an extinct species of capybara, based on a fragmented mandible from the "conglomerado osífero" in the Ituzaingó Formation, Huayquerian SALMA, late Miocene, which outcrops in the vicinity of the city of Paraná, Entre Rios Province, Argentina (Fig. 2.2). A large number of new genera and species, grouped in four subfamilies, were later nominated, following findings of additional fragmentary remains of different sizes and differing occlusal cheek teeth design (Mones 1991). Originally, capybaras were considered to be highly diversified, with a long, slow evolutionary history. Recent findings have, however, permitted a new interpretation of the fossil record. With this in mind, we describe the evolutionary history of capybaras and discuss their relationships with other South American hystricognath rodents based on paleontological data.

2.2 The Origin and Early Radiation of South American Hystricognath Rodents

Caviomorphs are immigrants to South America. It is generally believed that they are more closely related to the Old World hystricognaths than to any other group of rodents, whether they are considered monophyletic or not (Lavocat 1976; Patterson and Wood 1982; Martin 1994; Nedbal et al. 1994; Bryant and McKenna 1995; Marivaux et al. 2002; Poux et al. 2006; Blanga-Kanfi et al. 2009; Sallam et al. 2009; Vilela et al. 2009), and that they probably came from Africa by raft during the early Eocene (about 41 Ma; Vucetich et al. 2010a).

The oldest known hystricognath rodents of South America come from Contamana in Peru (Antoine et al. 2011), Tinguiririca in Chile (Figs. 2.1 and 2.2; Flynn et al. 2003 and references therein), and La Cantera site in Gran Barranca, Patagonia, Chubut Province, Argentina (Vucetich et al. 2010a), estimated at 31.5 Ma and 31.5–28 Ma, respectively (Figs. 2.1 and 2.2). The rodents found in Santa Rosa, Peru, were proposed as the oldest in South America, but estimations of the age of this site are dubious, dating between the late Eocene and late Oligocene (37–23 Ma; Figs. 2.1 and 2.2; Campbell 2004; Frailey and Campbell 2004; Shockey et al. 2004). The rodents from this site are abundant, brachyodont (Box 2.1) and relatively homogeneous in morphology when compared to those of other old fauna. In Tinguiririca and La Cantera, remains are scarcer but show great morphological diversity with a clear trend toward hypsodonty (high crown teeth; Box 2.1) in some taxa (Wyss et al. 1993; Flynn et al. 2003; Vucetich et al. 2010a).

The degree of diversification shown by these rodents suggests that the initial radiation occurred well before these records. Vucetich et al. (1999) suggest that caviomorphs colonized South America during the late or even early Eocene (ca. 55–34 Ma; Fig. 2.1). Taking the same line, Poux et al. (2006) suggest that caviomorphs radiated during the late Eocene, before the early Oligocene (33.9 Ma) cooling (Fig. 2.1; Zachos et al. 2001). Contamana findings confirm these hypotheses. Opazo (2005) disagrees, suggesting that radiation took place a little later in the early Oligocene. The earliest differentiation probably did occur before the early Oligocene cooling (Antarctic Glaciation; Fig. 2.1), but differentiation of the first high-crowned caviomorphs, the primitive cavioids and chinchilloids, appears to have occurred in association with this cooling.

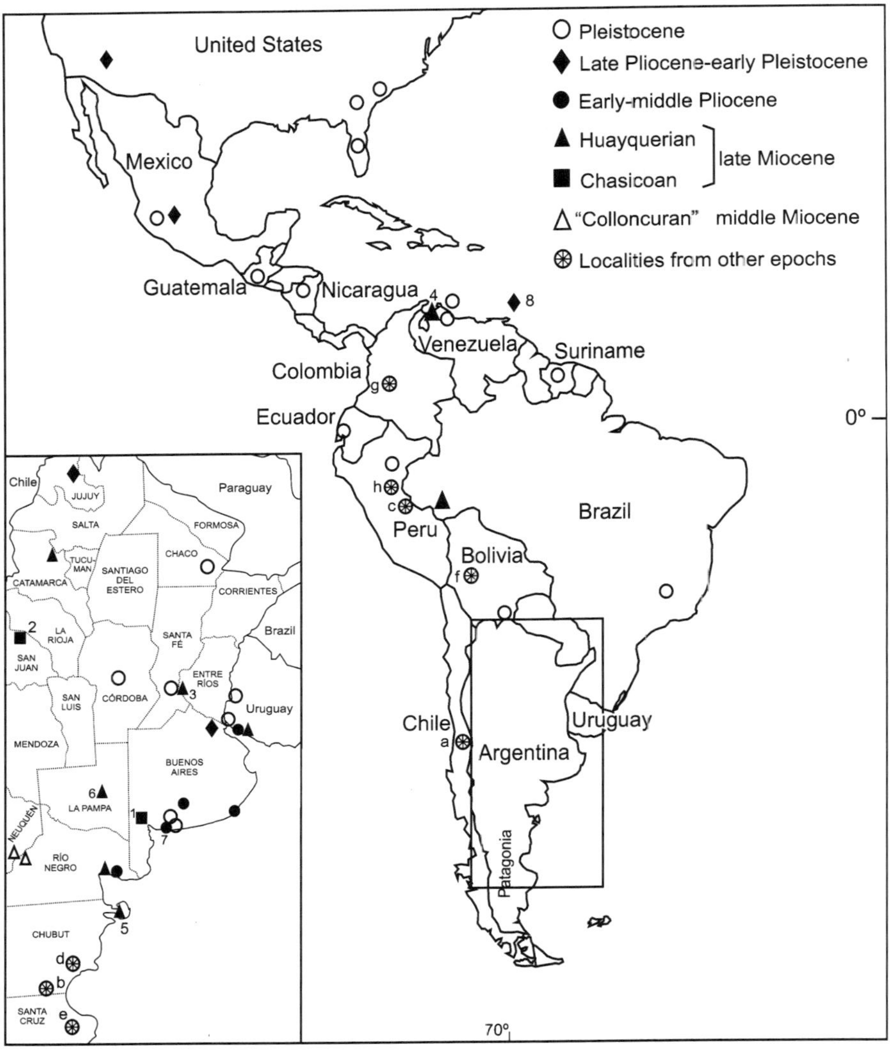

Fig. 2.2 Main localities of fossil capybaras. Locations mentioned in the text are referenced as follows: (**1**) Arroyo Chasicó, Buenos Aires Province, Argentina; (**2**) Albardón, San Juan Province, Argentina; (**3**) "conglomerado osífero" of the Ituzaingó Formation, Paraná City, Entre Ríos Province, Argentina; (**4**) Estado Falcón, Venezuela; (**5**) Estancia Rincón Chico, Península Valdés, Chubut Province, Argentina; (**6**) Laguna Chillhué and Laguna Guatraché, La Pampa Province, Argentina; (**7**) Farola Monte Hermoso, Buenos Aires Province, Argentina; (**8**) Grenada, Lesser Antilles. Symbols labeled (**a–g**) indicate other localities mentioned in the text: (**a**) Tinguiririca, Chile; (**b**) La Cantera, Chubut Province, Argentina; (**c**) Santa Rosa, Peru; (**d**) Cabeza Blanca, Chubut Province, Argentina; (**e**) La Flecha, Santa Cruz Province, Argentina; (**f**) Salla-Luribay, Bolivia; (**g**) La Venta, Colombia; (**h**) Contamana, Peru

Box 2.1 Different Types of Mammalian Teeth

Generalized mammals have teeth with roots and a short crown that do not change throughout life. Other mammals, like capybaras, horses, and rabbits have high-crowned teeth usually named hypsodont teeth. In these teeth, roots delay their development and the crown keeps on growing throughout life or at least part of it. This phenomenon is called hypsodonty. Hypsodonty is an adaptation to extend the life of teeth and, thus, the life of the animal, facing increasing rates of tooth wear resulting either from eating more abrasive plant tissues (phytoliths) or wear-inducing particles, such as wind-blown grit, that adhere to plant surfaces. Hypsodonty is also related to open environments in which animals feed closer to the ground (Janis and Fortelius 1988; Janis et al. 2000).

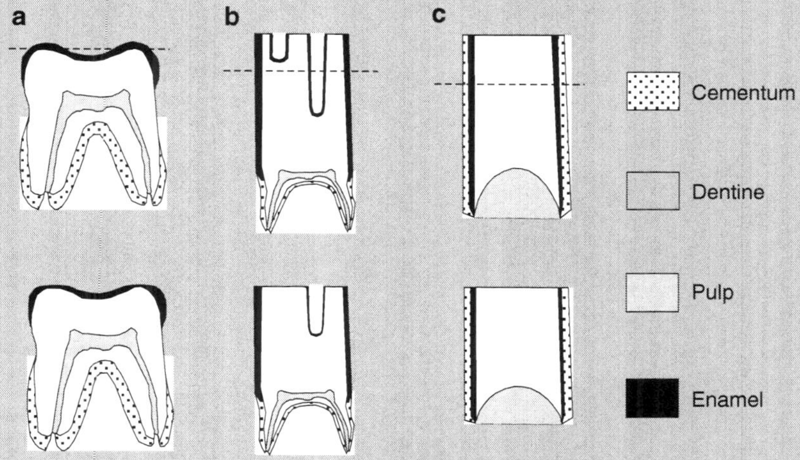

The diagrams show (**a**) brachyodont (low-crowned), (**b**) protohypsodont (high-crowned), and (**c**) euhypsodont (ever growing) teeth and their change with wear (lower row). The occlusal morphology varies accordingly.

Southern South America provided special conditions for the early development of hypsodonty. On the one hand, there was a general trend toward climatic deterioration after the Eocene-Oligocene transition (33.9 Ma), which would have been more intense at high latitudes. On the other hand, during the early Cenozoic (60–20 Ma), periods of intense volcanism affected Patagonia, providing large volumes of volcanic glass in the sediments (Mazzoni 1985; Bellosi 2010). These two elements, climatic deterioration and a large amount of abrasive material, deposited on plants that early rodents ate, favored the development of hypsodonty (Kay et al. 1999). These conditions could explain why rodents with a tendency to hypsodonty are recorded in Tinguiririca and La Cantera, but not further north in Santa Rosa (Fig. 2.2), which probably saw milder climatic conditions and no volcanic glass in the sediments.

2.3 Divergence of the Hydrochoeridae

Although the oldest caviomorphs derive from the middle Eocene (Fig. 2.1), it is not until the late Oligocene (Deseadan SALMA – ca. 28–25 Ma; Fig. 2.1) that the caviomorph record becomes abundant and continuous (Vucetich et al. 1999). Eocardiids (Box 2.2; Figs. 2.1 and 2.3), considered the stem group for Caviidae and Hydrochoeridae (Box 2.2), are first recorded in Argentinean Patagonia at Cabeza Blanca, Chubut Province, and La Flecha, Santa Cruz Province (Fig. 2.2), during the Deseadan SALMA (Fig. 2.1). No eocardiids have been recorded in the Deseadan of Salla-Luribay, Bolivia (Fig. 2.2), or in Santa Rosa. From this, it can be inferred that the diversification of the Cavioidea s.s. (Box 2.2) took place in Patagonia, or at least in the southern part of the continent. The diversification and later evolution of the

Box 2.2 Proposed Relationships Among the Cavioidea

Eocardiidae, Caviidae, and Hydrochoeridae are a very cohesive group sharing heart-shaped cheek teeth, among other characters. All three families are grouped together in the superfamily Cavioidea, although some authors include others in this superfamily, such as Dasyproctidae, Cuniculidae, and Dinomyidae. Patterson and Wood (1982) considered the three first families as the Cavioidea s.s. excluding the Dasyproctidae and other putative cavioids from this informal group. In the scheme below, we illustrate proposed relationships among the Cavioidea. Relationships with Octodontoids and Chinchilloids are controversial and not discussed in this entry.

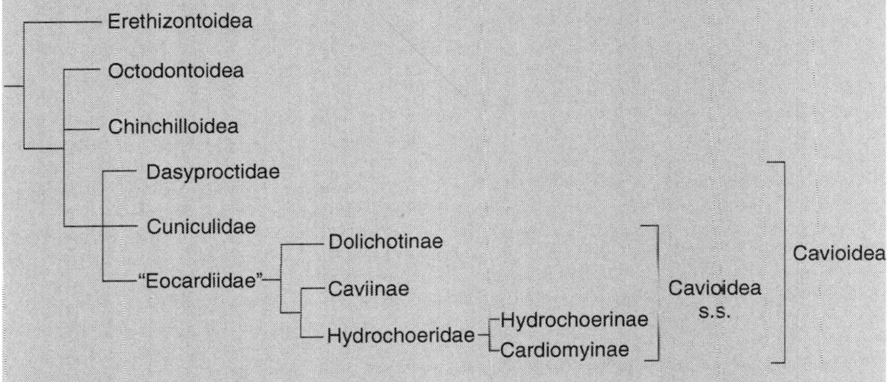

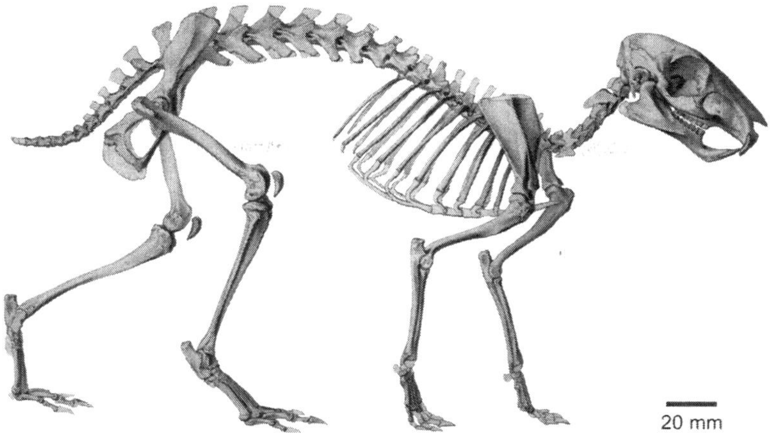

Fig. 2.3 Reconstruction of fossil *Eocardia* (Drawing by Bruce Horsfall, for Scott (1905))

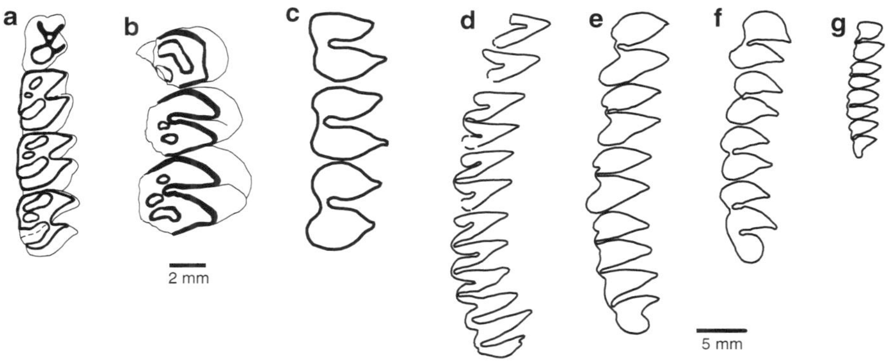

Fig. 2.4 Dental series in occlusal view. (**a**) *Asteromys* (right p4-m3); (**b**) *Luantus* (right p4-m2); (**c**) *Eocardia* (right p4-m2); (**d–g**) right P4-M3 of (**d**) *Caviodon*; (**e**) *Cardiomys*; (**f**) *Dolichotis*; (**g**) *Kerodon*. Anterior above

Cavioidea s.s. involved the development of high-crowned cheek teeth. The evolution of hypsodonty in this group can be followed from the Deseadan SALMA onward, starting with the mesodont (slightly higher-crowned than brachyodont) *Asteromys* (Fig. 2.4a), then the protohypsodont species of *Luantus* from the Colhuehuapian – "Pinturan" SALMAs (Figs. 2.1 and 2.4b), up to the euhypsodont *Eocardia* (Figs. 2.3 and 2.4c) of the Santacrucian SALMA (late early Miocene – 16.5–15.5 Ma; Fig. 2.1). This was a rather slow process of change that took about 12 million years, during which euhypsodonty and heart-shaped teeth, characteristic of the Cavioidea s.s., arose. In fact, it is very difficult to separate the most derived species of eocardiids from the first Caviidae only based on tooth morphology (but see Pérez 2010b; Pérez and Vucetich 2011).

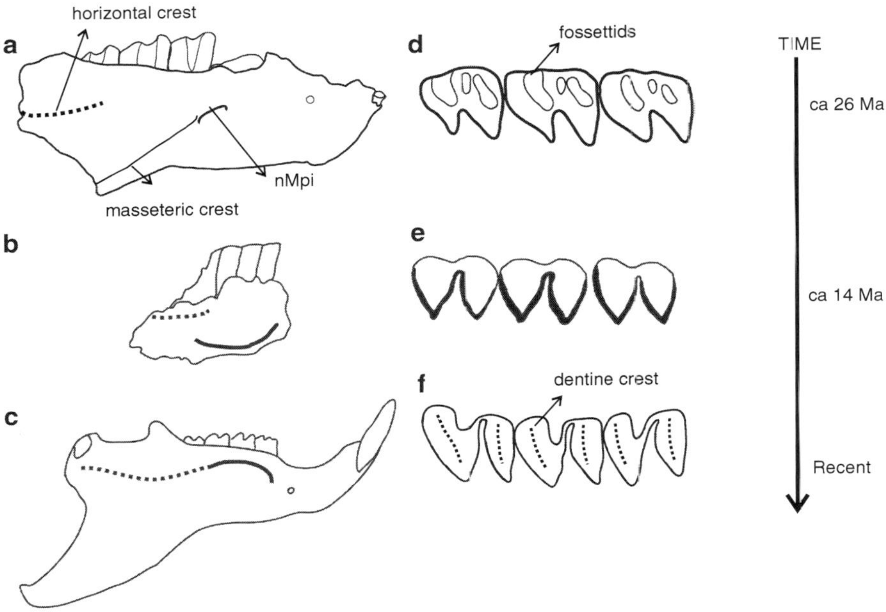

Fig. 2.5 Evolution of selected characters in Cavioidea s.s. (**a–c**) mandible; (**d–f**) m1–m3 in occlusal surface (anterior to the right). (**a**, **d**) *Asteromys*; (**b**, **e**) *Guiomys*; (**c**, **f**) *Galea*. nMpi notch for the insertion of the masseter medialis pars infraorbitalis. Not to scale. The nMpi changes its relation to other crests of the mandible: in *Asteromys* it is joined to the masseteric crest, in *Guiomys* it is isolated, and in *Galea* it is joined to the horizontal crest. Cheek teeth in *Asteromys* are mesodont, retain fossettids, and have different sizes; in *Guiomys* they are euhypsodont without fossettids, and similar in size; and in *Galea*, to these features they add a central dentine crest (Pérez 2010a, b; Pérez and Vucetich 2011)

The last eocardiids come from the "Colloncuran" (early middle Miocene; ca 15.5 Ma) of northern Patagonia (Figs. 2.1 and 2.2). The first modern Cavioidea s.s. (Caviinae, Dolichotinae, and Hydrochoeridae) come from the middle Miocene of La Venta in Colombia, Laventan SALMA, 11.8–13.5 Ma (Figs. 2.1 and 2.2; Fields 1957; Madden et al. 1997; Walton 1997) with the dolichotine *Prodolichotis*, but they were well diversified by the Chasicoan SALMA (Fig. 2.1). Species with intermediate morphology between eocardiids and modern Cavioidea s.s. (Fig. 2.5) were found in Patagonian sediments assigned to the "Colloncuran" and in the Laventan SALMA, Colombia (Pérez 2010a, b; Pérez and Vucetich 2011).

This temporal context allows at least two interpretations. On the one hand, all modern groups could have differentiated together between 13.5 and 15.5 Ma (Fig. 2.1), and the lack of hydrochoerids and caviines during that period could be an artifact of the paleontological record. On the other hand, dolichotines could have differentiated before the caviines and hydrochoerids. In accordance with the second hypothesis, dolichotines have the most primitive cheek teeth morphology and enamel microstructure among modern Cavioidea s.s. (Vieytes et al. 2001; Vieytes 2003), similar to those of eocardiids. Additionally, the Caviinae and

Hydrochoeridae share a more derived cheek teeth enamel microstructure (Vieytes et al. 2001; Vieytes 2003).

The cardiomyines are a group of large cavioids that were first recorded during the "Mayoan" (Fig. 2.1; Vucetich and Pérez, 2011). Like capybaras, they have an increasing number of laminae in the M3 and p4 (Figs. 2.4d–e), some deep flexids (see Sect. 2.5.1. Dental Structure, below), and a wide palate. Cardiomyines were previously considered Caviids, but Vucetich et al. (2005) postulated them to be a sister group to capybaras (Box 2.2 and Appendix 1). Recent phylogenetic analyses of Cavioidea (Pérez 2010a, b) support the latter hypothesis.

The differentiation of hydrochoerids would have been relatively rapid, explaining the absence of potential close ancestors from the middle Miocene strata of Patagonia (Vucetich 1984; Vucetich et al. 1993) and Colombia (Walton 1997).

2.4 Geographic Distribution of Fossil Capybaras

The earliest capybaras (*Cardiatherium chasicoense*) derive almost exclusively from Arroyo Chasicó, Buenos Aires Province, Argentina (Deschamps et al. 2007); however, one specimen has recently been described in San Juan Province (Fig. 2.2; Deschamps et al. 2009). By the Huayquerian SALMA (late Miocene; Fig. 2.1), capybaras were distributed throughout most of South America, from northern Venezuela (Linares 2004; Vucetich et al. 2010b) to western Brazil (Frailey 1986), south-western Uruguay (Francis and Mones 1965), and several sites in Argentina (Fig. 2.2). *Cardiatherium patagonicum* represents the most southern hydrochoerid, from Península Valdés, northern Patagonia, Argentina, and it is more derived than the remaining Huayquerian hydrochoerids.

Pliocene (5.3–2.5 Ma) representatives (*Phugatherium*) have been found in several sites in Argentina (Fig. 2.2). MacPhee et al. (2000) reported a new species, *Hydrochoerus gaylordi*, from the late Pliocene (dated 2.6–3.7 Ma) of the Antilles (Fig. 2.2), which may be the oldest record of this genus, but see below. During the Pleistocene-Holocene period (2.5 Ma to Recent), capybaras (*Neochoerus* and *Hydrochoerus*) ranged from southern North America to central Argentina (Fig. 2.2).

2.5 Dental Structure and Its Bearing on the Systematics of Capybaras

2.5.1 Dental Structure

As for most extinct mammals, the taxonomy of extinct capybaras is based mainly on tooth morphology, because teeth are the most frequent remains in the mammal fossil record. The morphology of the occlusal surface is so peculiar in this group

Box 2.3 Specific Nomenclature of the Occlusal Morphology of Capybara Cheek Teeth

Mammalian cheek teeth are intensively studied because they display a number of characters with high systematic and phylogenetic value. Thus, a very detailed nomenclature has been developed, especially for the occlusal surface. In the case of capybaras the occlusal morphology of cheek teeth is so peculiar that it was necessary to develop a specific nomenclature for them (see below). Capybara teeth are euhypsodont (Box 2.1), formed by a prism partially (or completely) divided by re-entrant folds which are also named flexi (uppers) or flexids (lowers). It is traditional in capybara literature to maintain the abbreviated Spanish name of dental characters to avoid further complications in this already complex nomenclature. The length of re-entrant folds varies among species, but also with the age of the animal, making the interpretation of occlusal morphology a complicated but fascinating task. In primitive capybaras, re-entrant folds are comparatively shallow, whereas in the most derived species, such as *H. hydrochaeris*, some of these folds cross to the opposite side, completely dividing the tooth into independent smaller prisms (Fig. 2.6h).

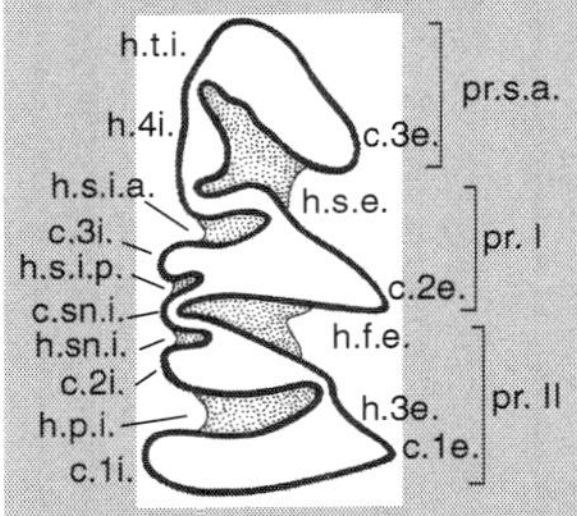

right 4th lower premolar (p4)

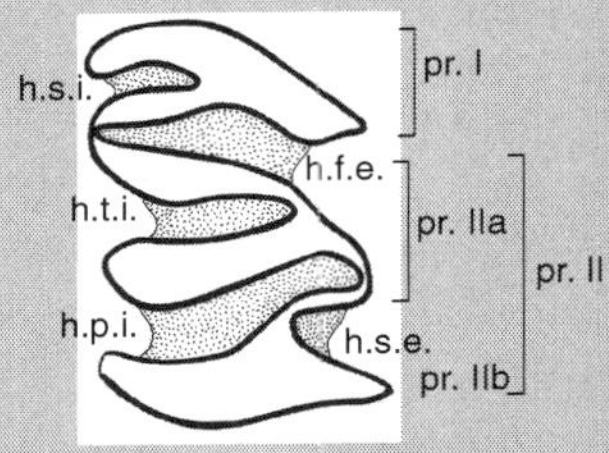

right 1st or 2nd lower molar (m1 or m2)

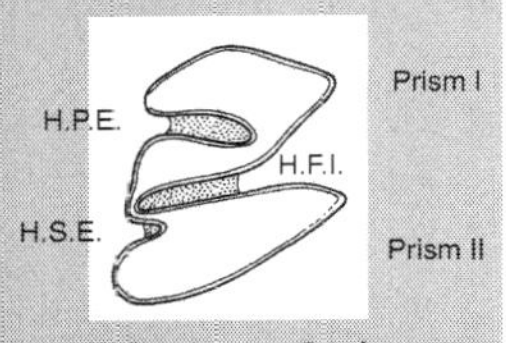

right 1st or 2nd upper molar (M1 or M2)

Nomenclature of cheek teeth. Lower teeth: h.t.i., tertiary internal flexid; h.4i., 4th internal flexus; h.s.i.a., secondary anterior internal flexid; c.3.i., third internal column; h.s.i.p., secondary posterior internal flexid; c.sn.i., supernumerary internal column; h.sn.i., supernumerary internal flexid; c.2i., second internal column; h.p.i., primary internal flexid; c.1i., 1st internal column; pr.s.a., anterior secondary prism; c.3e., third external column; h.s.e., secondary external flexid; c.2e., second external column; h.f.e., fundamental external flexid; h.3e., third external flexid; c.1e., first external column; h.s.i., secondary internal flexid; pr. I-II, II a-b, prisms I-II, II a-b. Upper teeth: H.S.E., secondary external flexus; H.P.E., primary external flexus; H.F.I., fundamental internal flexus.

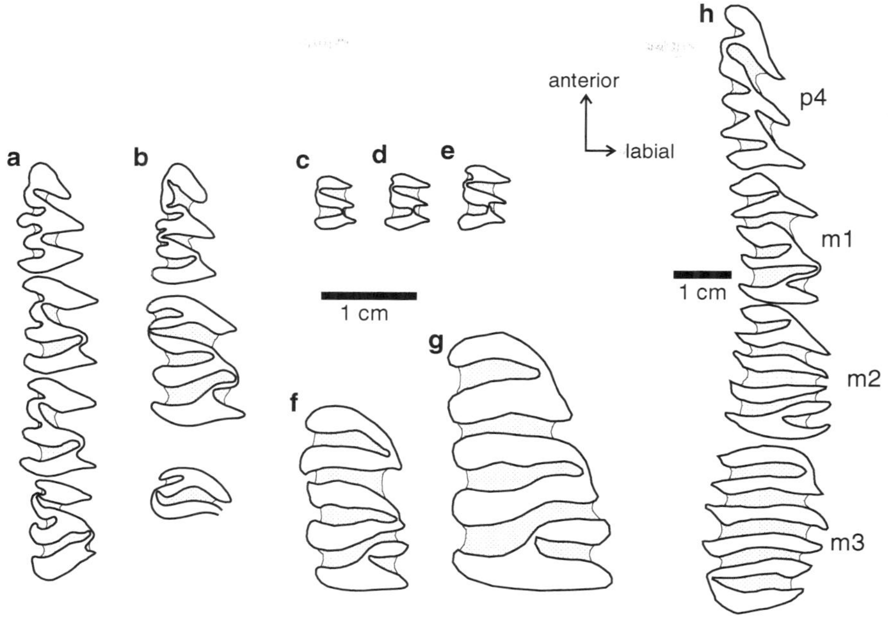

Fig. 2.6 Lower cheek teeth of hydrochoerines in occlusal view. (**a**) right p4-m3 of *Cardiatherium chasicoense*; (**b**) right p4, m1 or m2 and fragment of m3 of *Cardiatherium patagonicum*; (**c–g**) *Phugatherium cataclisticum* (different ontogenetic stages). (**c–e**) right juvenile m1 or m2; (**f–g**) right adult m1 or m2; (**h**) right p4-m3 of *Hydrochoerus hydrochaeris*

that it was necessary to develop a specific nomenclature (Box 2.3). Capybara teeth are easily characterized by the following: increasing number of laminae, especially in the last upper (M3) and lower (m3) molars; increasing depth of the reentrant folds (flexi/flexids; see Box 2.3) already present in more primitive cavioids (especially h.p.i.; see Box 2.3); and the development of a new flexid, the h.t.i. (Box 2.3), which is a novelty for the superfamily Cavioidea (Vucetich et al. 2005), and the one that grows most in ontogeny. Several of these flexi/flexids cross to the opposite side in adult *Hydrochoerus*. Systematics at genus and species level is based largely on the number and depth of flexi/flexids (e.g., Pascual and Bondesio 1982; Mones 1991).

The complicated dental morphology of capybaras developed from a simpler, bilobed one, seen in the early Miocene (20 Ma) *Eocardia* and the living *Dolichotis* (Fig. 2.4c, f). The number of laminae and depth of flexids increased from the oldest *C. chasicoense* (Fig. 2.6a), through the Huayquerian species of *Cardiatherium* (Fig. 2.6b) and the Pliocene *Phugatherium* (Fig. 2.6c–g), to the Pleistocene–Holocene *Neochoerus* and *Hydrochoerus* (Fig. 2.6h).

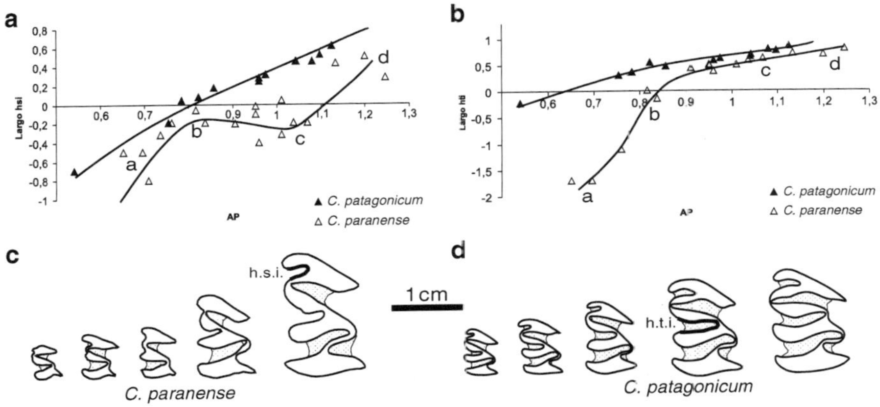

Fig. 2.7 Relationship between (log_{10}) anteroposterior length (AP) and (log_{10}) h.s.i. and h.t.i. lengths of *Cardiatherium patagonicum* and *Cardiatherium paranense*. Growth lines were fitted by eye. (**a–d**) point holotypes of some species now regarded as different ontogenetic stages of *C. paranense*. (**a**) "*Eucardiodon marshi*"; (**b**) "*Procardiatherium simplicidens*"; (**c**) "*Kiyutherium scillatoyanei*"; (**d**) *C. paranense*. Ontogenetic stages of m1 or m2 of *C. paranense* and *C. patagonicum* are shown below

2.5.2 Morphological Tooth Variation and Ontogenetic Trajectories

The fossil record of capybaras shows a large diversity in size and morphology. For example, the anteroposterior diameter of m1 ranges from 5.12 mm in "*Anchimys leidyi*" (Appendix 1) to 24.82 mm in *Neochoerus aesopi*. This diversity was classically interpreted as systematic richness. In contrast, Rocha and Montalvo (1999) interpreted this morphological diversity as individual variation in the analysis of one population of late Miocene capybaras from Guatraché and Chillhué lagoons (Fig. 2.2), in La Pampa Province, Argentina. Recently, Vucetich et al. (2005) described how cheek teeth are modified in structure through a lifetime in a fossil population from Estancia Rincón Chico (Fig. 2.2), in Chubut Province, also in Argentina.

Capybaras are born with all cheek teeth already erupted, even with occlusal wear, and as in other euhypsodont mammals, teeth continue growing in all dimensions throughout life. In fossil capybaras, the design of the occlusal surface becomes more complex with individual age (Vucetich et al. 2005), because these teeth grow allometrically, deepening flexids at different rates (Fig. 2.7). This growth pattern is unique among rodents, and was originally misinterpreted: small specimens with simpler morphology were considered primitive taxa, whereas large specimens with more complicated morphology were considered more derived (Figs. 2.6c–g and 2.7). Descriptions of morphological variants as specific characters have suggested a highly diversified taxonomy, with about 23 genera and 56 species (Mones 1991).

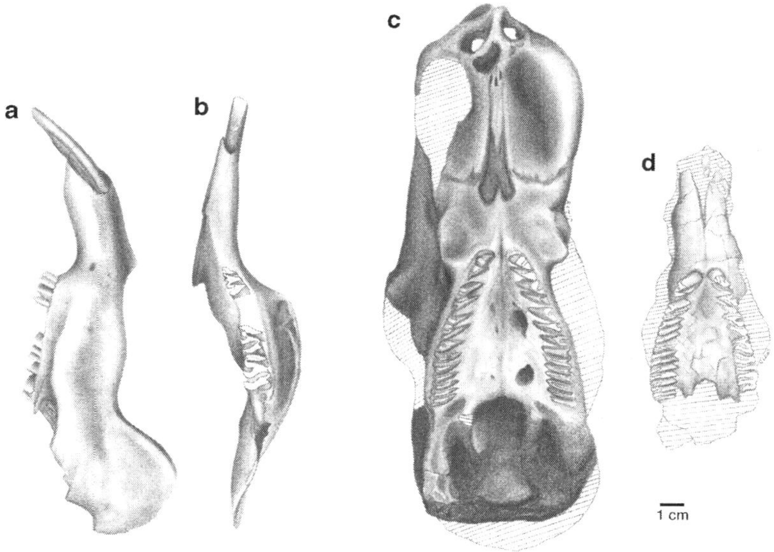

Fig. 2.8 Late Miocene capybaras. (**a–b**) right mandible in lateral and occlusal views of *Cardiatherium chasicoense*; (**c**) skull in occlusal view of *Cardiatherium paranense* (="*Anatochoerus inusitatus*"); (**d**) skull in occlusal view of *Cardiatherium orientalis* ((**a–b**) are Modified from Deschamps et al. (2007))

The most recent discovery of morphological change in ontogenetic tooth development (Vucetich et al. 2005) has resulted in the synonymy of many taxa, because some small species have been reclassified as the juveniles of larger ones. Thus, the high specific diversity suggested by typological criteria used in classical systematics may actually derive from high intraspecific morphological diversity caused by individual variation and extensive morphological change during ontogeny. In this context, the morphological difference of the species *H. gaylordi* (MacPhee et al. 2000) from the other species of *Hydrochoerus* may be due to ontogenetic variation, since it is based on a small specimen. Misinterpretation of morphological differences in teeth had in fact been so extreme that juveniles of a species were placed in one separate subfamily and the adults of that species in another. The new interpretation of ontogenetic morphological change does not support the classical subdivision of capybaras into four subfamilies. Three particularly illustrative cases are discussed in detail below.

2.5.2.1 Wide Rostrum Versus Narrow Rostrum: *Cardiatherium paranense* Versus *Cardiatherium orientalis*

Peculiar skull remains of capybaras found in the "conglomerado osifero" of Ituzaingó Formation, Paraná City, were assigned to the subfamily Anatochoerinae and named "*Anatochoerus*" (Fig. 2.8c; Mones 1991), in view of the great width of the rostrum. Although not associated with mandibles assigned to *C. paranense*, the single species to which all the mandibles from these sediments were supposed to have belonged (Vucetich et al. 2005), we attribute these peculiar skulls to this species because they have a similar dental pattern and are the only skull remains of capybaras in this site (Deschamps et al. 2010). However, *C. paranense* was considered a Cardiatherinae, and therefore presents a beautiful example of materials that belong to the same species having been wrongly attributed to different subfamilies. *C. paranense* is very close to *C. orientalis* (Fig. 2.8d), which is known through associated mandibles and skulls found in the late Miocene of La Pampa Province (Fig. 2.2), but the rostrum of these skulls is much narrower than that of *C. paranense* (compare Fig. 2.8c, d). In addition, the palate of *C. paranense* extends up to the end of the M3, whereas that of *C. orientalis* is shorter, reaching to about the middle of the M3 (Fig. 2.8c, d). Consequently, both species are considered valid despite the fact that the mandibles and occlusal morphology of teeth are quite similar (Deschamps et al. 2010).

2.5.2.2 The Case of the Early Pliocene Capybaras

The holotypes of the species *Phugatheriun cataclisticum* and "*Anchimysops villalobosi*" of the Monte Hermoso Formation (early Pliocene – 5 Ma; Fig. 2.2) are juvenile specimens; the former is represented by a mandible and the latter by a skull fragment. These taxa were originally assigned to the subfamily Cardiatheriinae. The only other hydrochoerid known from these levels is *Chapalmaterium perturbidum*, assigned to the subfamily Protohydrochoerinae. This species is represented by several skulls with larger mandibles and more derived cheek teeth morphology (Fig. 2.6f–g) than the specimens assigned to *Phugatherium* (Fig. 2.6c–e) and "*Anchimysops*". It is quite plausible that *Phugatherium* and "*Anchimysops*" are, in fact, juveniles of the protohydrochoerine *Chapalmatherium*. If this were so, the correct name for the single Montehermosan genus would be *Phugatherium* (Appendix 1).

2.5.2.3 *Kerodon* as a Hydrochoerid

Recently, there has been some debate on cavioid classification involving the position of the small rock-dwelling *Kerodon*, an extant species traditionally considered a Caviidae. Based on molecular analyses, Rowe and Honeycutt (2002) suggested close relationships between *Hydrochoerus* and *Kerodon* (see also Pérez 2010a, b), and on this basis, *Kerodon* was transferred to Hydrochoeridae (Wilson and Reeder 2005; Honeycutt 2012). Some morphological characters of *Kerodon*, such as a palate with a deep mesopterygoid fossa and very simple tooth morphology (Fig. 2.4g), appear quite different from those of the other hydrochoerids, both hydrochoerines and cardiomyines. Da Silva Neto (2000) also considered

Kerodon as a caviid on the basis of other skull characters. Hence, if we accept close relationships between *Hydrochoerus* and *Kerodon*, the split of *Kerodon* from Hydrochoerinae + Cardiomyinae should have occurred at least during the middle Miocene (Fig. 2.1; Opazo 2005; Pérez and Vucetich 2011). The short fossil record of *Kerodon* (late Pleistocene – about 40 Ky = 400 hundred years) gives little information about its evolutionary history, and it is thus not yet possible to corroborate this statement.

2.5.3 *Macrosystematics of Hydrochoeridae*

The topics discussed above outline the current lack of support for four distinct subfamilies of caviomorph rodents as originally described: the Cardiatherinae, Protohydrochoerinae, and Anatochoerinae with exclusively fossil representatives, and Hydrochoerinae including the living capybara and *Neochoerus*, its closest fossil representative. We suggest that the members of all four subfamilies belong within a single subfamily, Hydrochoerinae, which in turn should be grouped together with the subfamily Cardiomyinae within the family Hydrochoeridae (Box 2.2 and Appendix 1). If *Kerodon* is more closely related to this group than to the Caviinae perhaps it deserves its own subfamily.

2.6 Paleobiology

What can we infer from the fossil record about the natural history of capybaras? When did they attain their modern characteristics? Three main topics are of interest: size, semiaquatic habits, and aggregation in herds.

Gigantism in caviomorphs is more common than in other rodents and had reached its pinnacle by the end of the Miocene in the Huayquerian SALMA (6 Ma; Fig. 2.1). Capybaras in particular were already large in the previous Chasicoan SALMA (Fig. 2.1) when they were first recognized, but continued to increase in size, peaking during the Plio-Pleistocene (5,3 Ma to 10 Ky); examples are *Chapalmatherium* (200 kg) and *Neochoerus* (110 kg) (Vizcaíno et al. 2012).

Modern capybaras are semiaquatic, and in the fossil record capybaras have always been found in lithological units interpreted as water-related settings: *C. chasicoense*, in a swampy environment (Zárate et al. 2007), *C. paranense* in a fluvial deposit (Cione et al. 2000; Herbst 2000), *C. patagonicum* in levels deposited by a freshwater channel (Cione et al. 2005), and *C. orientalis* in the only two sites of the Cerro Azul Formation interpreted as deposits of lacustrine environments, Laguna Chillhué and Laguna Guatraché (Fig. 2.2; Verzi et al. 2008). In many other localities

Fig. 2.9 Reconstruction of fossil *Phugatherium cataclisticum* (=*Protohydrochoerus perturbidus*) (Drawing by Martín Barrios for the exhibitions of the Museo de La Plata, Argentina)

of La Pampa Province, in which eolian facies that are not related to water settings of the Cerro Azul Formation are exposed, vertebrate remains are abundant (Verzi et al. 2008 and references therein) but no capybara have been found. It was generally thought that Pliocene capybaras, however, were not adapted to semiaquatic habitats. Kraglievich (1930: 509) described the genus "*Protohydrochoerus*" (=*Phugatherium*; Appendix 1) from Monte Hermoso (Fig. 2.2) as a large running capybara (Fig. 2.9), based on the limb proportions; the humerus and femur were both longer than the radius and tibia, respectively. However, in a morphofunctional study of the limbs, Candela et al. (2006, and personal communication) concluded that they were not significantly different from those of the modern capybara, especially in the articular regions.

The body size range of the capybaras found in the Ituzaingó Formation ("conglomerado osífero") and the late Miocene of Península Valdés is very large (Vucetich et al. 2005). These deposits cover a short time span and, therefore, we believe that they reflect a community structure in which several generations were living together, as in modern capybara herds. In sum, it appears that the basic biological characteristics of capybaras were attained before the late Miocene.

Acknowledgments The authors thank the valuable suggestions of anonymous reviewers, A. Mones, and the editors that greatly improved this manuscript. Financial support was provided by ANPCyT Grant 38112, and Facultad de Ciencias Naturales y Museo, UNLP Grant N568 (to MGV).

Appendix 1 Capybara Revised Taxonomy

Synonymies already published in Mones (1991), Prado et al. (1998), Vucetich et al. (2005), and Deschamps et al. (2007) are not listed

Order Rodentia Bowdich (1821)
Suborder Hystricomorpha Brandt (1855)
Infraorder Hystricognathi Tullberg (1899)
Superfamily Cavioidea (Gray 1821; Kraglievich 1930)
Family Hydrochoeridae (Gray 1825; Gill 1872)
Subfamily Hydrochoerinae (Gray 1825; Weber 1928)

Genus *Cardiatherium* Ameghino (1883)
= *Contracavia* Burmeister (1885); *Anchimysops* Kraglievich (1927), partim; *Anatochoerus* Vucetich and Mones (1991)
Cardiatherium chasicoense (Pascual and Bondesio 1968)
Cardiatherium isseli Rovereto (1914)
Cardiatherium orientalis (Francis and Mones 1965)
Cardiatherium paranense (Ameghino 1883)
= *Anchimysops radicei* Kraglievich (1940); ?*Anchimysops dubius* Kraglievich (1940); *Contracavia minuta* (Ameghino 1885); *Anatochoerus inusitatus* Vucetich and Mones (1991) in Mones (1991)
Cardiatherium patagonicum Vucetich et al. (2005)
Cardiatherium rosendoi (Bondesio 1985)
Cardiatherium talicei Francis and Mones (1965)

Genus *Phugatherium* Ameghino (1887)
= *Chapalmatherium* Ameghino (1908); *Protohydrochoerus* Rovereto (1914); *Anchimysops* Kraglievich (1927), partim
Phugatherium cataclisticum Ameghino (1887)
= *Anchimysops ultra* Kraglievich (1940); *Chapalmatherium perturbidum* (Rovereto 1914).
Phugatherium novum (Ameghino 1908)

Genus *Hydrochoeropsis* Kraglievich (1930)
Hydrochoeropsis dasseni Kraglievich (1930)

Genus *Neochoerus* Hay (1926)[a]
Neochoerus dichroplax Ahearn and Lance (1980)
Neochoerus fontanai (Rusconi 1933)
Neochoerus tarijensis (Ameghino 1902)
Neochoerus sulcidens (Lund 1839)
Neochoerus aesopi (Leidy 1853)

Genus *Hydrochoerus* Brisson (1762)[a]
Hydrochoerus hydrochaeris (Linnaeus 1766)
Hydrochoerus ballesterensis (Rusconi 1934)
Hydrochoerus isthmius Goldman (1912)
Hydrochoerus gaylordi MacPhee et al. (2000)[b]

Subfamily Cardiomyinae Kraglievich (1930)[c]

Genus *Cardiomys* Ameghino (1885)
=*Neoprocavia* Ameghino (1889); *Parodimys* Kraglievich (1932)
Cardiomys cavinus Ameghino (1885)

(continued)

Appendix 1 (continued)

Genus *Procardiomys* Pascual (1961)	
	Procardiomys martinoi Pascual (1961)
Genus *Caviodon* Ameghino (1885)	
	Caviodon multiplicatus Ameghino (1885)
Genus *Xenocardia* Pascual and Bondesio (1963)	
	Xenocardia diversidens Pascual and Bondesio (1963)

[a]Differentiation of *Neochoerus* from *Hydrochoerus* on the basis of fragmentary material is, as it currently is in most cases, dubious. Thus, the stratigraphic range of these genera has been taken from literature (e.g., Mones 1991) without further analysis

[b]Characters used to distinguish this species are within the range of individual or ontogenetical variation of *H. hydrochaeris*

[c]Only the type species of cardiomyines are listed because a revision of the species of the whole group is still needed

References

Ameghino F (1883) Sobre una nueva colección de mamíferos fósiles recogidos por el Profesor Pedro Scalabrini en las barrancas del Paraná. B Acad N Cienc Córdoba 5:257–306

Antoine P-O, Marivaux L, Croft DA, Billet G, Ganerød M, Jaramillo C, Martin T, Orliac MJ, Tejada J, Altamirano AJ, Duranthon F, Fanjat G, Rousse S, Salas Gismondi R (2011) Middle Eocene rodents from Peruvian Amazonia reveal the pattern and timing of caviomorph origins and biogeography. Proc R Soc B. doi10.1098/rspb.2011.1732

Bellosi ES (2010) Physical stratigraphy of the Sarmiento Formation (middle Eocene – coger Miocene) at Gran Barranca, central Patagonia. In: Madden RH, Carlini AA, Vucetich MG, Kay RF (eds) The paleontology of Gran Barranca: evolution and environmental change through the middle Cenozoic of Patagonia. Cambridge University Press, Cambridge, UK, pp 19–31

Blanga-Kanfi S, Miranda H, Penn O, Pupko T, DeBry RW, Huchon D (2009) Rodent phylogeny revised: analysis of six nuclear genes from all major rodent clades. BMC Evol Biol 9:71

Bryant JD, McKenna MC (1995) Cranial anatomy and phylogenetic position of *Tsaganomys altaicus* (Mammalia, Rodentia) from the Hsanda Gol Formation (Oligocene), Mongolia. Am Mus Novitates 3156:1–42

Campbell KE (2004) The Santa Rosa local fauna: a summary. In: Campbell KE Jr (ed) The Paleogene mammalian fauna of Santa Rosa, Amazonian Peru, vol 40, Science series. Natural History Museum of Los Angeles County, Los Angeles, pp 156–163

Candela AM, García Esponda CM, Gainza MA (2006) Estudio morfofuncional del esqueleto postcraneano de los grandes carpinchos extintos. In: 9° Congreso Argentino de Paleontología y Bioestratigrafía Resúmenes:125. Córdoba

Cione AL, Azpelicueta MM, Bond M, Carlini AA, Casciotta J, Cozzuol M, de la Fuente MS, Gasparini Z, Goin FJ, Noriega JI, Scillato-Yané GJ, Soibelzon L, Tonni EP, Verzi DH, Vucetich MG (2000) Miocene vertebrates from Entre Ríos, eastern Argentina. In: Aceñolaza F, Herbst R (eds) El Mio-Plioceno Argentino, vol 14, Serie de Correlación Geológica. INSUGEO, San Miguel de Tucumán, pp 191–237

Cione AL, Azpelicueta MM, Casciotta JR, Dozo MT (2005) Tropical freshwater teleosts from Miocene beds of eastern Patagonia, southern Argentina. Geobios 38:29–42

da Silva Neto EJ (2000) Morphology of the regiones ethmoidalis and orbitotemporalis in *Galea musteloides* Meyen 1832 and *Kerodon rupestris* (Wied-Neuwied 1820) (Rodentia: Caviidae) with comments of the phylogenetic systematics of the Caviidae. J Zoolog Syst Evol Res 38:219–229

Deschamps CM, Olivares AI, Vieytes EC, Vucetich MG (2007) The oldest capybaras (Rodentia, Hydrochoeridae; Late Miocene of Argentina): ontogeny and diversity. J Vert Paleontol 27:683–692

Deschamps CM, Vieytes EC, Olivares AI, Vucetich MG (2009) Primer registro de *Cardiatherium chasicoense* (Rodentia, Hydrochoeridae) fuera del área pampeana (Argentina) y su valor bioestratigráfico. Ameghiniana 46:295–305

Deschamps CM, Vucetich MG, Montalvo CI (2010) Rostro ancho vs. rostro angosto: *Cardiatherium paranense* Ameghino vs. *C. orientalis* (Francis y Mones) (Rodentia, Hydrochoeridae). X Congreso Argentino de Paleontología y Bioestratigrafía y IV Congreso Latinoamericano de Paleontología. Actas, p 157, La Plata, Sept 2010

Fields RW (1957) Hystricomorph rodents from the Late Miocene of Colombia, South America, vol 32, Geological sciences. University of California Publications, Berkeley, pp 273–399

Flynn JJ, Wyss AR, Croft DA, Charrier R (2003) The Tinguiririca fauna of Chile: biochronology, paleoecology, biogeography, and a new earliest Oligocene South American Land Mammal "Age". Palaeogeogr Palaeoclim Palaeoecol 195:229–259

Frailey CD (1986) Late Miocene and Holocene mammals, exclusive of the Notoungulata, of the Río Acre region, Western Amazonia, vol 374, Contribution in sciences. Natural History Museum of Los Angeles County, Los Angeles, pp 1–46

Frailey CD, Campbell KE (2004) Paleogene rodents from Amazonian Peru: the Santa Rosa Local Fauna. In: Campbell KE Jr (ed) The Paleogene mammalian fauna of Santa Rosa, Amazonian Peru, vol 40, Science series. Natural History Museum of Los Angeles County, Los Angeles, pp 71–130

Francis JC, Mones A (1965) Sobre el hallazgo de *Kiyutherium orientalis* n. g., n. sp. (Rodentia Hydrochoeridae) en la Formación Kiyú, de Barrancas de San Gregorio, Departamento de San José, República Oriental del Uruguay. Kragleviana 1:45–54

Herbst R (2000) La Formación Ituzaingó (Plioceno). Estratigrafía y distribución. In: Aceñolaza FG, Herbst R (eds) El Neógeno de Argentina, vol 14. Instituto Superior de Correlación Geológica INSUGEO, San Miguel de Tucumán, pp 181–190

Honeycutt RL (2012) Phylogenetics of Caviomorph Rodents and genetic perspectives on the evolution of sociality and mating systems in the Caviidae. In: Moreira JR, Ferraz KMPMB, Herrera EA, Macdonald DW (eds) Capybara: biology, use and conservation of an exceptional neotropical species. Springer, New York, pp 61–81

Janis CM, Fortelius M (1988) On the means whereby mammals achieve increased functional durability of their dentitions, with special reference to limiting factors. Biol Rev 63:197–230

Janis CM, Damuth J, Theodor JM (2000) Miocene ungulates and terrestrial primary productivity: where have all the browsers gone? Evolution 97(14):7899–7904

Kay RF, Madden RH, Vucetich MG, Carlini AA, Mazzoni MM, Re GH, Heizler M, Sandeman H (1999) Revised geochronology of the Casamayoran South American land mammal age: climatic and biotic implications. Proc Natl Acad Sci 96(23):13235–13240

Kraglievich L (1930) La Formación Friaseana del río Frías, río Fénix, Laguna Blanca, etc., y su fauna de mamíferos. Physis 10:127–161

Lavocat R (1976) Rongeurs caviomorphes de l'Oligocéne de Bolivie. II Rongeurs de Bassin Déséadien de Salla-Luribay. Palaeovertebrata 7(3):21–90

Linares OJ (2004) Bioestratigrafia de la fauna de mamíferos de las Formaciones Socorro, Urumaco y Codore (Mioceno medio-Plioceno temprano) de la región de Urumaco, Falcón, Venezuela. Paleobiología Neotropical, No 1. http://www.paleobio.labb.usb.ve/Paleobio01.pdf

Macdonald DW (1981) Dwindling resources and the social behaviour of capybaras, (*Hydrochoerus hydrochaeris*) (Mammalia). J Zool 194:371–391

Macdonald DW, Herrera EA, Taber AB, Moreira JR (2007) Social organization and resource use in capybaras and maras. In: Wolff JO, Sherman PW (eds) Rodent societies: an ecological & evolutionary perspective. The University of Chicago Press, Chicago, pp 393–402

MacPhee RDE, Singer R, Diamond M (2000) Late Cenozoic Land Mammals from Grenada, Lesser Antilles Island-Arc. Am Mus Novitates 3302:1–20

Madden RH, Guerrero J, Kay RF, Flynn JJ, Swisher CC III, Walton AH (1997) The Laventan stage and age. In: Kay RF, Madden RH, Cifelli RL, Flynn JJ (eds) Vertebrate paleontology in the neotropics: the Miocene fauna of La Venta, Colombia. Smithsonian Institution Press, Washington, pp 499–519

Marivaux L, Vianey-Liaud M, Welcomme J-L, Jaeger J-J (2002) The role of Asia en the origin and diversification of hystricognathous rodents. Zoolog Scripta 31:225–239

Martin T (1994) African origin of caviomorph rodents is indicated by incisor enamel microstructure. Paleobiology 20:5–13

Mazzoni MM (1985) La Formación Sarmiento y el vulcanismo paleógeno. Rev Asoc Geol Argent 40:60–68

Mones A (1991) Monografía de la familia Hydrochoeridae (Mammalia, Rodentia): sistemática–paleontología–bibliografía. Courier Forschungsinstitut Senckenberg 134:1–235

Nedbal MA, Allard MW, Honeycutt RL (1994) Molecular systematics of hystricognath rodents: evidence from the mitochondrial 12S rRNA gene. Mol Phylogenet Evol 3:206–220

Ojasti J (1973) Estudio biológico del chigüire o capibara. Fondo Nacional de Investigaciones Agropecuarias, Editorial Sucre, Caracas, pp 1–275

Opazo JC (2005) A molecular timescale for caviomorph rodents (Mammalia, Hystricognathi). Mol Phylogenet Evol 37:932–937

Pascual R, Bondesio P (1982) Un roedor Cardiatheriinae (Hydrochoeridae) de la Edad Huayqueriense (Mioceno tardío) de La Pampa: sumario de los ambientes terrestres en la Argentina durante el Mioceno. Ameghiniana 19:19–35

Patterson B, Wood AE (1982) Rodents from the Deseadan Oligocene of Bolivia and the relationships of the caviomorpha. Bull Mus Comp Zool 149:371–543

Pérez ME (2010b). Sistemática, ecología y bioestratigrafía de Eocardiidae (Rodentia, Hystricognathi, Cavioidea) del Mioceno temprano y medio de Patagonia. Ph.D. thesis, Universidad Nacional de La Plata, La Plata

Pérez ME (2010b) A new rodent (Cavioidea, Hystricognathi) from the middle Miocene of Patagonia. Mandibular homologies and the origin of the crown group Cavioidea sensu stricto. J Vert Paleontol 30(6):1848–1859

Pérez ME, Vucetich MG (2011) A new extinct genus of Cavioidea (Rodentia, Hystricognathi) from the Miocene of Patagonia (Argentina) and the evolution of Cavioid Mandibular morphology. J Mamm Evol. doi:10.1007/s10914-011-9154-1

Poux C, Chevret P, Huchon D, de Jong WW, Douzery EJP (2006) Arrival and diversification of caviomorph rodents and platyrrhine primates in South America. Syst Biol 55:228–237

Prado JL, Cerdeño E, Roig-Juñent S (1998) The giant rodent *Chapalmatherium* from the Pliocene of Argentina: new remains and taxonomic remarks on the Family Hydrochoeridae. J Vert Paleontol 18:788–798

Rinderknecht A, Blanco E (2008) The largest fossil rodent. Proc R Soc 275:923–928

Rocha AM, Montalvo CI (1999) Variabilidad morfológica de los molariformes inferiores de *Kiyutherium* (Rodentia, Hydrochoeridae, Cardiatheriinae). VII Jornadas Pampeanas de Ciencias Naturales Actas, pp 279–287

Rowe DL, Honeycutt RL (2002) Phylogenetic relationships, ecological correlates, and molecular evolution within the Cavioidea (Mammalia, Rodentia). Mol Biol Evol 19:263–277

Sallam HM, Seiffert ER, Steiper ME, Simons EL (2009) Fossil and molecular evidence constrain scenarios for the early evolutionary and biogeographic history of hystricognathous rodents. Proc Natl Acad Sci 106(39):16722–16727

Sánchez-Villagra MR, Aguilera O, Horovitz I (2003) The anatomy of the world's largest extinct rodent. Science 301:1708–1710

Scott WB (1905) Mammalia of the Santa Cruz beds. Part III: Glires. Reports of Princeton University expeditions to Patagonia 5:348–489

Shockey B, Hitz R, Bond M (2004) Paleogene notoungulates from the Amazon Basin of Peru. In: Campbell KE Jr (ed) The Paleogene mammalian fauna of Santa Rosa, Amazonian Peru, vol 40, Science series. Natural History Museum of Los Angeles County, Los Angeles, pp 61–70

Verzi DH, Montalvo CI, Deschamps CM (2008) Biostratigraphy and biochronology of the Late Miocene of central Argentina: evidence from rodents and taphonomy. Geobios 41:145–155
Vieytes EC (2003) Microestructura del esmalte de roedores Hystricognathi sudamericanos fósiles y vivientes. Significado morfofuncional y filogenético. Unpublished Ph.D. thesis, Universidad Nacional de La Plata, La Plata, p 295
Vieytes EC, Verzi DH, Vucetich MG (2001) Enamel evolution in molariforms of Cavioidea (Rodentia, Caviomorpha). In: VI international congress of vertebrate morphology Abstracts: 296. Friedrich-Schiller- Universität. Jena
Vilela RV, Machado T, Ventura K, Fagundes V, Silva MJ, Yonenaga-Yassuda Y (2009) The taxonomic status of the endangered thin-spined porcupine, *Chaetomys subspinosus* (Olfers, 1818), based on molecular and karyologic data. BMC Evol Biol 9:29
Vizcaíno SF, Cassini GH, Toledo N, Bargo MS (2012) On the evolution of large size in mammalian herbivores of Cenozoic faunas of Southern South America. In: Patterson BD, Costa LP (eds) Bones, Clones, and Biomes. The History and Geography of Recent Neotropical Mammals. The University of Chicago Press, Chicago, USA, pp 76–101
Vucetich MG (1984) Los roedores de la Edad Friasense (Mioceno medio) de Patagonia. Revista del Museo de La Plata (Nueva Serie) 8. Paleontología 50:47–126
Vucetich MG, Mazzoni MM, Pardiñas UFJ (1993) Los roedores de la Formación Collón Cura (Mioceno medio) y la Ignimbrita Pilcaniyeu. Ameghiniana 30:361–381
Vucetich MG, Pérez ME (2011) The putative cardiomyines (Rodentia, Cavioidea) of the middle Miocene of Patagonia (Argentina) and the differentiation of the Family Hydrochoeridae. Journal of Vertebrate Paleontology 31:1382–1386
Vucetich MG, Verzi DH, Hartenberger JL (1999) Review and analysis of the radiation of South American Hystricognath Rodents. Comptes Rendus de l'Académie des Sciences: series IIA: earth and planetary science, vol 329, pp 763–769
Vucetich MG, Deschamps CM, Olivares AI, Dozo MT (2005) Capybaras, size, shape, and time: a model kit. Acta Palaeontol Pol 50:259–272
Vucetich MG, Vieytes EC, Pérez ME, Carlini AA (2010a) The rodents from La Cantera and the early evolution of caviomorphs in South America. In: Madden RH, Carlini AA, Vucetich MG, Kay RF (eds) The paleontology of Gran Barranca: evolution and environmental change through the middle Cenozoic of Patagonia. Cambridge University Press, Cambridge, UK, pp 189–201
Vucetich MG, Carlini AA, Aguilera O, Sánchez Villagra MR (2010b) The Tropics as reservoir of otherwise extinct mammals: the case of rodents from a new Pliocene faunal assemblage from northern Venezuela. J Mamm Evol. doi:10.1007/s10914-010-9142-x
Walton AH (1997) Rodents. In: Kay RF, Madden RH, Cifelli RL, Flynn JJ (eds) Vertebrate paleontology in the neotropics: the Miocene Fauna of La Venta, Colombia. Smithsonian Institution Press, Washington, pp 392–409
Wilson D, Reeder D (eds) (2005) Mammal species of the world: a taxonomic and geographic references, vol 2. Johns Hopkins University Press, Baltimore, pp 1–1600
Wyss AR, Flynn JJ, Norell MA, Swisher CC III, Charrier R, Novacek MJ, McKenna MC (1993) South America's earliest rodent and recognition of a new interval of mammalian evolution. Nature 365:434–437
Zachos J, Pagani M, Sloan L, Thomas E, Billups K (2001) Trends, rhythms, and aberrations in global climate 65 Ma to present. Science 292:686–693
Zárate MA, Schultz PH, Blasi A, Heil C, King J, Hames W (2007) Geology and geochronology of the type Chasicoan (late Miocene) mammal-bearing deposits of Buenos Aires Province (Argentina). J S Am Earth Sci 23(1):81–90

Chapter 3
Phylogenetics of Caviomorph Rodents and Genetic Perspectives on the Evolution of Sociality and Mating Systems in the Caviidae

Rodney L. Honeycutt

3.1 Overview of Caviomorph Rodents

The diversity of terrestrial mammals in South America is the result of both isolation and continental interchanges, preceded and followed by extinction and speciation (Simpson 1980; Marshall 1988). Mammalian biodiversity in South America can be partitioned into three major phases, beginning in the Paleocene (65.5–55.8 million years ago – Ma) with xenarthrans, notoungulates, marsupials, and ending with the "Great American Interchange" that occurred with the influx of North American mammals upon formation of the Panamanian land bridge approximately 3 Ma (Simpson 1980; Marshall 1988; Flynn and Wyss 1998). South America's first rodents, the New World Hystricognathi (Caviomorpha), termed "old native rodents" by Simpson (1980), appear in the fossil record around 37.5–31 Ma in the Eocene/Oligocene transition (Wyss et al. 1993; Vucetich et al. 1999; Flynn et al. 2003). The origin of South American caviomorph rodents is apparently a consequence of over-water dispersal from Africa during the Late Paleocene (c.58 Ma) to Middle Eocene (c.40 Ma; Lavocat 1969, 1980; Huchon and Douzery 2001; Rowe 2002; Rowe et al. 2010). The Late Eocene to Early Oligocene of South America represents a period of transition in terms of climatic and environmental changes (Flynn and Wyss 1998). According to both Mares and Ojeda (1982) and Simpson (1980), caviomorph rodents during this time period experienced an adaptive radiation that resulted in these "hypsodont herbivores" (herbivores with high-crowned cheek teeth) filling niches previously occupied by archaic ungulates from the first phase of mammalian history in South America (Flynn and Wyss 1998). One particular Pliocene/

R.L. Honeycutt (✉)
Natural Science Division, Pepperdine University, 24255 Pacific Coast Highway, Malibu, CA 90263-4321, USA
e-mail: rodney.honeycutt@pepperdine.edu

J.R. Moreira et al. (eds.), *Capybara: Biology, Use and Conservation of an Exceptional Neotropical Species*, DOI 10.1007/978-1-4614-4000-0_3,

Pleistocene (4–2 Ma) form, *Josephoartigasia monesi*, was the size of a rhinoceros and adapted for a semiaquatic life style (Rinderknecht and Blanco 2008). By the Late Oligocene (24 Ma), representatives of all four superfamilies and 8 of the 14 recent families of caviomorph rodents were present in the fossil record (Simpson 1980; Vucetich et al. 1999).

Today, caviomorph rodents occur throughout South America, occupying a diversity of habitats including deserts, grasslands, tropical forests, the Andes, coastal shrub, rocky outcrops, and many other habitats. Relative to the order Rodentia, caviomorph rodents represent only 12.5% of all species and 15% of all genera (Wilson and Reeder 2005), yet nearly half of all species of rodents in South America (Reig 1986), and approximately 40% of all families of rodents, are caviomorphs (Carleton and Musser 2005). The large number of families of caviomorphs is indicative of their ecomorphological and behavioral diversity. Both body size and basal metabolic rate of caviomorphs vary by as much as 400–800 and 65-fold (Arends and McNab 2001), respectively, with the largest being the capybara (*Hydrochoerus hydrochaeris*), averaging 50 kg. In addition to their morphological and ecological diversity, many caviomorphs are highly social and span the entire spectrum of mammalian mating behaviors.

The caviomorph rodents are an integral and diverse component of South America's biota, and many species are important to various regions of the continent for reasons related to both economics and conservation. Genetic markers in combination with analytical methods used in phylogenetics and population genetics provide a unique framework for interpreting various evolutionary processes associated with the origin and diversification of caviomorph rodents. This chapter provides an overview of recent phylogenetic studies of caviomorph rodents, including a discussion of how phylogenetic methods can be used to evaluate the evolution of mating systems as they relate to the capybara and its closest relatives. At the end of the chapter I also provide a discussion of the usefulness of genetic markers for studying the ecology and behavior of capybaras and offer some ideas for further investigations.

3.2 Phylogenetics

3.2.1 Phylogenetic Position of Caviomorphs Relative to Other Rodents

Resolving relationships among the major families of rodents remains a challenge, even with new molecular data (Honeycutt et al. 2007; Honeycutt 2009). Nevertheless, morphological and molecular studies do support some associations and help interpret the evolution of morphological features previously used in rodent classifications (Luckett and Hartenberger 1985; Meng 1990; Nedbal et al. 1996; Landry 1999; Huchon et al. 2000, 2002, 2007; Huchon and Douzery 2001; Adkins et al. 2001, 2003;

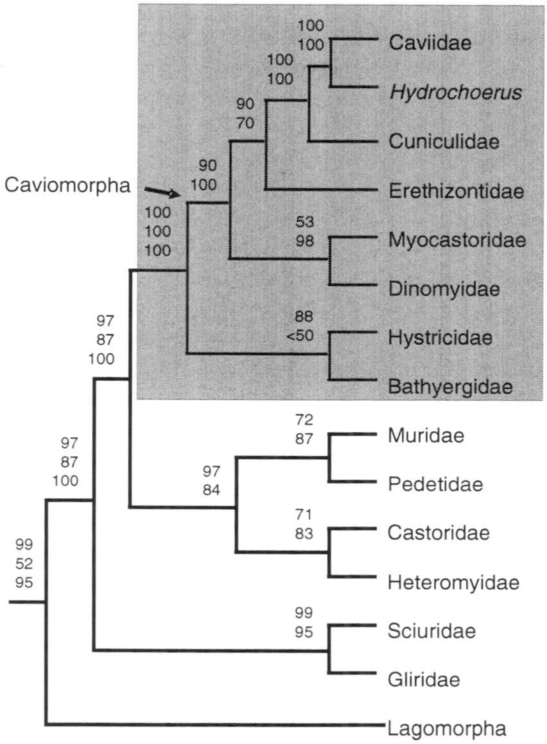

Fig. 3.1 Phylogeny, modified from Murphy et al. (2001), based on data from 18 orthologous gene segments. The numbers represent bootstrap support derived from neighbor-joining (*top*), maximum parsimony (*middle*), and maximum likelihood or ML (*lower*). Those with only two numbers are the result of missing values for ML because these values were obtained for only a subset of all taxa. The *shaded area* represents the Hystricognathi

Marivaux et al. 2004; Honeycutt et al. 2007; Rowe et al. 2010). Based on molecular data and some morphological analyses, families of rodents can be grouped into a minimum of five major clades (Huchon et al. 2002; Adkins et al. 2003; Honeycutt et al. 2007; Honeycutt 2009). Although relationships among, and contents of, the major rodent clades are not completely resolved, there is considerable support for a monophyletic Hystricognathi (Fig. 3.1). Morphologically, hystricognath rodents share many derived characteristics (Patterson and Wood 1982; Bugge 1985; Lavocat and Parent 1985; Luckett 1985; Luckett and Hartenberger 1993; Landry 1999; Marivaux et al. 2004). In addition, several recent molecular phylogenetic analyses, using both nuclear and mitochondrial sequences, consistently support hystricognath monophyly (Nedbal et al. 1994, 1996; Huchon et al. 2000, 2002; Murphy et al. 2001; Rowe 2002; Honeycutt et al. 2007; Kjer and Honeycutt 2007).

Establishment of the sister group to hystricognath rodents is complicated by interpretations of morphological features (Wood 1974, 1985; Bugge 1985; George 1985; Lavocat and Parent 1985; Luckett 1985; Sahni 1985; Meng 1990; Martin 1993). Recently, however, detailed molecular studies based on sequences from the nuclear genome have strongly supported a sister-group relationship

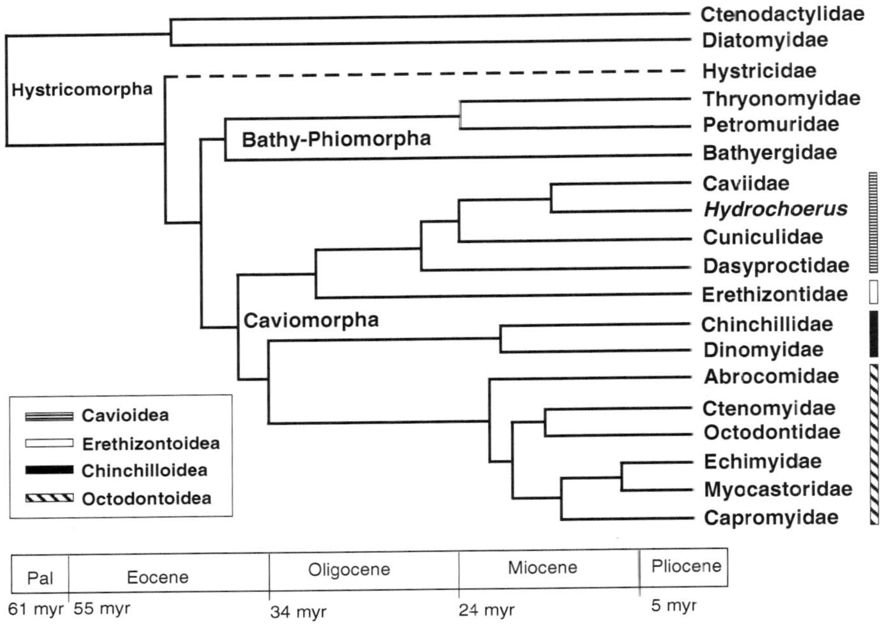

Fig. 3.2 Phylogeny of hystricognath rodent families and *Hydrochoerus* based on several molecular studies (Huchon et al. 2000; Walton et al. 2000; Adkins et al. 2001; Huchon and Douzery 2001; Rowe 2002; Rowe and Honeycutt 2002; Honeycutt et al. 2003; Ingram et al. 2004). Divergence times are based on molecular dates derived from several sources (Honeycutt et al. 2003; Ingram et al. 2004; Opazo 2005; Huchon et al. 2007). The *dashed line* represents the controversial placement of the Hystricidae

between the Ctenodactylidae and Hystricognathi, denoted as the crown group Hystricomorpha (Fig. 3.2; Huchon et al. 2000, 2002, 2007; Adkins et al. 2003), and this result is highly consistent with the morphological evidence (Marivaux et al. 2004). The recently discovered living fossil, *Laonastes aenigmamus*, representing the Diatomyidae, appears sister to the Ctenodactylidae (Dawson et al. 2006; Huchon et al. 2007), rather than the Hystricognathi as proposed by Jenkins et al. (2004).

3.2.2 Relationships Among Families of Hystricognath Rodents

Hystricognath rodents consist of 17 families (with Heptaxodontidae extinct) occurring primarily in South America, Africa, and Asia. The Old World forms include three families, Bathyergidae (mole-rats), Thryonomyidae (cane rats), Petromuridae (rock rats), occurring in Africa, and one family, Hystricidae (porcupine), found in both Africa and Asia. These families are collectively known as the Phiomorpha (sensu Lavocat 1973), but some place the Hystricidae in a separate group known as the Hystricomorpha and retain the other families in the Bathy-Phiomorpha (Wood 1965). The Caviomorpha (Wood 1955) are most diverse in South America with just one extant species (*Erethizon dorsatum*, porcupine) occurring in North America and some fossil forms found in North America and parts of the Caribbean (Marshall 1988; Webb and Barnosky 1989; MacPhee et al. 2000). Caviomorph families include: Abrocomidae (chinchilla rat), Cuniculidae (= Agoutidae – pacas), Capromyidae (hutias), Caviidae (cavies, guinea-pig, capybaras, and relatives), Chinchillidae (chinchillas), Ctenomyidae (tuco-tucos), Dasyproctidae (agoutis), Dinomyidae (pacarana), Echimyidae (spiny rats), Erethizontidae (American porcupine), Heptaxodontidae (giant hutias), Myocastoridae (coypu), and Octodontidae (degus and relatives; see Fig. 3.2).

Divergence times for both Old World and New World hystricognath rodents are similar. Based on fossil evidence, hystricognath rodents in Africa appear in the Late Eocene (37–34 Ma) and in South America by the Late Eocene/Early Oligocene, approximately 37.5–31 Ma (Lavocat 1980; Patterson and Wood 1982; Wyss et al. 1993; Flynn and Wyss 1998). Estimates of time since divergence based on molecular data (Nedbal et al. 1994; Huchon and Douzery 2001; Honeycutt et al. 2003; Opazo 2005; Poux et al. 2006; Huchon et al. 2007; Honeycutt 2009; Rowe et al. 2010) also indicate that separation of phiomorphs and caviomorphs does not extend past the Eocene (Fig. 3.2). The most recent molecular calibration places the divergence time at approximately 50 Ma (Rowe et al. 2010). It appears that the caviomorphs are derived from African ancestors that colonized South America, yet the timing of this event is complicated by geological evidence that indicates a separation of 1,700 km by the Late Eocene (Holroyd and Maas 1994). Therefore, this colonization event involved overwater dispersal as suggested by Hoffstetter (1972).

Although the Caviomorpha are monophyletic, relationships of the four Old World families are only partially substantiated. Both molecular and morphological data support the monophyly of Bathy-Phiomorpha, represented by a sister-group relationship between the families Thryonomyidae and Petromuridae followed by the Bathyergidae (Fig. 3.2; Lavocat 1973; Nedbal et al. 1994; Adkins et al. 2001; Huchon and Douzery 2001; Poux et al. 2006; Huchon et al. 2007). The placement of the Old World porcupines (family Hystricidae) is still uncertain (Patterson and Wood 1982; Adkins et al. 2001; Murphy et al. 2001; Huchon et al. 2002, 2007; Rowe 2002; Poux et al. 2006; Rowe et al. 2010). Based on the most recent analyses including extensive taxon sampling and two nuclear genes as well as an assessment of fewer taxa and eight genes, the Hystricidae appear sister to the Caviomorpha (Rowe et al. 2010). This association would suggest a somewhat more complicated biogeographic scenario for the Old World hystricognath lineages and their South American descendants.

Box 3.1 Various Classification Schemes for Caviomorph Rodents

Four major classification schemes (Simpson 1945[1]; Patterson and Wood 1982[2]; Wood 1955[3]; McKenna and Bell 1997[4]) of caviomorph rodents are discussed. Caviomorpha is a name provided by all but one classification[1] to denote South American hystricognath rodents. The most recent classification (Carleton and Musser 2005[5]) recognizes the suborder Hystricomorpha and two infraorders, with the Hystricognathi containing the Old World Phiomorpha, Hystricidae, and Caviomorpha. Diversity within this group is shown in parentheses (number of genera/number of species).

Most classifications recognize four superfamilies, whereas one[4] recognizes the distinction of Erethizontidae but does not assign it to a specific superfamily	*Superfamilies* Erethizontoidea (4/12) Cavioidea Chinchilloidea Octodontoidea
Most traditional classifications recognize four families in the Cavioidea, whereas one[4] also recognizes Cuniculidae (=Agoutidae). In contrast to these classifications, recent molecular data suggest placing Hydrochoeridae within the Caviidae[5] and the Dinomyidae within the superfamily Chinchilloidea	*Cavioidea* Cuniculidae (1/2) Caviidae (5/14) Dasyproctidae (2/13) Dinomyidae (1/1) Hydrochoeridae (1/1)
All classifications recognize the Chinchillidae as being a separate superfamily Chinchilloidea. Nevertheless, they vary in the assignment of other families to this superfamily. Molecular data and all but one classification clearly place Capromyidae and Abrocomidae in Octodontoidea. The association of Dinomyidae with the Chinchilloidea is based on molecular data alone	*Chinchilloidea* Chinchillidae (3/7) Plus Capromyidae[3] Plus Abrocomidae[4]
Six families are currently placed in the superfamily Octodontoidea. The primary difference among classifications relates to the recognition of Capromyidae, Ctenomyidae, and/or Myocastoridae as separate families and the placement of Abrocomidae and Capromyidae. Heptaxodontidae is extinct	*Octodontoidea* Abrocomidae (1/3) Capromyidae (12/25) Ctenomyidae (1/38) Echimyidae (21/80) Heptaxodontidae (4/4) Myocastoridae (1/1) Octodontidae (6/9)

3.2.3 *Relationships Among Caviomorph Rodents*

Most classifications recognize four superfamilies of caviomorph rodents including the Cavioidea, Chinchilloidea, Erethizontoidea, and Octodontoidea (Box 3.1; Fig. 3.2). The primary difference in these classifications relates to either the placement of particular families (e.g., Abrocomidae) or the recognition of families (e.g., Capromyidae, Ctenomyidae, and Myocastoridae). Even the recognition of the Hydrochoeridae (capybaras) as a separate family is questionable (Rowe and Honeycutt 2002), and the most recent classification (Woods and Kilpatrick 2005) places capybaras in the family Caviidae.

Various molecular studies contribute to an understanding of relationships among families of caviomorphs (Fig. 3.2). For instance, evidence for Abrocomidae being sister to the Octodontoidea is well supported (Honeycutt et al. 2003). Independent analyses of molecular data show strong support for the placement of Dinomyidae as a member of the Chinchilloidea instead of the Cavioidea (Adkins et al. 2001; Huchon and Douzery 2001; Rowe 2002). Molecular data also provide phylogenetic resolution for relationships among genera and families within the larger superfamily Cavioidea (Rowe and Honeycutt 2002) and Octodontoidea (Honeycutt et al. 2003). There are still some relationships that are not well resolved, such as the position of the Erethizontoidea and relationships among the major superfamilies (Honeycutt et al. 2007).

3.3 Cavioidea

The superfamily Cavioidea is represented by 9 genera and 31 species distributed throughout most of South America, with some occurring in Central America and Mexico (Redford and Eisenberg 1992; Wilson and Reeder 2005). Members of this superfamily are diverse in terms of behavior, choice of habitats, morphology, locomotion, and other life-history traits. In addition, they reveal a broad range in body size and metabolic rate (Arends and McNab 2001).

With the exception of the Dinomyidae (Box 3.1; Fig. 3.2), traditional members of the superfamily Cavioidea share a common ancestry (Patterson and Wood 1982; Nedbal et al. 1996; Huchon et al. 1999; Huchon and Douzery 2001; Rocha-Barbosa et al. 2007). Nevertheless, the number of families within this superfamily varies from three to four, depending on which classification scheme is used (Box 3.1). This variation results from whether or not the families Cuniculidae, Dasyproctidae, and Hydrochoeridae are separate families. Some classifications combine dasyproctids and cuniculids into a single family, either Dasyproctidae or Cuniculidae (Corbet and Hill 1991; McKenna and Bell 1997), and others recognize these as separate families (Woods 1984; Wilson and Reeder 2005). Familial-level status for capybaras also varies. Woods (1984) recognizes the Hydrochoeridae as a separate family on the basis of chromosomal and morphological differences, whereas Woods and Kilpatrick (2005) do not.

The most diverse family of cavioid rodents is the Caviidae, a family traditionally subdivided into two subfamilies (Woods 1984), Caviinae (*Cavia*, *Galea*, *Microcavia*, and *Kerodon*) and Dolichotinae (*Dolichotis*). The family occurs in South America and is most diverse in the Southern Cone (Redford and Eisenberg 1992; Honeycutt 2003). Most morphological comparisons consider capybaras (Hydrochoeridae) to be closely related to the Caviidae (Patterson and Wood 1982; Woods 1984; Rocha-Barbosa et al. 2007), and some place capybaras in the family Caviidae (Walton 1997). Surprisingly, few detailed phylogenetic studies of the various genera and species of the Cavioidea are available. Both Rocha-Barbosa et al. (2007) and Quintana

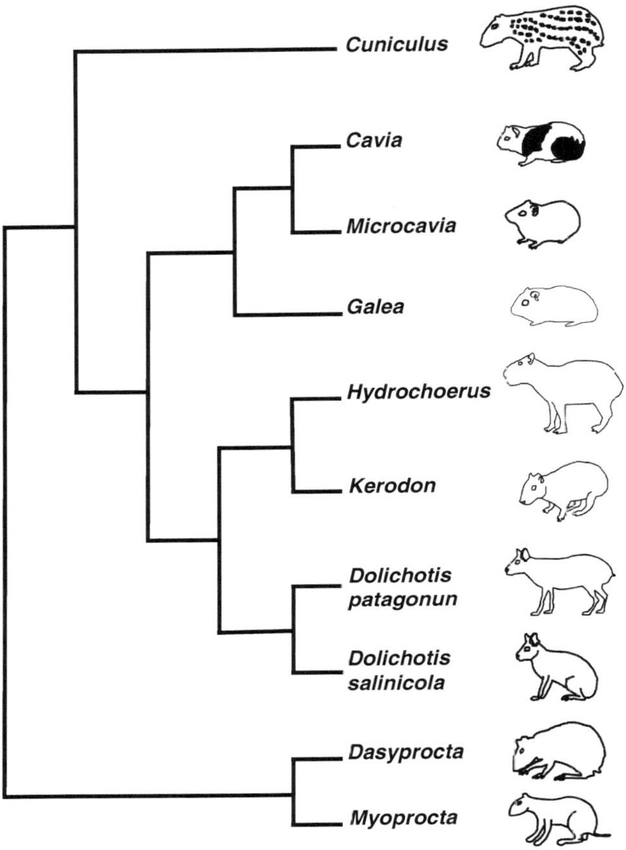

Fig. 3.3 Molecular phylogeny of the Cavioidea based on nucleotide sequences examined by Rowe and Honeycutt (2002)

(1998) use morphological features of foot musculature and skeletal features, respectively, to construct phylogenies for subsets of cavioids. Both cases are restricted in their ability to assess clearly either the relationships among all the major genera and families of cavioids or the placement of the capybara. In addition, Hugot's (2003) interpretation of hystricognath relationships based on comparisons to a phylogeny of pinworms is highly incongruent with most recent morphological and molecular data. The most thorough assessment of cavioid relationships is the molecular study by Rowe and Honeycutt (2002) that presents a phylogeny based on sequences from nuclear and mitochondrial genes (Fig. 3.3). Several conclusions can be drawn from this phylogeny. First, the traditional subfamily Caviinae is not a monophyletic group because the genus *Kerodon* is closer to members of the Dolichotinae. There is some justification for this arrangement based on behavioral and reproductive patterns (Lacher 1981; Woods 1984). Second, rather than grouping sister to the Caviidae, the Hydrochoeridae resides within the Caviidae, being closest to *Kerodon*. This particular finding makes some sense when mating systems and social behavior are considered (below). Finally, the family Cuniculidae is more closely aligned with the Caviidae rather than grouping sister to Dasyproctidae. Therefore, these data support the recognition of both Cuniculidae and Dasyproctidae as separate families.

3.3.1 Placement of Capybaras in the Cavioidea

As Patterson and Wood (1982) indicate, the capybara is a "good cavioid," and it is morphologically most similar to members of the Caviidae (Woods 1984). More recent assessments of both foot musculature (Rocha-Barbosa et al. 2007) and nucleotide sequences (Rowe and Honeycutt 2002) support this conclusion. The remaining issues regarding the phylogenetic placement of the capybara are twofold. Is the capybara sister to a clade containing other members of the Caviidae? Alternatively, does the capybara group belong within the Caviidae? In the most recent classification of the Hystricognathi, Woods and Kilpatrick (2005) place the capybara in a separate subfamily Hydrochoerinae within the Caviidae. They place *Kerodon* within this subfamily, rather than its traditional placement in the Caviinae. Their decision to align *Kerodon* with the capybara is based on the molecular study by Rowe and Honeycutt (2002) that shows strong support for a sister-group relationship between *Hydrochoerus* and *Kerodon* (Fig. 3.3). This particular association is supported by a more recent phylogenetic analysis based on sequence data from the 12S and 16S rRNA genes (Trillmich et al. 2004). The morphological precedent for this arrangement is lacking. Most morphological studies (e.g., Quintana 1998; da Silva Neto 2000; Vucetich et al. 2012) assume a priori that *Hydrochoerus* is sister to a clade containing the remaining genera of Caviidae. This assumption is clearly not valid based on the molecular data. A reanalysis of da Silva Neto's (2000) data based on features of the fetal cranium and the inclusion of *Hydrochoerus* as an ingroup taxon is presented in Fig. 3.4. Unlike the molecular results, *Hydrochoerus* groups outside a clade containing the other caviids, with the remainder of caviids representing an unresolved trichotomy represented by *Kerodon*, *Dolichotis*, and a group containing the remaining genera of Caviinae.

How does one explain the strong molecular association of *Kerodon* with *Hydrochoerus*? Although the molecular data contradict the traditional placement of *Kerodon* within the Caviinae and *Hydrochoerus* as sister to the other genera of Caviidae, the existing morphological data are not compelling. No detailed morphological analysis, based on a large number of characters and an appropriate outgroup, exists. For instance, the only detailed phylogenetic analysis by da Silva Neto (2000) relies on nine characters and lacks two of the caviid genera. As indicated by Dos Reis (1994), the association of *Kerodon* with the subfamily Caviinae may represent morphological similarity as a result of convergence. In addition, morphological specializations associated with particular life styles and overall differences in size

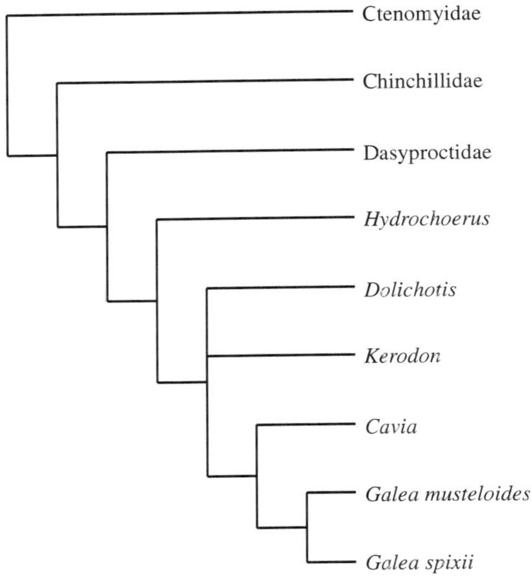

Fig. 3.4 Reanalysis of da Silva Neto's (2000) data from fetal regions of the cranium. The tree is based on a strict consensus of three trees representing a tree length of 11 and a rescaled consistency index of 0.7091. The families Ctenomyidae, Chinchillidae, and Dasyproctidae were used as outgroups and *Hydrochoerus* was treated as a member of the ingroup. Nine characters were used in an analysis utilizing maximum parsimony and the exhaustive search option. These characters were treated as either present (1) or absent (0)

complicate the morphological association of *Kerodon* and *Hydrochoerus*. Therefore, I suggest that the molecular results be used as a prompt for more thorough analysis of morphological data.

3.4 Sociality and Mating Systems in the Caviidae

A phylogenetic framework provides a means of testing hypotheses pertaining to the evolution of behavior, morphology, and various ecological characteristics (Dobson 1985; Harvey and Pagel 1991; Ebensperger and Cofré 2001; Ebensperger and Blumstein 2006). For instance, behavioral and ecological traits shared among species may be the result of convergent evolution in response to similar environmental conditions. Alternatively, species sharing particular traits may represent a monophyletic group, reflecting shared ancestry. A phylogeny provides a means of distinguishing between these two explanations.

Members of the Caviidae, including the capybara, display differences in life-history associated with both the utilization of habitat and mating strategies. For instance, *Cavia*, *Galea* (cui), and *Microcavia* (mountain cavy) are habitat generalists, occupying a diversity of habitats with evenly distributed resources (Lacher 1981). In contrast, several species are habitat specialists. The rock cavy, *Kerodon rupestris*, for example, is a specialist, preferring den sites associated with rocky outcrops, which are not evenly distributed (Lacher 1981). Capybaras are semiaquatic herbivores, dependent on "low-lying grassy patches" and "bushy scrub" adjacent to water, and these components of habitat are not evenly dispersed (Herrera and Macdonald 1989). Maras (*Dolichotis patagonum*) are cursorial and prefer open areas with low vegetation

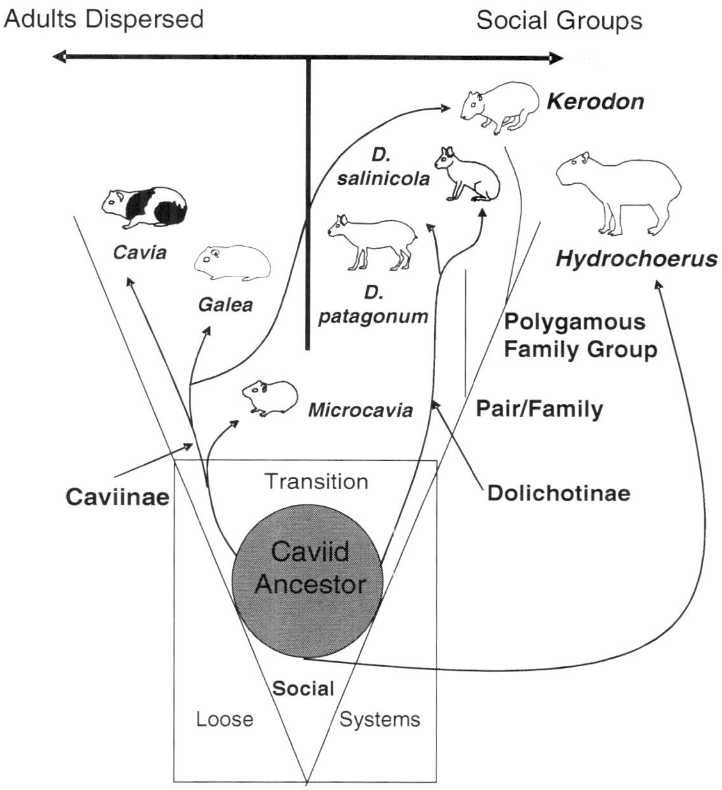

Fig. 3.5 Evolution of social behavior in the family Caviidae based on Lacher (1981). *Microcavia* is considered more like the ancestral caviid in displaying a loose social system and promiscuity. The *lines with arrows* show development of these trends, with *Kerodon* and *Hydrochoerus* showing the formation of social groups more like members of the Dolichotinae

for breeding and more barren sites for the construction of communal dens (Taber and Macdonald 1992; Baldi 2007). These areas tend to maximize visibility, thus reducing predation. Another cursorial species, *Dolichotis salinicola* (dwarf Patagonian cavy), also prefers open areas in dry forests and thorn scrub (Tabeni and Ojeda 2003; Macdonald et al. 2007).

"The evolution of morphological and behavioral differences in the family Caviidae is strongly linked to the different habitat requirements" Lacher (1981, p. 55). This hypothesis suggests that ecological constraints, such as the distribution of resources and susceptibility to predation, influence patterns of sociality and mating in different species of caviids. Species occupying more evenly dispersed resources tend to show dispersion as adults, have less group cohesion, and lack the formation of pair bonds (Fig. 3.5). For instance, *Galea* and *Microcavia* have promiscuous mating systems, and exhibit a linear dominance hierarchy (Rood 1972; Sachser et al. 1999). *Cavia* has a more skewed breeding system that tends to be polygynous with dominant males being more successful at monopolizing females during breeding (Rood 1972; Sachser et al. 1999). In contrast to more evenly dispersed resources, patchily distributed resources tend to clump, allowing one or a few individuals to dominate resources. Species occupying such clumped resources form more stable social groups and display more complex and stable mating systems, such as resource defense polygyny in *Kerodon* (Lacher 1981) and *Hydrochoerus* (Herrera and Macdonald 1989, 1994). The rather homogeneous grazing of the Patagonian steppes leads to intense mate-guarding and monogamy in *Dolichotis* (Taber and Macdonald 1992).

Lacher (1981, Fig. 38 legend p. 59) states that "environmental variables were more important than phylogenetic constraints in determining trends in the Caviidae." This assumes that *Kerodon* groups within the Caviinae rather than the Dolichotinae. The placement of the capybara as a family separate from the Caviidae appears to strengthen Lacher's statement. Nevertheless, the phylogeny derived by Rowe and Honeycutt (2002) implies that at least in terms of social group formation and the development of more complex mating systems, *Kerodon*, members of the Dolichotinae, and *Hydrochoerus* share a common ancestry (Fig. 3.3). Therefore, behavioral and environmental variables associated with these more complex social structures may be bounded by history, rather than reflecting convergence. The position of the capybara as sister to *Kerodon* strengthens this position. Capybaras also form "stable linear dominance hierarchies" consisting of a dominant male, subordinate males, females, and young (Herrera and Macdonald 1993), and this social system is somewhat similar to that seen in *Kerodon* (Lacher 1981).

Trillmich et al. (2004, p. 522) contradict the above conclusions by stating that "the social and mating systems of the Caviidae are not primarily phylogenetically constrained." These authors misinterpret the intent of the conclusions drawn by Rowe and Honeycutt (2002). Contrary to the insinuations of Trillmich et al. (2004) and Asher et al. (2008), these phylogenetic results do not negate the idea that ecology plays a role in the evolution of complex social systems. Rather, the phylogenetic data support monophyly of *Kerodon*, *Dolichotis*, and *Hydrochoerus*, thus allowing for a more focused study of factors responsible for the evolution of behaviors observed in this group of rodents. Having made this statement, I agree with Trillmich et al. (2004) in that further testing of various hypotheses associated with the evolution of behavior in the Caviidae requires a more detailed study based on a more precise designation of categories of both habitat types and social behavior. I strongly suggest that any future analysis use a phylogenetic framework for such investigations.

3.5 Future Genetic Investigations

3.5.1 Dispersal Patterns and Kinship

Capybara social groups contain a mixture of adult males and females as well as young. These groups appear stable and are characterized by a dominant male that, along with subordinate males and females, defends territories against intruding males (Herrera and Macdonald 1994). Presumably, dispersal involves young of both sexes and a subordinate male (Herrera and Macdonald 1987; Herrera 1992; Tang-Martínez 2003). In addition, there is some evidence to suggest that different age groups of young are maintained in "crèches," where they receive care from communally nursing females (Macdonald 1981; Alho and Rondon 1987; Tang-Martínez 2003; Macdonald et al. 2007).

Although basic information on group composition, behavior, and patterns of dispersal is available, little is known about the genetic structure either within or between groups of capybaras. This information is essential for addressing questions related to the behavioral patterns observed for capybaras. For instance, what role does kinship play in the behavior and group structure of capybara populations? Kinship is important in many complex social systems, and to a great extent influences many social interactions (Hamilton 1963). At least 14 species of rodents communally nurse, and several, including the capybara, are caviomorphs (Hayes 2000). In some cases, communal nursing occurs along kin lines, with sisters more likely to provision milk to each other's offspring (Jesseau et al. 2009). This appears to be the case for capybaras (Nogueira et al. 2000). Microsatellite data demonstrate that nest sharing in the caviomorph rodent, *Octodon degus*, occurs among close kin (Ebensperger et al. 2004). Given the lack of aggression among female capybaras and cooperation in provisioning of young, one might hypothesize that females within these stable groups are more closely related. Some support for this hypothesis comes from a captive breeding study that showed higher infanticide among unfamiliar females in comparison to females associated since weaning (Nogueira et al. 1999).

The general pattern of dispersal in mammals is sex-biased, with males dispersing and females demonstrating philopatry (adherence within or near the natal group; Greenwood 1980). As indicated by Dobson (1982), juvenile male dispersal is common in mammals for species demonstrating a polygynous or promiscuous mating system. Gender-biased dispersal can influence the genetic structure of populations. For instance, in the red deer (*Cervus elaphus*), a polygynous species that demonstrates male-biased dispersal, comparisons of genetic structure based on estimates of F_{ST} values reveal higher values for mtDNA (a maternal marker) than seen for microsatellite loci, nuclear gene markers inherited by both sexes (Pérez-Espona et al. 2010). This suggests that one can tease apart the contributions to genetic cohesiveness within and between populations by a detailed examination of patterns of gene flow for both males and females. Although the dispersal pattern in capybaras is somewhat different from that seen in red deer in that both sexes disperse in capybaras, the pattern of dispersal may provide a genetic footprint. For instance, females that disperse may be either full or half siblings, whereas the subordinate male may be unrelated. Assuming that juveniles of both sexes disperse, do newly formed groups consist of males and females that are related, and if so, do these juvenile males become subordinates to the older male?

Clearly, genetic studies are necessary to understand completely the overall structure of capybara populations. A combination of nuclear and mitochondrial markers would provide an effective means of addressing questions related to both kinship and dispersal patterns, as well as the influence of current behavioral patterns on the partitioning of genetic variation between groups.

3.5.2 Mating Systems

Genetic markers are powerful tools for detailed investigations of mating systems because they provide a more accurate means of determining parentage, especially paternity (Hughes 1998). How successful is the dominant male in terms of monopolizing females in a polygynous mating system? Generally, dominance is assumed to be an indicator of reproductive success. As Dewsbury (1982) indicates, observed copulations imply differences in reproduction, yet these observations do not necessarily establish paternity. Rather than the mating system, it appears that mixed paternity involving additional males relates more to the "dispersion of females" and the degree to which a male can guard a female (Clutton-Brock and Isvaran 2006). In mammals, nearly half of all species examined show 20% "extra-group paternity," and according to Isvaran and Clutton-Brock (2007), this relates to the length of the breeding season and number of females in a group.

Many of the more recent conclusions about presumed reproductive success not being totally predictable are based on the establishment of parentage with nuclear microsatellite loci. These genetic markers are highly polymorphic, and variation is inherited as codominant alleles. Therefore, a panel of microsatellite loci provides a powerful means of assessing parentage (for reviews see DeWoody 2005; DeYoung and Honeycutt 2005). For instance, many ungulates reveal a polygynous mating system, similar to caviids and the capybara, and recent genetic studies provide an interesting perspective that sometimes contrasts with assumptions regarding dominance and mating success. Based on behavioral observations, dominant male pronghorn antelope (*Antilocapra americana*) appear to have high reproductive success, yet genetic data reveal 44% mixed paternity in twins (Carling et al. 2003). Dominant males of Bighorn sheep (*Ovis canadensis*) tend females in estrus and guard against intruders. Nevertheless, Coltman et al. (2002, p. 165) state that "Although a few large-horned, mature rams had very high reproductive success, younger rams sired *ca.* 50% of the lambs." In genetic comparisons of twins in white-tailed deer (*Odocoileus virginianus*), multiple paternity occurs at a frequency of 25%, even though previous behavioral studies imply a mating bias toward dominant individuals (DeYoung et al. 2002).

Microsatellite loci also provide information about reproductive success and the mating system in social rodents. The yellow-pine chipmunk (*Tamias amoenus*) appears to have a promiscuous mating system, and recent genetic analysis confirms that both males and females mate with more than one partner, resulting in over 90% of all litters being sired by multiple fathers (Schulte-Hostedde et al. 2004). In addition, the alpine marmot (*Marmota marmota*) forms social groups and is monogamous, yet extra-pair paternity increases as the number of subordinate males in the group increases (Cohas et al. 2006). Two caviomorph rodents, *Galea musteloides*, a species with a more promiscuous mating system, and the cavy (*Cavia aperea*), a presumably polygynous species, display multiple paternity (Keil et al. 1999; Asher et al. 2004, 2008). In the case of *Cavia aperea*, multiple paternity ranges between 13% and 27% (Asher et al. 2008).

As in the alpine marmot, capybaras form harem-like social groups, containing several adult males, females, and juveniles (Herrera and Macdonald 1993). Individual dominant males are larger in body size and appear to perform more copulations than subordinate males, but as a group, subordinates still perform more copulations than individual dominant males (Herrera et al. 2011). Groups appear to be stable and usually consist of approximately 10–15 adults (Herrera and Macdonald 1987, 1989). Herrera et al. (2004) describe a recently developed panel of microsatellite loci for the capybara. Although exclusion probabilities are not available for these loci, the level of polymorphism in at least four loci is reasonable. It would be interesting to test for multiple paternity in the capybara, especially in light of the skewed number of copulations observed. Furthermore, if statements from other studies of social mammals with a range of mating systems are accurate, one might expect the degree of multiple paternity to coincide with the overall group dynamics of the population examined.

3.6 Conclusions

The behavioral, ecological, morphological, and physiological diversity within the Caviomorpha provides a treasure trove for those interested in the evolution of life-history traits and the influence of social structure on the partitioning of genetic variation within and between species. As indicated with the Caviidae, prior to proposing ecological constraints as potentially important to the evolution of social groups and complex mating systems, one must determine whether or not similarities are the result of shared ancestry. Even when a phylogenetic perspective is used to evaluate the evolution of sociality, one must be careful not to make generalities. Bennett and Faulkes (2000) propose the aridity-food distribution hypothesis (AFDH) as an explanation for the evolution of sociality in African mole-rats. Phylogenetically, the two eusocial species, *Heterocephalus glaber* and *Cryptomys damarensis*, do not share a common ancestry, suggesting independent origins of this trait (Allard and Honeycutt 1992). Nevertheless, both species occupy arid regions with intermittent rainfall and a patchy distribution of food. This implies that risk of predation and the uncertainty of food provide preconditions for convergent evolution of group-living in these two species (see Burda et al. 2000 for a somewhat different interpretation). Even though ecological constraints may explain the system seen in African mole-rats, the AFDH appears to provide a limited explanation of group-living in hystricognath rodents. For instance, Lacey and Wieczorek (2003) present data for *Ctenomys sociabilis*, a colonial species of caviomorph, and *Ctenomys haigi*, a solitary species. These two species occur sympatrically in southwestern Argentina, and the solitary species appears to reside in a habitat more conducive to group-living, a finding contrary to the AFDH.

Results from two other studies on hystricognath rodents provide tests of hypotheses pertaining to the evolution of group-living and conclude that group-living relates primarily to body size, diurnality, and digging burrows, with larger species that are diurnal and dig burrows being more likely to form large groups (Ebensperger and Cofré 2001; Ebensperger and Blumstein 2006). Unfortunately, neither uses the most accurate phylogeny for caviomorph rodents, and it would be interesting to reevaluate their conclusions with a more recently derived molecular phylogeny.

In conclusion, phylogenies and genetic markers are available for detailed studies of hystricognath rodents in general and the capybara in particular. Generic and species-level phylogenies provide a framework for detailed studies of the behavior and ecology of several major groups, including the Bathyergidae, Caviidae, Ctenomyidae, Echimyidae, and Octodontidae. In addition, many species of caviomorph rodents have a social structure and geographic distribution that may result in considerable phylogeographic structure. The capybara is no exception. Mones and Ojasti (1986) indicate that past taxonomic treatments of *Hydrochoerus* recognize as many as four species, and these authors recognize two. Nevertheless, no detailed phylogeographic study exists for the capybara. In addition, microsatellite loci are available for several species of hystricognath rodents, and many of these rodents display an ecology that may result in restricted gene flow between groups (Schroeder et al. 2000; Walker et al. 2000; Burland et al. 2001), and one comparative study of solitary and social species of *Ctenomys* provides evidence of genetic differences that may be related to different population histories (Lacey 2001). It is apparent from the preliminary information presented in this chapter that the current level of genetic and phylogenetic information provides an excellent starting point for more detailed studies of the evolution of hystricognath rodents. Such information in parallel with more detailed studies of the ecology and behavior of various species, especially caviomorphs, will provide the necessary starting point for comparisons of parameters important to the evolution of complex rodent societies.

Acknowledgments I thank the editors of this book for soliciting my contribution. Emilio Herrera and David Macdonald made many helpful comments. In addition, I thank my daughter, Heather A. Wakefield, for the animal drawings in Figs. 3.3 and 3.5.

References

Adkins RM, Gelke EL, Rowe D, Honeycutt RL (2001) Molecular phylogeny and divergence time estimates for major rodent groups: evidence from multiple genes. Mol Biol Evol 18:777–791

Adkins RM, Walton AH, Honeycutt RL (2003) Higher-level systematics of rodents and divergence time estimates based on two congruent nuclear genes. Mol Phylogenet Evol 26:409–420

Alho CJR, Rondon NL (1987) Habitats, population densities, and social structure of capybaras (*Hydrochoerus hydrochaeris*, Rodentia) in the pantanal, Brazil. Evita Bras Zoo S Paulo 4:139–149

Allard MW, Honeycutt RL (1992) Nucleotide sequence variation in the mitochondrial 12S rRNA gene and the phylogeny of African mole-rats (Rodentia: Bathyergidae). Mol Biol Evol 9:27–40

Arends A, Mcnab BK (2001) The comparative energetics of 'caviomorph' rodents. Comp Biochem Physiol A 130:105–122

Asher M, Spinelli De Oliveira E, Sachser N (2004) Social system and spatial organization of wild guinea pigs (*Cavia aperea*) in a natural population. J Mammal 85:788–796

Asher M, Lippmann T, Epplen JT, Kraus C, Trillmich F, Sachser N (2008) Large males dominate: ecology, social organization, and mating system of wild cavies, the ancestors to guinea pig. Behav Ecol Sociobiol 62:1509–1521

Baldi R (2007) Breeding success of the endemic mara *Dolichotis patagonum* in relation to habitat selection: conservation implications. J Arid Environ 68:9–19

Bennett NC, Faulkes CG (2000) African mole-rats: ecology and eusociality. Cambridge University Press, Cambridge, UK

Bugge J (1985) Systematic value of the carotid arterial pattern in rodents. In: Luckett WP, Hartenberger JL (eds) Evolutionary relationships among rodents: a multidisciplinary analysis. Plenum, New York, pp 381–402

Burda H, Honeycutt RL, Begall S, Locker-grütjen O, Scharff A (2000) Are naked and common mole-rats eusocial and if so, why? Behav Ecol Sociobiol 47:293–303

Burland TM, Bishop JM, O'ryan C, Faulkes CG (2001) Microsatellite primers for the African mole-rat genus *Cryptomys* and cross species amplification within the family, Bathyergidae. Mol Ecol 1:311–314

Carleton MD, Musser GG (2005) Order rodentia. In: Wilson DE, Reeder DM (eds) Mammal species of the world: a taxonomic and geographic reference, 3rd edn. Johns Hopkins University Press, Baltimore, pp 745–752

Carling MD, Wiseman PA, Byers JA (2003) Microsatellite analysis reveals multiple paternity in a population of wild pronghorn antelopes (*Antilocapra americana*). J Mammal 84:1237–1243

Clutton-Brock T, Isvaran K (2006) Paternity loss in contrasting mammalian societies. Biol Lett 2:513–516

Cohas A, Yoccoz NG, da Silva A, Goossens B, Allainé D (2006) Extra-pair paternity in the monogamous alpine marmot (*Marmota marmota*): the roles of social setting and female mate choice. Behav Ecol Sociobiol 59:597–605

Coltman DW, Festa-bianchet M, Jorgenson JT, Strobeck C (2002) Age-dependent sexual selection in bighorn rams. Proc R Soc Lond B 269:165–172

Corbet GG, Hill JE (1991) A World list of mammalian species, 3rd edn. Oxford University Press, Oxford, UK

da Silva Neto EJ (2000) Morphology of the regions ethmoidalis and orbitotemporalis in *Galea musteloides* Meyen 1832 and *Kerodon rupestris* (Wied-Neuwied 1820) (Rodentia: Caviidae) with comments on the phylogenetic systematics of the Caviidae. J Zoolog Syst Evol Res 38:219–229

Dawson MR, Marivaux L, Li C-K, Beard KC, Métais G (2006) *Laonastes* and the "lazarus effect" in recent mammals. Science 311:1456–1458

Dewoody JA (2005) Molecular approaches to the study of parentage, relatedness and fitness: practical applications for wild animals. J Wildl Manag 69:1400–1418

Dewsbury DA (1982) Dominance rank, copulatory behavior, and differential reproduction. Q Rev Biol 57:135–159

Deyoung RW, Honeycutt RL (2005) The molecular toolbox: genetic techniques in wildlife ecology and management. J Wildl Manag 69:1362–1384

Deyoung RW, Demarais S, Gonzales RA, Honeycutt RL, Gee KL (2002) Multiple paternity in white-tailed deer (*Odocoileus virginianus*) revealed by DNA microsatellites. J Mammal 83:884–892

Dobson FS (1982) Competition for mates and predominant juvenile male dispersal in mammals. Anim Behav 30:1183–1192

Dobson SF (1985) The use of phylogeny in behavior and ecology. Evolution 39:1384–1388

Dos Reis SF (1994) Allometry, convergence of ontogenetic trajectories and cranial evolution in caviine rodents (Rodentia: Caviinae). J Morphol 220:342

Ebensperger LA, Blumstein DT (2006) Sociality in New World hystricognath rodents is linked to predators and burrow digging. Behav Ecology 17:410–418

Ebensperger LA, Cofré H (2001) On the evolution of group-living in the New World cursorial hystricognath rodents. Behav Ecol 12:227–236

Ebensperger LA, Hurtado MJ, Soto-Gamboa M, Lacey EA, Chang AT (2004) Communal nesting and kinship in degus (*Octodon degus*). Naturwissenschaften 91:391–395

Flynn JJ, Wyss AR (1998) Recent advances in South American mammalian paleontology. Trends Ecol Evol 13:449–454

Flynn JJ, Wyss AR, Croft DA, Charrier R (2003) The Tinguiririca Fauna, Chile: biochronology, paleoecology, biogeography, and a new earliest Oligocene South American Land Mammal 'Age'. Palaeogeogr Palaeoclim Palaeoecol 195:229–259

George W (1985) Reproductive and chromosomal characters of ctenodactylids as a key to their relationships. In: Luckett WP, Hartenberger J-L (eds) Evolutionary relationships among rodents: a multidisciplinary analysis. Plenum, New York, pp 453–474

Greenwood PJ (1980) Mating systems, philopatry and dispersal in birds and mammals. Anim Behav 28:1140–1162

Hamilton WD (1963) The evolution of altruistic behavior. Am Nat 97:354–356

Harvey PH, Pagel M (1991) The comparative method in evolutionary biology. Oxford University Press, Oxford, UK

Hayes LD (2000) To nest communally or not to nest communally: a review of rodent communal nesting and nursing. Anim Behav 59:677–688

Herrera EA (1992) Growth and dispersal of capybaras (*Hydrochaeris hydrochaeris*) in the Llanos of Venezuela. J Zool 228:307–316

Herrera EA, Macdonald DW (1987) Group stability and the structure of a capybara population. Symp Zoolog Soc Lond 58:115–130

Herrera EA, Macdonald DW (1989) Resource utilization and territoriality in group-living capybaras (*Hydrochoerus hydrochaeris*). J Anim Ecol 58:667–679

Herrera EA, Macdonald DW (1993) Aggression, dominance, and mating success among capybara males (*Hydrochaerus hydrochaeris*). Behav Ecol 4:114–119

Herrera EA, Macdonald DW (1994) Social significance of scent marking in capybaras. J Mammal 75:410–415

Herrera EA, Chemello ME, Lacey EA, Salas V, Sousa BF (2004) Characterization of microsatellite markers from capybaras, *Hydrochoerus hydrochaeris* (Rodentia: Hydrochoeridae). Mol Ecol Notes 4:541–543

Herrera EA, Salas V, Congdon ER, Corriale MJ, Tang-Martínez Z (2011) Capybara social structure and dispersal patterns: variations on a theme. J Mammal 92:12–20

Hoffstetter R (1972) Relationships, origins, and history of the ceboid monkeys and caviomorph rodents: a modern reinterpretation. Evol Biol 6:322–347

Holroyd PA, Maas MC (1994) Paleogeography, paleobiogeography, and anthropoid origins. In: Fleagle JG, Kay RF (eds) Anthropoid origins. Plenum, New York, pp 297–334

Honeycutt RL (2003) Cavies and maras. In: Kleiman DG, Geist V, Hutchins M, McDade MC (eds) Grzimek's animal life encyclopedia, mammals V, vol 16. Thomson Gale Group/Farmington Hills, Minnesota, pp 389–400

Honeycutt RL (2009) Rodentia. In: Hedges SB, Kumar S (eds) The time tree of life. Oxford University Press, Oxford, UK, pp 490–494

Honeycutt RL, Rowe DL, Gallardo MH (2003) Molecular systematics of the South American caviomorph rodents: relationships among species and genera in the family Octodontidae. Mol Phylogenet Evol 26:476–489

Honeycutt RL, Frabotta LJ, Rowe DL (2007) Rodent evolution, phylogenetics, and biogeography. In: Wolff JO, Sherman PW (eds) Rodent societies: an ecological & evolutionary perspective. The University of Chicago Press, Chicago, pp 8–23

Huchon D, Douzery EJP (2001) From the old World to the New World: a molecular chronicle of the phylogeny and biogeography of hystricognath rodents. Mol Phylogenet Evol 20:238–251

Huchon D, Catzeflis FM, Douzery EJP (1999) Molecular evolution of the von Willebrand factor gene in mammals and the phylogeny of rodents. Mol Biol Evol 16:577–589

Huchon D, Catzeflis FM, Douzery EJP (2000) Variance of molecular datings, evolution of rodents, and the phylogenetic affinities between Ctenodactylidae and Hystricognathi. Proc R Soc Lond B 276:393–402

Huchon D et al (2002) Rodent phylogeny and a timescale for the evolution of Glires: evidence from an extensive taxon sampling using three nuclear genes. Mol Biol Evol 19:1053–1065
Huchon D et al (2007) Multiple molecular evidences for a living mammalian fossil. Proc Natl Acad Sci USA 104:7495–7499
Hughes C (1998) Integrating molecular techniques with field methods in studies of social behavior: a revolution results. Ecology 79:383–399
Hugot JP (2003) New evidence for hystricognath rodent monophyly from the phylogeny of their pinworms. In: Page RDM (ed) Tangled trees: phylogeny, cospeciation, and coevolution. University of Chicago Press, Chicago, pp 144–173
Ingram CM, Burda H, Honeycutt RL (2004) Molecular phylogenetics and taxonomy of the African mole-rats, genus *Cryptomys* and the new genus *Coetomys* Gray, 1864. Mol Phylogenet Evol 31:997–1014
Isvaran K, Clutton-Brock T (2007) Ecological correlates of extra-group paternity in mammals. Proc R Soc Lond Series B 274:219–224
Jenkins PD, Kilpatrick CW, Robinson MF, Timmins RJ (2004) Morphological and molecular investigations of a new family, genus and species of rodent (Mammalia: Rodentia: Hystricognatha) from Lao PDR. Syst Biodivers 2:419–454
Jesseau SA, Holmes WG, Lee TM (2009) Communal nesting and discriminative nursing by degus, *Octodon degus*. Anim Behav 78:1183–1188
Keil A, Epplen JT, Sachser N (1999) Reproductive success of males in the promiscuous-mating yellow-toothed cavy (*Galea musteloides*). J Mammal 80:1257–1263
Kjer KM, Honeycutt RL (2007) Site specific rates of mitochondrial genomes and the phylogeny of Eutheria. BMC Evol Biol 7:8
Lacey EA (2001) Microsatellite variation in solitary and social tuco-tucos: molecular properties and population dynamics. Heredity 86:628–637
Lacey EA, Wieczorek JR (2003) Ecology of sociality in rodents: a ctenomyid perspective. J Mammal 84:1198–1211
Lacher TE Jr (1981) The comparative social behavior of *Kerodon rupestris* and *Galea spixii* and the evolution of behavior in the Caviidae. Bull Carnegie Mus Nat Hist 17:1–71
Landry SO (1999) A proposal for a new classification and nomenclature for the Glires (Lagomorpha and Rodentia). Mitt Zoologischen Mus Berlin 75:283–316
Lavocat R (1969) La systématique des rongeurs hystricomorphes et la dérive des continents. CRAcad Sci Paris D 269:1496–1497
Lavocat R (1973) Les rongeurs du Miocène d'Afrique Orientale. Mémoires et Travaux de l'Institut de Montpellier de l'École Pratique des Hautes Études, Institut de Montpellier 1, pp 1–284
Lavocat R (1980) The implications of rodent paleontology and biogeography to the geographical sources and origin of platyrrhine primates. In: Ciochon RL, Chiarelli AB (eds) Evolutionary biology of the New World monkeys and continental drift. Plenum, New York, pp 93–102
Lavocat R, Parent JP (1985) Phylogenetic analysis of middle ear features in fossils and living rodents. In: Luckett WP, Hartenberger JL (eds) Evolutionary relationships among rodents: a multidisciplinary analysis. Plenum, New York, pp 211–227
Luckett WP (1985) Superordinal and intraordinal affinities of rodents: developmental evidence from dentition and placentation. In: Luckett WP, Hartenberger JL (eds) Evolutionary relationships among rodents: a multidisciplinary analysis. Plenum, New York, pp 227–276
Luckett WP, Hartenberger J-L (1985) Evolutionary relationships among rodents: comments and conclusions. In: Luckett WP, Hartenberger JL (eds) Evolutionary relationships among rodents: a multidisciplinary analysis. Plenum, New York, pp 685–712
Luckett WP, Hartenberger J-L (1993) Monophyly or polyphyly of the order Rodentia: possible conflict between morphological and molecular interpretations. J Mamm Evol 1:127–147
Macdonald DW (1981) Dwindling resources and the social behaviour of capybaras, (*Hydrochoerus hydrochaeris*) (Mammalia). J Zool 194:371–391
Macdonald DW, Herrera EA, Taber AB, Moreira JR (2007) Social organization and resource use in capybaras and maras. In: Wolff JO, Sherman PW (eds) Rodent societies, an ecological and evolutionary perspective. University of Chicago Press, Chicago, pp 393–402

Macphee RDE, Singer R, Diamond M (2000) Late Cenozoic land mammals from Grenada, Lesser Antilles island-arc. Am Mus Novit 3302:1–20
Mares MA, Ojeda RA (1982) Patterns of diversity and adaptation in South American hystricognath rodents. Pymatuning Lab Ecol Spec Publ Ser 6:393–431
Marivaux L, Vianey-liaud M, Jaeger J-J (2004) High-level phylogeny of early tertiary rodents: dental evidence. Zoolog J Linn Soc 142:105–134
Marshall LG (1988) Land mammals and the great American interchange. Am Sci 76:380–388
Martin T (1993) Early rodent incisor enamel evolution: phylogenetic implications. J Mamm Evol 1:227–254
Mckenna MC, Bell SK (1997) Classification of mammals: above the species level. Columbia University Press, New York
Meng J (1990) The auditory region of *Reithroparamys delicatissimus* (Mammalia, Rodentia) and its systematic implications. Am Mus Novit 2972:1–35
Mones A, Ojasti J (1986) *Hydrochoerus hydrochaeris*. Mamm Species 264:1–7
Murphy WJ, Eizirik E, Johnson WE, Zhang YP, Ryder A, O'Brien SJ (2001) Molecular phylogenetics and the origins of placental mammals. Nature 409:614–618
Nedbal MA, Allard MW, Honeycutt RL (1994) Molecular systematics of hystricognath rodents: evidence from the mitochondrial 12S rRNA gene. Mol Phylogenet Evol 3:206–220
Nedbal MA, Honeycutt RL, Schlitter DA (1996) Higher-level systematics of rodents (Mammalia, Rodentia): evidence from the mitochondrial 12S rRNA gene. J Mamm Evol 3:201–237
Nogueira SSC, Nogueira-Filho SLG, Otta E, Dias CTS, Carvalho A (1999) Determination of infanticide causes in capybara (*Hydrochoerus hydrochaeris*) groups in captivity. Appl Anim Behav Sci 62:351–357
Nogueira SSC, Nogueira-filo SLG, Dias CTS (2000) Alloparental behavior in capybara (*Hydrochoerus hydrochaeris*). Rev Etologia 2:17–21
Opazo JC (2005) A molecular timescale for caviomorph rodents (Mammalia, Hystricognathi). Mol Phylogenet Evol 37:932–937
Patterson B, Wood AE (1982) Rodents from the Deseadan Oligocene of Bolivia and the relationships of the Caviomorpha. Bull Mus Comp Zool 149:371–543
Pérez-Espona S, Pérez-Barberia FJ, Jiggins CD, Gordon IJ, Pemberton JM (2010) Variable extent of sex-biased dispersal in a strongly polygynous mammal. Mol Ecol 19:3101–3113
Poux C, Chevret P, Huchon D, de Jong WW, Douzery EJP (2006) Arrival and diversification of caviomorph rodents and platyrrhine primates in South America. Syst Biol 55:228–244
Quintana CA (1998) Relaciones filogenéticas de roedores Caviinae (Caviomorpha, Caviidae) de América Sur. B Soc Esp Hist Nat Biol 94:125–134
Redford KH, Eisenberg JF (1992) Mammals of the Neotropics: the southern cone, vol 2. University of Chicago Press, Chicago
Reig OA (1986) Diversity patterns and differentiation of high Andean rodents. In: Vuilleumier F, Monasterio M (eds) High altitude tropical biogeography. Oxford University Press, Oxford, UK, pp 404–439
Rinderknecht A, Blanco RE (2008) The largest fossil rodent. Proc R Soc B 1645:1–6
Rocha-Barbosa O, Loguercio MFC, Renous S, Gasc J-P (2007) Comparative study on the forefoot and hindfoot intrinsic muscles of some Cavioidea rodents (Mammalia, Rodentia). Zoology 110:58–65
Rood JP (1972) Ecological and behavioural comparisons of three genera of Argentine cavies. Anim Behav Monogr 5:1–83
Rowe DL (2002) Molecular phylogenetics and evolution of hystricognath rodents. Ph.D. thesis, Texas A&M University, College Station
Rowe DL, Honeycutt RL (2002) Phylogenetic relationships, ecological correlates, and molecular evolution within the Cavioidea (Mammalia, Rodentia). Mol Biol Evol 19:263–277
Rowe DL, Dunn KA, Adkins RM, Honeycutt RL (2010) Molecular clocks keep dispersal hypotheses afloat: evidence for trans-Atlantic rafting by rodents. J Biogeogr 37:305–324
Sachser N, Schwarz-Weigi E, Keil A, Epplen JT (1999) Behavioural strategies, testis size, and reproductive success in two caviomorph rodents with different mating systems. Behaviour 136:1203–1217

Sahni A (1985) Enamel structure of early mammals and its role in evaluating relationships among rodents. In: Luckett WP, Hartenberger JL (eds) Evolutionary relationships among rodents: a multidisciplinary analysis. Plenum, New York, pp 133–150
Schroeder JW, Honeycutt RL, Rooney AP, Han G, Begall S, Gallardo MH (2000) Microsatellites from the South Americaan coruro, *Spalacopus cyanus*. Mol Ecol 9:1447–1449
Schulte-Hostedde AI, Millar JS, Gibbs HL (2004) Sexual selection and mating patterns in a mammal with female-biased sexual size dimorphism. Behav Ecol 15:351–356
Simpson GG (1945) The principles of classification and a classification of mammals. Am Mus Nat Hist Bull 85:1–350
Simpson GG (1980) Splendid isolation: the curious history of South American mammals. Yale University Press, New Haven
Tabeni S, Ojeda RA (2003) Assessing mammal responses to perturbations in temperate aridlands of Argentina. J Arid Environ 55:715–726
Taber AB, Macdonald DW (1992) Spatial organization and monogamy in the mara *Dolichotis patagonum*. J Zool Lon 227:417–438
Tang-Martínez Z (2003) Emerging themes and future challenges: forgotten rodents, neglected questions. J Mammal 84:1212–1227
Trillmich F et al (2004) Species-level differentiation of two cryptic species pairs of wild cavies, genera Cavia and Galea, with a discussion of the relationships between social systems and phylogeny in the Caviinae. Can J Zool 82:516–524
Vucetich MG, Verzi DH, Hartenberger J-L (1999) Review and analysis of the radiation of the South American Hystricognathi (Mammalia, Rodentia). CR Acad Sci Paris Earth Plant Sci 329:763–769
Vucetich MG, Deschamps CM, Pérez ME (2012) Paleontology, evolution and systematics of capybara. In: Moreira JR, Ferraz KMPMB, Herrera EA, Macdonald DW (eds) Capybara: biology, use and conservation of an exceptional neotropical species. Springer, New York, pp 39–59
Walker RS, Farmerie WG, Branch LC (2000) Characterization of microsatellite loci from the mountain vizcacha *Lagidium viscacia* and their use for the plains vizcacha *Lagostomus maximus*. Mol Ecol 9:1672–1674
Walton AH (1997) Rodents. In: Kay RF, Madden RH, Cifelli RL, Flynn JJ (eds) Vertebrate paleontology in the neotropics. The Miocene fauna of La Venta, Colombia. Smithsonian Institution Press, Washington, DC, pp 392–409
Walton AH, Nedbal MA, Honeycutt RL (2000) Evidence from intron 1 of the nuclear transthyretin (prealbumin) gene for the phylogeny of African mole-rats (Bathyergidae). Mol Phylogenet Evol 16:467–474
Webb SD, Barnosky AD (1989) Faunal dynamics of Pleistocene mammals. Annu Rev Earth Planet Sci 17:413–438
Wilson DE, Reeder DM (eds) (2005) Mammal species of the world: a taxonomic and geographic reference, vol 3. Johns Hopkins University Press, Baltimore
Wood AE (1955) A revised classification of the rodents. J Mammal 36:165–187
Wood AE (1965) Grades and clades among rodents. Evolution 19:115–130
Wood AE (1974) The evolution of the Old World and New World Hystricomorphs. Symp Zoolog Soc Lond 34:21–60
Wood AE (1985) The relationships, origin and dispersal of the Hystricognathous rodents. In: Luckett WP, Hartenberger J-L (eds) Evolutionary relationships among rodents: a multidisciplinary analysis. Plenum, New York, pp 475–513
Woods CA (1984) Hystricognath rodents. In: Anderson S, Jones JK Jr (eds) Orders and families of recent mammals of the world. Wiley, New York, pp 389–446
Woods CA, Kilpatrick CW (2005) Infraorder hystricognathi. In: Wilson DE, Reeder DM (eds) Mammal species of the world: a taxonomic and geographic reference, 3rd edn. Johns Hopkins University Press, Baltimore, pp 1538–1600
Wyss AR et al (1993) South America's earliest rodent and recognition of a new interval of mammalian evolution. Nature 365:434–437

Chapter 4
Foraging Strategies and Feeding Habits of Capybaras

Guillermo R. Barreto and Rúben D. Quintana

4.1 Introduction

The capybara (*Hydrochoerus hydrochaeris*) is a herbivorous semiaquatic mammal that grazes near water bodies. A number of physiological and morphological adaptations of the capybara's digestive system (Herrera 2012) allow the species to meet its energy requirements from a diet with a high fiber and low nutritional content and silica deposits. Capybara molars, for example, are structured to reduce particle size to 0.001–0.3 mm, a chewing effectiveness comparable to that achieved by bovids (Ojasti 1973). The capybara cecum contains cellulolytic microorganisms (Borges et al. 1996) from a wide variety of taxa, and its digestive efficiency is comparable to that of ruminants (González-Jiménez 1977). Fermentation in the cecum is associated with cecotrophy (Herrera 2012; Mendes and Nogueira-Filho 2012) making digestion highly efficient.

Increasing its efficiency still more, the capybara's digestive tract shows seasonal modifications that allow it to adjust intake quantity as the nutritional quality of available food varies (Borges et al. 1996). During the dry season, the relative weight of the small intestine (plus the cecum) increases, indicating a larger surface area to absorb nutrients at a time when food quality is in decline. These characteristics make capybaras efficient grazers and ecologically equivalent to the medium-sized ungulates of Africa (Ojasti 1983).

G.R. Barreto (✉)
Departamento de Biología de Organismos, Universidad Simón Bolívar,
Apdo. 89000, Caracas 1080-A, Venezuela
e-mail: guibarre@usb.ve

R.D. Quintana
Instituto de Investigación e Ingeniería Ambiental (3iA), Universidad Nacional de San Martín,
Peatonal Belgrano 3563, San Martín, Pcia. de Buenos Aires B1650, Argentina
e-mail: rubenq@ege.fcen.uba.ar; rquintana@unsam.edu.ar

J.R. Moreira et al. (eds.), *Capybara: Biology, Use and Conservation of an Exceptional Neotropical Species*, DOI 10.1007/978-1-4614-4000-0_4,

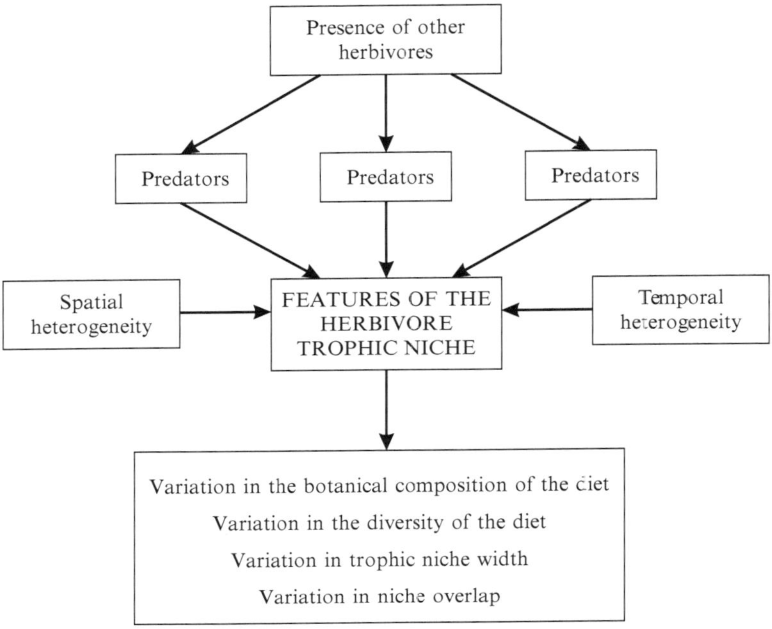

Fig. 4.1 The main factors that influence the foraging patterns of a herbivore (From Quintana et al. (2002))

Studying the feeding habits of herbivores such as capybaras, and indeed of any consumer, requires the researcher to consider different scales (Charnov 1976; Pyke 1984; Stephens and Krebs 1986; Senft et al. 1987). Resource selection takes place at a variety of levels, and in decreasing hierarchical order these levels would be: (a) the species' geographical range, (b) the home range of the individual or group within the species' geographical range, (c) habitats within the home range, and (d) patches within the preferred feeding habitat (Manly et al. 2002). Thus, at the most general level, foraging studies describe the habitat type chosen for foraging, and at a more detailed level they describe the choice of patches within that habitat. Particular foraging strategies used within the patch, together with the species consumed, constitute the most detailed level of study.

Grazing mammals in patchy habitats face a landscape in which both quantity and quality of food resources vary over space and time. Such consumers must adjust their feeding habits and foraging tactics to match the type of food present in the different patches within their home range (Sibly 1981). So a herbivore is able to select high-quality food in periods of abundance while opting to consume large quantities of low-quality food in leaner seasons (Hume 1989). Herbivore foraging patterns are also influenced by factors such as the presence of competitors, the proximity of predators and specific physical and physiological characteristics of the plant species consumed (Fig. 4.1).

In this chapter, we describe the feeding ecology of capybaras, starting with habitat use and patch selection and moving on to foraging strategies and diet. We also discuss competition for food between capybaras and cattle.

4.2 Habitat and Space Use

Capybaras are rarely found far from water bodies. They inhabit several types of wetland, including gallery forests along rivers (Soini and Soini 1992), mangroves, and marshes, but capybaras reach their highest densities in the seasonally flooded savannas of the Llanos of Venezuela and Colombia, the Pantanal of the Mato Grosso, and on Marajó Island in Brazil (Fig. 4.2; Mones and Ojasti 1986). There are also important populations in the marshes of Esteros del Iberá in the Argentinian province of Corrientes and in the agricultural landscapes of southeastern Brazil (Fig. 4.2). Water is a vital resource for capybaras; it is used not only for drinking, but also to

Fig. 4.2 Map of South America with the locations of the places referred to in the text

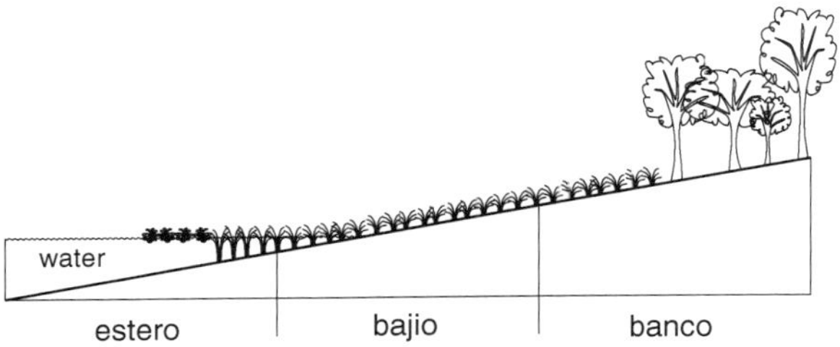

Fig. 4.3 Physiographic units of neotropical savannas: esteros, inundated areas; bajios, 0.5 m higher; and bancos, 2 m higher, rarely flooded

control their body temperature and as an escape from predators (Macdonald 1981). Capybaras usually mate in the water (Azcárate 1981; Macdonald 1981) and, as semiaquatic grazers, most of their food is found near or within water bodies.

Capybaras establish territories in which water is the key resource (Herrera and Macdonald 1989), and it is around these water holes where the most palatable forage occurs (Escobar and González Jiménez 1979). In the Venezuelan Llanos and the Brazilian Pantanal, capybara density is much higher at less than 300 m from a water hole. Capybara groups near the Paraguay River, however, have been reported as far as 1 km away from the nearest water hole looking for the bushy areas they favor for resting (Schaller and Crawshaw 1981).

In the seasonally flooded savannas (low Llanos) of Venezuela, capybaras change their use of space as water levels vary: they need water in the dry season and dry land in the wet. In both situations, food availability is crucial (Herrera and Macdonald 1989). These savannas are characterized by three well-defined physiographic units (Fig. 4.3): (a) *esteros*, inundated areas which can dry up completely in the dry season; (b) *bajios*, 0.5 m higher than esteros, comprising close to 70 % of the low Llanos, where palatable and nutritious grasses grow; and (c) *bancos*, 2 m higher than esteros, rarely flooded and containing tall grasses, bushes, and, occasionally, forested patches. Observational studies by Herrera and Macdonald (1989) in this type of habitat showed seasonal differences in the use of these physiographic units. Capybaras fed in the bajios throughout most of the year, but spent less time feeding there during peak flooding at the end of the wet season. Bancos were used increasingly toward the end of the wet season as they became the only available dry land. During the dry season, capybara groups remained in the bajios or on dry reed beds (Cyperacea: *Eleocharis* sp.) along the edges of esteros. These changes in habitat use did not entail migratory movements, as the group's center of activity rarely moved more than 300 m from the edge of whatever water was present during the dry season. In the tropics, according to Jorgenson (1986), capybaras spend 31 % of their time grazing during the wet season and 42 % in the dry.

Habitat use also varies throughout the day. In the early morning, capybara groups rest in the bajios. Toward mid-morning they move to tall grass bancos as the temperature begins to rise and then to esteros, around midday, where they wade. In the afternoon, and as the sun goes down, groups forage in the bajios and continue to do so until well into the night. In the Llanos of Colombia, Jorgenson (1986) found that capybaras started feeding at around 10:00 h for 1.5–2 h in the dry season, while in the wet season they set out to graze 1 h later. Similar observations of diurnal foraging have been recorded in the Brazilian Pantanal (Alho et al. 1987). However, although capybaras are considered predominantly diurnal, groups have been observed grazing during the night, with a resting period around midnight, on Marajó Island, Pará, Brazil (José R. Moreira personal communication), in the Pantanal (Rodiney A. Mauro personal communication) and also in the agricultural habitats in southeastern Brazil (Katia M.P.M.B. Ferraz personal communication). Additionally, capybaras appear to change the grazing patch within their territory every day or two (Barreto and Herrera 1998).

4.3 Foraging Decisions

Capybaras have been described as selective consumers that choose highly palatable grasses (Ojasti 1973; Gonzalez-Jimenez 1977; Herrera and Macdonald 1989). In a study of capybara foraging behavior, Barreto and Herrera (1998) showed that capybaras are also able to adapt their foraging strategies in response to varying seasonal conditions. During the wet season, capybaras were more selective, spending more time grazing on *Hymenachne amplexicaulis* (27 % of observation time), an aquatic grass of high caloric and low fiber content, than on the less palatable reeds, *Eleocharis interstincta*, which were all but ignored by the animals (3 % of observation time). Comparison of effective feeding time (Barreto and Herrera 1998) on the grasses *H. amplexicaulis* and *Reimarochloa acuta* showed that capybaras were less selective when feeding on the lower quality grass (*R. acuta*, Table 4.1).

During the dry season, availability and quality of food is drastically reduced and, as expected, capybaras were less selective, spending roughly the same amount of time grazing on *E. interstincta* reeds and on the grass *Paratheria prostrata*. The latter grass provides more energy than reeds but has similar protein and fiber values (Table 4.2). Individuals feeding on reeds interspersed with water hyacinth

Table 4.1 Effective feeding time (EFT) of capybaras, caloric content (CC), and protein and fiber of three plants (green biomass) consumed in the wet season in a seasonally flooded savanna of the Venezuelan Llanos. Different letters next to values indicate significant differences (Barreto 1994)

Species	EFT (%)	CC (Kjoules/g)	Protein (%)	Fiber (%)
Hymenachne amplexicaulis	69b	19.26 ± 0.03a	10.31 ± 0.86a	65.0 ± 1.3ab
Eleocharis interstincta	65b	18.72 ± 0.11b	8.07 ± 0.30b	67.6 ± 1.1a
Reimarochloa acuta	81a	19.54 ± 0.13a	7.29 ± 2.58b	62.8 ± 1.4b

Table 4.2 Effective feeding time (EFT) of capybaras, caloric content (CC), and protein and fiber of three plants (green biomass) consumed in the dry season in a seasonally flooded savanna of the Venezuelan Llanos. Different letters next to values indicate significant differences (Barreto 1994)

Species	TEA (%)	CC (Kjoules/g)	Protein (%)	Fiber (%)
Hymenachne amplexicaulis	84a	19.54 ± 0.35a	16.62 ± 0.30a	54.5 ± 1.3b
Eleocharis interstincta	68b	18.64 ± 0.08b	11.32 ± 1.0b	64.4 ± 0.2a
Paratheria prostrata	86a	19.44 ± 0.10a	12.25 ± 1.58b	63.8 ± 3.3a

(*Eichhornia* sp.) ignored the latter. Indeed, consumption of water hyacinth has only been observed during severe drought, and even then only stalks were eaten and the leaves rejected, leading Ojasti (1973) to remark that this plant is an emergency resource used only in times of extreme scarcity.

The foregoing discussion suggests that capybaras are able to adapt and optimize their foraging patterns in response to spatial and temporal variation in the quality of food resources. However, in a study in the Lower Delta of the Paraná River (Argentina; Fig. 4.2), Corriale et al. (2011) found that calorific content was the only variable quality that determined capybaras' preference for the plants they consumed in greater proportion than their availability. Other variables, such as ash (minerals) and nitrogen content (protein) were not different between plants selected and those not chosen. Corriale et al. (2011) suggest that other factors such as the presence of chemical defenses in plants or physiological constraints may influence plant selection by capybaras.

4.4 Diet Composition

The earliest reports of capybara diet date back to the nineteenth century. Humboldt (1826) described capybaras as herbivore grazers but also suggested that they consume fish. He probably based the latter on hearsay, and merely on popular misconception or folklore at that (Ojasti 1973). Later studies include observations by Osgood (1912), who described *Hydrochoerus isthmius* (the capybara species found west of the Andes) feeding on algae; by Mondolfi (1957) on *Hydrochoerus hydrochaeris*, who mentioned grasses, particularly *Paspalum fasciculatum*; and by Barlow (1969), who reported Uruguayan capybaras eating *Digitaria* sp., *Cynodon dactylon*, and *Oxalis* sp. It was not until the 1970s that a systematic study of capybara biology, including diet, was carried out (Ojasti 1973). Based on field observations, Ojasti (1973) described capybaras consuming 21 species of grasses; 5 aquatic plants; and the bark, stems, and leaves of 4 bushes; he also suspected the consumption of 3 species of Cyperaceae. Ojasti suggested that *Axonopus purpusii*, *P. prostrata*, and *R. acuta* were important dietary items in the dry season, while *Luziola pittieri* and the aquatic *Heteranthera limosa* were consumed mainly in the wet season.

More recent studies are based on microhistological techniques (see Box 4.1). Escobar and González-Jiménez (1976) found that capybaras in the Venezuelan Llanos consume at least 17 plant species, five of which make up more than 80 % of the diet, both in frequency of consumption and in proportion of weight consumed. Three of these species, the grasses *H. amplexicaulis*, *Leersia hexandra*, and an unidentified Cyperaceae, possibly *E. interstincta*, are associated with esteros, while *P. prostrata* is abundant in bajios and *Sporobolus indicus* grows in bancos. *S. indicus* was the least consumed of all five species, revealing the preference of capybaras for bajios and esteros as feeding habitats in the seasonally flooded savannas of Venezuela.

Much farther south in Argentina's Paraná River Delta, Quintana et al. (1994, 1998a) described consumption of between 12 and 18 plant species, 3–5 of which

Box 4.1 Characterization of the Diet of Capybaras Using the Microhistological Method

The microhistological technique is based on the fact that the characteristic microstructures of the epidermis (type of stomata and trichomes, thickness and shape of the cell wall, presence and types of hairs, etc.) of many plants survive digestion intact, thus providing a diagnostic character for the identification of plant species consumed by herbivores (Fig. 4.4). Examination of these undigested remains in feces can yield both qualitative and quantitative descriptions of a herbivore's diet. All plant groups possess characteristic features, but grasses attain the greatest degree of differentiation using these methods, which means that microhistological techniques are especially useful for studying the foraging habits of grazers, such as capybaras. Several authors (Metcalfe 1960; Metcalfe and Chalk 1979; Johnson et al. 1983; Monge 1989) have published detailed descriptions of various structures useful for the identification of epidermal fragments.

Fecal analysis does not interfere with the activities of the animal and samples are almost unlimited. In contrast, direct observations of consumed species are often impractical, while the analysis of stomach contents involves killing the animal, an unacceptable practice when the study species is endangered.

Fecal studies have been used in recent years for the study of the diet in a wide range of herbivores. By 1978, in the USA alone, this technique had been used for at least 16 large mammals, 20 small mammals, 4 birds, and a few insects. Since the 1970s, interest in the microhistological analysis of vegetation fragments has been increasing, and ever more effective techniques to describe the patterns of the herbivore's diet have appeared, making it a widely accepted method.

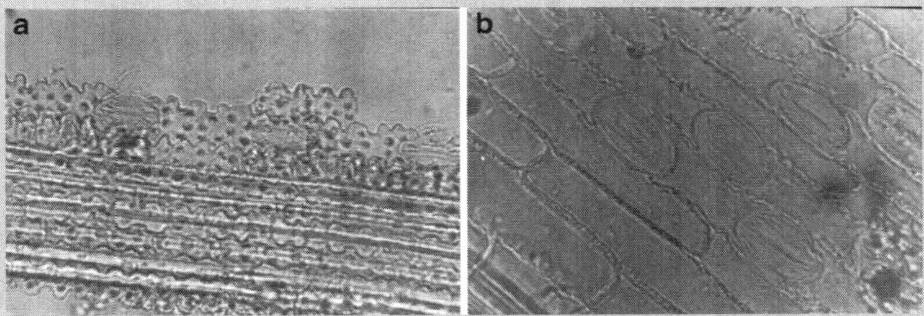

Fig. 4.4 Microhistology of some neotropical plant species: (**a**) grass *Luziola peruviana*; (**b**) Cyperacea *Rhynchospora* sp.

Table 4.3 Diet of capybaras (% by weight) in insular areas of the Parana River Delta, Argentina. Only those species which made up more than 75 % in each season are shown (Data from a microhistological analysis of feces (Quintana et al. 1994))

	Rains		Drought	
Species	Summer	Autumn	Winter	Spring
Axonopus fissifolius (=*Axonopus affinis*)		18.00		
Carex fuscula (=*Carex riparia*)	37.33	48.00	61.00	46.50
Cynodon dactylon	30.33	9.00	7.33	13.50
Hymenachne grumosa (=*Panicum grumosum*)	11.83	7.33	12.50	14.83
Paspalum sp.		6.33	5.17	
Several cyperaceae				12.67
Subtotal	79.49	88.66	86.00	87.50

Table 4.4 Diet of capybaras (% by weight) in non-insular areas of the Parana River Delta in Argentina. Only those species which made up more than 6 % in each season are shown (Data from a microhistological analysis of feces (Quintana et al. 1998a))

	Rains		Drought	
Species	Summer	Autumn	Winter	Spring
Carex fuscula (=*Carex riparia*)		8.67		9.34
Cynodon dactylon	26.33	28.00		16.78
Eleocharis spp.			14.85	19.00
Luziola peruviana	18.33	8.34		16.56
Hymenachne grumosa (=*Panicum grumosum*)	10.34	14.00	26.51	9.33
Zizaniopsis bonariensis	6.00	12.67	16.90	
Several cyperaceae			13.40	
Subtotal	61.00	71.68	71.66	71.01

constituted more than 70 % of the total weight consumed (Tables 4.3 and 4.4). The high proportion (up to 40 % on the mainland and 60 % on the islands) of Cyperaceae that these authors found in capybaras' diet is noteworthy, contrasting with Venezuela where a single unidentified Cyperacea contributed only 16 % of the diet (Escobar and González-Jiménez 1976). This is probably due to the abundance of these plant families in their respective ecosystems. In fact, in the savanna habitat of the Entre Ríos Province of Argentina (Fig. 4.2), where grasses dominate the herbaceous stratum, the main species eaten by capybaras is the grass *Paspalum* sp., with Cyperacea making up only 14 % in weight of the diet (Quintana et al. 1998a; Quintana 2003). Furthermore, the greatest proportion of Cyperaceae consumption was found in winter, indicating that such plants are consumed mainly in times of scarcity. The range of plant species reportedly consumed in Brazil's Pantanal swamps in Nhecolandia is much wider (Mauro and Pott 1996). Among the 26 species eaten by capybaras in the dry season and 32 during the rains, the most important grasses were *A. purpusii*,

Panicum repens, *H. amplexicaulis* and the aquatic herb *Pontederia cordata* (=*Pontederia lanceolata*), which together constituted 48.9 % and 54.7 % of dry weight in the dry and wet seasons, respectively. Although some beard algae were found in feces collected during the rains, the authors considered this finding accidental. As for Cyperaceae, Mauro and Pott (1996) found moderate consumption of *Cyperus brevifolius*, with 8.1 % of dry weight in the dry season and 3.4 % in the rains. Borges and Colares (2007) found that only 14.8 % of 133 species of herbs were present in capybaras' fecal samples from the wetlands of Taim in southern Brazil, indicating a marked preference for Poaceae (grasses) which were present in 52.9 % of fecal samples analyzed. *Zizaniopsis bonariensis* was the most common species in the diet in all seasons (Borges and Colares 2007) and diet diversity was highest in the summer.

In Colombia, recent observations in the Llanos region (Forero-Montaña et al. 2003) and in Amazonia (Arteaga and Jorgenson 2007) mirrored those described above. In the Llanos, from a total of 89 plants, seven including the grasses *H. amplexicaulis*, *Digitaria bicornis*, *P. fasciculatum*, *C. dactylon*, *Urochloa arrecta*, *Urochloa maxima* (=*Panicum maximum*), and the Cyperaceae *Rhynchospora corymbosa* constituted 60 % of the diet. In the Colombian Amazon, the study was limited to observations of plants associated with capybara footprints, but this was the only study of capybaras in forested habitats, albeit a nonsystematic one. The findings indicate that capybaras adapted their diet to agricultural crops in altered habitats.

In conclusion, capybaras are grazers that feed mainly on grasses in savanna habitat, while including Cyperacea in times of scarcity or in places where these are predominant. Their diet has been documented in only a small part of the species' distribution.

4.5 Potential Competition with Cattle

In the open areas of seasonally flooded savannas where capybaras may occur at particularly high densities, they share the habitat with deer. Three species, pampas deer (*Ozotoceros bezoarticus*), marsh deer (*Blastocerus dichotomus*), and white-tailed deer (*Odocoileus virginianus*) occur, depending on the region, and they comprise the bulk of the herbivore biomass. Cervids are browsers, survivors from an important group of ungulates that invaded South America during the Pliocene (3 Ma), when the emergence of the Isthmus of Panama connected this continent with North America. Capybaras, in contrast, are grazers, descendents of rodents that invaded South America probably during the late Eocene (ca 40 Ma; Vucetich et al. 2012), which evolved during the Miocene (23.03–5.33 Ma) and prospered in the South American savannas, filling the herbivore niche that bovids occupy in the African savannas (Ojasti 1983).

Increasingly, however, capybaras face competition from cattle. Now farmed across much of South America, they were introduced together with horses, pigs, and sheep in the sixteenth century and therefore have only recently faced adaptation to

Table 4.5 Diet overlap (% overlap index; Quintana 2003) among herbivore species in each season. Ca=Capybara; Co=Cow; Sh=Sheep (From Quintana (2003))

Comparison	Summer %	Autumn %	Winter %	Spring %
Ca-Co	53.7	64.8	64.5	66.9
Ca-Sh	51.8	70.2	77.3	69.4
Co-Sh	70.3	77.8	63.5	66.1

the native grasses and to coexistence with capybaras. In a matter of only centuries, however, cattle may have altered the feeding patterns of native grazers such as capybaras. In Brazil alone, there are an estimated 200 million head of cattle, turning over 67 billion reais (about US $42 billion in 2011). Farmers certainly view capybaras as competitors, because they feed mostly on grasses which are also consumed by cattle and in some cases by sheep.

Exclusion plot experiments have revealed that capybaras' consumption alone (i.e., without the other two major grazers, cattle and white-tailed deer), had the same effect on grass growth as that of all three grazers (Box 4.2; Barreto 1994). However, in Venezuela, from an analysis of feces collected at the end of a wet season, Escobar and González-Jiménez (1976) found important differences between the diet of white-tailed deer and that of capybaras, cattle, and horses. Although the diet of these animals showed overlap, especially in consumption of plants growing on flooded patches, capybaras in Venezuelan savannas consumed plant species that were rejected by the other herbivores, such as *P. prostrata* and some Cyperaceae. *P. prostrata* is a very short grass, which cannot easily be reached by cattle. Although cattle graze mainly on palatable and nutritious grasses found on bajios, it is to be expected that during the dry season they concentrate on the esteros where the most nutritious plants are found as ponds recede. In the dry season, with its more limited choices, capybaras and cattle compete for these decreasing resources.

In the savannas of Argentina, capybaras appear to compete with sheep as well as cattle. Two habitats were investigated by Quintana (2002, 2003) and Quintana et al. (1998a, b), comparing what capybaras ate when cattle were absent, when they coexisted with cattle, and with both cattle and sheep. Over a 2-year period, Quintana (2003) found capybaras and cattle consuming similar diets in summer in non-insular areas in the Paraná River Delta (Table 4.5). Also in summer, sheep and capybaras had a similar diet in the Entre Rios province. However, the diet of capybaras was altered by the presence of these domestic herbivores, as evidenced by their consuming a greater number of plant species in shared areas when compared to areas where cattle were absent. This may be due to the fact that in the presence of other herbivores capybaras are forced to consume less palatable (suboptimal) items in order to satisfy their energy requirements (Quintana 2002).

In Brazil, a 2-year study comparing the diets of pampas deer, cattle, and capybaras (Desbiez et al. 2011) found that all three species preferred *L. hexandra* and *H. amplexicaulis* (a semiaquatic and an aquatic grass), both of which are common in the diet of capybaras in Venezuela (Escobar and Gonzalez-Jimenez 1979). However, the authors conclude that the presence of cattle is not significantly detrimental to capybaras or deer in terms of diet competition. In general, Desbiez et al. (2011)

Box 4.2 Effects of Grazing by Capybaras on Savanna Vegetation: Field Experiments Based on Exclusion Plots

Exclusion plot experiments have been used to evaluate the impact of predators on prey species (Macdonald et al. 1999) as well as the effect of herbivores on their target vegetation. Barreto (1994) carried out exclusion experiments to assess the impact of capybaras on their main resource, including in the experimental design total exclusion (TE, i.e., neither cattle nor capybaras could have access to the grasses), and capybara-only exclusion (only capybaras have access to the grass, PE). No significant differences were found for the height and aerial biomass of three grass species (*Hymenachne amplexicaulis*, *Reimarochloa acuta*, and *Paratheria prostrata*) between the PE plots and the TE plots. This indicates that capybaras grazing alone have the same effect on those three grasses as consumption by all large herbivores (capybaras, cattle, and white-tailed deer, *Odocoileus virginianus*). Furthermore, grazing by capybaras affected not only grass growth but species composition as well (Table 4.6). Thus, the grass *Heliotropium indicum* exhibited greater densities in grazed plots than in TE plots, while *Cyperus* sp. and *P. prostrata* were particularly affected by the grazing of capybaras. Capybara grazing on *P. prostrata* also altered the production of inflorescences. Although the mean number of inflorescences did not vary among experimental plots (Barreto 1994), the frequency distribution of inflorescences per individual plant did vary. The density of inflorescences of *H. amplexicaulis* was greater in TE and PE plots, that is, grasses react to grazing by reproducing sexually. A germination assay with seeds collected on the experimental plots revealed that the percentage of germinated seeds from TE plots was significantly greater than that on plots subjected to herbivory (Barreto 1994).

Table 4.6 Density (ind/m^2) and presence/absence of five plant species in experimental total exclusion plots (TE, capybaras and cattle are excluded), partial exclusion (PE, cattle excluded, capybaras allowed) and control plots in flooded savannas of Venezuela (Barreto 1994)

Species	TE	PE	Control
Heliotropium indicum	0.25	3.62	6.83
Cyperus sp.	1.62	6.75	6.12
Cyperaceae	98	8	4
Paratheria prostrata	Present	Absent	Absent
Eleocharis sp.	Absent	Absent	Present

argue that cattle ranching, as carried out in the Pantanal (very similar to Venezuela), is a low-impact form of land utilization. In a study in Venezuela, Escobar and González-Jiménez (1979) found low diet overlap between capybaras and cattle and attributed that to two main reasons: (1) Some plants, such as *P. prostrata*, were so short as to be out of reach of cattle. (2) Cattle and capybaras used available land differently, that is, capybaras mainly used patches next to water holes while cattle used drier and higher patches a longer distance from the water. Thus, in the present patterns of land use of the flooded savannas of Brazil and Venezuela, cattle ranching appears to be compatible with the maintenance of healthy capybara populations.

Acknowledgments We thank Emilio Herrera, Katia Ferraz, José Roberto Moreira, and two anonymous referees for comments that helped to substantially improve the manuscript. Juhani Ojasti inspired the exclusion experiment (among other things) and helped with its design. Susana Monge authorized the use of photographs in Box 4.1. Results from Argentina were partially funded by grants PICT 1849 and X-817 of the National Agency for the Scientific and Technological Promotion and the University of Buenos Aires. Results for Box 4.2 were funded by CONICIT (now FONACIT, Venezuela) and the Graduate Studies Deanery of Universidad Simón Bolívar, Caracas.

References

Alho CJR, Campos ZMS, Goncalves HC (1987) Ecologia de capivara (*Hydrochoeris hydrochaeris*, Rodentia) do Pantanal: II. Actividade, sazonalidade, uso do espaço e manejo. Rev Brasilera Biol 47:99–110

Arteaga MC, Jorgenson JP (2007) Hábitat de desplazamiento y dieta del capibara (*Hydrochoerus hydrochaeris*) en la Amazonia colombiana. Mastozoología Neotrop 14:11–17

Azcárate T (1981) Sociobiología y manejo del capibara. Doñana Acta Vertebrat 7:1–228

Barlow JC (1969) Observations on the biology of rodents in Uruguay. Life science contributions. R Ont Mus 75:1–59

Barreto GR (1994) Herbivoría en chigüires: un estudio de campo de las relaciones planta-animal. M.Sc. dissertation, Universidad Simón Bolívar, Sartenejas

Barreto GR, Herrera EA (1998) Foraging patterns of capybaras in a seasonally flooded savanna of Venezuela. J Trop Ecol 14:87–98

Borges LV, Colares IG (2007) Feeding habits of capybaras (*Hydrochoerus hydrochaeris,* Linnaeus 1766), in the Ecological Reserve of Taim (ESEC – Taim) – South of Brazil. Braz Arch Biol Technol 50:409–416

Borges PA, Domínguez-Bello MG, Herrera EA (1996) Digestive physiology of wild capybara. J Comp Physiol B 166:55–60

Charnov EL (1976) Optimal foraging, the marginal value theorem. Theor Popul Biol 9:129–136

Corriale MJ, Arias SM, Quintana RD (2011) Nutritional value of the capybara's diet in the Paraná River delta. Argent Rangeland Ecol Manage 64:257–263

Desbiez ALJ, Santos SA, Alvarez JM, Tomas WM (2011) Forage use in domestic cattle (*Bos indicus*), capybara (*Hydrochoerus hydrochaeris*) and pampas deer (*Ozotoceros bezoarticus*) in a seasonal Neotropical wetland. Mamm Biol 76:351–357

Escobar A, González-Jiménez E (1976) Estudio de la competencia alimenticia de los herbívoros mayores del llano inundable con referencia especial al chigüire (*Hydrochoerus hydrochaeris*). Agron Trop 23:215–227

Escobar A, González-Jimenez E (1979) La producticn primaire de la savane inondable d'Apure (Venezuela). Geogr Ecol Trop 3:53–70

Forero-Montaña J, Betancur J, Cavelier J (2003) Dieta del capibara *Hydrochoerus hydrochaeris* (Rodentia: Hydrochaeridae) en Caño Limón, Arauca, Colombia. Rev Biol Trop 51:579–590

González-Jimenez E (1977) Digestive physiology and feeding of capybaras. In: Rechligl M (ed) Handbook series in nutrition and food. CRC Press, Cleveland, pp 163–177

Herrera EA (2012) Capybara digestive adaptations. In: Moreira JR, Ferraz KMPMB, Herrera EA, Macdonald DW (eds) Capybara: biology, use and conservation of an exceptional neotropical species. Springer, New York, pp 97–106

Herrera EA, Macdonald DW (1989) Resource utilization and territoriality in group-living capibaras (*Hydrochoerus hydrochaeris*). J Anim Ecol 58:667–679
Hume ID (1989) Optimal digestive strategies in mammalian herbivores. Physiol Zool 62:1145–1163
Johnson MK, Wofford H, Pearson HA (1983) Microhistological techniques for food habits analyses. United States Department of Agriculture, Forest Service, Research paper SO-199, New Orleans, p 40
Jorgenson J (1986) Notes on the ecology and behaviour of capybaras in Northeastem Colombia. Vida Silvestre Neotrop 1:31–40
Macdonald DW (1981) Dwindling resources and the social behavior of capybaras (*Hydrochoerus hydrochaeris*) (Mammalia). J Zool (Lond) 194:371–391
Macdonald DW, Mace GM, Barreto GR (1999) The effects of predators on fragmented prey populations: a case study for the conservation of endangered prey. J Zool (Lond) 247:487–506
Manly B, McDonald LL, Thomas DL, McDonald TL, Erikson WP (2002) Resource selection by animals: statistical design and analysis for field studies. Kluwer, Dordrecht
Mauro RA, Pott A (1996) Dieta de capibara (*Hydrochoerus hydrochaeris*) basado en análisis microhistológico de las heces. Vida Silvestre Neotrop 5:151–153
Mendes A, Nogueira-Filho SLG (2012) Feeds and nutrition of farmed capybaras. In: Moreira JR, Ferraz KMPMB, Herrera EA, Macdonald DW (eds) Capybara: biology, use and conservation of an exceptional neotropical species. Springer, New York, pp 261–274
Metcalfe CR (1960) Anatomy of the monocotyledons: I. Gramineae. Oxford University Press, London, p 731
Metcalfe CR, Chalk L (1979) Anatomy of the dicotyledons, vol 1, 2nd edn. Oxford University Press, London, p 745
Mondolfi E (1957) El Chigüire. El Farol 168:38–40
Mones A, Ojasti J (1986) *Hydrochoerus hydrochaeris*. Mamm Species 264:1–7
Monge S (1989) Descripción de las características epidérmicas de gramíneas del centro-este de Mendoza (Ñacuñán, Santa Rosa). Rev Argent Produccion Anim 9:57–68
Ojasti J (1973) Estudio biológico del chigüire o capibara. Fondo Nacional de Investigaciones Agropecuarias, Caracas
Ojasti J (1983) Ungulates and large rodents of South America. In: Bourlière F (ed) Tropical savannas. Elsevier Scientific, Amsterdam, pp 427–439
Osgood WH (1912) Mammals from Western Venezuela and Eastern Colombia. Field Mus Nat Hist Zool S 10:29–66
Pyke GH (1984) Optimal foraging theory: a critical review. Ann Rev Ecol Syst 15:523–575
Quintana RD (2002) Influence of livestock grazing on the capybara's trophic niche and forage preferences. Acta Theriol 47:175–183
Quintana RD (2003) Seasonal effects on overlap trophic niche between capybara (*Hydrochoerus hydrochaeris*) and livestock, and on trophic niche breadths in a rangeland of Central Entre Ríos, Argentina. Mammalia 67:33–40
Quintana RD, Monge S, Malvarez AI (1994) Feeding habits of capybara (*Hydrochoerus hydrochaeris*) in afforestation areas of the lower delta of the Paraná River, Argentina. Mammalia 58:569–580
Quintana RD, Monge S, Malvarez AI (1998a) Feeding patterns of capybara *Hydrochoerus hydrochaeris* (Rodentis Hydrochaeridae) and cattle in the non-insular area of the lower delta of the Paraná River, Argentina. Mammalia 62:37–52
Quintana RD, Monge S, Malvarez AI (1998b) Composición y diversidad de las dietas del capibara (*Hydrochoerus hydrochaeris*) y del ganado doméstico en un agroecosistema de la región central de Entre Ríos, Argentina. Ecotropicos 11:33–44
Quintana RD, Puig S, Rosati V, Arias SM (2002) Selección de dieta de mamíferos herbívoros silvestres a escala de hábitat de alimentación. In: Cid MS, Bonno N, Cassini M, Anchorena J, de Pelliza Sbriller A, Arriaga M (eds) Selección de dieta por grandes herbívoros mamíferos: procesos y escalas, vol 1, Revista del Museo Argentino de Ciencias Naturales (Ediciones Especiales). Museo Argentino de Ciencias Naturales, Buenos Aires, pp 59–77
Schaller GB, Crawshaw PG Jr (1981) Social organization in a capybara population. Säugetierkundliche Mitt 29:3–16

Senft RL, Coughenour MB, Baily DW, Rittenhouse LR, Sala OE, Swift DM (1987) Large ecological hierarchies. Bioscience 37:789–799

Sibly RM (1981) Strategies of digestion and defecation. In: Townsend CR, Calow P (eds) Physiological ecology. Blackwell Scientific, Oxford, UK, pp 109–139

Soini P, Soini M (1992) Ecología del ronsoco o capibara (*Hydrochoerus hydrochaeris*) en la reserva nacional Pacaya-Samiria, Perú. Folia Amazonica 4:135–150

Stephens DW, Krebs JR (1986) Foraging theory. Princeton University Press, New Jersey

von Humboldt A (1826) Voyages aux regions équinoxiales du Nouveau Continent fait en 1799, 1800, 1801, 1802, 1803 et 1804 par Al. Von Humboldt et A. Bompland. VI. Chez N. Maze, Paris, p 318

Vucetich MG, Deschamps CM, Pérez ME (2012) Paleontology, evolution and systematics of capybara. In: Moreira JR, Ferraz KMPMB, Herrera EA, Macdonald DW (eds) Capybara: biology, use and conservation of an exceptional neotropical species. Springer, New York, pp 39–59

Chapter 5
Capybara Digestive Adaptations

Emilio A. Herrera

5.1 Introduction

Structural polymers of plants (mainly cellulose and hemicellulose) are possibly the most widely available sources of energy from primary producers for herbivores (Parra 1978). However, such highly fibrous diet components are extremely difficult to digest and, therefore, herbivores possess specific adaptations for the digestion of these materials. The best-known, and undoubtedly the most common, adaptation to a high-fiber diet among mammals is fermentation by symbionts (bacteria, fungi, and protozoa), coupled with mechanisms for the digestion and absorption of the products of fermentation. Symbionts themselves thrive on the large amounts of vegetation swallowed by the herbivore.

Among mammals, there are two distinct types of symbiotic digestion of plant material, which differ in the position within the digestion system where fermentation occurs: (1) foregut fermentation, present in Artiodactyla (bovids, including antelopes, cervids, etc.), and (2) hindgut fermentation, adopted by Perissodactyla (horses, tapirs), Lagomorpha (rabbits and hares), and many rodents (Hirakawa 2001, 2002). Different anatomical, physiological, and even behavioral adaptations are observed in each of these two kinds of herbivores, allowing them to thrive on the plants they eat. Thus, foregut fermenters usually perform rumination, where foodstuffs processed by symbionts in a modified stomach (the rumen) are regurgitated, chewed a second time (ruminants "chew the cud"), swallowed, and then moved directly to the regular stomach, bypassing the rumen and continuing to the small intestine, where the digested products of the symbionts' fermentation are absorbed.

Hindgut fermenters use the cecum, located between the small and large intestines, as a fermentation chamber, which precludes regurgitation and re-swallowing of

E.A. Herrera (✉)
Departamento de Estudios Ambientales, Universidad Simón Bolívar,
Caracas 1080-A, Caracas Apdo. 89000, Venezuela
e-mail: eherre@usb.ve

J.R. Moreira et al. (eds.), *Capybara: Biology, Use and Conservation of an Exceptional Neotropical Species*, DOI 10.1007/978-1-4614-4000-0_5,

the fermented plants as a strategy for the absorption of nutrients. So, these herbivores either have adaptations in the large intestine (horses) for the absorption of nutrients or they resort to cecotrophy (lagomorphs and rodents). In the latter, there is a daily cycle of feeding and reingestion: food goes once along the digestive tract, entering the cecum where it is fermented for a certain amount of time, and then excreted. These excreted feces are taken directly from the anus by the herbivore and they pass one more time through the entire digestive tract. After digestion in the stomach, nutrients are absorbed in the small intestine. The waste products bypass the cecum and move on to the large intestine, where hard, dry feces are voided (but not reabsorbed this time). The two processes occur within a 24-h cycle. It has been argued that, since hindgut fermenters can take advantage of any available, directly digestible (i.e., non-fiber) nutrients before the bacterial fermentation takes place, they are more efficient at extracting nutrients from food than are foregut fermenters (González-Jiménez and Escobar 1975).

Capybaras (*Hydrochoerus hydrochaeris*) are 50 kg grazing rodents with a fully herbivorous diet, consisting mainly of grasses, with a varying proportion of sedges (Cyperacea) and just a few other plants (Barreto and Quintana 2012). They are hindgut fermenters possessing a large cecum and performing cecotrophy (sometimes called somewhat incorrectly coprophagy; Herrera 1985; Mendes et al. 2000). In this chapter, I will review the digestive adaptations of capybaras in order to complete the picture of this rodent's overall feeding strategy as outlined elsewhere in this book (Barreto and Quintana 2012; Mendes and Nogueira-Filho 2012).

5.2 Morphology

The cranium of the capybara is box-shaped and the mouth, with highly labile lips, is located in the lower front part of the head, thus placing the teeth directly in contact with the plants as the animal lowers its head to graze (Fig. 5.1). The dental formula of capybaras is incisors 1/1, canines 0/0, premolars 1/1, and molars 3/3 (20 in total), where the third upper molar is as long as the premolar and other two molars put together (Ojasti 1973). The large, sharp incisors are separated from the premolars and molars by a wide diastema. Top and bottom incisors produce a scissor effect which allows animals to crop grasses literally to the ground, enabling them to feed on very short grasses (Fig. 5.1) and take advantage of whatever is available (Ojasti 1973). This is especially important in the dry season when food can be extremely scarce and the few available plants are composed of almost pure structural carbohydrates (Ramia 1967).

All teeth grow continually, which means they are always healthy, tough, and sharp (although from a researcher's point of view, this precludes the use of teeth for age determination, a reliable method in many other mammals). The convoluted enamel pattern (Fig. 5.2) and the back and forth movement of jaws, coupled with strong musculature, allow capybaras to grind grasses virtually to a paste (Ojasti 1973). Particle size after chewing can be very small, down to 0.001–0.3 mm wide (Ojasti 1973). In a review of particle size among mammalian herbivores, Fritz et al. (2009) found

Fig. 5.1 A capybara grazing, showing how the mouth virtually touches the ground when grass is short (Photo by E.A. Herrera)

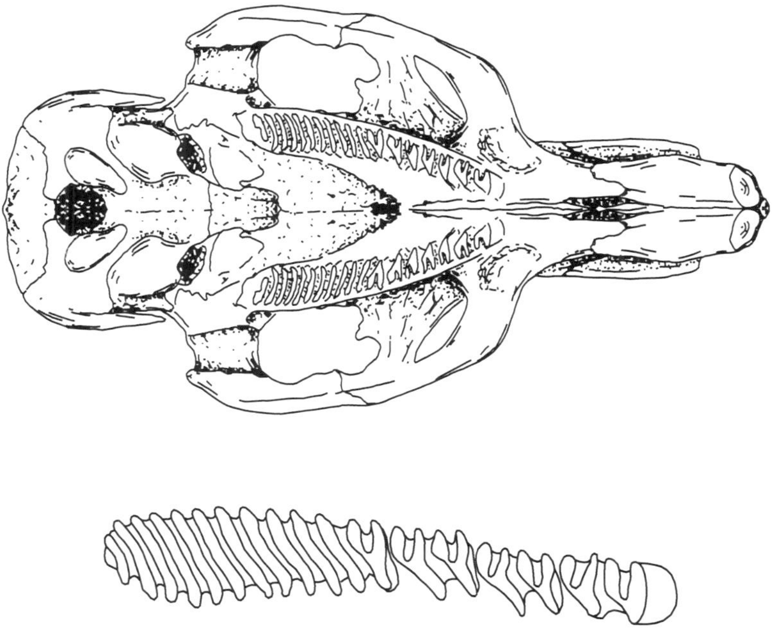

Fig. 5.2 *Top*: View from below capybara skull showing diastema. *Bottom, Left to right*: largest molar, two other molars, and premolar. Note transverse enamel plates and size of the third molar as long as other three cheek teeth put together

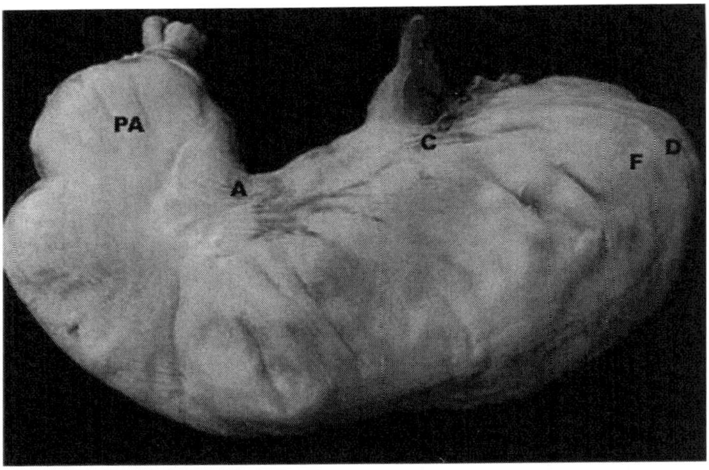

Fig. 5.3 Stomach of a capybara showing the angular incisures (*a*), pyloric antrum (*pa*), cardiac incisure (*c*), gastric fundus (*f*) and gastric diverticulum (*d*) (From Barros Moraes et al. (2005))

that capybaras' mean particle size was comparable to that of similarly sized ruminants. This corroborates the proposed convergence between capybaras and this group of mammalian herbivores (Eisenberg and Mackay 1974).

The lumen of the capybara esophagus is covered with a longitudinally folded membrane and is somewhat cornified, possibly an adaptation to withstand the passage of tough fibers (Mones and Ojasti 1986). The stomach is a kidney- or inverted J-shaped organ and, although it is not compartmentalized (i.e., it is a "simple" stomach), it has cardiac, fundic, gastric diverticulum, body, and pyloric sections (Fig. 5.3), with a volume varying from 850 to 2,010 mL (average 1,500 mL approximately; Barros Moraes et al. 2002). It has numerous folds throughout its internal surface and a complex and well-defined muscular architecture that differs from that of other rodents (Barros Moraes et al. 2002, 2005). The weight of the stomach contents is 10% of the total digestive tract (1,175 ± 315 g dry weight; Gonzalez-Jimenez 1977), with a pH of 2.5–3.5 (Gonzalez-Jimenez 1977; Baldizán et al. 1983). It is filled with a dark and a lighter mass of chewed vegetation, the former probably corresponding to reingested feces (see below, "physiology") and the latter to freshly swallowed grasses.

The stomach is followed by the small intestine, which can be 10 m long with a mean volume of 2,302 mL (Rodrigues et al. 2006). The largest and arguably the most interesting part of the digestive tract is the cecum, located between the small intestine and the colon (Fig. 5.4). The capybara cecum takes up 63–74% of the total volume of the digestive tract with a volume of up to 5 L (Ojasti 1973; Borges et al. 1996) and is the site of bacterial fermentation (see below). In captivity, the volume of the cecum is 34% greater when the animals are fed forage than when they are fed concentrate (Baldizán et al. 1983). The rest of the large intestine is the colon where water absorption and fecal production occurs.

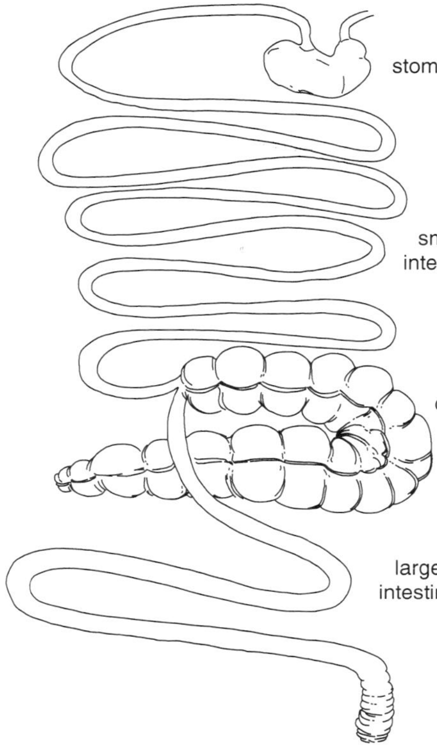

Fig. 5.4 Diagrammatic drawing of the digestive tract of a capybara (Drawing by Luis Miguel Márquez)

5.3 Physiology

Capybaras are the largest among the cecum fermenters and cecotrophous mammals (Herrera 1985; Borges et al. 1996; Hirakawa 2001, 2002). Chewed grasses and other plants are first digested in the stomach, where protein and other digestible materials are broken down in an acid enzymatic process. The digested particles flow into the small intestine where absorption of amino acids and other nutrients takes place. However, a significant proportion of the foodstuffs, namely, high-molecular-weight hydrocarbons such as cellulose, are at this point undigested and unabsorbed. These important components of the capybara diet then enter the cecum, the fermentation chamber in capybaras (Baldizán et al. 1983). The cecum hosts a large variety of bacterial "species" (formally, operational taxonomic units or OTUs) from 21 phyla, mainly Firmicutes (34.5%) and Proteobacteria (32.3%; García-Amado et al. 2012), and at least 17 protozoan species (Tengler de McClure 1970). Firmicutes contain several species of cellulolytic bacteria while Proteobacteria are known to cause a number of gut diseases in humans. The concentrations of bacteria and protozoa in the cecum are comparable to those of ungulates (Borges et al. 1996).

Fig. 5.5 Piles of capybara droppings of different ages. On the *left (arrow)* are larger droppings, similar to what are called "cecotrophs" (Photo by E.A. Herrera)

The digestive capacity (relative volume) of the capybara cecum appears to be equivalent to that of the rumen of similarly sized ruminants such as sheep (Gonzalez-Jimenez 1977). Microscopically, it presents numerous nervous ganglions and endocrine cells (Soares Bressan et al. 2005). The pH inside the cecum is 6.3–6.6 (Baldizán et al. 1983; Mendes et al. 2000), a neutral pH that allows a diverse bacterial flora to develop. The fermentation process inside the cecum digests the structural polymers and generates volatile fatty acids, vitamins, and microbial proteins. These nutrients need to be absorbed, but the part of the digestive tract immediately following the cecum is the large intestine where only water is absorbed and feces produced (although volatile fatty acids can be absorbed in the cecum; Démigné and Rémésy 1985). Thus, the adaptive strategy present in capybaras for the absorption of the products of bacterial fermentation is cecotrophy.

Capybara droppings are oval shaped and shiny from being damp when just dropped, but dark and matte when dry (Fig. 5.5). Piles contain as many as 100 pellets averaging 24.7 ± 2.27 mm in length (Ojasti 1973), which varies in relation to the size of the animal. In fact, the body weight of a capybara can be estimated from the size of its pellets, since there is a statistically significant regression between pellet size and body mass (body mass in kg = 15.55 + 0.17 pellet length in mm, for animals between 10 and 60 kg; Ojasti 1973). Whether capybaras produce "cecotrophs" or larger softer feces to be reingested and regular pellets to be voided (as rabbits do) is as yet unclear (Borges et al. 1996).

Retention time for 68% of the total daily ration was 72 h (Ojasti 1973) and 60 h for the recovery of 80% of a chemical marker (Gonzalez-Jimenez 1977). These values are close to those found for sheep and rabbits (Gonzalez-Jimenez 1977).

However, they were obtained from a captive study where cecotrophy was not observed, probably because the animals were fed concentrate and kept in "metabolic cages." Cecotrophy implies a separation mechanism between feces to be voided and feces to be reingested. The existence of such a colonic separation mechanism is suspected (Hirakawa 2002) and believed to be associated with selective retention of fluid digesta in the cecum (Borges et al. 1996).

5.4 Behavior

Capybaras have a clear daily routine: they are most active in the early afternoon and evening when they set out to graze; they then alternate between grazing and resting through the night (Macdonald 1981; Herrera 1985). During the night, as they graze, they void most of their feces (Ojasti 1973). From dawn until midday they rest and in the hot midday hours they wade in the shallow water of the pond next to their grazing patch.

Perhaps the most remarkable event in the capybara's daily routine is the reingesting of their feces, which they do during the morning resting period (Herrera 1985). To do this, the semi-recumbent capybara props itself up slightly on its forelegs, arches its neck slightly to one side and reaches back between its hind legs toward the anus, from which it takes the feces directly as they emerge (Fig. 5.6). Capybaras reingest their feces on average once per hour between 0700 and 1400 h and significantly more frequently during the dry season than during the wet (Herrera 1985). An increase in the frequency of cecotrophy is expected during the dry season given

Fig. 5.6 Cecotrophy (Photo by José Eduardo Moreira)

the severe scarcity of fresh grass and its high concentration of cellulose at this time (Ramia 1967). Cecotrophy has not been observed during the night (Herrera 1985). At the beginning of the wet season, capybaras take advantage of the increased concentration of protein in the growing grasses and tend to feed at all times of the day (Herrera 1985), apparently relying less on cecotrophy and utilizing a larger proportion of the soluble materials within plant cells. This digestive flexibility probably allows capybaras to forage without cecotrophy when it is not necessary. In fact, studying capybara digestion in captivity where the animals were fed a mixture of grass and concentrate, González-Jiménez (1977) concluded that this peculiar adaptation was not present in capybaras. There is evidence that in the dry season, when food is of poorer quality (i.e., more fibrous), the cecum increases in weight, suggesting greater fermentation (Borges et al. 1996) and hence more cecotrophy activity.

In another captivity study, however, cecotrophy was observed, and in a similar pattern to that in a natural situation (Mendes et al. 2000), occurring between 10 and 11 h after the animals were fed. These authors found that reingested feces were softer than voided ones and contained 37% more protein than the latter. Borges et al. (1996) found no differences in the composition of the digesta in the various compartments of the digestive tract at different times of day, although the daily routine of feeding and cecotrophy should have produced differences. However, as mentioned by Hirakawa (2001), a certain amount of mixing can occur between fecal material to be reingested and that to be voided, and this may preclude detection of differences.

Capybaras tend to deposit their feces as they walk during the evening and nocturnal grazing period (Ojasti 1973); they do not have a particular location (a "latrine") in which to defecate, as do some other mammals (such as felines). Sometimes they walk into the water to defecate, but the function of this behavior is unknown.

5.5 Final Remarks

Capybaras have a digestive system that is as efficient as that of similarly sized ruminants (González-Jiménez and Escobar 1975; Gonzalez-Jimenez 1977). They are able to take advantage of very short and dry grasses, an obvious adaptation to the extremely harsh dry season of the Llanos (Venezuela and Colombia) or the Pantanal (Brazil). In fact, capybaras lose weight (as much as 4.5 kg or about 10% of their body mass; Ojasti 1973) and can become thin and weak at the height of the dry season (Fig. 5.7), but they make a very quick recovery as soon as they start feeding on the nutritious new growth of grasses available at the start of the wet season (Emilio A. Herrera personal observation). Indeed, capybaras are most commonly born at the end of the rainy season (Ojasti 1973), so they face the very harsh dry period almost immediately after birth, and they still manage to gain 92 g on average per day (Herrera 1992). Thus, they are exquisitely adapted to survive, grow, and reproduce in the markedly seasonal habitat where they live, undoubtedly thanks – at least in part – to their highly efficient digestive system.

Fig. 5.7 A thin, weakened capybara at the end of the dry season (March) (Photo taken at Hato El Frío, Venezuela. Photo by E.A. Herrera)

Acknowledgments I would like to thank Carlos Bosque and María Gloria Domínguez-Bello, as well as the editors of the book, for their comments on the manuscript.

References

Baldizán A, Dixon RM, Parra R (1983) Digestion in the capybara (*Hydrochoerus hydrochaeris*). S Afr J Anim Sci 13:27–28

Barreto GR, Quintana RD (2012) Foraging strategies and feeding habits of capybaras. In: Moreira JR, Ferraz KMPMB, Herrera EA, Macdonald DW (eds) Capybara: biology, use and conservation of an exceptional neotropical species. Springer, New York, pp 83–96

Borges PA, Domínguez-Bello MG, Herrera EA (1996) Digestive physiology of wild capybara. J Comp Physiol B 166:55–60

de Barros Moraes PT, Pacheco MR, de Souza WM, da Silva RA, Neto PBS, de Figuereido Barretto CS, Ribeiro AACM (2002) Morphological aspects of the capybara stomach (*Hydrochoerus hydrochaeris*): gross and morphological structure. Anat Histol Embryol 31:362–366

de Barros Moraes PT, de Souza WM, Neto PBS, de Figueiredo Barretto CS, Ribeiro AACM (2005) The muscular organization of the stomach of capybara (*Hydrochoerus hydrochaeris*): an architectural view. Ann Anat 187:51–56

Démigné C, Rémésy C (1985) Stimulation of absorption of volatile fatty acids and minerals in the cecum of rats adapted to a very high fiber diet. J Nutr 115(1):53–60

Eisenberg J, Mackay GM (1974) Comparison of ungulate adaptations in the New World and Old World tropical forests with special reference to Ceylon and the rainforests of Central America. In: Geist V, Walther F (eds) Behavior of ungulates in relation to management, vol 24, IUCN series. IUCN, Morges, pp 585–602

Fritz J, Hummel J, Kienzle E, Arnold C, Nunn C, Clauss M (2009) Comparative chewing efficiency in mammalian herbivores. Oikos 118:1623–1632

Garcia-Amado MA, Godoy-Vitorino F, Piceno I, Andersen G, Herrera EA, Dominguez-Bello MG (2012) Bacterial diversity in the cecum of the Capybara (*Hydrochoerus hydrochaeris*) Microbial Ecology 63:719–725

Gonzalez-Jimenez E (1977) Digestive physiology and feeding of the capybara. In: Rechcigl M (ed) Handbook series in nutrition and food Section G, vol 1. CRC press, Cleveland, pp 167–177

González-Jiménez E, Escobar A (1975) Digestibilidad comparada entre chiguire (*Hydrochoerus hydrochaeris*) conejos y ovinos con raciones de diferentes proporciones de forrajes y concentrado. Agron Trop 25:283–290

Herrera EA (1985) Coprophagy in the capybara, *Hydrochoerus hydrochaeris*. J Zool Lond 207A:616–619

Herrera EA (1992) Growth and dispersal of capybaras (*Hydrochoerus hydrochaeris*) in the Llanos of Venezuela. J Zool 228:307–316

Hirakawa H (2001) Coprophagy in leporids and other mammalian herbivores. Mamm Rev 31:61–80

Hirakawa H (2002) Supplement: coprophagy in leporids and other mammalian herbivores. Mamm Rev 32:150–152

Macdonald DW (1981) Dwindling resources and the social behaviour of capybaras, *Hydrochoerus hydrochaeris* (Mammalia). J Zool Lond 194:371–391

Mendes A, Nogueira-Filho LG (2012) Feeds and nutrition of farmed capybaras. In: Moreira JR, Ferraz KMPMB, Herrera EA, Macdonald DW (eds) Capybara: biology, use and conservation of an exceptional neotropical species. Springer, New York, pp 261–274

Mendes A, da Nogueira SSC, Lavorenti A, Nogueira-Filho SLG (2000) A note on the cecotrophy behavior in capybara (*Hydrochoerus hydrochaeris*). Appl Anim Behav Sci 66:161–167

Mones A, Ojasti J (1986) *Hydrochoerus hydrochaeris*. Mamm Species 247:1–7

Ojasti J (1973) Estudio biológico del chigüire o capibara. FONAIAP, Caracas, 2nd Ed. 2011, Editorial Equinoccio, USB, Caracas, p 273

Parra A (1978) Comparison of foregut and hindgut fermentation in herbivores. In: Montgomery GG (ed) The ecology of arboreal folivores. Smithsonian Institution Press, Washington, DC, pp 205–229

Ramia M (1967) Tipos de sabanas en los llanos de Venezuela. Boletín de la Sociedad. Venezolana de Ciencias Naturales 27:264–288

Rodrigues SS, Fonseca CC, de Paula TAR, Peixoto JV (2006) Aspectos biométricos corporais e do intestino delgado da capivara *Hydrochoerus hydrochaeris* Linnaeus, 1766 (Mammalia, Rodentia, Hydrochaeridae). Biotemas 19:79–86

Soares Bressan M, Fonseca CS, de Menin E, Paula TAR (2005) Aspectos anátomo-histológicos e neuroendócrinos do ceco da capivara *Hydrochoerus hydrochaeris* Linnaeus, 1766 (Mammalia, Rodentia). Arquivos de Ciência Veterinária e Zootecnia UNIPAR 8:197–203

Tengler de Mcclure M (1970) Algunas observaciones sobre los ciliados cecales del chigüire (*Hydrochoerus hydrochaeris*) en Venezuela. Actas del IV Congreso Latinoamericano de Zoología 1:313–319

Chapter 6
Reproductive Morphology and Physiology of the Male Capybara

Tarcízio A.R. Paula and Nicholas J. Walker

6.1 Introduction

The interrelations between morphology, physiology, and behavior are inseparable, and together provide the key to the evolutionary success and failure of each species. Within this triad, experimentation and the complicity of nuances determine, and in turn are determined by, the evolutionary process. The capybara (*Hydrochoerus hydrochaeris*), like every other species, has undergone many adaptations throughout its evolutionary history. Thus, abiotic components and inter- and intra-specific relations have defined both the individual strategies and group behaviors necessary for its survival. Reproduction is certainly one of the most important aspects in determining evolution. In the capybara the morpho-functional and behavioral aspects involved are extremely rich and in some cases unique, especially considering the adaptations of males.

Parental investment by the male capybara in the production of progeny is much less than that of the females: contrasting the energy expenditure limited to fertilization with that of weeks of gestation and more of nursing. However, male capybaras can father more offspring than females can conceive and nurse (Alho 1989), and receptive females are therefore the main ecological resource to which dominant males have priority access (Herrera and Macdonald 1993). Thus, as the male's reproductive investment is limited to herding and keeping the largest number of actively reproductive females as possible, behaviors such as aggressiveness, fighting,

T.A. Paula (✉)
Departamento de Veterinária, Universidade Federal de Viçosa, Av. PH Rolfs s/n, Campus Universitário, Viçosa, Minas Gerais 36570-000, Brazil
e-mail: tarcizio@ufv.br

N.J. Walker
Laboratório de Sistemática Molecular, Departamento de Biologia Animal, Universidade Federal de Viçosa, Av. PH Rolfs s/n, Campus Universitário, Viçosa, Minas Gerais 36570-000, Brazil
e-mail: nicholas.walker@ufv.br

J.R. Moreira et al. (eds.), *Capybara: Biology, Use and Conservation of an Exceptional Neotropical Species*, DOI 10.1007/978-1-4614-4000-0_6,

and glandular marking abilities are costly components of the reproductive investment, which causes males to make more effort than females in gene propagation.

In addition to maintaining spermatogenesis and male secondary sexual characteristics (O'Donnel et al. 2001), testosterone also affects aggressiveness, glandular marking, social status, and libido (Hadley 1988; Herrera and Macdonald 1994). The main source of serum testosterone is the endocrine part of the testicle, composed of Leydig cells. Along with connective cells, leucocytes, and blood and lymphatic vessels, Leydig cells form the testicular intertubular tissue. The arrangement and proportions of these components vary between mammal species and form mechanisms to maintain the level of testosterone 2–3 times higher in the interstitial fluid than in the testicular blood vessels and 40–250 times higher than in the peripheral blood (Sharpe 1994). In the adult capybara male, over 50% of the testicular parenchyma is composed of intertubular tissue, which is the highest proportion recorded among mammals (Moreira et al. 1997a; Paula 1999), and Leydig cells correspond to approximately one third of the testicular volume (Costa and Paula 2006; Paula et al. 2007).

Glandular scent marking is an integral component of capybara behavior (Macdonald and Herrera 2012). Although urine is used routinely in marking, both males and females have specific scent glands. One gland, located in the dorsal region of the snout, is more developed and functional in males than females. Both sexes also have anal glands with their own characteristics. According to Macdonald et al. (1984), different individuals secrete the same chemical components but in different quantities and frequencies, suggesting that individual recognition based on chemical markings may be possible.

Within a group, the social behavior of capybaras obeys a rigidly established pattern in which males are more competitive than females. The male hierarchy is maintained through aggressive interactions, with the dominant male being characterized by greater body size and a higher proportion of mounts than other males achieve (Herrera and Macdonald 1993). Glandular marking behavior is seen more in dominant males than in subordinates (Herrera and Macdonald 1994). According to Herrera (1992), there is a positive correlation between the size of the nasal gland and testicular size. Paula (1999), working with adult males, also described a significant positive correlation of nasal gland volume in relation to spermatozoid production, as well as with age. Costa and Paula (2006) concluded that the increase in nasal glandular volume in adult male capybaras is associated with a proportional increase in the serum level of testosterone.

Thus, the size of the nasal gland is a predictor of social status and reproductive success. However, even in dominant animals, capybara sperm production lies in the lower threshold of values observed among studied mammals, and its gonadosomatic index (the gonad mass as a proportion of total body mass) is one of the lowest recorded among rodents (Kenagy and Trombulak 1986; Moreira et al. 1997b; Paula 1999). Interrelation between behavioral and morpho-functional aspects is therefore complex, and clearly inseparable from the male capybara's contribution to the reproductive success of this species.

6.2 Male Reproductive Organs

6.2.1 Testicular Positioning and Gonadosomatic Index

The male capybara does not have a properly defined scrotum, and the testicles are located subcutaneously in the inguinal region (Fig. 6.1). The testicular coverings are the same as those observed in other mammals, although like most rodents and lagomorphs, the capybara presents a thick layer of cremaster muscle which composes the cremasteric tunic in a continuous manner. Due to the large size of this tunic in the capybara and the separate disposition of the testicles in the subcutaneous space, two distinct muscular inguinal sacs are seen during dissection, lacking the presence of a septum or the formation of cutaneous folding. The well-developed cremasteric tunic combined with a wide inguinal channel allows testicular movement to the interior of the abdominal cavity. Due to capture stress and manipulation, it is common to see the intra-abdominal position of the testicle in animals slaughtered for meat.

Consequently, capybara testicles are located on each side of the median sagittal plane. The longitudinal axis of the testicle is parallel to the animal's body, with the

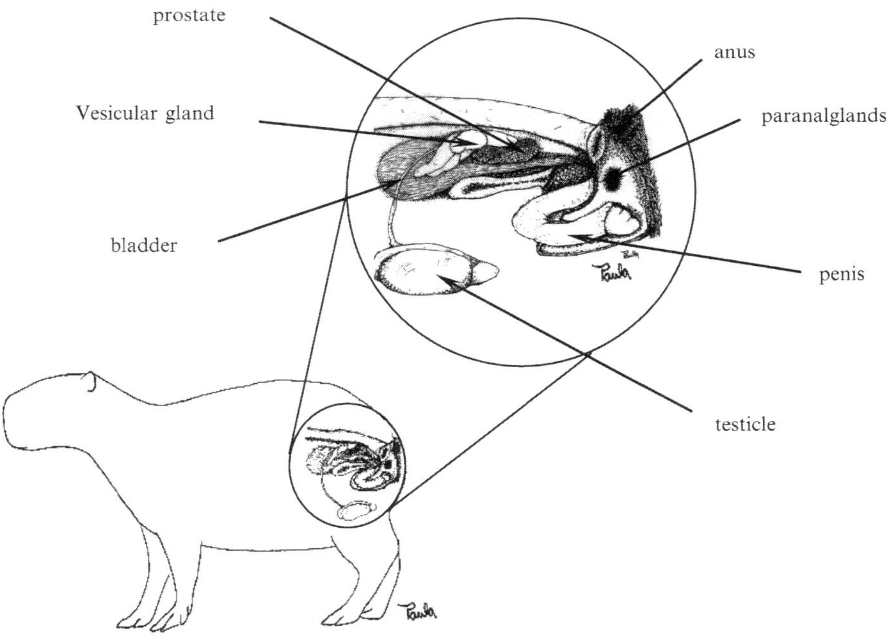

Fig. 6.1 Schematic of the positioning and relationship of reproductive organs in male capybaras. Presence of anogenital invagination containing penis, anus, and paranal gland

caudata extremity positioned caudally and dorso-medially to the epididymal edge. The epididymis seems to be small; therefore even the epididymis tail is not evident as is the case with some rodents, such as the rat (*Rattus norvegicus*).

In mammals, there is generally no correlation between the size and permanent location of the testicle (abdominal or scrotal), or between the size of the testicles, the shape of the body, and how the animals move (in land, water, and air; Kenagy and Trombulak 1986). However, it is known that in most mammals studied, a temperature of around 4–5 °C below regular body temperature is necessary to ensure normal spermatogenesis. The testicular thermal-regulation mechanisms in the capybara probably consider the testicular position during displacement and permanence in land or water.

Within the same mammal species, it is common to record up to a 50% difference in testicular weight among sexually mature individuals, even in species which have been domesticated and genetically selected such as bovines (Berndtson et al. 1987). However, in adult male capybaras, an average testicular weight of 32 g was observed, with only small variations seen among individuals from distinct populations (Paula 1999). Body size does not seem to be the primary determining factor in predicting testicular size. The gonadosomatic index in capybaras is approximately 0.12% (Moreira et al. 1997b; Paula 1999). In studies involving 133 different species of mammals, among them 62 rodents, Kenagy and Trombulak (1986) found that the order Rodentia had the greatest variation in relative testicular weight, ranging from 0.05% in the beaver (*Castor canadensis*) to 8.41% in the African gerbil (*Taterona afra*). These authors also observed that animals with a lower body weight allocate a higher proportion of body mass and energy expenditure to testicular tissue compared with larger animals. In this respect, it is evident that the gonadosomatic indices in capybaras (0.12%) and in the beaver (0.05%) are the lowest currently known for rodents, while their average body weights (54 kg in capybaras and 19 kg in the beaver) are the highest.

In many different species of mammals, the size of the testicles appears to vary depending on the mating system adopted (Kenagy and Trombulak 1986; Breed and Adams 1992; Short 1997). Animals that have promiscuous or polyandrous systems, in which different males mate with the same female in the same estrus cycle, tend to have larger testicles than those that have monogamous or polygynous systems, in which only one male is responsible for mating. In the first case, the competition for progeny production depends on the quality and quantity of semen deposited in the female genital tract, making higher levels of spermatic production advantageous. On the other hand, in monogamous and polygynous systems, the competition between males is for mating rights, in which behavioral dominance over opponents is more important than spermatic production itself (Short 1997). The capybara fits the second case, having a low gonadosomatic index and presenting reproductive behavior in which one male is responsible for most copulation after precopulation competitions (Herrera and Macdonald 1993). Thus, due to the strong male hierarchy seen in these animals, sperm competition may not be of primary importance in the general context of capybara reproduction, although it may play some role in the mating strategy of subordinate males (López et al. 2008).

6.2.2 Reproductive Organs

In addition to having larger body size in relation to other rodents, the morphology of the capybara's reproductive organs is unique. Although capybaras are not sexually dimorphic, and male and female external genitals appear extremely similar, genders can often, but not infallibly, be distinguished visually by the size of the nasal gland.

6.2.2.1 Penis and Foreskin

The foreskin of the adult male capybara is connected to the anus in a unique way, forming a wide anogenital invagination. This invagination is comprised of the flaccid penis, the dorsally located anus and a pair of scent glands (paranal glands) located on each antimere (Fig. 6.1). The parallel anal sacs differ between males and females in both shape and depth (Macdonald and Herrera 2012). In females they appear as an invagination, while in males they appear as a small circular plane, easily reversible to the external environment. In both genders, the parallel anal sacs are covered with fur and oily secretions. In contrast to the nasal gland, the anal sacs are used by both sexes in glandular marking by rubbing them on vegetation.

On first analysis, the resting position of the capybara penis seems unusual since it is flexed caudally. However, a caudal direction of the penis is seen in some species (such as felines), and the presence of flexure is an adaptation present in suids and ruminants. In the capybara, the base of the penis is directed cranially and next to the middle third of the penis, which presents a ventro-caudal curvature of 180°, pointing its distal extremity caudally (Fig. 6.1). Throughout its cranial trajectory, including the flexure, the penis presents a subcutaneous position when at rest. In this caudal trajectory, it is totally coated by free mucosa in the interior of the anogenital cavity (Fig. 6.1). For the accurate identification of subadult males, touching and exposure of the penis are necessary. This is easily done with external cranial-caudal manipulation of the anogenital cavity (Fig. 6.2a).

The penile gland of the capybara is externally well-marked by the presence of the *colo* of the gland. Its opening is sub-apical and has an inverted "T" shape, which is composed of a longitudinal median sulcus ending in a transversal sulcus (Fig. 6.3a). The unique shape of this opening is related to the fact that the external urethral ostium is connected to the opening of the terminal-invagination, which runs parallel and ventral to the urethra (Fig. 6.3b). This invagination is also found as a homologous structure in some species (such as equids), described as a urethral sinus (Dyce et al. 1996; Schaller 1999). Although this urethral sinus is not responsible for any relevant physiological role in equids, in capybaras the invagination seems to be morphologically related to the process of penile erection. During attempts to collect capybara semen through electro-ejaculation in anesthetized animals, it was possible to stimulate an erection with a considerable increase of penile rigor. However, what stood out during this process was the huge increase in volume seen in the gland,

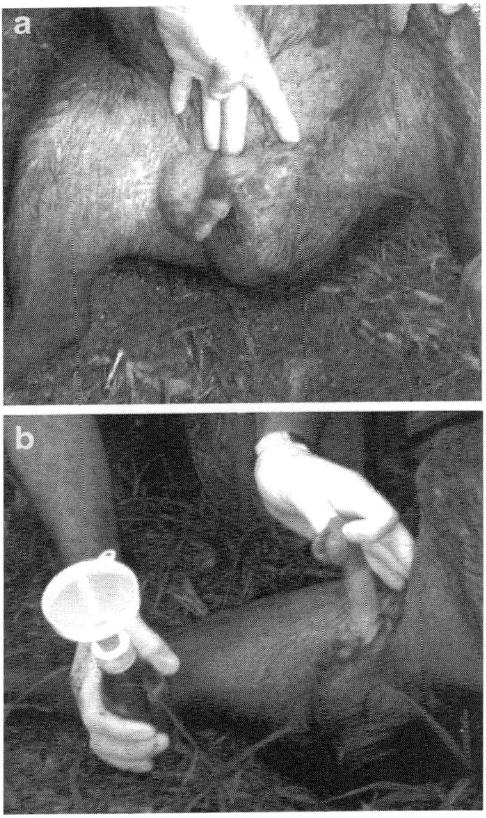

Fig. 6.2 (**a**) exposure of the penis from the anogenital cavity by external manipulation. (**b**) swelling of the glans penis in erection during electroejaculation

which swelled to form two large round masses isolated to each antimere by a continuous median sulcus with an external urethral ostium (Figs. 6.2b and 6.3c). These expansions are formed as a result of blood flow into the corpus spongiosum of the penis and are concentrated mainly in the lateral labia of the longitudinal median sulcus at the opening of the gland. These labia present continuity to the interior of the urethral sinus in the flaccid penis (Fig. 6.3b), while during erection they promote the reversal of the urethral sinus of the gland (Fig. 6.3c).

Large expansions in the corpus spongiosum in the penis gland constitute an erection strategy also seen in Canidae, in which all empty spaces of the penis are filled. The expansion caused by the filling of these corpora spongiosa causes the appearance of a large bulb (the bulb of the gland). This bulb causes a temporary fixation of the penis in the interior of the female genital during copulation, which is necessary for its long ejaculation process. Since the corpora spongiosa do not supply enough rigidity for intravaginal penetration, capybaras have a penile bone (bacculum) for this purpose. In adult male capybaras, the distal third of the free part of the penis lacks resilient cavernous bodies, and this portion is occupied by the penile bone (Fig. 6.4). Shorter than those seen in Canidae, the penile bone of the capybara has

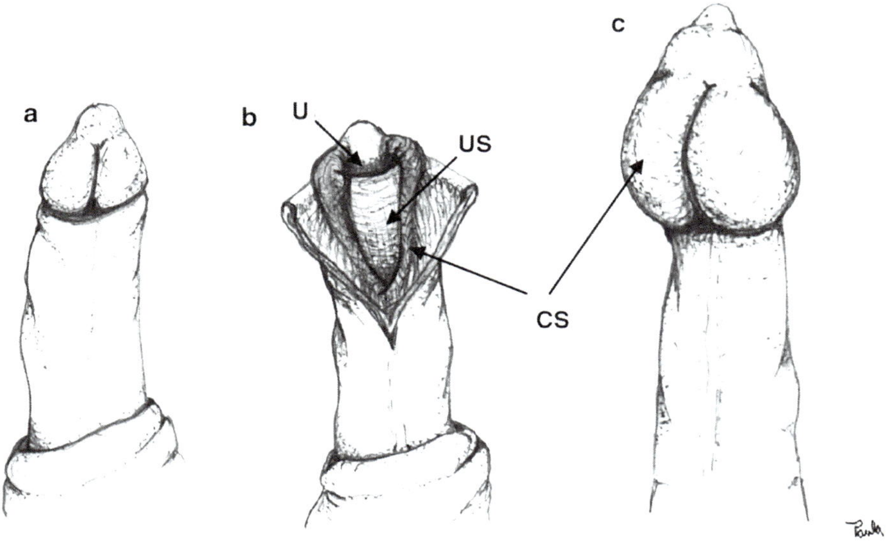

Fig. 6.3 Schemes showing ventral view of penis of capybara. (**a**) appearance of the penis at rest, (**b**) dissection of the urethral sinus through ventral median section: (U) urethra and (US) urethral sinus, (**c**) enlargement of the bilateral reversal urethral sinus during penile erection in capybara: (CS) corpora spongiosa

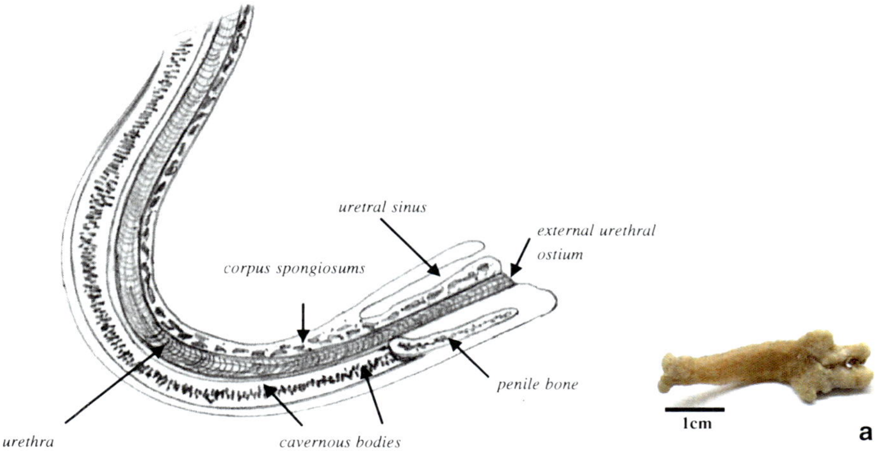

Fig. 6.4 Schema of a longitudinal section of the penis of adult capybara. (**a**) penile bone

similar mechanisms to promote rigidity and swelling, but it does not cause the female and male genitals to lock together during copulation.

Dissection reveals a pair of retractor muscles in the penis, firmly connected to the external portion of the penile curvature by a firm tendon and directed cranially to

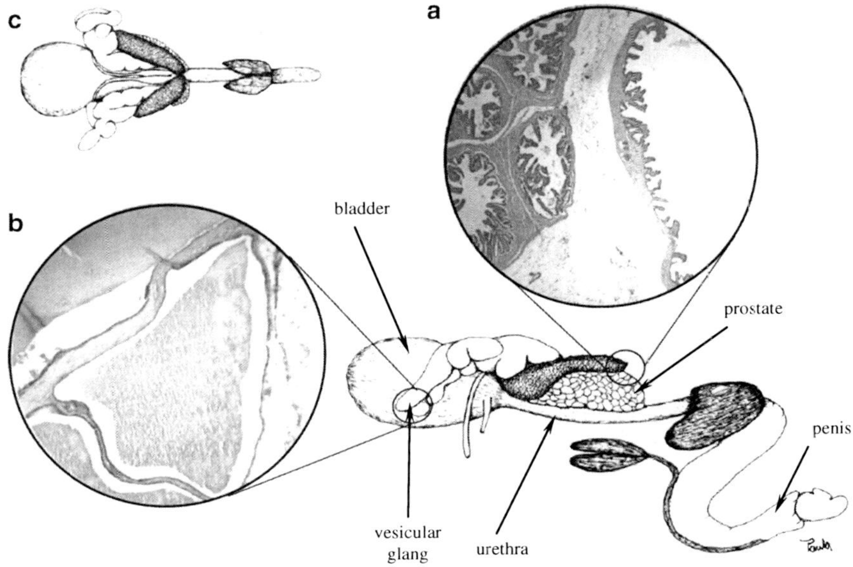

Fig. 6.5 Scheme of a left-side view of the bladder, genital glands dorsally attached to the pelvic urethra, and penis in adult capybaras. Details: histological sections (**a**) from the vesicular gland acini, (**b**) transition between two portions of the prostate (dorsolateral portion and left portion ventro-lateral to the right), (**c**) dorsal view

their origin, close to the ventral aspect of the coxal bone (Fig. 6.5). At their origin, the retracting muscles of the penis are hypertrophied with the formation of two small dilations. A pair of well-developed ischiocavernosus muscles holds the cross of the penis in its fixation to the ischiatic arch.

6.2.2.2 Glands

As seen in other mammals, the pelvic segment of the male reproductive organs in capybaras contains glands linked to seminal production, located dorsally to a long, thin pelvic urethra which is covered by the urethral muscle. The accessory genital glands are composed of well-developed vesicular glands and prostate, whereas bulbourethral glands were not detected (Fig. 6.5).

The ductus deferens of the capybara culminates directly in the seminal colliculus close to the ostium of the vesicular glands (Ojasti 1973). Before the opening, the ductus deferens presents a dilation forming two small but noticeable ampoules of the ductus deferens. Histologically, these ampoules are coated by a pseudostratified epithelium without microvillosities. The mucosa tunic has no glands, while the muscle tunic is composed of three types of layers: internal longitudinal, medium circular, and external longitudinal.

The vesicular glands are well developed, with one on each antimere, located dorsally to the *colo* of the bladder (Fig. 6.5) and presenting a number of digiform ramifications. According to Ojasti (1973), the number of projections ranges from 2 to 14 in each gland, registering great variation even between the vesicular glands of a given individual. Ojasti (1973) cites the presence of vesicular glands with only one ramification per antimere in an individual of the species *Hydrochoerus isthmius*. The vesicular glands culminate dorsally in the seminal colliculus through isolated ostium. Histologically, these glands are formed by a few large acini containing great quantities of secretion (Fig. 6.5a). The coating epithelium of the vesicular glands is pseudostratified in the folds that limit the lobes, as well as in its limiting external portions, with a well-vascularized *tunica propria*. The muscle tunic consists of at least two layers of smooth muscle, one being circular internal and the other longitudinal external.

The prostate gland is a glandular mass formed by lobules coated with a layer of connective serosa (Ojasti 1973). These lobules are grouped in at least two distinct portions on each antimere: dorsomedial and ventro-lateral, both located dorsal to the pelvic urethra (Fig. 6.5). As seen by Ojasti (1973), the dorsomedial portions of the prostate present smaller lobes than the ventro-lateral portion (Fig. 6.5b). In both portions, the internal coating of the lobes is composed of a simple cuboidal prismatic epithelium. Folding of the mucosa is observed in the interior of the lumen lobule, where each fold is filled by loose vascularized connective tissue. In their ventro-lateral portion, the prostatic lobules present a greater accumulation of secretion, causing an increase in lumen and a reduction in the height of their folding (Fig. 6.5b).

6.2.3 Testicles

6.2.3.1 Capybara Testicle Structure

Despite a difference in location, the testicles of capybaras present architecture and structure very similar to those observed in other mammal species. The testicular albuginea in capybaras has an average width of approximately 250 μm, and is very rich in blood and lymph vessels. The albuginea penetrates the interior of the testicle at its capitata extremity, following an inverted trajectory from the efferent ducts and continuing with the testicular mediastinum. Several large connective septa are internalized from the testicular albuginea, converging at the mediastinum and defining lobes that contain the seminiferous tubules and a vast intertubular space. The testicular mediastine is easily observed and is located longitudinally in the central portion of the testicle (Fig. 6.6). The testicular albuginea and mediastinum, respectively, represent an average of 6.5% and 5.5% of testicular weight in adult capybaras (Paula 1999).

The most prominent aspects of the histological and histometric analyses of adult capybara testicles is the presence of large quantities of Leydig cells (Fig. 6.7) and the disposition of the lymphatic spaces in the intertubular compartment. The lymphatic spaces form an abundant network adjacent to the seminiferous tubules and to the rich mass of Leydig cells (Fig. 6.7).

According to Fawcett et al. (1973), the arrangement and proportion of the elements that constitute the intertubular space of different mammal species investigated so

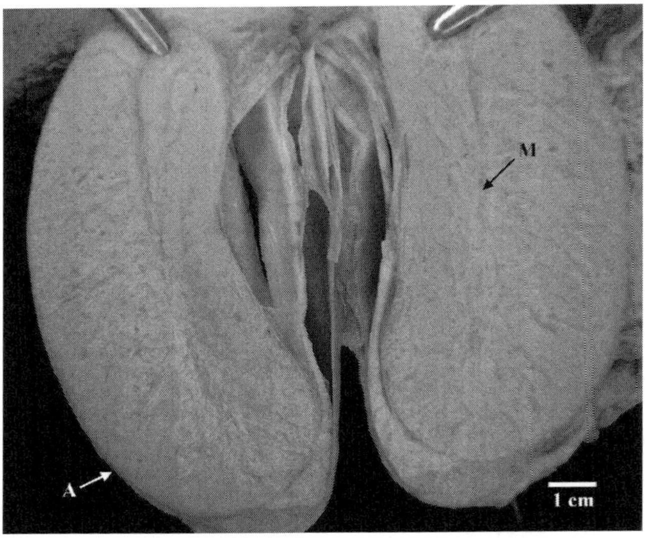

Fig. 6.6 Longitudinal section in the middle portion of the testis of capybaras: testicular albuginea (*A*) and mediastinum (*M*)

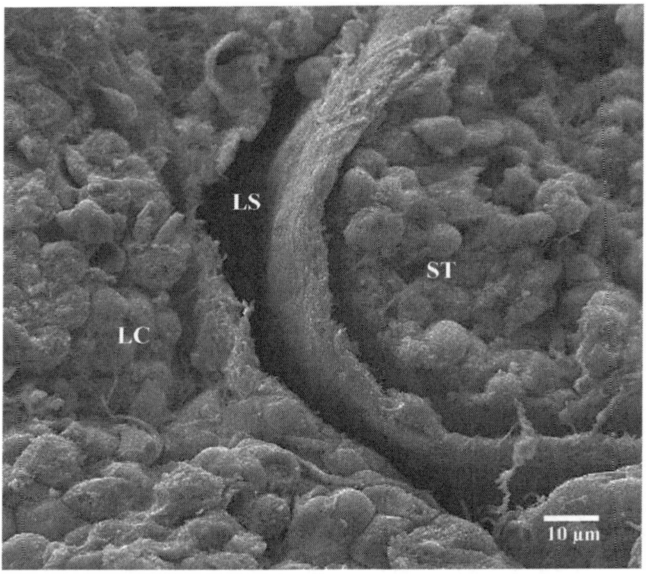

Fig. 6.7 Photomicrograph of scanning electron microscopy of a piece of testicular parenchyma from an adult capybara. Lymphatic space (*LS*) observed between the seminiferous tubule (*ST*) and the mass of Leydig cells (*LC*)

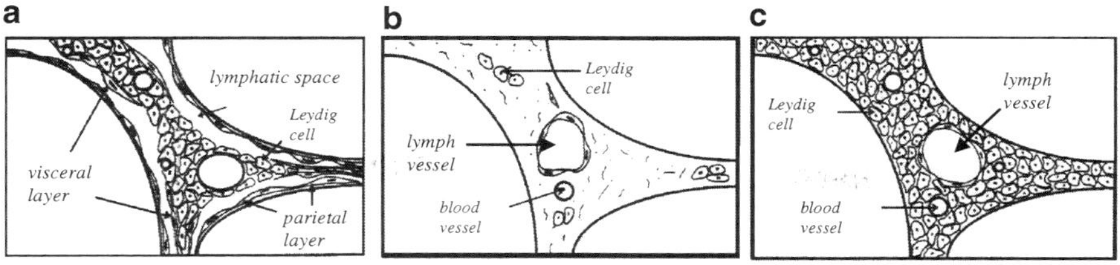

Fig. 6.8 Organization of key elements of the intertubular compartment, according to Fawcett et al. (1973): (**a**) standard type I, (**b**) type II, and (**c**) type III

far, all generally follow three distinct patterns (Fig. 6.8): (1) species in which Leydig cells and connective tissue occupy a very small area in the intertubular compartment, contrasting with extensive lymphatic sinusoids or lymphatic spaces; (2) species that present groups of Leydig cells spread in abundant loose, swollen connective tissue, which is drained by a lymph vessel located in the center or eccentrically in the intertubular space; and (3) species in which abundant groupings of Leydig cells occupy practically the entire intertubular compartment, presenting few connective and lymphatic tissues. The models described above indicate a great variation in relation to the quantity of Leydig cells and the disposition of lymphatic vessels/spaces. The main function of these different arrangements is apparently to maintain high concentrations of androgens in the testicle while simultaneously providing physiological seric concentrations.

Generally, rodents present the type I pattern described by Fawcett et al. (1973). The lymphatic space seen in these animals is composed of a parietal endothelial layer in contact with the seminiferous tubules, and by a visceral layer in contact with the Leydig cells and blood vessels. Based on further studies, Clark (1976) and Russell (1996) observed that the visceral endothelium in the rat, hamster (*Mesecricetus auratus*), and mouse (*Mus nusculus*) is discontinuous in the meeting regions between three seminiferous tubules, placing the Leydig cells in direct contact with the lymph in these regions (Fig. 6.9a). On the other hand, in the guinea pig (*Cavis porcellus*) and chinchilla (*Chinchilla lanigera*) this endothelium is continuous (Fig. 6.9b). Using electronic transmission microscopy (Fig. 6.10), the presence of a continuous endothelium in capybaras can be seen, despite the huge volumetric proportion of Leydig cells, with the arrangement of intertubular components being similar to those seen in the guinea pig and chinchilla (Fig. 6.9c).

6.2.3.2 Components of the Testicular Parenchyma

In functional terms, mammalian testicles can be divided into two basic compartments: the tubular or spermatogenic and the intertubular or androgenic. The tubular compartment is composed of seminiferous tubules, ranging in number from several dozen to several hundred. These are divided into three portions: one well-developed intermediate portion comprising very contorted grips (loops) and two smaller

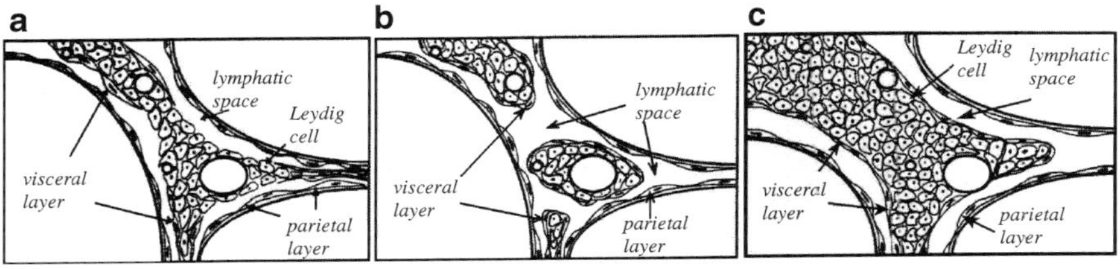

Fig. 6.9 Subdivisions of the architectural pattern of type I: (**a**) rat (Clark 1976), mice, and hamsters (Russell 1996); (**b**) guinea pig and chinchilla (Russell 1996); and (**c**) capybara (Paula 1999)

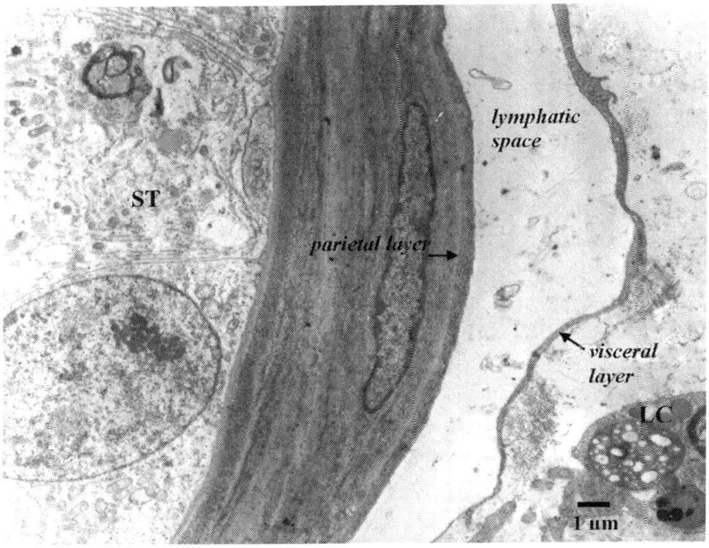

Fig. 6.10 Capybara testis observed through transmission electron microscopy showing the disposition of lymphatic space, which is bordered by the parietal endothelium and visceral endothelium. Also indicated are the seminiferous tubule (*ST*) and Leydig cell (*LC*)

peripherals called straight tubules, which are connected to the testicular network in the testicular mediastinum. The intertubular compartment contains the Leydig cells, blood and lymphatic vessels, nerves, cells and fibers from the connective tissue, and various other cell types such as macrophages and mastocytes. The high level of variation in the proportions of these compartments is one of the factors responsible for interspecific differences in spermatic production (Russell et al. 1990a; França and Russell 1998).

6.2.3.3 The Intertubular Compartment

In addition to some singular morphological aspects, the male genital organs in capybaras have quantitative biometric parameters unique among other adult animals.

Fig. 6.11 Photomicrograph of scanning electron microscopy of a piece of testicular parenchyma from an adult capybara. Cross-section of a bundle of tubules (*ST*) surrounded by a large amount of intertubular tissue, particularly the Leydig cell (*LC*)

While in domestic mammals the percentage of the testicles occupied by the intertubular compartment can vary from 10% in canids to 30% in equids (França and Russell 1998), male adult capybaras are notable in having around 50% of the testicular parenchyma volume contained in the intertubular compartment (Fig. 6.11). It seems that phylogenetic factors are not responsible for the percentage occupied by the intertubular compartment since enormous variation is observed between species, even among those in the same genus (Fawcett et al. 1973; Breed 1982).

The Leydig cells are the main element that composes the intertubular compartment. Nearly a third of the testicle is occupied by these cells in adult capybaras (Paula et al. 2007) – this is the highest volumetric proportion of this cellular type seen in all mammals studied (Moreira 1995; Moreira et al. 1997a; França and Russell 1998; Paula 1999; Costa et al. 2006; Costa and Paula 2006; Paula et al. 2007). The quantity of Leydig cells varies among different mammal species, with this variation not found in other endocrine glands (Fawcett et al. 1973). However, the other components of the intertubular space present fewer evident variations.

Additionally, the number of Leydig cells quantified per gram of testicle is noticeable in capybaras. With approximately 126.4 million Leydig cells per gram of testicle (Costa et al. 2006), the capybara exceeds the values registered in domestic animals (between 20 and 60 million; França and Russell 1998). When compared to other rodents, the exceptional abundance of Leydig cells in the testicular parenchyma in capybaras becomes increasingly evident, since guinea pigs, mice, and rats present 6, 10.5, and 24.9 million Leydig cells per gram of testicle, respectively (Mori et al. 1980; Zirkin and Ewing 1987).

Leydig cells are principally responsible for steroid production, and many factors can contribute to their enormous interspecific quantitative variation. The Sertoli cell acts as the main regulator of the spermatogenic process. During testicular development, and especially during puberty and seasonal testicular recrudescence, the Sertoli cell is influenced first by the follicle-stimulating hormone (FSH) and then by testosterone, which gradually replaces FSH (Means et al. 1976; Jégou et al. 1983; Sharpe 1994). Thus, as preparation for this substitution, the Leydig cells must be numerically and physiologically adequate to provide support to the new testosterone requirement (Sharpe 1994). In sexually mature animals, studies correlating the structure and function of the Leydig cells showed that variations in testosterone secretion are more often the result of individual capacity than of the differences in their total volume in the testicle (Ewing et al. 1979; Costa and Paula 2006). This capacity is most correlated to the individual volume of Leydig cells (Costa and Paula 2006), especially their smooth endoplasmic reticulum content (Zirkin et al. 1980).

Individual variations in the need for testosterone for spermatic production are not the only factors that regulate the quantity of Leydig cells. The maintenance of extra testicular blood levels, and adequate levels of testosterone in particular, are key to secondary sexual characterization and especially reproductive behavior (Hadley 1988), and are of fundamental importance to the capybara (Herrera and Macdonald 1994). According to some researchers (Fawcett et al. 1973; Russell 1996), the greater quantity of Leydig cells in some species may be related to the synthesis of other types of steroids like estrogens, pheromones, or other substances yet to be determined. Estrogens are very important in the functional control of spermatic pathways, being responsible for regulating and reabsorbing fluids produced in the seminiferous epithelium, efferent ducts and the initial segment of the epididymis head (Hess et al. 1997). Pheromones are widely used in social and sexual communication, and field data suggest that the pheromones of dominant capybara males may have a strong influence on the inhibition of subordinate male development (Tarcízio A.R. Paula personal observation). The removal of the dominant male from a determined group begins a quick process (approximately 40 days) for the maximum development of the nasal gland and reproductive behavior of the sub-dominant male.

A number of different factors can influence the quantity of Leydig cells per animal, among which are the quantity of available luteinizing hormone (LH); the number of LH receptors per cell; the quantity of testosterone that the Leydig cell is capable of secreting at a given time; the speed at which testosterone leaves the testicles via lymphatic vessels, blood vessels, and seminal fluid; the blood volume of the animal; and the testosterone's metabolic rate (Russell et al. 1994; Russell 1996). Therefore, considering that only 2.2% of rat testicles are occupied by Leydig cells (Mendis–Handagama and Ariyaratne 2001), it is evident that this value is much lower than the 33% seen in the testicle of capybaras (Paula et al. 2007). However, if these values are related to the percentage of body weight allocated to the testicle (gonadosomatic index) for the rat, 4.26% (Kenagy and Trombulak 1986), and for the capybara, 0.125% (Paula et al. 2007), the body weight index allocated in Leydig cells (Leydigosomatic index) for the rat is 0.094% and for the capybara only 0.041%. Therefore, proportional to body weight, the rat presents twice the quantity of Leydig cells seen in capybaras, suggesting that testicular volumetric proportion does not reflect the actual physiological requirement of Leydig cells per animal.

6.2.3.4 Seminiferous Tubules

The seminiferous tubules in capybaras generally do not differ in their morphofunctional aspects from those seen in other mammals. The seminiferous tubules do not have any vascularization or innervation of their own and are constituted by only a single tunica propria, a seminiferous epithelium and the tubular lumen. The proper tunic coats the epithelium externally, consisting of myoid or peritubular cells and acellular elements that together form the limiting membrane of the seminiferous tubule. The gametogenic function of the seminiferous tubule is carried out by the seminiferous epithelium, which is formed by two cellular types of embryological origin and distinct function (Russell et al. 1990a), but equally important in the spermatogenic process. The productive process is coordinated by Sertoli cells, while germinative cells control proliferation and differentiation in spermatozoids. The spermatozoids, like a considerable amount of secreted fluid by Sertoli cells, are released to the tubular lumen and propelled toward the testicular mediastinum through the contraction of myoid cells.

In addition to the well-known endocrine control of the hypothalamic-pituitary-testicular axis, an elaborate intercellular communication system is located in the testicle to ensure the proper functioning of the spermatogenic process (Roser 2000). Thus, the cells directly and indirectly involved in the production of male gametes (Sertoli cells, germinative cells, Leydig cells, Peri-tubular myoid cells, leucocytes, etc.), maintain a paracrine/autocrine system to modulate this intrinsic network of cellular interaction, which is fundamental for the proper functioning of the testicle (Schlatt et al. 1997).

The seminiferous tubules in the capybara are arranged individually, forming tubular bundles in the midst of a great mass of intertubular tissue (Fig. 6.11). The variable abundance of intertubular mass seen among individuals determines the degree of isolation of these tubular bundles through the testicular parenchyma. It is well known that the seminiferous tubule compartment physically dominates the testicle, in most mammals occupying from 70% to 90% of its parenchyma (França and Russell 1998), thus greatly influencing weight and spermatic production (Amann 1970). In adult male capybaras, the proportion occupied by the seminiferous tubules is only 50% of the testicular parenchyma, which is reflected by low spermatic production.

To estimate the efficiency of sperm production in a certain species, quantitative aspects other than the somatic indices and volumetric proportion of the testicular components must be considered, such as the width of the seminiferous epithelium as well as the tubular diameter and length. However, qualitative aspects such as the intrinsic yield of the spermatogenic process and the cycle duration of the seminiferous tubules are also determining factors. The tubular diameter remains relatively constant in sexually mature animals, and is not affected by seasonal variation. However, the tubule diameter can present significant variations across several species and among different lineages or races (França and Russell 1998). Also, the average tubular diameter can reach up to 550 μm in some species of marsupials (Woolley 1975), whereas the values seen for most of the amniotes range from 180 to 300 μm (Roosen-Runge 1977). Adult male capybaras have an average tubular diameter of 213 μm. The average height of the seminiferous epithelium in capybaras is 79.1 μm, which is within the values cited for domestic animals (60 to 100 μm; França and Russell 1998). However, while Moreira (1995) has observed seasonal variation in the tubular diameter of capybaras studied on Marajó Island in northern Brazil, no variation was seen in animals from southeastern and southern Brazil, nor were differences observed in the tubular diameter and height of the seminiferous epithelium at different times of the year (Paula 1999).

The tubular length is obtained from the volume of seminiferous tubules and the tubular diameter. The capybara presents an average of 374 m of seminiferous tubules per testicle, which represents around 11 m of tubules per gram of testicle. Generally, between 10 and 15 m of seminiferous tubules per gram of testicle are seen in most of the mammals studied (Wing and Christensen 1982; Neaves and Johnson 1985; Sinha-Hikim et al. 1988; França and Russell 1998; Guião Leite and Paula 2003; Bittencourt et al. 2007; Costa et al. 2007). However, some variations have been reported, like the small marsupial *Antechinus stuartii* which has only 6 m per gram of testicle (Woolley 1975), and the rabbit (*Oryctolagus cuniculus*) with approximately 20 m per gram of testicle (Amann 1970). Thus, even with only 50% of the testicle occupied by seminiferous tubules, the value obtained for seminiferous tubules per gram of testicle in capybaras (11 m) is within the observed range for mammals.

6.2.3.5 The Spermatogenic Process

In mammals, the seminiferous epithelium has four or five layers of germinative cells. The spermatogenic process is a series of events that results from proliferation and differentiation during the development of such cells. Under the guidance of Sertoli cells, a continuous flow of germinative cells is established from the base of the lumen of the seminiferous epithelium. Generations of germinative cells recently produced in the base migrate through the epithelium as they develop, while at the same time preceding generations reach the tubular lumen and are released as spermatozoids. So each time a new generation is produced an old generation is released, establishing the cycle of the seminiferous epithelium. Until one spermatogonium generation is released as spermatozoids, at least four cycles must take place. Thus, a transversal section in a seminiferous tubule reveals different combinations of germinative cells at different stages of development. A total of eight combinations that repeat in each cycle can be seen, and these eight are denominated by the cycle stages of the seminiferous epithelium. This epithelial classification is known as the tubular morphology method and is applicable to all generally studied mammals.

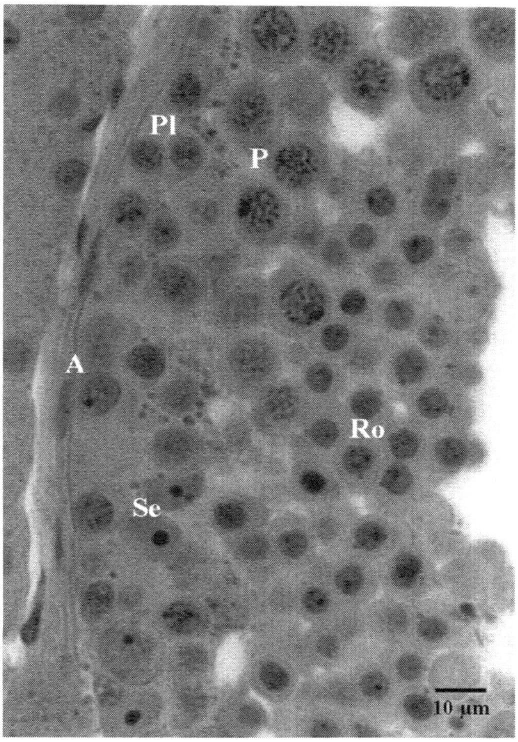

Fig. 6.12 Stage 1 of the seminiferous epithelium cycle of capybaras. *Pl* primary spermatocytes in pre-leptotene, *P* primary spermatocytes in pachytene, *A* type A spermatogonia, *Ro* round spermatids, *Se* Sertoli Cells

In this aspect, spermatogenesis in capybaras is very similar to that described for other rodents (Russell et al. 1990b; De Rooij 1998; Paula 1999).

Using the transverse section of seminiferous tubules in several cycle stages, population quantification of the different types of germinative cells and Sertoli cells allows functional analysis of the spermatogenic process and estimates of spermatic production, in addition to providing kinetic information on the spermatogenesis of a species (Clermont 1972; Cardoso 1981; França 1991; Castro 1995). Stage 1 of the seminiferous epithelium cycle is traditionally used to quantify the intrinsic yield of the spermatogenic process, since the seminiferous epithelium strategically contains generations of spermatogonia, primary spermatocytes in preleptotene, primary spermatocytes in pachytene, and round spermatids (Fig. 6.12). The ratios between the populations of different types of germinative cells provide a very accurate way to analyze the efficiency of the spermatogenic process. This approach enables comparisons between different species and serves as a correction factor for the numbers obtained from different methodologies and histological cuts of varied widths (Russell et al. 1990a; Paula 1999; Guião Leite and Paula 2003).

In capybaras, around nine primary spermatocytes are formed for every type A_1 spermatogonium. This quotient is one of the lowest found in the literature when compared with other species. For purposes of illustration, between 15 and 22 primary spermatocytes are formed from a type A_1 spermatogonium in domestic ruminants (França and Russell 1998), in dogs (Ibach et al. 1976; Paula 1992), and in the Chinese hamster (*Cricetulus barabensis*; Oud and De Rooij 1977). Thus, the analysis of this index proves that the spermatogonial phase is one of extreme cellular loss in capybaras.

Normally, it is in the meiotic phase that the highest numerical cell loss occurs in most studied mammals (Roosen-Runge 1973; França and Russell 1998; França et al. 1999). Most authors state that the apoptoses that occur during meiosis are part of a cell elimination mechanism with abnormal or aneuploidy chromosomes (Roosen-Runge 1973; Sharpe 1994), which is the probable reason for infertility in most hybrid animals such as the mule (Benirschke et al. 1962; Hernández-Jáuregui and Monter 1977). In mammals, this loss is generally around 25%, which means that for every four expected round spermatids, three are formed (França and Russell 1998). However, in capybaras only around 50% of the round spermatids were seen, causing a 2:1 meiotic index, which is slightly higher than the index for humans (1.3:1; Sinha-Hikim et al. 1985). Analysis of the apoptotic index determines that this index is relatively stable throughout the eight different stages of the cycle, but slightly higher in the meiotic division stage (stage 4). Analysis of primary spermatocytes and round spermatid populations in the different cycle stages shows that the populations of these cell types are relatively stable. This reinforces the fact that the spermatogonial proliferation and meiotic division stages are the most susceptible to apoptosis in capybaras.

The general yield of spermatogenesis is also low in capybaras (21:1): two to four times lower than most domestic animals and laboratory animals (França and Russell 1998; França et al. 1999). However, it is worth mentioning that in all the investigated mammals, only 15–30% of the expected spermatozoids are produced (Huckins 1978; França and Russell 1998; Swerdloff et al. 1998), meaning that cell losses are an integral part of the spermatogenic process.

The interactions between Sertoli and germinative cells are crucial to maintaining normal spermatic production (Griswold 1995). The population of Sertoli cells is stable after puberty and throughout the different stages of the cycle, so it serves as a reference to quantify and functionally evaluate the spermatogenic process. Nevertheless, the support capacity of Sertoli cells varies and is specific to each species (Russell and Peterson 1984; França and Russell 1998; França et al. 1999). The number of spermatids per Sertoli cell is considered a very accurate index of the efficiency and functioning of the Sertoli cell, being a determining factor in spermatic production (Russell and Peterson 1984; Sharpe 1994). In capybaras, the number of round spermatids per Sertoli cell is 5.6, a level considered low when compared to most investigated rodents (Wing and Christensen 1982; Parreira 1990; Patil and Saidapur 1991; Rocha et al. 1999) and domestic mammals (França and Russell 1998). It is practically half the number seen for species with high reproductive efficiency, such as the pig and the rabbit (Russell and Peterson 1984; França and Russell 1998).

The duration of the seminiferous epithelium cycle is a biological constant specific to each species, which is under the control of the germinative cell genotype

(França et al. 1998), and not influenced by any known factor (Clermont 1972; Amann and Schandbacher 1983). Approximately 4.5 cycles are required to complete the spermatogenic process in mammals, which means that spermatozoids are released in the seminiferous tubule lumen of a spermatogonium (Amann and Schandbacher 1983; França and Russell 1998). Among the studied placental mammals, both the shortest and longest durations of the seminiferous epithelium cycles are found in the order Rodentia, family Cricetidae. The shortest duration is found in the bank vole (*Myodes glareolus*) of the Muridae family: 6.7 days (Grocock and Clark 1976), and the longest in the Chinese hamster: 17 days (Oud and de Rooij 1977). The duration of one seminiferous epithelium cycle in capybaras, 11.9 days, is one of the highest among rodents. Compared to other non-rodent species, spermatogenesis duration in capybaras is close to the rabbit (11.2 days; Orgebin-Crist 1968), the banteng (11.8 days; *Bos javanicus*; McCool 1989), the horse (12.2 days; Swierstra et al. 1975), the blue fox (12.0 days; *Vulpes lagopus*; Berg et al. 1990), and the stump-tailed macaque (11.6 days; *Macaca arctoides*; Clermont and Antar 1973).

6.2.3.6 Quantification of Spermatic Production

By using the population of more advanced germinative cells in a tubular section in stage 1 of the seminiferous epithelium cycle, the total length of the seminiferous tubules, the duration of a seminiferous epithelium cycle, and the testicular weight, it is possible to estimate the daily spermatic population per gram of testicle. This is the principal parameter in determining the efficiency of spermatogenesis, since all the interspecific oscillations are eliminated and all productive variables are considered. In further analysis of the literature, it is possible to group daily spermatic production per gram of testicle into three levels: (1) species with high spermatogenic efficiency and that produce around 20–30 million spematozoids, including animals such as the domestic pig, horse, sheep, rabbit (França and Russell 1998), the rat (Robb et al. 1978), the rhesus monkey (*Macaca mulatta*; Amann et al. 1976), the cougar (*Puma concolor*; Guião Leite and Paula 2003), and the maned wolf (*Chrysocyon brachyurus*; Bittencourt et al. 2007); (2) species with medium spermatogenic efficiency, producing from 10 to 20 million spermatozoids, among them cattle and the buffalo (*Bubalus bubalis*; França and Russell 1998); and, lastly, (3) species that produce below 10 million spermatozoids, including, for example, humans (Amann 1981) and the white-eared opossum (*Didelphis albiventris*; Queiroz and Nogueira 1992). The capybara, with approximately 10 million spermatozoids produced per day for every gram of testicle, is placed between the medium and low levels.

A significant and positive correlation was seen between the daily spermatic production per gram of testicle and the volume of the nasal gland in adult male capybaras (Paula 1999). Since the size and usage rate of the nasal gland are significantly higher in animals that are more physically and physiologically apt for dominance (Herrera and Macdonald 1993; Herrera and Macdonald 1994), we conclude that those males considered dominant are also the largest producers of spermatozoids.

References

Alho CJR (1989) Comportamento no habitat natural como fator limitante de criação e manejo de capivaras. In: de Animais E, Ades C (eds) Etologia de Animais e de homens. Edicon, São Paulo, pp 75–85

Amann RP (1970) Sperm production rates. In: Johnson AD, Gomes WR, Vandemark NL (eds) The testis. Academic, New York, pp 433–482

Amann RP (1981) A critical review of methods for evaluation of spermatogenesis from seminal characteristics. J Androl 2:37–58

Amann RP, Schandbacher BD (1983) Physiology of male reproduction. J Anim Sci 57(Suppl 2): 380–403

Amann RPL, Johnson DL, Thompson JR, Pickett BW (1976) Daily spermatozoal production, epididymal spermatozoal reserves and transit time of spermatozoa through the epididymis of the Rhesus monkey. Biol Reprod 15:586–592

Benirschke K, Brownhill LE, Beath MM (1962) Somatic chromosomes of the horse, the donkey and their hybrids, the mule and the hinny. J Reprod Fertil 4:319–326

Berg AK, Clausen OP, Huitfield HS (1990) The spermatogenic cycle in the blue fox (Alopex lagopus): relative frequency and absolute duration of the different stages. Int J Androl 13(4):315–326

Berndtson WE, Igboeli G, Pickett BW (1987) Relationship of absolute number of Sertoli cells to testicular size and spermatogenesis in young beef bulls. J Anim Sci 64:241–246

Bittencourt VL et al (2007) The seminiferous epithelium cycle and daily spermatic production in the adult maned wolf (*Chrysocyon brachyurus*, Illiger, 1811). Micron 38(6):584–589

Breed WG (1982) Morphological variation in the testes and accessory sex organs of Australian rodents in the genera *Pesudomys* and *Notomys*. J Reprod Fertil 66:607–613

Breed WG, Adams M (1992) Breeding systems of spinifex hopping mice (*Notomys alexis*) and plains rats (*Pseudomys australis*): a test for multiple paternity within the laboratory. Aust J Zool 40:13–20

Cardoso FM (1981) Morfologia, cinética e quantificação da espermatogênese em zebus (*Bos indicus*). Ph.D. thesis, Instituto de Ciências Biológicas, Universidade Federal de Minas Gerais, Belo Horizonte

Castro ACS (1995) A proposed acrossomal system for identifying stages of the cycle of the seminiferous epithelium and a model for the kinetics of spermatogenesis in the rabbit. Ph.D. thesis, University of New Hampshire, Durham

Clark RC (1976) Three-dimensional organization of testicular interstitial tissue and lymphatic space in the rat. Anat Rec 184:203–226

Clermont Y (1972) Kinetics of spermatogenesis in mammals, seminiferous epithelium cycle and spermatogonial renewal. Physiol Rev 52(1):198–236

Clermont Y, Antar M (1973) Duration of the cycle of the seminiferous epithelium and the spermatogonial renewal in the monkey (*Macaca arctoides*). Am J Anat 136:153–166

Costa DS, Paula TAR (2006) Testosterone level, nasal gland volume and Leydig cell morphometry in capybaras (*Hydrochoerus hydrochaeris*). Arquivos Brasileiros de Medicina Veterinária e Zootecnia 58(6):1086–1091

Costa DS, Paula TAR, Matta SLP (2006) The intertubular compartment morphometry in capybaras (*Hydrochoerus hydrochaeris*) testis. Anim Reprod Sci 91:173–179

Costa DS, Menezes CMC, Paula TAR (2007) Spermatogenesis in White-lipped peccaries (*Tayassu pecari*). Anim Reprod Sci 98:322–334

de Rooij DG (1998) Stem cells in the testis. Int J Exp Pathol 79:67–80

Dyce KM, Sack WO, Wensing CJG (1996) Tratado de Anatomia Veterinária, 2ath edn. Editora Guanabara Koogan S.A, Rio de Janeiro

Ewing LL, Zirkin BR, Cochran RC, Kromann N, Peter C, Ruiz-Bravo N (1979) Testosterone secretion by rat, rabbit, guinea pig, dog and hamster testes perfused in vitro: correlation with Leydig cell mass. Endocrinology 105(5):1135–1142

Fawcett DW, Neaves WB, Flores MN (1973) Comparative observations on intertublar lymphatic and the organization of the interstitial tissue of the mammalian testis. Biol Reprod 9:500–532
França LR (1991) Análise morfofuncional da espermatogênese de suínos adultos da raça Piau. Ph.D. thesis, Universidade Federal de Minas Gerais, Belo Horizonte
França LR, Russell LD (1998) The testis of domestic animals. In: Regadera J, Martinez-Garcia F (eds) Male reproduction. A multidisciplinary overview. Churchill Livingstone, Madrid, pp 197–219
França LR, Ogawa T, Avarbock MR, Brinster RL, Russell LD (1998) Germ cell genotype controls cell cycle during spermatogenesis in the rat. Biol Reprod 59:1371–1377
França LR, Becker-Silva SC, Chiarini-Garcia H (1999) The length of the cycle of seminiferous epithelium in goats. Tissue Cell 31(3):274–280
Griswold MD (1995) Interaction between germ cells and Sertoli cells in the testis. Biol Reprod 52:211–216
Grocock CA, Clark JR (1976) Duration of spermatogenesis in the vole (*Microtus agrestis*) and the bank vole (*Cleterionomys glareolus*). J Reprod Fertil 47:133–135
Guião Leite FL, Paula TAR (2003) Intrinsic yield of spermatogenesis, Sertoli cell index and daily sperm production in cougar (*Puma concolor*). Revista Brasileira de Reprodução Animal 27(1):21
Hadley ME (1988) Endocrinology, 2nd edn. Prentice Hall, New Jersey
Hernández-Jáuregui P, Monter HM (1977) Fine structure of mule testes: light and electron microscopy study. Am J Vet Res 38(4):343–347
Herrera EA (1992) Size of testes and scent glands in capybaras *(Hydrochoerus hydrochaeris Rodentia: Caviomorpha)*. J Mammal 73(4):871–975
Herrera EA, Macdonald DW (1993) Aggression, dominance and mating success among capybara males *(Hydrochoerus hydrochaeris)*. Behav Ecol 4:114–119
Herrera EA, Macdonald DW (1994) Social significance of scent marking in capybaras. J Mammal 75(2):410–415
Hess RA et al (1997) A role for oestrogens in the male reproductive system. Nature 390(4):509–512
Huckins C (1978) The morphology and kinetics of spermatogonial degeneration in normal adult rats: an analysis using a simplified classification of the germinal epithelium. Anat Rec 190(4):905–926
Ibach B, Weissbach L, Hilscher B (1976) Stages of the cycle of the seminiferous epithelium in the dog. Andrologia 8(4):297–307
Jégou B, Le Gac F, Irby DC, de Krestser DM (1983) Studies on seminiferous tubule fluid production in the adult rat: effect of hypophysectomy and treatment with FSH, LH and testosterone. Int J Androl 6:249–260
kenagy GJ, Trombulak SC (1986) Size and function of mammalian testes in relation to body size. J Mammal 67(1):1–22
López M, Muñoz MG, Herrera EA (2008) Reproductive morphology of capybaras, *Hydrochoerus hydrochaeris* (*Rodentia: Hystricognathi*): no evidence for sperm competition ? Mamm Biol 73:241–244
Macdonald DW, Herrera EA (2012) Capybara scent glands and scent-marking behavior. In: Moreira JR, Ferraz KMPMB, Herrera EA, Macdonald DW (eds) Capybara: biology, use and conservation of an exceptional neotropical species. Springer, New York, pp 185–193
Macdonald DW, Krantz K, Aplin RT (1984) Behavioral, anatomical and chemical aspects of scent marking amongst capybaras, *Hydrochoerus hydrochaeris* (*Rodentia: Caviomorpha)*. J Zool Lond 202:341–360
Mccool CJ (1989) The cycle of the seminiferous epithelium in bali cattle (Bos sondaicus). J Reprod Fertil 87:327–330
Means AR, Fakunding JL, Huckins C, Tindall DJ, Vitale R (1976) Follicle-stimulating hormone, the Sertoli cell and spermatogenesis. Recent Prog Horm Res 32:477–527
Mendis-Handagama SMLC, Ariyaratne S (2001) Differentiation of adult Leydig cell population in the postnatal testis. Biol Reprod 65:660–671
Moreira JR (1995) The reproduction, demography and management of capybaras (Hydrochaeris hydrochaeris) on Marajó Island- Brazil. D.Phil., thesis. University of Oxford, Oxford, UK

Moreira JR, Clarke JR, Macdonald DW (1997a) The testis of capybaras (*Hydrochaeris hydrochaeris* – Rodentia). J Mammal 78:1096–1100

Moreira JR, Macdonald DW, Clarke JR (1997b) Correlates of testis mass in capybaras (*Hydrochoerus hydrochaeris)*: dominance assurance or sperm production? J Zool Lond 241:457–463

Mori H, Shimizu D, Takeda A, Takioka Y, Fikinishi R (1980) Stereologic analysis of Leydig cells in normal guinea pig testis. J Electron Microsc 29:8–21

Neaves WB, Johnson L (1985) Seminiferous tubules and daily sperm production (DSP) in older adult men with varied numbers of Leydig cells. Biol Reprod 32(Suppl):86

O'Donnel L, Robertson KM, Jones ME (2001) Estrogen and spermatogeneis. Endocrinol Rev 22:280–318

Ojasti J (1973) Estudio biológico del chigüire o capibara. Fondo Nacional de investigaciones agropecuarias, Caracas

Orgebin-Crist MC (1968) Gonadal and epididimal sperm reserves in the rabbit: estimation of the daily sperm production. J Reprod Fertil 15:15–25

Oud JL, de Rooij DG (1977) Spermatogenesis in the Chinese hamster. Anat Rec 187(1):113–124

Parreira GG (1990) Morfologia e variação sazonal da atividade dos testículos e órgãos genitais acessórios de *Bolomys lasiurus* Lund, 1841 (*Rodentia, Cricetidae*). M.Sc., Dissertation, Universidade Federal de Minas Gerais. Belo Horizonte

Patil SB, Saidapur SK (1991) Kinetics of spermatogenesis in the wild squirrel *Funambulus palmarum* (Linnaeus). Acta Anat (Basel) 141(4):352–363

Paula TAR (1992) Estudo histológico quantitativo da atividade espermatogênica de cães s.r.d. em diferentes faixas etárias após a puberdade. M.Sc., dissertation, Universidade Federal de Minas Gerais. Belo Horizonte

Paula TAR (1999) Avaliação histológica e funcional do testículo de capivaras adultas (*Hydrochoerus hydrochaeris*). Ph.D. thesis, Universidade Federal de Minas Gerais. Belo Horizonte

Paula TAR, Matta SLP, Silva VA Jr, Costa DS, Fonseca CC, Neves MTD (2007) Intertubular space characterization in adult capybara (*Hydrochoerus hydrochaeris*) testis. Braz Arch Biol Technol 50(2):289–297

Queiroz GF, Nogueira JC (1992) Duration of the cycle of the seminiferous epithelium and quantitative histology of the testis of the South American white belly opossum (*Didelphis albiventris*), Marsupialia. Reprod Fertil 4:213–222

Robb GE, Amann RP, Killian GJ (1978) Daily sperm production and epididymal sperm reserves of pubertal and adult rats. J Reprod Fertil 54:103–107

Rocha DCM, Debeljuk L, França LR (1999) Exposure to constant light during testis development increases daily sperm production in adult Wistar rats. Tissue Cell 31(3):128–136

Roosen-Runge EC (1973) Germinal-cell loss in normal metazoan spermatogenesis. J Reprod Fertil 35:339–348

Roosen-Runge EC (1977) The process of spermatogenesis in animals. Cambridge University Press, Cambridge, UK

Roser JF (2000) Reproductive endocrinology of the stallion. In: Samper JC (ed) Equine breeding management and artificial insemination. Saunders Company, Philadelphia, pp 123–132

Russell LD (1996) Mammalian Leydig cell structure. In: Payne AH, Hardy MP, Russell LD (eds) The Leydig cell. Cache River Press, Vienna

Russell LD, Peterson RN (1984) Determination of the alongate spermatid-Sertoli cell ratio in various mammals. J Reprod Fertil 70:635–641

Russell LD, Ettlin RA, Sinha-Hikim AP, Clegg ED (1990a) Histological and histopathological evaluation of the testis. Cache River Press, Clearwater

Russell LD, Ren HP, Sinha-Hikim I, Schulze W, Sinha-Hikim AP (1990b) A comparative study in twelve mammalian species of volume densities, volumes and numerical densities of selected testis components, emphasizing those related to the Sertoli cell. Am J Anat 188(1):21–30

Russell LD, Chandrashekar V, Bartke A, Sinha-Hikim AP (1994) The hamster Sertoli cell in early testicular regression and early recrudescence: a stereological and endocrine study. Int J Androl 17(2):93–106

Schaller O (1999) Nomenclatura Anatômica Veterinária Ilustrada. Editora Manole Ltda, São Paulo

Schlatt S, Meinhardt A, Nieschlag E (1997) Paracrine regulation of cellular interactions in the testis: factors in search of a function. Eur J Endocrinol 137:107–117

Sharpe RM (1994) Regulation of spermatogenesis. In: Knobil E, Neil JD (eds) The physiology of reproduction, vol 2. Raven, New York, pp 1363–1434

Short RV (1997) The testis: the witness of the mating system, the site of mutation and the engine of desire. Acta Paediatr 422(Suppl):3–7

Sinha-Hikim AP, Chakraborty J, Jhunjhunwala JS (1985) Germ cell quantitation in human testicular biopsy. Urol Res 13:111–115

Sinha-Hikim AP, Bartke A, Russell LD (1988) Morphometric studies on hamster testes in gonadally active and inactive states: light microscope findings. Biol Reprod 39:1225–1237

Swerdloff RS, Lue Y, Wang C, Rajavashisth T, Sinha-Hikin A (1998) Hormonal regulation of germ cell apoptosis. In: Zirkin BR (ed) Germ cell development, division, disruption and death. Springer, New York

Swierstra EE, Pickett BW, Gebaure MR (1975) Spermatogenesis and duration of transit of spermatozoa through the excurrent ducts of stallions. J Reprod Fertil 23(Suppl):53–57

Wing TY, Christensen AK (1982) Morphometric studies on rat seminiferous tubule. Am J Anat 165:13–25

Woolley P (1975) The seminiferous tubules in dasyurid marsupials. J Reprod Fertil 45(2):255–261

Zirkin BR, Ewing LL (1987) Leydig cell differentiation during maturation of the rat testis: a stereological study of cell number and ultrastructure. Anat Rec 219:157–163

Zirkin BR, Ewing LL, Kromann N, Cochran RC (1980) Testosterone secretion by rat, rabbit, guinea pig, dog, and hamster testes perfused in vitro: correlation with Leydig cell ultrastructure. Endocrinology 107:1867–1874

Chapter 7
Morphology and Reproductive Physiology of Female Capybaras

Maria Angélica Miglino, Tatiana Carlesso dos Santos, Claudia Kanashiro, and Rosa Helena dos Santos Ferraz

7.1 Introduction

Despite an abundant and diverse literature on the biology and reproductive management of capybaras, there is no detailed description of the morphophysiology of the urogenital system of the species in particular or, indeed, the hystricomorph rodents in general. This rodent group, which includes the guinea pig (*Cavia porcellus*), chinchilla (*Chinchilla lanigera*), cane rat (*Thryonomys* spp.), paca (*Cuniculus paca*), rock cavy (*Kerodon* spp.), and agouti (*Dasyprocta* spp.), could be regarded as monophyletic, based on placental structure alone (Luckett and Mossman 1981).

Rodents of this suborder are embryologically interesting, not least because of their extraordinarily long gestation periods relative to other rodents. They are also characterized by having precocious young and a moderate to small litter size when compared to myomorph rodents (Weir 1974). The capybara (*Hydrochoerus hydrochaeris*), for example, with an adult weight of about 40–55 kg, has a gestation of 147–156 days to produce 4–8 young, while rock cavies weighing 0.7–0.8 kg carry a litter of 1–2 young for 70 days (Miglino et al. 2004).

M.A. Miglino (✉) • C. Kanashiro
Universidade de São Paulo, Av. Prof. Dr. Orlando Marques de Paiva, 87, Cidade Universitária, São Paulo, São Paulo 05508-270, Brazil
e-mail: miglino@usp.br; claudiakanashiro@uol.com.br

T.C. dos Santos
Departamento de Zootecnia, Universidade Estadual de Maringá Av. Colombo 5790, Maringá 87020-900, Paraná, Brazil
e-mail: tcsantos@uem.br

R.H. dos Santos Ferraz
Universidade Federal de Mato Grosso, Av. Fernando Correa da Costa, s/n Bairro Coxipó, Cuiabá, Mato Grosso 78060-600, Brazil
e-mail: rhsferraz@ufmt.br

J.R. Moreira et al. (eds.), *Capybara: Biology, Use and Conservation of an Exceptional Neotropical Species*, DOI 10.1007/978-1-4614-4000-0_7,

In general, the structure of the hystricomorph female genital organs is similar to that of other rodents, but the characteristic subplacenta is unique to this group (Miglino et al. 2004). Indeed, very little was known about hystricomorph placentation before studies were published by Miglino and her group (Miglino et al. 2002, 2004). In this chapter, we describe the morphophysiology of the sexual organs of female capybaras, including a detailed description of the form and function of the capybara placenta. We also discuss the female capybara's reproductive potential and describe the development of the embryo in the first weeks of gestation.

7.2 Reproductive Morphology of Female Capybaras

The female sexual organs of capybaras comprise a pair of ovaries; a pair of uterine tubes (oviducts); a bipartite uterus with a Y-shaped cervix; a long, wide vagina; and a vulva (Fig. 7.1).

7.2.1 Ovaries

Capybara ovaries are covered by the well-developed ovarian bursa. This is composed of the mesosalpinx and the mesovarium; the opening of the ovarian bursa, the ostium, is positioned on the medial side near the uterine horns. The ostium is two-thirds of the length of the ovary. A deposit of fatty tissue is observed in the mesosalpinx (Fig. 7.2).

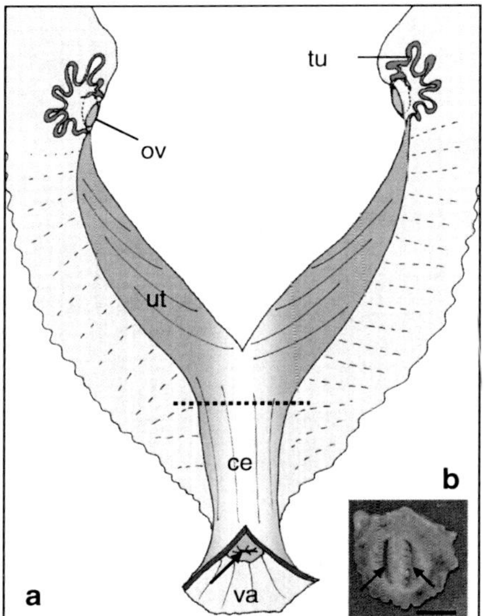

Fig. 7.1 (**a**) A schematic diagram of the genital organs of the adult capybara. Ovaries (*ov*), oviducts (*tu*), uterus composed of uterine horns (*ut*) and uterine cervix (*ce*), and vagina (*va*) opened dorsally to demonstrate the vaginal canal of the cervix (*arrow*). (**b**) A photograph of a cross section of the cranial portion of the uterine cervix (*dotted line*) showing a septum in the cranial-most area, which divides the cervix into two cervical canals (*arrows*). Bar: 1 cm

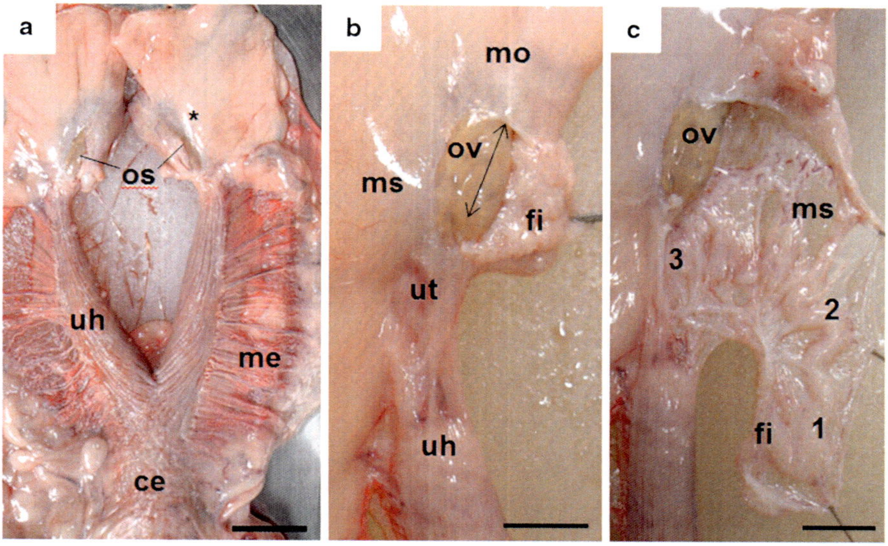

Fig. 7.2 Photographs of the genital organs of a female capybara. (**a**) The uterine horns (*ut*) are cranially joined to the uterine tubes and the ovaries (*ov*) contained in the ovarian bursa (*), which is surrounded by fatty tissue, and caudally to the cervix (*ce*). Laterally, the uterus is suspended by the mesometrium (*me*). (**b**) Detail of the ovary (*ov*) observed through the opening (↔) of the ovarian bursa (*in natura*), the ostium (*os*) and the mesos forming the bursa, the mesosalpinx (*ms*), and mesovarium (*mo*). (**c**) The uterine tube (*tu*) exposed after dissection of the mesosalpinx evidencing the fimbriae (*fi*) in the infundibulum (*1*), the ampoule (*2*), and the isthmus (*3*), and copious adjacent fatty tissue. Bars: (**a**) 3 cm, (**b**) and (**c**) 1.5 cm

In young females, the ovaries are smooth and ovoid (López-Barbella 1993), but in adult females, the size and surface become irregular (Ojasti 1973). The dimensions of the ovaries vary according to reproductive state. In young animals, the weight, length, and width of the ovaries increase, respectively, from 0.1 g, 1.0 cm, and 0.5 cm at 2 months of age, to 0.6 g, 1.8 cm, and 1.2 cm at 10 months of age (López-Barbella 1993). In adult females, measurements reach 2.9 g, 1.6 cm, and 0.9 cm and 2.8 g, 1.9 cm, and 1.0 cm for the weight, length, and width of the right and left ovary, respectively (Ojasti 1973), or 2.8 g, 1.5 cm, and 0.5 cm (Silva and Perdomo 1983).

The ovaries of 2-month-old females have primary follicles in the cortex but scarce secondary follicles. At 4 months, tertiary follicles appear, and at 6 months there is hypertrophy of follicles, and the granulosa cells are visible. By 8 months of age, the ovaries contain follicles in the cortical region, and by 12 months, folliculogenesis (the process of follicle maturation) begins and corpora lutea of different sizes, which characterize puberty, can be seen (López-Barbella 1993). Females at reproductive age have ovaries composed of a covering layer of simple cuboidal epithelium, a cortical region containing follicles and corpora lutea at different stages of development, and a medullar region composed of blood vessels, lymphatic vessels, and nerves (Fig. 7.3; Silva and Perdomo 1983).

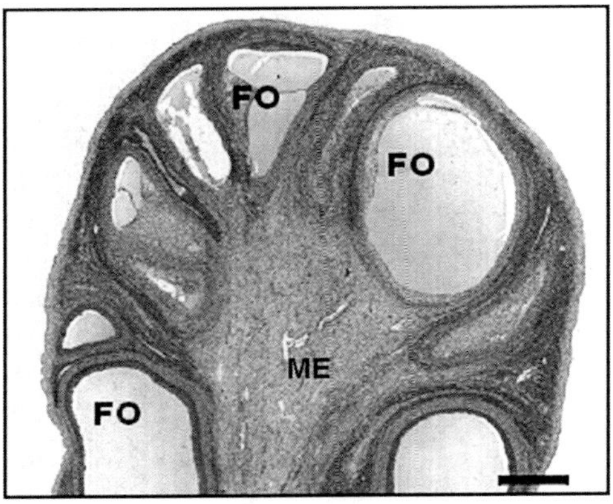

Fig. 7.3 Photomicrographs of the cross section in an ovary from an adult capybara. Note the cortical region with follicles (*FO*) and medullar region (*ME*). Hematoxylin and Eosin (*HE*). Bar 500 μm

7.2.2 Uterine Tubes

The capybara uterine tube is a tortuous organ, bilateral, located between the ovary and the apex of the uterine horns, reaching up to 17 cm in length (Ojasti 1973). It runs along the mesosalpinx and opens cranially on the medial surface of the ovary, the area where slender fimbriae are found next to the opening of the ovarian bursa. These fimbriae are the mucosal folds of the first segment of the uterine tube, the infundibulum, which is followed caudally by the ampulla and isthmus (Fig. 7.2c).

Histologically, the uterine tube is a folded mucosa that consists of a simple cylindrical epithelium with ciliated cells and a clear mucosal membrane. The region of the submucosa is nongranular and comprises loose connective tissue, blood vessels, and lymphatic vessels. The muscular coat has two layers of smooth muscle, which are externally covered by the serosa coat (Silva and Perdomo 1983).

7.2.3 The Uterus

The uterus of adult capybaras has two straight uterine horns, which externally resemble a Y-shaped form (Fig. 7.1; Kanashiro 2006). The uterine horns join at their caudal ends, forming the double uterine body, divided internally by a well-vascularized fibrous septum (Silva and Perdomo 1983). Because of this conformation, the uterus of the capybara includes two uterine bodies, one for each uterine horn; both end independently in the cervix. This division is also observed in the cranial portion of the cervix, where a dorsal-ventral septum is present, dividing the cavity into two cervical channels. Caudally, the septum does not reach the dorsal wall of the cervix, and only one external uterine ostium is observed in the vagina (Fig. 7.1). This conformation gives the capybara a bipartite uterus (Ojasti 1973). The uterine horns are supported in the abdominal cavity by the mesometrium and vascularized by branches of the uterine artery, which branch into a "half-feathered" shape before penetrating the uterine wall. The weight of the uterus in nulliparous and pluriparous females

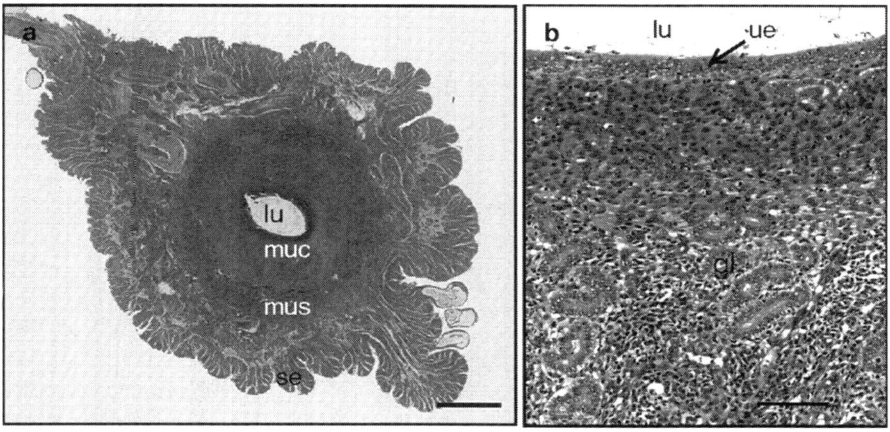

Fig. 7.4 (**a**) Overview of the transversal sections of the uterine horn in a nonpregnant capybara. Note the layers of the uterus: the tunica mucosa (*muc*), the tunica muscularis (*mus*), and the tunica serosa (*se*). (**b**) Detail of the uterine epithelium (*ue*) and glands (*gl*) of the mucosa. Lumen (*lu*). Hematoxylin and Eosin (*HE*). Bar: (**a**) 1 cm; (**b**) 100 μm

varies from 6.5 to 57.0 g and 40 to 765 g, respectively (Ojasti 1973). Longitudinal grooves can be seen on the wall of the uterine horns in nonpregnant females. In the early stage of pregnancy, rounded areas are visible in the uterine horns. These areas are a set of fetal membranes and the embryo or fetus, and they develop into large dilations as pregnancy advances.

Histologically, as in every mammal, the uterus of the capybara is composed of three layers: the tunica mucosa, the tunica muscularis, and the tunica serosa (Fig. 7.4). The tunica mucosa or endometrium comprises a pseudostratified epithelium and the lamina propria, which is composed of loose connective tissue and simple tubular glands. The tunica muscularis or myometrium is thicker and contains three distinct layers: the innermost stratum submucosum; the middle stratum vasculare, which is thicker and more vascularized; and the outermost stratum serosum, with its longitudinally arranged muscle bundles. The tunica serosa covers the outside of the uterine wall and is composed of a simple squamous epithelium resting on a strip of loose connective tissue (Kanashiro 2006). In the body of the uterus, the cylindrical stratified epithelium possesses periodic acid Schiff (PAS) positive mucus-secreting cells, suggesting the activity of these cells in cervical secretion. In the cervix, the mucosa has a stratified epithelium that extends through the vagina (Silva and Perdomo 1983).

7.2.4 The Vagina and Vulva

The vagina (Fig. 7.1) is a tubular muscle-membranous organ located between the uterine cervix and the vulva. Its surface contains a deep median groove in which the external urethral orifice opens. In adult females, the vagina is wide, has longitudinal folds, and can reach 14–15 cm in length (Ojasti 1973; Gonzalez-Jimenez 1995). The vaginal mucosa is lined by a stratified epithelium with cuboidal basal cells and a surface varying from polyhedral or flat to cylindrical with PAS-positive cells (Silva and Perdomo 1983). Intraepithelial adenomers can be found in the epithelium. The submucosal connective tissue of the vagina has elastic fibers, many blood and lymphatic vessels, nerves, and lymph nodes.

After copulation, a clot of yellowish seminal fluid called the vaginal plug is formed within the vagina. This plug, 41 mm long and 17 mm wide, remains in the vagina for some time and is then eliminated (Ojasti 1973).

The vulva is the most caudal portion of the female genital organ. It connects the vagina with the external environment; therefore, besides being the female copulatory organ, it is also the birth canal. It measures 6 cm and is bounded laterally by the perianal glands (Gonzalez-Jimenez 1995). In capybaras, it is not possible to distinguish the labia, and researchers disagree as to whether capybaras have a clitoris; Ojasti (1973) states that capybaras do not, whereas Costa et al. (2002) have described one. The latter authors suggest that the "membrane of occlusion" described by Clark and Oflert (1986) is in fact the hymen. Externally, the skin of the vulva is pigmented, while internally, the mucosa is composed of a stratified squamous epithelium (Silva and Perdomo 1983).

7.3 The Estrus Cycle

The age of sexual maturity in capybaras varies, apparently influenced by local climate and resource availability. Zoo-bred animals reach sexual maturity at 15 months (Zara 1973; Soiron 1993), while wild-bred animals seem to reach sexual maturity between 6 and 12 months (López-Barbella 1984).

Puberty takes place in female capybaras between 10 and 12 months. During puberty, the uterus doubles in weight, and the ovaries undergo a high degree of follicular differentiation, with the appearance of the first luteal bodies accompanied by a significant increase in the concentration of progestogens in the blood. At 4 months of age, the concentration of progestogens fluctuates between 3.50 and 4.30 nmol/l, increasing abruptly at 6 months to 9.54 nmol/l. At around 1 year of age, progestogen concentrations reach 13.36 nmol/l (López-Barbella 1993).

The estrus cycle of capybaras was studied in captivity by López-Barbella (1982) using exfoliative cytology, changes in body temperature and blood concentrations of luteinizing hormones and progestogens. The average duration of the estrus cycle was 7.5 ± 1.2 days; ovulation was spontaneous and the receptive period the last 8 h. Blood progesterone levels varied from 0.8 to 3.5 mg/ml on the third day of estrus, fell drastically to 0.9 mg/ml on the fourth day, and increased again during metestrus. On the third day of estrus, body temperature, normally 36.0–36.2°C, rose to 36.6°C, probably associated with the ovulation process.

In the first phases of estrus, the capybara exhibits vaginal smears with parabasal dark cells and some polymorphic cells, followed by a mucous secretion with

Fig. 7.5 Capybaras mating in water (Photo by Rita Barreto)

few leucocytes (López-Barbella 1982). In the next stages (24–30 h) of mucous secretion, the vaginal epithelium peels off keratinized cells, which may have pyknotic nuclei or no nuclei. Subsequently, these keratinized cells disappear from the vaginal smear, and there is a considerable increase in the keratin content in polymorphic cells for approximately 2–5 days. When the estrus cycle restarts, parabasal cells appear again in the vaginal cytology of the capybara.

Capybaras produce no external physical signs of being in estrus (Alho et al. 1989). During estrus, the female becomes receptive to the male, and copulation almost always takes place in water (Fig. 7.5). The female moves in and out of the water, always followed by the male, until she demonstrates receptivity (Ojasti 1973; Zara 1973; Alho et al. 1987b; Herrera and Macdonald 1993). The male initiates courtship by scent marking and sniffing the female's external sexual organ (López-Barbella 1984). Subsequently, the pair circle one another for around 5–10 min, culminating in the first intention of coition, and this is when the female exposes the vulva. The male makes about six attempts before introducing his penis and ultimately ejaculates for about 3 s.

7.4 Mating and Gestation

Capybaras may reproduce at any time of the year and there is no specific reproductive season (Alho et al. 1987a), although the frequency of mating tends to be more intense at the beginning of the wet season (April-May in Venezuela,

October-November in Brazil; Mones and Ojasti 1986). Each dominant male secures significantly more matings than each subordinate, but subordinate males, as a class, are responsible for more matings than each dominant male (Herrera and Macdonald 1993).

Capybaras have the longest period of gestation known amongst rodents. In a study by López-Barbella (1987), in which coitus was observed, and the presence of spermatozoid confirmed inside the vagina, the average duration of gestation was 150.6 ± 2.8 days. In Venezuela, delivery is concentrated mainly in two periods of the year: one at the beginning of the rainy season and another at the beginning of the dry period (Parra et al. 1978). This allows capybaras to have two pregnancies per year (Mones and Ojasti 1986).

In the wild, females tend to breed when they reach a body weight of 30–40 kg, which usually occurs at 1.5–2 years of age (Ojasti 1973). In general, capybaras give birth to between one and seven young per litter (Ojasti 1973; Zara 1973). Studies carried out in Venezuela show that capybaras in captivity can give birth to 4.4 ± 1.3 (López-Barbella 1987) or 3.7 (Parra et al. 1978) young per litter. Older females usually have larger litters (Soiron 1993). Several studies show that, in the wild, the litters on average contain four young (Ojasti 1973; Herrera and Macdonald 1987; Moreira and Macdonald 1996). Studies in Argentina with different captive breeding systems have noted that capybaras can give birth twice per year, with 6.5 young per year (Alvarez and Kravetz 2006) or 3.4 ± 0.2 young per year (Cueto 1999).

There are few reports of litter sizes in Brazil; those on captive-bred capybaras describe values between 3.28 and 3.6 young per birth (Lavorenti 1989; Silva Neto et al. 1990; Nogueira 1997). Captive capybaras that were monitored for 5 years in Piracicaba (São Paulo State) showed an average of six young per year (Lavorenti 1989). For wild capybaras on Marajó Island (Pará State), the fertility rate was found to be 4.2 young per year with an incidence of gestation of one litter per female per year (Moreira and Macdonald 1996).

7.5 The Placenta

The general characteristics of the placenta of capybaras are shared with some other hystricomorph rodents (Miglino et al. 2002), such as pacas (Bonatelli et al. 2005), agoutis (Rodrigues et al. 2006), and rock cavies (Oliveira et al. 2006). The capybara placenta is located on the mesometrial face of the uterus, discoid in shape (López-Barbella 1981), prominently lobulated (Soiron 1993), labyrinthine, and hemomonochorial (Miglino et al. 2002); within the labyrinth its blood flow runs countercurrent (Miglino et al. 2004).

The first visible sign of gestation in capybaras is the presence of conceptuses (the fertilized embryo and associated parts), which at this stage are spherical and whitish bodies around 8 mm in diameter, joined to the endometrium and separated from each other by the transversal folds of the uterine wall Ojasti (1973). As pregnancy progresses, these masses become conical on the mesometrial face.

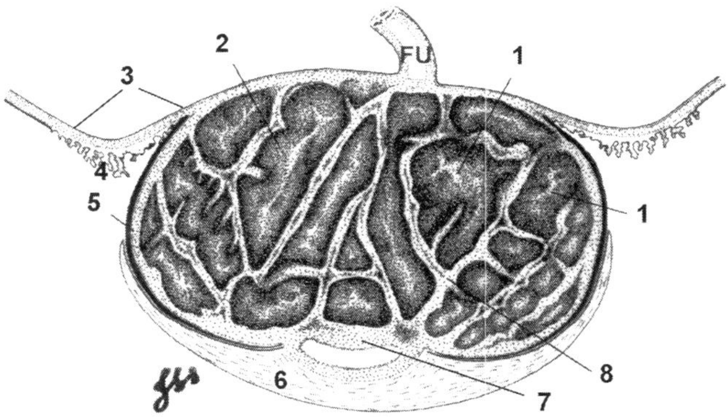

Fig. 7.6 A schematic diagram of the median sagittal section of the placenta of the capybara. The center of the lobe (*1*), labyrinth (*2*), amnion (*3*), visceral vitelline placenta (*4*), parietal vitelline placenta (*5*), decidua in the region of the placenta peduncle (*6*), subplacenta (*7*), interlobar area (*8*), and umbilical cord (FU)

7.5.1 Description of the Placenta and the Decidua

The *placenta* is a transitory organ responsible for the exchange of nutrients between the mother and the fetus. In capybaras, two types of placenta (Fig. 7.6) are found: the chorioallantoic or principal placenta (Fig. 7.7c) and the vitelline placenta; both are associated with the decidua (Fig. 7.6). Between the principal placenta and the decidua there is the subplacenta, which is as yet scantily described and whose true function is poorly understood.

Prior to birth, the *principal placenta* can reach a diameter of 7–8 cm, a thickness of 3–3.5 cm, and a weight of 100–150 g (Soiron 1993). In studies by Ferraz (2001), seven placentas were examined, being on average 7.3 cm long, 5.6 cm wide, and 3.7 cm thick, and weighing 172.5 g. The shape of the principal placenta is discoid; however, at the beginning of gestation, it takes on the shape of a chalice, which grows considerably throughout gestation (Fig. 7.7).

The principal placenta is made up of *lobes*, which can be visualized macroscopically and are mainly composed of a labyrinth separated by interlobe areas (Miglino et al. 2002). Each lobe is composed of two distinct regions: the center of the lobe and the area of the labyrinth (Fig. 7.8a). The principal placenta is also known as the labyrinthine. In the center of the lobe, thick blood vessels can be seen immersed in the mesenchyme. In the placenta, the labyrinth consists of approximately 150 parallel cones, each one approximately 3 cm long and 4–6 mm wide (sSoiron MI 1993). It is this labyrinth that allows exchange of nutrients between the mother and the fetus (Miglino et al. 2002). Within the labyrinth, the fetal blood, flowing through the capillary network, is separated from the maternal blood only by the wall of the capillary itself and by the trophoblast. This single layer of trophoblastic cells between maternal and fetal blood borders the irregular channels along which the maternal blood

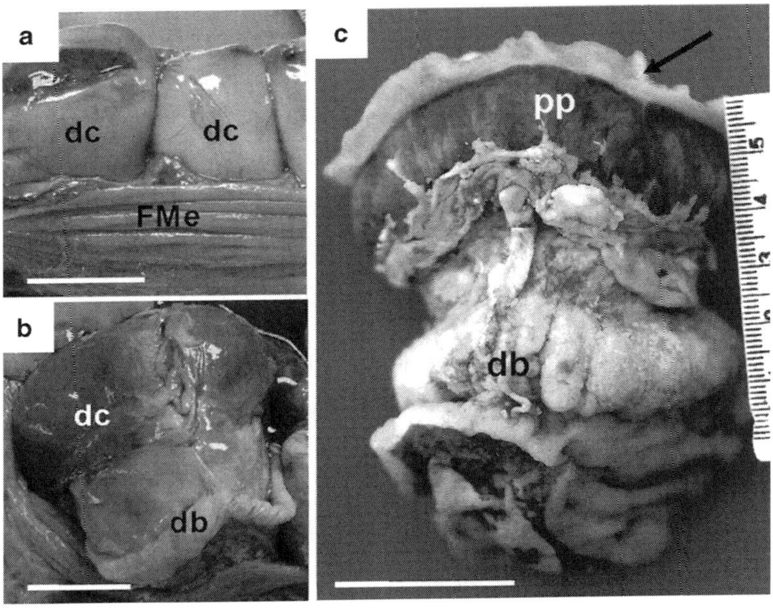

Fig. 7.7 Photographs of fresh placentas from capybaras in the first third (**a, b**) and middle of pregnancy (**c**). (**a**) A side view of the uterine horn after opening it to expose the placenta inserted in the mesometrial side of the uterus (*FMe*) still covered by thick capsular decidua (*dc*). (**b**) Note the significant change in the capsular decidua (*dc*) that becomes transparent as gestation progresses, and the basal decidua (*db*). (**c**) From the middle to the end of pregnancy, the principal placenta (*pp*) increases considerably and the basal decidua shrinks. The margin of insertion of the vitelline placenta (*arrow*) in the uterus. Bars: 3 cm

flows, and is the characteristic that gives the principal placenta its denomination "hemomonochorial" (Soiron 1993; Ferraz 2001). The *interlobe* region, composed of trophoblastic cells, is less organized (Fig. 7.8a). At the beginning of gestation, the interlobe is predominant, occupying most of the principal placenta. With the growth of the lobes and the consequent development of the labyrinth in each lobe, the interlobe areas become reduced and limited to the regions between the lobes.

The *junction zone* is positioned around the subplacenta and represents the transition between it and the decidua. This zone is comprised of a large band of amorphous, eosinophil, PAS-positive tissue containing scattered or grouped trophoblastic cells, the latter characterized by their clear cytoplasm and numerous round nuclei.

The *vitelline placenta* of rodents is inverted and covers the lateral surface of the principal placenta (parietal vitelline placenta) or folds over the amnion (visceral vitelline placenta; Fig. 7.6). In the first third of gestation, the *visceral vitelline* placenta is a fine and transparent membrane with clear blood vessels distributed over its entire surface. This membrane is inserted on the fetal side of the principal placenta, projecting out completely over the embryo and the amnion (Kanashiro 2006). At the end of gestation, next to the principal placenta, the visceral vitelline

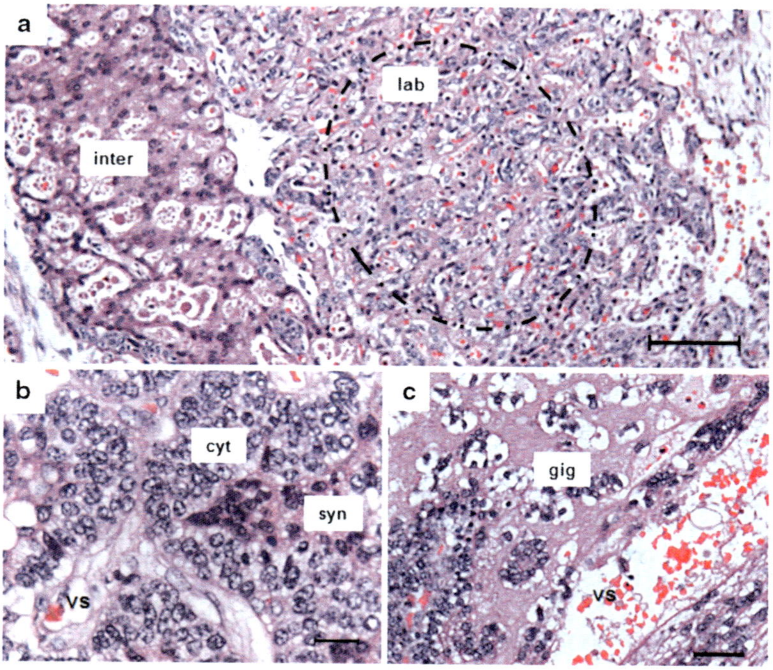

Fig. 7.8 Photomicrographs of the labyrinth of the main placenta in an adult capybara (early pregnancy). (**a**) Details of the lobe where the labyrinth (*lab*) and interlobular (*inter*) regions are observed. (**b**) Details of the subplacenta composed of syncytiotrophoblasts (*sin*) and cytotrophoblasts (*cit*) supported by the vascularized mesenchyme (*vs*). (**c**) The area of the basal decidua and giant cells (*gig*) and vessels (*vs*) probably invaded by maternal syncytiotrophoblasts. Inclusion in the paraplast, Hematoxylin and Eosin (*HE*). Bars: (**a**) 200 μm; (**b**) 40 μm; and (**c**) 50 μm

placenta develops many finger-like projections, giving it, macroscopically, a "plush-like" aspect (Ferraz 2001). Histologically, the folds of the visceral vitelinic placenta are composed of a pseudostratified endodermic epithelium, supported on a thin mesenchymal layer with many collagen fibers in which blood vessels are distributed.

The region of the vitelline placenta that covers the periphery of the principal placenta is the *parietal vitelline* placenta, composed of a prismatic pseudostratified epithelium that can exhibit branched folds, especially close to folds of the decidua. The parietal vitellinic placenta is positioned on a band of connective tissue that constitutes Reichert's membrane.

Within the vitelline placenta, the *amnion,* a fine and transparent membrane, wraps itself around the embryo or fetus completely. The amnion is formed by a squamous epithelium, with cells containing flattened nuclei arranged on a loose mesenchyme with collagen fibers (Kanashiro 2006).

The *decidua* is formed from maternal tissue. In the early gestation stages, the embryo and its associated parts are completely enclosed by the *capsular decidua*, which is thick and whitish; however, it becomes delicate and transparent as gestation progresses, disappearing late in gestation as the fetus develops and the volume of liquid increases

(Fig. 7.7a, b). The basal decidua forms the connection between the placenta and the endometrium of the uterus. This area is easily removed from the remainder of the uterus and is connected to it only via blood vessels. In the early stages of gestation, the basal decidua corresponds to a large portion of the placenta (Kanashiro 2006), so that in the placenta of pregnant capybaras, the remains of decidual tissue are observed to form a peduncle between the placenta and the womb (Miglino et al. 2002).

Another prominent region in the placenta of hystricomorph rodents is the subplacenta, whose function is still being studied. The subplacenta occupies the central area between the principal placenta and the basal decidua (Fig. 7.6); microscopically, it is distinguished from the adjacent regions by its whitish coloration. Weighing approximately 1.5 g, it is surrounded by a necrotic zone and by uteroplacental vessels (Soiron 1993). By the end of gestation, the subplacenta makes up part of the peduncle of the placenta. It is divided into small irregular lobes, covered by cytotrophoblasts and syncytiotrophoblasts (Fig. 7.8b). The cytotrophoblast forms a layer of two or more cells supported by mesodermic lamellas, which contain fetal blood vessels (Miglino et al. 2002). Although the syncytiotrophoblasts contain closely packed nuclei with poorly defined aspects and cellular limits, they also contain basophile cytoplasm and highly PAS-positive cytoplasm (Kanashiro 2006).

7.5.2 *Placental Vascularization*

Microcirculation within capybara placenta is similar to that of other hystricomorph rodents such as the paca, the agouti, the rock cavy, and the guinea pig (Miglino et al. 2004). The fetal arteries pass through the interlobular areas, becoming capillaries in the region of the labyrinth and draining into the fetal veins at the center of the lobes. The maternal blood penetrates the placenta through spiral arteries, which pass through the perimeter of the subplacenta toward the main placenta (Fig. 7.8b, c). At the center of the lobes, the maternal arteries branch out, and the blood circulates in the region of the labyrinth delimited by the trophoblast (maternal cavity) toward the periphery. Here, these cavities drain into others located in the interlobular areas, which flow together to form the maternal veins. The fetal capillaries and the maternal cavities run in parallel and thus form the morphological basis for the countercurrent exchange mechanism (Ferraz 2001; Miglino et al. 2004). The region of the subplacenta is vascularized by a single fetal artery, and the capillaries in the mesenchyme harbor the trophoblast (Miglino et al. 2004).

7.6 Embryonic Development

Information on capybara embryonic and fetal development is scarce. The morphological characteristics of the blastocyst, details of the implantation period, and the mechanisms for pregnancy recognition are lacking, especially for the first 3 weeks of gestation.

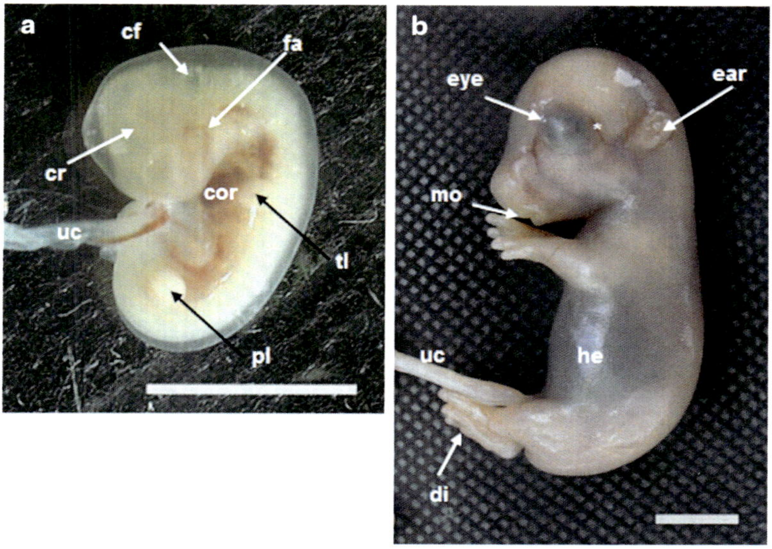

Fig. 7.9 Photographs of the embryo and fetus of the capybara in the first third of pregnancy after fixation in formalin. (**a**) An embryo (12.4 mm) showing the cervical flexure (*fc*), the pharyngeal arches (*af*), the cardiac region (*cor*), the buds of the thoracic limbs (*mt*) and the pelvic limbs (*mp*), the cephalic region (*rc*), and the umbilical cord (*cd*) encased by the amnion (*am*). (**b**) A fetus (58 mm) with external morphology in which we can observe the eyelid covering the pigmented eye (*ol*), the pinna of the external ear (*or*), the liver area (*he*), a developed face with an evident external nose and mouth (*bo*), developed limbs with individual digits (*di*), dermal vessels seen by transparency (*), and the umbilical cord (*cd*). Bars: 10 mm

At 30 days of gestation, capybara embryos are spherical, 2.2 cm long, and weigh 4.4 g (López-Barbella 1987). At 60 days of gestation, the embryos are 4.5 cm long and weigh 7.6 g, and at 90 days, they measure 15 cm, weigh 252.1 g, and have visible fur on the face and back. By 100 days, fetuses measure 17 cm and weigh 425.4 g. Hair on other parts of the body appears at 120 days of gestation, when the fetuses measure 19.5 cm and weigh approximately 800 g. By 140 days (close to birth), pelage is complete, the hooves are no longer membranous, and fetuses are about 21.5 cm long and weigh approximately 1 kg (Table 1; López-Barbella 1987).

Detailed descriptions of embryonic and fetal development in relation to gestational age are not available, but it is possible to relate morphological development to various measurements of the size or weight of the embryo or fetus, producing correlations that are useful for comparative studies. Embryos with an average crown-rump length of 11.5 mm possess cervical flexure, pharyngeal arches, a cardiac region, buds of the thoracic and pelvic members, a cephalic region, and an umbilical cord (Fig. 7.9a; Kanashiro 2006). Fetuses with a crown-rump length of 55–58 mm possess pigmentation of the eye but no eyelids, the pinna, and the openings of the external ear and the external nose. The mouth, the ribs, the liver area, and buds of the thoracic and pelvic limbs with formed digits can also be seen at this stage (Fig. 7.9b). Fetuses at 24.9 cm in length along the cerebrospinal-coccygeal axis have fur over the entire body with keratinized nails (Ferraz 2001). The birth weight of capybara fetuses is around 1,300–2,010 g, and the sexual ratio is 1:1 (Ojasti 1973).

Not all capybara embryos reach term. Degenerated embryos, recognized as a tissue mass contained within fetal membranes but of smaller size than those of normal placentas at the same stage, can be found in the uterus in the early stages of gestation (Kanashiro 2006). "Abnormal embryos" may be accompanied by "normal embryos" and this phenomenon can also be observed in more advanced phases of gestation (Rosa H.S. Ferraz personal observation). Embryonic mortality rates are estimated to be 16–17% (Ojasti 1973; Moreira et al. 2001). These decidual capsules without live embryos are retained in the uterus, usually up to the time of delivery, probably because they remain the same size throughout the pregnancy. The weight of these clusters in advanced gestations suggests that there may be an accumulation of decidual tissue in the capsular decidua during pregnancy.

7.7 Final Remarks

Many questions still remain regarding the reproductive behavior of capybaras. Capybara production, management, and conservation will certainly benefit from clarifying the adaptive significance of some of the histological/physiological features of the species. To choose just one example, identifying the evolutionary reason behind the 17% embryonic loss during capybara gestation could have a remarkable impact on capybara use and conservation. The economic importance of the capybara, given its productive potential, and its ecological importance as a neotropical species, makes it an excellent model for further studies.

Acknowledgment The authors thank CNPq, FAPESP, and CAPES for financial support.

References

Alho CJR, Campos ZMS, Gonçalves HC (1987a) Ecologia da capivara (*Hydrochaeris hydrochaeris, Rodentia*) do pantanal: I habitats, densidades e tamanho de grupo. Rev Bras Biol 47:87–97

Alho CJR, Campo ZMS, Gonçalves HC (1987b) Ecologia da capivara (*Hydrochaeris hydrochaeris, Rodentia*) do pantanal: II atividade, sazonalidade, uso do espaço e manejo. Rev Bras Biol 47(2):99–110

Alho CJR, Campos ZM, Gonçalves HC (1989) Ecology, social behavior, and management of the capybara (*Hydrochoerus hydrochaeris*) in the Pantanal of Brazil. In: Redford KH, Eisenberg EF (eds) Advances in neotropical mammalogy. Sanches Crane Press, Gainesville, pp 163–194

Alvarez MR, Kravetz FO (2006) Reproductive performance of capybaras (*Hydrochoerus hydrochaeris*) in captivity under different management systems in Argentina. Anim Res 55:153–164
Bonatelli M, Carter AM, Machado MRF, Oliveira MF, Lima MC, Miglino MA (2005) Placentation in the paca (*Agouti paca* L). Reprod Biol Endocrinol 28:3–9
Clark JD, Oflert ED (1986) Rodents. In: Fowler ME (ed) Zoo and wild animal medicine. Saunders, Philadelphia, pp 727–747
Costa DS, Paula TAR, Fonseca CC, Neves MTD (2002) Reprodução de capivaras. Arquivos de Ciências Veterinárias e Zoologia da UNIPAR 5(1):111–118
Cueto GR (1999) Biología reproductiva y crescimiento del carpincho (*Hydrochoerus hydrochaeris*) em cautiverio: Una interpretación da las estrategias poblacionales, Ph.D. thesis, Universidad de Buenos Aires, Buenos Aires
Ferraz RH (2001) Estudo macro e microscópico da placenta de termo de capivara (*Hydrochaeris hydrochaeris* – Carleton MD 1984). 75 f. Doctoral thesis, Faculdade de Medicina Veterinária e Zootecnia, Universidade de São Paulo, São Paulo
Herrera EA, Macdonald D (1987) Group stability and the structure of a capybara population. Symp Zoo Soc Lond 58:115–130
Herrera EA, Macdonald DW (1993) Aggression, dominance and mating success among capybara males (*Hydrochaeris hydrochaeris*). Behav Ecol 4(2):114–119
González-Jiménez E (1995) El Capibara (*Hydrochoerus hydrochaeris*); estado actual de su producción, vol 122, Estudio FAO Producción y Sanidad Animal. FAO, Roma
Kanashiro C (2006) Estudo morfológico da placenta de capivara (*Hydrochaeris hydrochaeris*): terço inicial, 105 f. M.Sc. dissertation, Faculdade de Medicina Veterinária e Zootecnia, Universidade de São Paulo, São Paulo
Lavorenti A (1989) Domestication and potential for genetic improvement of capybara. Revista Brasileira de Genética 12(3):137–144
López-Barbella S (1981) Caracterización preliminar de la gestación en chiguire (*Hydrochoerus hydrochaeris*). Boletín del Informe Anual del IPA (Instituto de Producción Animal), Facultad de Agronomía, Universidad Central de Venezuela, Maracay, pp 18, 19
López-barbella S (1982) Determinación del ciclo estral en chiguires (*Hydrochoerus hydrochaeris*). Acta Cient Venez 33(6):497–501
López-Barbella S (1984) Una contribución al conocimiento de la reproducción del chiguire (*Hydrochoerus hydrochaeris*). Boletin del Informe Anual del IPA (Instituto de Producción Animal), Facultad de Agronomía, Universidad Central de Venezuela, Maracay, pp 109–117
López-barbella S (1987) Consideraciones generales sobre la gestación del chiguire (*Hydrochoerus hydrochaeris*). Acta Cient Venez 38:84–89
López-barbella S (1993) Puberdad en hembras chiguires (*Hydrochoerus hydrochaeris*). Revista de la Facultad de Agronomia. Venezuela 19:121–127
Luckett WP, Mossman HW (1981) Development and phylogenetic significance of the fetal membranes and placenta of the African hystricognathous rodents *Bathyergus* and *Hystrix*. Am J Anat 16(3):265–285
Miglino MA, Carter AM, Dos Santos Ferraz RH, Fernandes Machado MR (2002) Placentation in the capybara (*Hydrochaerus hydrochaeris*), agouti (*Dasyprocta aguti*) and paca (*Agouti paca*). Placenta 23:416–428
Miglino MA, Carter AM, Ambrosio CE, Bonatelli M, de Oliveira MF, Dos Santos Ferraz RH, Rodrigues RF, Santos TC (2004) Vascular organization of the hystricomorph placenta: a comparative study in the agouti, capybara, guinea pig, paca and rock cavy. Placenta 25:438–448
Mones A, Ojasti J (1986) *Hydrochoerus hydrochaeris*. Mammalian species, vol 264. The American Society of Mammalogists, U.S.A, pp 1–7
Moreira JR, Macdonald DW (1996) Capybara use and conservation in South America. In: Taylor VJ, Dunstone N (eds) The exploitation of mammal populations. Chapman & Hall, London, pp 88–101
Moreira JR, Macdonald DW, Clarke JR (2001) Alguns aspectos comportamentais da reprodução da capivara. Revista Brasileira de Reprodução Animal 25:120–122

Nogueira SS (1997) Manejo reprodutivo da capivara (*Hydrochoerus hydrochaeris hydrochaeris* L. 1766) sob sistema intensivo de criação. Ph.D. thesis, Universidade de São Paulo, São Paulo
Ojasti J (1973) Estudio biológico del chigüire o capibara. Fondo Nacional de Investigaciones Agropecuarias, Caracas, p 262
Oliveira MF, Carter AM, Bonatelli M, Ambrosio CE, Miglino MA (2006) Placentation in the Rock Cavy, *Kerodon rupestris* (Wied). Placenta 27:87–97
Parra R, Escobar A, Gonzalez JE (1978) El chiguire su potencial biologico y su cria en confinamiento. Boletin del Informe Anual del IPA (Instituto de Producción Animal), Facultad de Agronomía, Universidad Central de Venezuela, Maracay, pp 83–94
Rodrigues RF, Carter AM, Ambrósio CE, Santos TC, Miglino MA (2006) The subplacenta of the red-rumped agouti (*Dasyprocta leporina* L). Reprod Biol Endocrinol 4:1–8
Silva CL, Perdomo F (1983) Algunos aspectos anatomicos e histologicos del sistema genital femenino del chigüire (*Hidrochoerus hidrochaeris*). Revista de la Facultad de Ciencias Veterinarias 30(1–8):89–97
Silva Neto PB, Lavorenti A, Almeida AF, Marchi AP, Fernandes MB (1990) Parâmetros reprodutivos da capivara (*Hydrochoerus hydrochaeris hydrochaeris*) em cativeiro. In: Reunião Anual da Sociedade Brasileira de Zootecnia, vol 27. Campinas, Annais… FEALQ, Piracicaba, p 457
Soiron MI (1993) Das Südamerikanische wasserschwein (*Hydrochoerus hydrochaeris*, L. 1766). Physiologisch-anatomische und klinische untersuchungen unter dem aspekt der versuchstierkundlichen eignung. Aachen, Giessen, 133 f. Ph.D. thesis, Fachbereich Veterinärmedizin, Justus-Liebig-Universität Giessen, Germany
Weir BJ (1974) Reproductive characteristics of hystricomorph rodents. Symp Zoolog Soc Lond 34:265–301
Zara JL (1973) Breeding and husbandry of the capybara (*Hydrochoerus hydrochaeris*). Int Zoo Yearb 13:137–139

Chapter 8
Capybara Demographic Traits

José Roberto Moreira, Helga Wiederhecker, Katia Maria P.M.B. Ferraz, Juanita Aldana-Domínguez, Luciano M. Verdade, and David W. Macdonald

8.1 Introduction

An understanding of the structure, demography and dynamics of animal populations underpins species conservation and management. For a given population, a number of demographic parameters, which may be obtained from short- or long-term studies, are useful in this regard. These parameters explain the broad features of a species' life

J.R. Moreira (✉)
Embrapa Recursos Genéticos e Biotecnologia, Parque Estação Biológica, Final W5 Norte, Brasília, Distrito Federal 70770-900, Brazil
e-mail: alencar_moreira@yahoo.com.br

H. Wiederhecker
SHIS QI 23 CH 3, Brasília, Distrito Federal 71660-710, Brazil
e-mail: helgaw@abordo.com.br

K.M.P.M.B. Ferraz
Departamento de Ciências Florestais, Escola Superior de Agricultura "Luiz de Queiroz", Universidade de São Paulo, Av. Pádua Dias 11, Piracicaba, São Paulo 13418-900, Brazil
e-mail: katia.ferraz@usp.br

J. Aldana-Domínguez
Independent consultant on biodiversity and landscape, Calle 16, No, 1 F-12, Puerto Colombia, Atlantico, Colombia
e-mail: juanitaldana@gmail.com

L.M. Verdade
Laboratório de Ecologia Isotópica, CENA, Universidade de São Paulo, Caixa Postal 96, Piracicaba, São Paulo 13416-000, Brazil
e-mail: lmverdade@usp.br

D.W. Macdonald
Wildlife Conservation Research Unit Zoology Department, University of Oxford
The Recanati-Kaplan Centre Oxford, Tubney House, Abingdon Road, Tubney, Abingdon OX13 5QL, United Kingdom
e-mail: david.macdonald@zoo.ox.ac.uk

J.R. Moreira et al. (eds.), *Capybara: Biology, Use and Conservation of an Exceptional Neotropical Species*, DOI 10.1007/978-1-4614-4000-0_8,

cycle, and include birth, age at maturity, number of offspring, growth, reproductive investment, death and length of life (Stearns 1992). Most of these life history traits can be summarized in a "life table," an important tool in the study of populations (Sinclair et al. 2006). Demographic traits have important uses in conservation (e.g., in the reintroduction and recovery of species), agriculture (evaluating the effect of biological control or reduction of pest species), and human health (effects of epidemics or malnutrition). They are also useful for managing the sustainable use of species of economic interest.

The high natural productivity of the capybara (*Hydrochoerus hydrochaeris*) makes it a natural choice for a sustainable management program (Ojasti 1991). Knowledge and understanding of capybara demographics, and the key factors that affect them, are required for effective management. Age-specific mortality and fertility rates are particularly relevant: At what age do capybaras suffer the highest mortality rates? When do females have the highest rates of reproduction? These are examples of questions that can be answered with the study of demographic traits, and the answers are fundamental to a program for sustainable use of capybara populations.

This chapter presents a review of capybara life history traits and population dynamics collected from wild populations. Most of the data were obtained from two separate studies; one conducted by Ojasti (1973) in the Llanos, in the Venezuelan section of the Orinoco River basin (Fig. 8.1), and the other by Moreira (1995) on the floodplains of Marajó Island, at the mouth of the Amazon River, Brazil (Fig. 8.1). Both studies used data collected from live population censuses as well as from hunts. In Moreira's (1995) study on Marajó Island, capybaras were hunted throughout the year and, although hunting for capybaras in the Llanos is usually concentrated in the period before Holy Week (January to April; Ojasti 1973) also hunted animals throughout the year in his study.

8.2 Birth

Capybaras differ from most rodents in relation to several reproductive characteristics, starting with their conditions at birth. While rodents from the family Muridae (which encompass more than a quarter of the known mammal species) are born blind, furless, and needing extreme maternal care for their survival (Weir 1974), capybaras are born precocial, with open eyes; they are covered with fur, and are capable of moving within a few hours and eating solid food within a few days of birth (Ojasti 1973). And, of course, they are also born large in size. From a sample of nine neonatal wild-born capybaras, Ojasti (1973) found a mean weight of 1.5 ± 0.1 kg. This is the only birth weight recorded in the wild for capybaras, but in captivity the mean birth weight was even heavier: 2.0 ± 0.3 kg for males and 2.0 ± 0.4 kg for females (Nogueira 1997).

Fig. 8.1 Map of South America with the locations of the places referred to in the text

8.3 Growth

Ojasti (1973) defined age classes with ossification of sutures of the cranium and humerus as well as the weight of the eye lenses. Lord and Lord (1988) and later Moreira and Macdonald (1998) improved the eye lense method and made it a reliable tool for estimating age of hunted capybaras, using the following equation:

$$L = 53.9528\ x^{0.451588}$$

where L is the weight of the dried lens given in grams and x is the age of the animal in months.

Table 8.1 Age and weight ranges used for classification of capybaras into age classes (Ojasti 1973)

Age class	Class interval (months)	Class weight (kg)
Infant	0–3	1.5–7.8
Juvenile	3–12	7.8–22
Subadult	12–24	22–40
Adult	>24	>40

Using marked individuals, Ojasti (1973) modeled the correlation between weight and age and defined the age classes commonly used to describe the age structure of live capybara populations (Table 8.1). At 1 year of age, wild capybaras weigh 22 kg and at 2 years they reach 40 kg (Ojasti 1973). Their growth rate is on average 53.6 ± 1.9 g/day.

8.4 Maturity

Age and size at maturity are particularly important life history traits as they strongly influence fitness and, consequently, population model estimates of sustainable harvest rates (Sinclair et al. 2006). Maturation is the division in life between preparation and consummation (Stearns 1992).

Sperm can be found in wild male capybara testicles when they are from 20 kg of live weight (around a year old), but the sperm quantity is low until they reach 30–35 kg, at 1½ years of age (Ojasti 1973). Since some secondary sexual characteristics like nasal and anal glands are also fully developed when males are between 1½ and 2 years old, this is the age when they are considered to be sexually mature.

Female capybaras are mature and can be fertilized when between 30 and 40 kg of body weight, which happens between 1½ and 2 years of age. Ojasti (1973) noticed that every female capybara over 40 kg was either pregnant or parous.

8.5 Reproduction

As stated above, capybaras differ from most rodents in several reproductive characteristics. Most of these differences are actually phylogenetic: while Murids invest in short gestation of large litters of altricial infants, Hystricomorph rodents (the rodent suborder to which capybaras belong) have long gestations of small litters of precocial infants (Weir 1974; Kleiman et al. 1979). However, it is noteworthy that the capybara is a species whose reproductive characteristics differ from those of other closely related members of the Hystricomorph. While the correlation between litter size and body mass is negative among Hystricomorphs (Kleiman et al. 1979), capybaras have both the largest litter size (4.2) and greatest body mass (52 kg) of the suborder (Moreira and Macdonald 1997).

Table 8.2 Average litter size stratified by age of females and the average number of dead embryos observed in the uterus from capybaras of Marajó Island, Pará State, Brazil, calculated from pregnant females at 70 or more days of gestation (Moreira 1995)

Age class	Litter size	Standard error	Sample size
1	–	–	–
2	4.0	0.37	6
3	4.1	0.28	10
4	4.6	0.78	10
5	3.5	1.50	2
Mean	4.36	0.31	28
Dead embryos	0.86	0.18	28

For females of iteroparous species, a general pattern is that there will be a slow increase in litter size up to a certain age and following a reproductive peak there will be a rapid decrease (Bronson 1989). Capybaras follow the same pattern. On Marajó Island (Moreira 1995), the peak is reached by 4-year-old females with the largest litter sizes (4.6 ± 0.78 infants) followed by a decline (Table 8.2). Ojasti (1973) found a positive correlation between female cranium age and litter size ($r = 0.325$, $P < 0.01$) in the Llanos of Venezuela. In captivity, the largest litter sizes were reported in 5-year-old female capybaras (Hosken 1999). On Marajó Island, the sex ratio of capybaras at birth was 0.49 (expressed as the proportion of ; $N = 114$; Moreira 1995). Interestingly, this relatively small litter size (compared to Murids) is not due to a limitation of available teats for feeding newborn. The most frequent number of teats found per female capybara on Marajó Island was 11 ($N = 51$), followed by 10 ($N = 45$), and 12 ($N = 29$; Moreira 1995), well above the maximum number of infants in a litter (9; Ojasti 1973).

More embryos are implanted in the uterus of capybaras than become viable. Ojasti (1973) showed that 16.8% of visible implanted embryos in capybaras from the Llanos of Venezuela are lost in utero, probably during the first half of gestation. On Marajó Island, an average of 4.36 viable embryos and 0.86 dead embryos per litter was observed (Table 8.2). Overall 17.65% of the embryos implanted are lost during pregnancy (Moreira et al. 2001).

The same trend of a reproductive peak at 4 years was found in the prevalence and annual incidence of pregnancy in capybaras on Marajó Island (Table 8.3; Box 8.1; Moreira 1995). The drastic reduction in the number of females between 4 and 5 years in the population is evidence that corroborates the idea of a lesser physical (and hence reproductive) condition after 4 years. The average annual incidence of pregnancy found in capybaras on Marajó Island was 1.24 births per female per year and the average fertility rate was 2.7 females/female/year. The incidence of pregnancy found by Ojasti (1973) in the Venezuelan Llanos, when recalculated using a gestation time of 150.6 days (López-Barbella 1987), was 1.2 births per female per year and a fertility rate of 2.5 females/female/year (Moreira and Macdonald 1996). It is clear that during the peak of reproduction, some female capybaras breed more than once a year (Herrera 1998).

Table 8.3 Prevalence and incidence of pregnancy and fertility rates, stratified by age of female capybaras of Marajó Island, Pará State, Brazil, and their averages (calculated from the overall sample of adult females; Moreira 1995)

Age class	Prevalence of pregnancy $\overline{P}$ (%/year)	Incidence of pregnancy $\overline{I}$ (births/female/year)	Fertility rate (females/female/year)
1	–	–	–
2	0.37	0.80	1.60
3	0.43	0.95	1.94
4	0.64	1.39	3.21
5	0.50	1.09	1.91
6	0.05	0.11	–
Mean	0.57	1.24	2.70

Box 8.1 Calculations Used to Derive Capybara Birth Frequency and Fertility Rate

The calculation of the average number of litters produced per female in 1 year is carried out in two steps (Caughley 1977). First, the average prevalence of pregnancy $\overline{P}$ over the year is calculated from the average proportion of pregnant females every month:

$$\overline{P} = \frac{\sum_{t=1}^{n} \frac{g_t}{g_t + \overline{g}_t}}{n}$$

where g and $\overline{g}$ are the status of females that are pregnant or not and n is the number of months. The average incidence of pregnancy $\overline{I}$, which is the number of times a female becomes pregnant during the year, can be estimated from:

$$\overline{I} = \overline{P} / \overline{D}$$

where $\overline{D}$ is the average length of a visible pregnancy, defined as the average length of gestation in which pregnancy can be detected.

This method was developed for use with live animals, and uses visual recognition of pregnancies. However, it is also expected that gestation is visible in only a fraction of pregnant females killed. Both Moreira (1995) and Ojasti (1973) assumed that pregnancy in a dead female capybara will not be detected in the first 10 days, even allowing for different lengths of gestation.

Considering that $\overline{I}$ varies substantially with age, it is preferable that sampling be stratified by age class. It should be remembered that $\overline{I}$ is not strictly the average number of births undergone per female per year (Caughley 1977), as abortions are not included in the calculation.

8.6 Death

The life table calculated for capybaras on Marajó Island (Box 8.2; Moreira 1995) showed that the mortality of individuals in the first year of life was high (0.68, Table 8.4). Due to this high first-year mortality, the life expectancy of capybaras on Marajó Island was low (1.08 years for males and 1.28 years for females), resulting in rapid population turnover. Survival rate tended to increase with age until the third year, decrease in the fourth and fifth years, and decrease sharply in the sixth year.

Box 8.2 Calculating a Life Table for the Capybara

The life table for capybaras on Marajó Island (Pará state, Brazil) was calculated using "method 6" of Caughley (1977) to obtain mortality data. This method provides an approximation, where the age distribution is calculated from birth for a population with a stable age distribution and a known population growth rate (Caughley 1977). Data from hunted individuals recorded by Moreira (1995) were used to obtain population mortality parameters. The number of individuals in each age class *x* provided the frequency *fx*. The frequency of age class 1 was calculated assuming a fecundity rate of four pups/female/year and a sex ratio of 1:1 for all females of reproductive age. The age of individuals from the population was obtained using the mass of dry eye lenses, following the method described in Moreira and Macdonald (1998). Only animals slaughtered in the period from November to March (corresponding to the birth season on Marajó Island) were used in the calculations (Moreira 1995). The population was considered to be stable as hunting pressure (both frequency and intensity) has remained unchanged for over 10 years. The population growth rate *r* for the hunted population was estimated over 3 consecutive years to be –0.04. The growth rate of a population reveals not only the speed at which it grows, but also its general well-being (Caughley 1977).

Calculation of Population Survival and Mortality

Survival l_x was calculated by dividing the frequency of individuals in each age class by the starting vigor of the population (i.e., f_x calculated for age class 1). The calculations allow for the construction of a survival table (l_x), indicating the proportion of the population still alive at a given age in relation to the number of live births (age 0; Caughley 1977). All other parameters were calculated from l_x. Mortality d_x was calculated using the following equation (Caughley 1977):

$$d_x = l_x - l_{x+1}$$

(continued)

Box 8.2 (continued)

The equation used to calculate the mortality rate q_x was (Caughley 1977):

$$q_x = 1 - (l_{x+1} / l_x)$$

The survival rate p_x derived from q_x (Caughley 1977):

$$p_x = 1 - q_x$$

Survival l_x and all other parameters were then stratified by gender and presented in different life tables (Moreira 1995).

Methodological Limitations

A static life table, as used in these calculations, has some limitations. Unlike a dynamic life table, which follows a cohort from birth to death, static tables reconstruct a cohort using observations from a single moment in time, like a snapshot. Static tables can be used to calculate the population growth rate only if one assumption is valid – that the mortality at any age is constant over time. Thus, it is assumed that the birth rate and age-specific survival are independent of conditions in the year that sampling occurred (Caughley 1977). This assumption is rarely correct. Consequently, the conclusions that can be inferred from a static table must take into consideration the behavior of a population under conditions that are constant between years and similar to those at the moment of sampling. It is also important to note that a sample collected from animals slaughtered commercially, as was the case in the study on Marajó Island, present a biased age distribution (Caughley 1977 :95). Thus, the information and conclusions that can be obtained from the life table should be directed to general trends and patterns of survival and mortality for different age classes and sex.

Table 8.4 Static life table for capybaras of Marajó Island, Pará State, Brazil. The frequency of age class 1 was calculated assuming a fecundity rate of four pups/female/year for all females of reproductive age (Moreira 1995)

Age class	Frequency	Survival	Mortality	Mortality rate	Survival rate
x	f_x	l_x	d_x	q_x	p_x
1	227	1.000	0.680	0.680	0.320
2	76	0.320	0.129	0.403	0.597
3	47	0.191	0.073	0.382	0.618
4	30	0.118	0.049	0.415	0.585
5	18	0.069	0.036	0.522	0.478
6	9	0.033	0.028	0.848	0.152
7	1	0.005	–	–	–

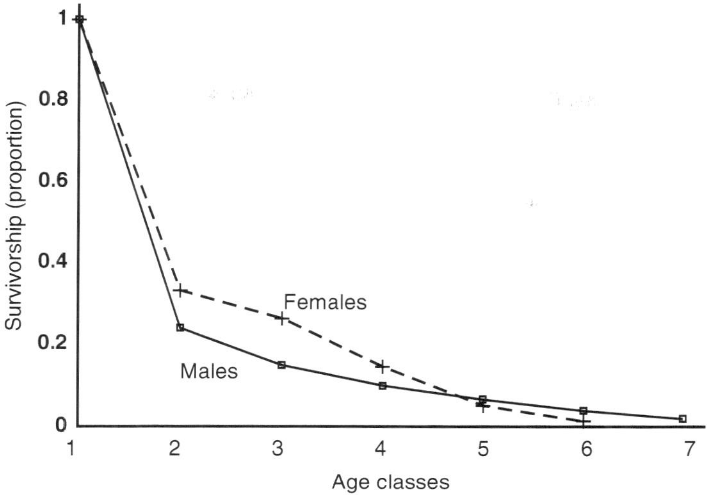

Fig. 8.2 Survivorship curves for male and female capybaras from Marajó Island, Brazil, based on hunting data (From Moreira (1995))

Table 8.5 Static life table for a population of female capybaras on Marajó Island, Pará State, Brazil. The frequency of age class 1 was calculated assuming a fecundity rate of two females/female/year for all females of reproductive age (Moreira 1995)

Age class x	Frequency f_x	Survival l_x	Mortality d_x	Mortality rate q_x	Survival rate p_x
1	114	1.000	0.670	0.670	0.330
2	39	0.330	0.072	0.218	0.782
3	32	0.258	0.117	0.453	0.547
4	18	0.141	0.096	0.681	0.319
5	6	0.045	0.038	0.844	0.156
6	1	0.007	–	–	–

In sum, the highest mortality rates were experienced by the infants, those less than 1 year old, and adults over 5 years.

There was no significant difference between the survival curves (Fig. 8.2) of males and females (Log-rank: $\chi^2_5=0.40$, $P=0.53$; Wilcoxon: $\chi^2_5=2.61$, $P=0.11$; –2Log(LR): $\chi^2_5=0.77$, $P=0.38$). However, a significantly higher number ($\chi^2_5=5.02$, $P=0.03$) of females (Table 8.5) survived the second year of life compared with males (Table 8.6; Moreira 1995) and, overall, the average mortality rate was slightly higher for males (0.63 per year) than for females (0.56 per year). Nevertheless, for those males that did survive, longevity was apparently higher (the oldest female captured was 6 years old, whereas five 6-year-old males and two 7-year-old males were captured). Several factors likely contribute to the difference in survival between sexes. Infant males may be unable to remain within the shelter of natal groups

Table 8.6 Static life table for a population of male capybaras on Marajó Island, Pará State, Brazil. The frequency of age class 1 was calculated assuming a fecundity rate of two males/female/year for all females of reproductive age presented in Table 8.5 (Moreira 1995)

Age class x	Frequency f_x	Survival l_x	Mortality d_x	Mortality rate q_x	Survival rate p_x
1	114	1.000	0.763	0.763	0.237
2	28	0.237	0.091	0.384	0.616
3	18	0.146	0.052	0.356	0.644
4	12	0.094	0.035	0.372	0.628
5	8	0.059	0.026	0.441	0.559
6	5	0.033	0.020	0.606	0.394
7	2	0.013	–	–	–

because of competition with adult males; outside the group, life is risky and males become subject to predation and vulnerable to starvation if they are unable to find/form new groups and establish new territories. In contrast, although females are able to remain within the group and thus might be expected to have greater survival in their first few years, the demands of reproduction are likely to be great.

8.6.1 Age Structure

Although there is generally an annual peak in births (Ojasti 1973), individuals of all age classes can be found in capybara groups throughout the year. In Venezuela, capybara populations typically include 70% adults and 30% juveniles (Ojasti 1973). In Colombia, a population studied comprised 43% adults, 28% subadults, 17% juveniles, and 12% infants (Aldana-Domínguez et al. 2002). In the Brazilian Pantanal (Fig. 8.1), the percentage of adults and infants is close to 50% (Schaller and Crawshaw 1981). In Taim Ecological Station, in the state of Rio Grande do Sul, Brazil (Fig. 8.1), groups included circa 58% adults, 25% juveniles, and 17% infants (Garcias and Bager 2009).

Age structures may also differ between the two capybara species: Aldana-Domínguez et al. (2002) found a predominance of adults in populations of *H. hydrochaeris* in Colombia, but a more balanced adult to juvenile ratio was found by Ballesteros (2001) for *Hydrochoerus isthmius*. In this case, the difference is most likely explained by the conservation status of the species and differences in the areas surveyed.

8.6.2 Sex Ratio

Capybara groups always contain more females than males (Table 8.7), because when male capybaras reach sexual maturity they are expelled from the group by the dominant adult male (Alho and Rondon 1987; Alho et al. 1987a; Herrera 2012).

Table 8.7 Sex ratios found in groups and adult populations of capybaras in different locations

Location	Sex ratio (proportion of ♀)	Unit	Reference
Llanos, Venezuela	0.67	Group	Macdonald (1981)[a]
Llanos, Venezuela	0.63	Group	Herrera and Macdonald (1987)[a]
Pantanal, Mato-Grosso do Sul, Brazil	0.75	Population	Alho and Rondon (1987)[1]; Alho et al. (1989)[a]
Pantanal, Mato-Grosso do Sul, Brazil	0.67	Group	Schaller and Crawshaw (1981)[a]
Llanos, Colombia	0.38	Population	Payan (2007)[b]
Marajó Island, Brazil	0.52	Population	Moreira (1995)[c]

[a]Gender identified from the presence of the nasal gland in wild individuals
[b]Gender identified from the upper incisor length from skulls of hunted individuals
[c]Gender identified from the visual inspection of genitalia of hunted individuals

Subordinate males live on the outskirts of groups, but do not maintain links with any specific group (Macdonald 1981).

The sex ratio of the capybara population on Marajó Island did not differ from 1:1 (0.52, $\chi_1^2=0.6$, $P=0.44$, $N=281$; Moreira 1995). Sex ratios evaluated from other capybara populations (Table 8.7) were probably misinterpreted since these studies used misleading methods for distinguishing gender, like the observation of the presence of nasal gland on males (Alho and Rondon 1987; Alho et al. 1989), or the length of the upper incisor from the skulls of dead animals (Payan 2007). The only infallible method of distinguishing the gender of capybaras is to examine their genitalia.

8.7 Population Dynamics

The way a population changes over time is determined by an often complex relationship with environmental resources (Sinclair et al. 2006). The obvious, but fundamental, generalization is that the greater the availability of limiting resources, the larger the population size. Primary production in ecosystems where rainfall is seasonal is subjected to production pulses and can influence the population size of animal species. This is plausible because the availability of grass, the main item in the capybara's diet, varies with rainfall in seasonal savannas. We addressed the association between rainfall and the size of three populations of free-living capybaras in different locations: Llanos of Venezuela (data from Ojasti 1973); Marajó Island, in the Brazilian Amazonia (data from Moreira 1995) and Pirassununga, in the state of São Paulo, Brazil (Fig. 8.1; data from Vargas 2005).

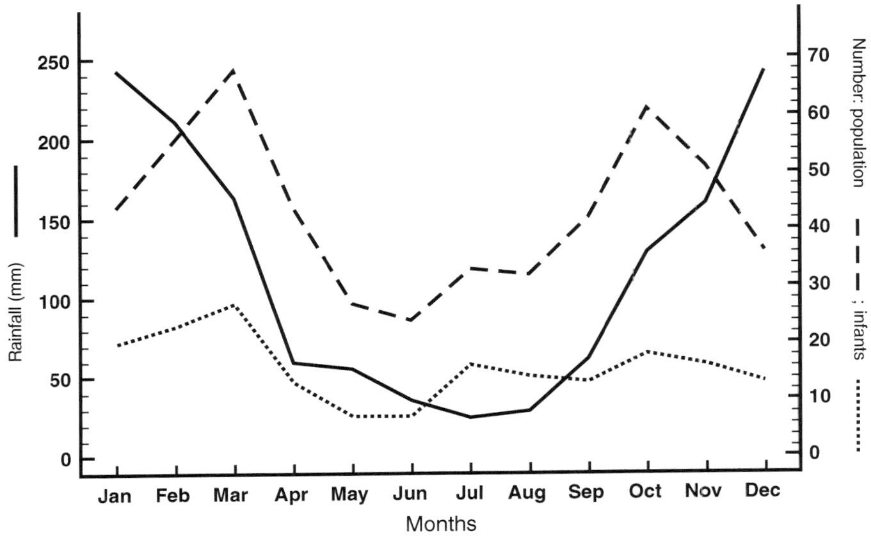

Fig. 8.3 Monthly variations in capybara populations and in the number of infants, and total monthly rainfall (mm) in Pirassununga, São Paulo state, Brazil (From Vargas (2005))

Since information on the seasonal variation in food resources in these locations was not available, we used rainfall as a proxy for primary production. If food availability is a limiting factor for these free-living capybara populations, we expect a positive correlation between population sizes and rainfall. For the capybaras of Pirassununga, the correlation between rainfall and the population size was positive (Pearson $r=0.627$, $N=12$, $P=0.02$). There is no significant correlation between rainfall and variation in the population size of capybaras of Marajó Island (Pearson $r=0.680$, $N=4$, $P=0.32$). In the Llanos, there is a negative correlation between rainfall and the population size (Pearson $r=-0.895$, $N=12$, $P<0.01$).

The population dynamics of capybaras in Pirassununga (Fig. 8.3) responded positively to the variation in rainfall (Vargas 2005) probably because the site is located in the Cerrado biome, a non-flooded savannah. In this biome, the peak production of green biomass (grasses) occurs during the rainy season (Silva and Klink 2001; Munhoz and Felfili 2005). This could be the case (Fig. 8.4) in capybaras on Marajó Island (Moreira 1995) because the population size peaks during the rainy season. It is noteworthy that the failure to detect a significant correlation between rain and population size might be due to the fact that the population on Marajó Island was sampled quarterly in just 1 year rather than monthly, as in other locations. Nevertheless, some population sizes might not respond to the variation in rainfall, as suggested by results found in the Llanos.

The very different population dynamics of capybaras in the Venezuelan Llanos (Fig. 8.5), where the population increases when there is less rainfall (and thus, presumably, lower availability of grazing) supports the idea that a density-independent factor – flooding – limits the population more stringently than a density-dependent

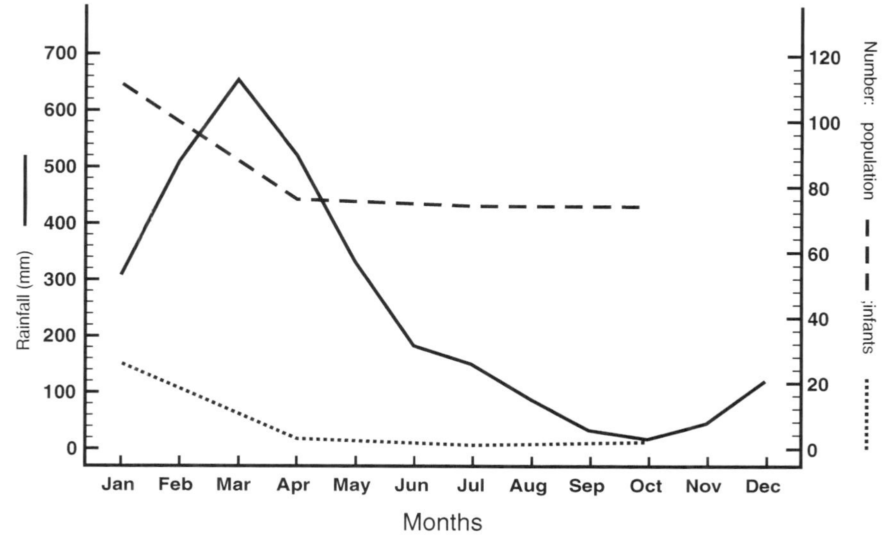

Fig. 8.4 Seasonal (three-monthly) variations in capybara populations and in the number of infants, and total monthly rainfall (mm) on Marajó Island, Pará state, Brazil (From Moreira et al. (2009))

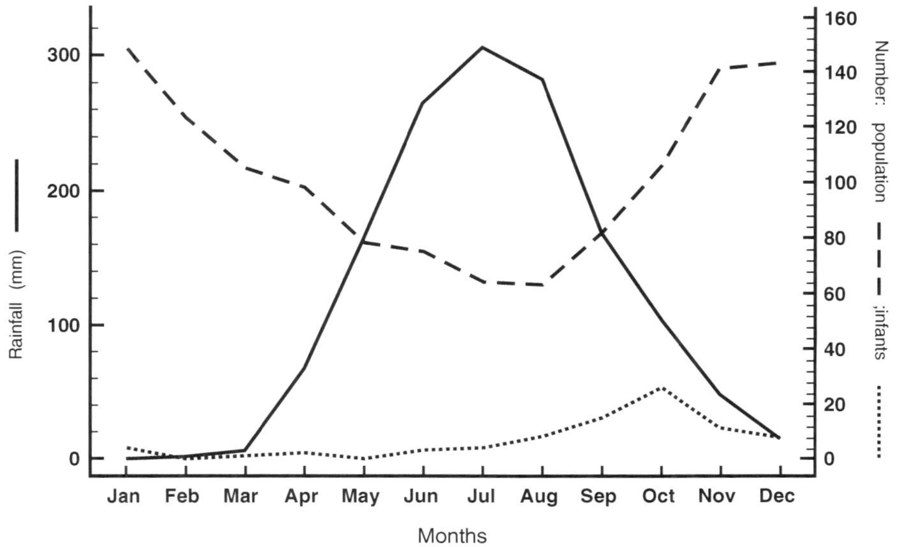

Fig. 8.5 Monthly variations in capybara populations and in the number of infants, and total monthly rainfall (mm) in the Llanos of Venezuela (From Ojasti (1973))

one, the availability of food. Some reasons for that may be an adjustment in birth timing to avoid high infant mortality due to floods caused by the rains (Azcárate et al. 1980) or a high incidence of pathogens (El-Kouba 2005).

8.7.1 Birth Season

Most mammals do not reproduce throughout the year, but have distinct seasonal peaks in reproductive activity across all latitudes, including the tropics (Brown and Shine 2006; Bronson 2009). The timing of reproduction is an important variable in mammalian life histories. Seasonal reproduction can be initiated by stimuli such as changes in day length, temperature, rainfall, food supply, or their interactions. However, the ultimate causes for the seasonality of reproduction, that is, the advantages and disadvantages associated with breeding in different seasons, are difficult to identify (Aung et al. 2001). Ignorance of proximate and ultimate factors that regulate reproduction is particularly acute in the tropics (Bronson 2009).

Biotic (predation and food availability) and abiotic (temperature, rainfall, water level) factors have been suggested as explanatory factors in the seasonality of reproduction in mammals (Aung et al. 2001). The energy costs of reproduction are high and when the ratio between the cost of foraging and the resulting gain varies seasonally, births should occur when the costs of obtaining food is lowest (Anderson et al. 2006; Martins et al. 2006; Dias et al. 2009). On the other hand, where predation (or other cause of mortality) of infants is high, the timing of the breeding season is predicted to reflect a balance between mortality risk and nutrition (Aung et al. 2001). The period from the middle to the end of lactation involves the greatest energy cost for females, so it might be expected that capybaras would have evolved adaptive mechanisms to synchronize the peak in energy requirements with seasonal peaks in the quality and quantity of forage and that the infants will be weaned during this period.

Many habitats in the tropics are as seasonal as those of temperate regions. Tropical savannas in particular are highly seasonal (Bourlière 1983) due to a profound, and predictable, variation in rainfall and hence food availability. Capybaras in Venezuela are fertile throughout the year (Ojasti 1973; Herrera 1998), but reproductive investment is not constant. Comparing the number of infants in a capybara population in the Venezuelan Llanos with the amount of rainfall in the region (Fig. 8.5), a clear peak in births is observed during the late rainy season (October), although there is no significant correlation between the number of births and the amount of rainfall per se (Pearson $r=0.277$, $N=12$, $P=0.38$). A similar pattern in the seasonality of reproduction was found in the Brazilian Pantanal (Schaller and Crawshaw 1981; Alho et al. 1987b), where a peak in capybara reproduction also occurs at the end of the rainy season.

However, this pattern is not found throughout the range of the capybara. On Marajó Island, capybaras also reproduce throughout the year but, contrary to the pattern observed in Venezuela and the Pantanal, capybaras on Marajó show a peak in births at the beginning of the rainy season (Fig. 8.4; Moreira et al. 2009).

Although the birth rate differed statistically among seasons ($F_{3,23}=87.29$, $P<0.01$), again, there was no significant correlation between the number of births and local rainfall (Pearson $r=0.491$, $N=4$, $P=0.52$). A peak in births at the beginning of the rainy season has also been observed in regions where savannas are not seasonally flooded, for example, in the Cerrados of Brasília (Moreira et al. 2002) and Pirassununga (Fig. 8.3; Vargas 2005), and in Piracicaba, state of São Paulo (Katia M.P.M.B. Ferraz personal observation).

In areas dependent on rainfall such as the flooded savannas of Marajó Island and the Venezuelan Llanos, grasses and sedges undergo an explosive but short-lived increase in biomass during the rainy season (Sarmiento 1984). Grasses and sedges comprise around 70% of the diet of the capybara (Escobar and González-Jiménez 1976; Quintana et al. 1994; Barreto and Quintana 2012). Thus, it is to be expected that the coincidence of birth peak with the rainy season, and the associated increase in food availability, maximizes the survival and growth of young and the future survival and reproductive success of females (Rutberg 1987). In regions where there is no distinct rainy season, such as the Atlantic rainforest on the coast of Bahia State, Brazil (Fig. 8.1), and where resources are relatively constant throughout the year, capybaras appear to have two birth peaks (Sérgio L.G. Nogueira-Filho personal communication). A similar situation occurs in captivity, where food is supplied at a constant rate, unaffected by seasonal changes (Mendes and Nogueira-Filho 2012).

The difference in the timing of the birth peak in relation to the timing of the rainy season in the Venezuelan Llanos and Marajó Island is puzzling. The peak in capybara births on Marajó Islands is in line with expectations – births occur at the onset of the rains, enabling exploitation of an abundance of grasses, and potential maximization of growth, prior to the hardships of an impending drought (Moreira et al. 2009). Why would births in the Venezuelan Llanos be delayed until the end of the rainy season? One possible explanation may be high mortality of young capybaras in Venezuela due to heavy rainfall and flooding during the early rainy season (Azcárate et al. 1980). The availability of a dry shelter is apparently an important prerequisite for the survival of young capybaras in Venezuela, and thus delaying births until the end of the rainy season when the frequency and intensity of floods diminish may be advantageous. The same situation occurs in the Pantanal of Mato Grosso do Sul state, Brazil, where extreme annual differences in water levels (from flooding to drought) have a strong influence on the behavior of local capybara populations. Furthermore, in the Venezuelan Llanos and the Brazilian Pantanal, where flooding is extensive, the floods recede at the end of the rainy season to expose a vast fertile plain renewed by the input of water and nutrients. So births in the late rainy season coincide here with a peak in resource quality and quantity (Ferraz 2003). Ojasti (1973:28) mentions that the green pastures and low water levels in the Llanos are seen at the ebb of the waters ("bajadas de aguas") during October and November.

Under some circumstances, a temporal synchrony in births may reduce predation (Estes 1976; Bertram 1978). The dilution effect (Hamilton 1971), confusion, and group defense (Macdonald 1981) can be advantageous when births occur in synchrony. The allonursing behavior (allowing non-filial offspring to suckle) of capybaras (Macdonald 1981; Nogueira et al. 2000; Herrera 2012) may also have favored selection for the synchronization of reproduction in this species. Herrera (1986) raised the possibility that allonursing can improve reproductive efficiency through cooperation, since females share the cost of milk production.

8.7.2 Rate of Population Growth

The intrinsic rate of population increase (r_{max}) is the highest possible rate of population growth for a particular species in a situation where there are unlimited resources. The r_{max} calculated for the capybara using data from Marajó Island was 0.69 (Box 8.3; Moreira and Macdonald 1996), and is one of the highest r_{max} values obtained for any neotropical mammal (Robinson and Redford 1986, 1991). Moreover, it reveals an intriguing exception to the generally inverse relationship between r_{max} and body size among nonflying mammals (Hennemann 1983). Unfortunately, there is little information on how r_{max} differs throughout the capybara's geographical range.

Box 8.3 Interpreting a Population's Demographic Data

Intrinsic Rate of Population Increase (r_{max})

Sustainable exploitation of a wild population subject to logistic growth is a function of its abundance, environmental carrying capacity, and intrinsic rate of population increase (r_{max}; Caughley 1977). The intrinsic rate of population increase is the maximum percentage by which a population can grow when food is plentiful and there are no predators, pathogens, or competitors. The value of r_{max} varies from 0 to 1 and the higher the value of r_{max} for a species, the greater its ability to be commercially exploited in a sustainable manner. The estimate of r_{max} is derived in an iterative manner, using an adaptation of the equation proposed by Cole (1954):

$$1 = e^{-r_{max}} + be^{-r_{max}(a)} - be^{-r_{max}(w+1)}$$

where a is the age at first reproduction, b fertility rate, and $w+1$ the age at last reproduction.

Moreira and Macdonald (1996) calculated r_{max} for capybaras from data collected from a population on Marajó Island (state of Pará; Brazil). They considered the age at first reproduction to be 2 years, a fertility rate of two females/female/year, and the age at last reproduction to be 7 years. The result, given as 0.67, was similar to that calculated for the species by Robinson and Redford (1986).

(continued)

Box 8.3 (continued)

Population Turnover

Population turnover measures the time required for the replacement of individuals in a population. There are several measures that provide information on the renewal of individuals in a population that may be useful for the sustainable management of a species. It identifies bottlenecks and pressure points that are crucial in controlling population growth.

Life expectancy is a measure that evaluates the average age at death of members of the population (Caughley 1977:103). If the calculated life expectancy for a population is low, it indicates that the mortality of infants is very high in this population. To achieve conservation objectives, management strategies for this population must seek to reduce the mortality of infants. Moreira and Macdonald (1996) calculated the life expectancy at birth (e_0), using the equation of Caughley (1966), where l_x is the survival of the population:

$$e_0 = \left(\sum l_x\right) - \frac{1}{2}$$

Values for l_x were calculated for capybaras on Marajó Island from life tables.

A measure that best describes how fast a population is renewed is the Weighted Mean Mortality Rate ($\overline{q}$), which can be calculated using the equation presented by Caughley (1977:104):

$$\overline{q} = 1 / \sum l_x$$

The Weighted Mean Mortality Rate presents the proportion of the population of animals that die between reproductive peaks (Caughley 1977). For the population of capybaras on Marajó Island, the Weighted Mean Mortality Rate was also calculated from life table data (Moreira 1995). High values found for this measure indicated a high turnover of individuals within the population. Management measures that aim to reduce mortality of age class that are most susceptible to mortality can have a marked effect on population growth.

8.8 Final Considerations

Knowledge of the age-specific survival patterns of the capybara is fundamental to understanding the evolution of the species' life history traits and its conservation. Both Ojasti's (1973) and Moreira's (1995) studies have shown that there were a greater number of individuals in the first year of life in the population and that mortality in this age class was high (0.68, Table 8.4). The parameter that best describes how fast a population is renewed is the weighted mean mortality rate ($\overline{q}$). The $\overline{q}$

is the proportion of animals alive during a given breeding season that die before the following breeding season (Caughley 1977:104). For the population of capybaras on Marajó Island, the weighted mean mortality rate was 0.58 per year (Box 8.3; Moreira 1995); that is, more than half the population of wild capybaras die between breeding seasons. In short, capybara populations have a rapid turnover, and first-year mortality is one, if not the, major factor that regulates them.

The population dynamics of large herbivorous mammals can be affected by stochastic environmental variation and density-dependent effects (Sæther 1997). The most important factors in which density dependence can operate are the survival of infants during the first year of life and the age at first reproduction (Gaillard et al. 1998). Amongst capybaras, there is a high mortality rate during the first year of life, but we do not know if it is density dependent. The operation of density dependence on capybara populations needs to be understood as a foundation for management plans, and generally requires detailed long-term studies.

All of the demographic parameters discussed are a direct result of the interactions between the species' genetic potential and the local environmental conditions. The available data concern only a small fraction of the capybara's extensive geographical range. Investigating intra-specific variation in the capybara's reproductive parameters offers a rich model for exploring the ecological processes affecting a mammalian herbivore's demography, and provides a foundation for the species' management.

Acknowledgments Part of the work presented was supported financially by the Conselho Nacional de Desenvolvimento Científico e Tecnológico (Brazil) in the form of a scholarship to JRM, the International Foundation for Science (Sweden), Wildlife Conservation International (USA), and the Fundação de Apoio à Pesquisa Família Alencar (Brazil). JRM thanks Matthew Eagle and Anthony Davidson for statistical support, J.R. Clarke for help in the reproductive studies and S. Casement for editorial suggestions. We also acknowledge the assistance and hospitality of the Lobato family, Belém (Brazil), and especially the support of employees from the Eco-Búfalos Farm on Marajó Island (Brazil). We thank the Fundação de Amparo à Pesquisa do Estado de São Paulo – FAPESP (Proc. No. 00/00180-9, 00/00602-0, 05/00405-4, 06/04878-7).

References

Aldana-Domínguez J, Forero-M J, Betancur J, Cavelier J (2002) Dinámica y estructura de la población de chigüiros (*Hydrochaeris hydrochaeris*: Rodentia, *Hydrochaeridae*) de Caño Limón, Arauca, Colombia. Caldasia 24:445–458

Alho CJR, Rondon NL (1987) Habitats, population densities and social structure of capybaras (*Hydrochaeris hydrochaeris*: Rodentia) in the Pantanal, Brazil. Rev Bras Zool 4(2):139–149

Alho CJR, Campos ZMS, Gonçalves HC (1987a) Ecologia de capivara (*Hydrochoerus hydrochaeris*, Rodentia) do Pantanal. I. Habitats, densidades e tamanho de grupo. Rev Bras Biol 47:88–97

Alho CJR, Campos MSC, Gonçalves HC (1987b) Ecologia de capivara (*Hydrochaeris hydrochaeris*, Rodentia) do Pantanal: II. Atividade, sazonalidade, uso do espaço e manejo. Rev Bras Biol 47:99–110

Alho CJR, Campos ZM, Gonçalves HC (1989) Ecology, social behavior and management of the capybara in the Pantanal of Brazil. In: Redford KH, Eisenberg JF (eds) Advances in neotropical mammalogy. Sandhill Crane, Gainesville, pp 163–194

Anderson DP, Nordheim EV, Boesch C (2006) Environmental factors influencing the seasonality of estrus in chimpanzees. Primates 47:43–50
Aung M, Mcshea W, Htung S, Than A, Soe TM, Monfort S, Wemmer C (2001) Ecology and social organization of a tropical deer (*Cervus eldi thamin*). J Mammal 82:836–847
Azcárate T, Alvarez F, Braza F (1980) Tendencias gregarias del capibara (*Hydrochoerus hydrochaeris*) en los Llanos de Venezuela. Reunión Iberoamericana de Zoologia Vertebrata 1:285–292
Ballesteros J (2001) Estado de conservación del chigüiro o ponche (*Hydrochaeris hydrochaeris isthmius*) en el departamento de Córdoba, Colombia. M.Sc. dissertation, Pontificia Universidad Javeriana, Bogotá
Barreto GR, Quintana RD (2012) Foraging strategies and feeding habits of capybaras. In: Moreira JR, Ferraz KMPMB, Herrera EA, Macdonald DW (eds) Capybara: biology, use and conservation of an exceptional neotropical species. Springer, New York, pp 83–96
Bertram BCR (1978) Living in groups: predators and prey. In: Krebs JR, Davies NB (eds) Behavioural ecology; an evolutionary approach. Blackwell Scientific, Oxford, UK, pp 64–96
Bourlière F (1983) Tropical savannas. Elsevier Scientific, Amsterdam
Bronson FH (1989) Mammalian reproductive biology. University of Chicago Press, Chicago
Bronson FH (2009) Climate change and seasonal reproduction in mammals. Phil Trans Royal Soc B 364:3331–3340
Brown GP, Shine R (2006) Why do most tropical animals reproduce seasonally? Testing hypotheses on an Australian snake. Ecology 87:133–143
Caughley G (1966) Mortality patterns in mammals. Ecology 47:906–918
Caughley G (1977) Analysis of vertebrate populations. Wiley, Chichester
Cole LC (1954) The population consequences of life history phenomena. Q Rev Biol 29:103–137
Dias BB, Santos LAD, Lara-Ruiz P, Cassano CR, Pinder L, Chiarello AG (2009) First observation of mating and reproductive seasonality in maned sloths *Bradypus torquatus* (Pilosa: Bradypodidae). J Ethol 27:97–103
El-Kouba MMAN (2005) Aspectos gerais da fasciolose e das endoparasitoses em capivaras (*Hydrochaeris hydrochaeris* – Linnaeus, 1766) e ratões-de-banhado (*Myocastor coypus* – Molina, 1782) residentes em três parques do estado do Paraná. M.Sc. dissertation, Universidade Federal do Paraná, Curitiba
Escobar A, González-Jiménez E (1976) Estudio de la competencia alimenticia de los herbivoros mayores del llano inundable con especial al chigüire (*Hydrochoerus hydrochaeris*). Agron Trop 23:215–227
Estes RD (1976) The significance of breeding synchrony in the wildebeest. East Afr Wildl J 14:135–152
Ferraz L (2003) Liberação de nutrientes em um campo inundável no Pantanal de Mato Grosso: o papel da inundação. M.Sc. dissertation. Universidade Federal de Mato Grosso, Cuiabá
Gaillard J-M, Festa-Bianchet M, Yoccoz NG (1998) Population dynamics of large herbivores: variable recruitment with constant adult survival. Trends Ecol Evol 13:58–63
Garcias FM, Bager A (2009) Estrutura populacional de capivaras na Estação Ecológica do Taim, Brasil, RS. Ciência Rural 39:2441–2447
Hamilton WD (1971) Geometry for the selfish herd. J Theor Biol 31:295–311
Hennemann WW III (1983) Relationship among body mass, metabolic rate and the intrinsic rate of natural increase in mammals. Oecologia 56:104–108
Herrera EA (1986) The behavioural ecology of the capybara, *Hydrochoerus hydrochaeris*. D.Phil. Thesis, University of Oxford, Oxford, UK
Herrera EA (1998) Reproductive strategies of female capybaras: dry season gestations. Symp Zool Soc Lond 71:381–392
Herrera EA (2012) Capybara social behavior and use of space: patterns and processes. In: Moreira JR, Ferraz KMPMB, Herrera EA, Macdonald DW (eds) Capybara: biology, use and conservation of an exceptional neotropical species. Springer, New York, pp 195–207
Herrera EA, Macdonald DW (1987) Group stability and the structure of a capybara population. Symp Zool Soc Lond 58:115–130

Hosken FM (1999) Criação de capivaras. SEBRAE-MT, Cuiabá, p 138. (Série Natureza e Negócios, 1)

Kleiman DG, Eisenberg JF, Maliniak E (1979) Reproductive parameters and productivity of caviomorph rodents. In: Eisenberg JF (ed) Vertebrate ecology in the Northern Neotropics. Smithsonian Institution Press, Washington, DC, pp 173–183

López-Barbella S (1987) Consideraciones generales sobre la gestacion del chigüire (*Hydrochoerus hydrochaeris*). Acta Cient Venez 38:84–89

Lord RD, Lord VR (1988) Cross checking censuses and a model of the annual cycle of mortality and reproduction in capybaras (*Hydrochaeris hydrochaeris*). Stud Neotrop Fauna Environ 23:213–224

Macdonald DW (1981) Dwindling resources and the social behaviour of capybaras, (*Hydrochoerus hydrochaeris*) (Mammalia). J Zool 194:371–391

Martins EG, Bonato V, Da-Silva CQ, Reis SF (2006) Seasonality in reproduction, age structure and density of the gracile mouse opossum *Gracilinanus microtarsus* (Marsupialia: Didelphidae) in a Brazilian cerrado. J Trop Ecol 22:461–468

Mendes A, Nogueira-Filho SLG (2012) Feeds and nutrition of farmed capybaras. In: Moreira JR, Ferraz KMPMB, Herrera EA, Macdonald DW (eds) Capybara: biology, use and conservation of an exceptional neotropical species. Springer, New York, pp 261–274

Moreira JR (1995) The reproduction, demography and management of capybaras (*Hydrochaeris hydrochaeris*) on Marajó Island – Brazil. D.Phil. thesis, University of Oxford, Oxford, UK

Moreira JR, Macdonald DW (1996) Capybara use and conservation in South America. In: Taylor VJ, Dunstone N (eds) The exploitation of mammal populations. Chapman and Hall, London, pp 88–101

Moreira JR, Macdonald DW (1997) Técnicas de manejo de capivaras e outros grandes roedores na Amazônia. In: Valadares-Pádua C, Bodmer RE (eds) Manejo e Conservação de Vida Silvestre. Conselho Nacional de Desenvolvimento Científico e Tecnológico, Brasília, pp 186–213

Moreira JR, Macdonald DW (1998) Métodos para a estimativa de idade e peso de capivaras (*Hydrochoerus hydrochaeris*). In Universidade Federal de Pernambuco. Resumos do 22° Congresso Brasileiro de Zoologia. Universidade Federal de Pernambuco, Recife, pp 350–351

Moreira JR, Macdonald DW, Clarke JR (2001) Alguns aspectos comportamentais da reprodução da capivara. Revista Brasileira de Reprodução Animal 25:120–122

Moreira JR, Cunha HJ, Pinha PRS, Carvalho JP, Hercos AP (2002) Estação de nascimentos de capivaras (*Hydrochoerus hydrochaeris*) no Cerrado. In: Simpósio Ecologia e Biodiversidade do Cerrado. Embrapa Sede, Brasília, p 34

Moreira JR, Eagle M, Gillespie OJ, Davidson A, Marriott FHC, Macdonald DW (2009) A model to search for birth probabilities of mammal populations using fertility data. Braz J Biol 69(4):1127–1131, ISSN 1519-6984

Munhoz CBR, Felfili JM (2005) Fenologia do estrato herbáceo-subarbustivo de uma comunidade de campo sujo na Fazenda Água Limpa no Distrito Federal, Brasil. Acta Bot Bras 19:981–990

Nogueira SSC (1997) Manejo reprodutivo da capivara (*Hydrochoerus hydrochaeris hydrochaeris*) em sistema intensivo de criação. Ph.D. thesis, Instituto de Psicologia – Departamento de Psicologia Experimental of the Universidade de São Paulo, São Paulo

Nogueira SSC, Nogueira-Filo SLG, Dias CTS (2000) Alloparental behavior in capybara (*Hydrochoerus hydrochaeris*). Revista de Etologia 2:17–21

Ojasti J (1973) Estudio Biologico del Chigüire o Capibara. FONAIAP, Caracas

Ojasti J (1991) Human exploitation of capybara. In: Robinson JG, Redford KH (eds) Neotropical wildlife use and conservation. University of Chicago Press, Chicago, pp 236–252

Payan E (2007) Análisis de proporciones de edad y sexo a partir de cráneos resultantes de cosechas de chigüiros en los Llanos Orientales de Colômbia, departamento del Casanere, para el uso y conservación de la espécie. In: Aldana-Domínguez J, Vieira-Muñoz MI, Ángel-Escobar DC (eds) Estudios sobre la ecología del chigüiro (*Hydrochoerus hydrochaeris*), enfocados a su manejo y uso sostenible en Colombia. Instituto Alexander von Humboldt, Bogotá D.C., pp 87–103

Quintana RD, Monge S, Malvárez AI (1994) Feeding habits of capybara (*Hydrochaeris hydrochaeris*) in afforestation areas of the lower delta of the Paraná River, Argentina. Mammalia 58:569–580

Robinson JG, Redford KH (1986) Intrinsic rate of natural increase in Neotropical forest mammals: relationship to phylogeny and diet. Oecologia 68:516–520
Robinson JG, Redford KH (1991) Sustainable harvest of Neotropical forest mammals. In: Robinson JG, Redford KH (eds) Neotropical wildlife use and conservation. University of Chicago Press, Chicago, pp 415–429
Rutberg AT (1987) Adaptive hypotheses of birth synchrony in ruminants: an interspecific test. Am Nat 130:692–710
Sæther B-E (1997) Environmental stochasticity and population dynamics of large herbivores: a search for mechanisms. Trends Ecol Evol 12:143–149
Sarmiento G (1984) The ecology of Neotropical Savannas. Harvard University Press, Cambridge
Schaller GB, Crawshaw PG Jr (1981) Social organization in a capybara population. Säugetierkundlichen Mitteilungen 29:3–16
Silva DA da, Klink CA (2001) Leaf and tiller dynamics of two C4 and one C3 native grass species of the Brazilian Cerrado vegetation. Rev Bras Bot 24:441–446
Sinclair ARE, Fryxell JM, Caughley G (2006) Wildlife ecology, conservation, and management, 2nd edn. Blackwell, Malden, 469p
Stearns SC (1992) The evolution of life histories. Oxford University Press, Oxford, UK
Vargas FC (2005) Estudo Comparativo de duas Populações de Capivaras (*Hydrochaeris hydrochaeris*) no Município de Pirassununga, SP. M.Sc., dissertation. Universidade de São Paulo, Pirassununga
Weir BJ (1974) Reproductive characteristics of Hystricomorph rodents. Symp Zool Soc Lond 34:265–301

Chapter 9
Diseases of Capybara

Gerardo Rubén Cueto

9.1 Introduction

Disease control is an important part of wildlife management both in the wild and in captivity (Caughley and Sinclair 1994). Capybaras, *Hydrochoerus hydrochaeris,* carry a wide range of parasites and other diseases including hemoparasites and gut parasites (Mones and Martinez 1982), but appear to be largely resistant to their effects and show few signs of ill-health (Emilio A. Herrera, personal communication). In this chapter, I describe the diseases that affect capybaras in the wild and in captivity. The information presented here is based on a decade of experience in capybara breeding and husbandry, based initially at the "Módulo Experimental de Cría de Carpinchos," located in the Experimental Station of the Paraná Delta (Province of Buenos Aires, Argentina; Fig. 9.1), and later on commercial breeding farms, with additional information obtained from the literature.

9.2 Wild and Captive Populations

Generally, capybaras are resilient animals and, in the wild, the main cause of death is not disease but predation, old age, and malnutrition (Ojasti 1973; González Jiménez 1995; Nogueira and da Cruz 2007). In captivity, however, a healthy capybara population is dependent on good sanitary management.

The large body size, high reproductive potential, and gregariousness of capybaras, and their potential economic value, make them ideal candidates for commercial

G.R. Cueto (✉)
Instituto de Ecología Genética y Evolución de Buenos Aires (IEGEBA), UBA-CONICET, Facultad de Ciencias Exactas y Naturales, Universidad de Buenos Aires, Ciudad Universitaria, Pabellón II, 4 Piso, Laboratorio 104, Buenos Aires C1428EHA, Argentina
e-mail: gcueto@ege.fcen.uba.ar

J.R. Moreira et al. (eds.), *Capybara: Biology, Use and Conservation of an Exceptional Neotropical Species*, DOI 10.1007/978-1-4614-4000-0_9,

Fig. 9.1 Map of South America with the locations of the places referred to in the text

captive breeding (Nogueira-Filho et al. 2012), and many breeders have attempted to produce capybara meat and hide on a commercial scale. Capybara breeding projects have been developed in various South American countries including Venezuela (González Jiménez 1995), Colombia (Fuerbringer 1974), Brazil (Alho 1986; Nogueira-Filho 1996; Pinheiro 2007), and Argentina (Allekotte 2003). Due to obvious limitations on space, captive stock is kept at much higher densities (between 200 and 600 individuals/ha depending on the type of pen) than capybaras in the wild, and thus the risk of disease transmission is higher in captivity than in a natural situation. The literature describing pathologies affecting capybaras in captivity, both in intensive and semi-intensive breeding, has therefore increased at a much greater rate than that for capybaras in the wild. For this reason, and because of the importance of sanitation in captive breeding, much of the review herein refers to captive studies.

In the following paragraphs, I present a description of the main diseases recorded for capybaras, enumerating clinical signs, pathologies, forms of diagnosis and, for some, their treatment.

9.3 Ectoparasites

9.3.1 *Scabies*

Sarcoptic scabies is caused by a mite (Acarina: *Sarcoptes scabiei*), a common parasite found on the skin of a large number of domestic and wild mammalian species. Waste products produced by the scabies mite cause severe dermatitis, characterized, in its acute phase, by erythema and edema, followed by intense hyperkeratosis (González Jiménez 1995). Symptoms include localized loss of hair and intense itching, made obvious by the animals' constant search for objects to scratch on. If untreated, the skin thickens and folds form, with hemorrhagic and exuding lesions caused by secondary bacterial infections. Affected animals become anorexic, losing weight to the point of death (Rivera 1983). According to González Jiménez (1995), scabies is the main pathological factor jeopardizing capybaras' high reproductive potential in captivity and in the wild. The economic impact of scabies comes from increased fattening time, reduced daily weight gain, cost of treatments and labor, and occasional mortality among infested animals, as has been recorded for cattle (Losson and Lonneux 1993). The skin lesions caused by this mite cause defects in raw leather and a negative effect on the quality of the commercially valuable hide.

Scabies is highly contagious, especially under high-density conditions in captive breeding facilities. Scabies can be detected in its early stages by microscopic examination of skin scrapings and treated with subcutaneous Ivermectin (1 mL per 33 kg), repeated after 2 weeks (Allekotte 2003). González Jiménez (1995) recommends dusting animals with anti-scabies medication twice a year as a preventive measure, while Giraldo and Ramírez Perilla (2001) recommend Asuntol (coumaphos) and Neguvon (Dimethyl: 2,2,2-trichloro-1-hydroxyethyl-phosphonate) in a 7.5 mg/L dose to prevent infection or to treat mild cases. For diagnosed cases, they recommend intramuscular injections of Ivoemec (Ivermectin 0.5 mg/kg) every 3 days.

9.3.2 *Ticks*

Ticks are obligate ectoparasites that affect a wide variety of vertebrates. Several species of the genus *Amblyomma* have been reported to parasitize both captive and wild capybaras (Nogueira and da Cruz 2007). Tick infestation may reduce red blood cell counts in captive capybaras, causing anemia in highly infested animals (Heijden et al. 2003) but ticks do not seem to affect the health of wild individuals (Nogueira and da Cruz 2007). In captive situations, Giraldo and Ramírez Perilla (2001) and

Allekotte (2003) recommend the use of Ivermectin (also used for scabies, above) to control ticks.

Amblyomma cajennense is the tick most frequently found on capybaras from commercial breeding farms in Brazil (Heijden et al. 2003). This species, distributed widely across South America, exhibits low species specificity. Horses act as the primary host for all parasitic stages of this tick, and are capable of maintaining high tick populations in pastures (Szabó et al. 2004). *A. cajennense* is the main vector in Central and South America of *Rickettsia rickettsii*, the causal agent of Brazilian spotted fever, transmitted to humans by the bite of larvae or nymphs of ticks (Lemos et al. 2001). *R. rickettsii* may infect capybaras asymptomatically while maintaining blood titers of the parasite that are capable of infecting other ticks, in effect turning capybaras into amplifying hosts for *R. rickettsii* (Souza et al. 2009; Labruna 2012). Indeed, recent increases in the capybara population in São Paulo state have been blamed for a reemergence of Brazilian spotted fever there (Fig. 9.1; Souza et al. 2009). Capybaras might also play an important role in the maintenance of *Rickettsia*-caused disease in the same state (Lemos et al. 1996; Labruna 2012).

9.4 Endoparasites

9.4.1 Helminths

In an exhaustive literature compilation of helminths (nematodes, cestodes, and trematodes) recorded in capybaras (and of the organs infected), Nogueira and da Cruz (2007) list 17 nematode, 4 cestode, and 10 trematode genera.

Parasite prevalence varies considerably between captive and wild animals depending on the parasite species and the peculiarities of its specific life cycle. Most trematodes, for example, have complex life cycles with stages affecting several host species: as adults, they are vertebrate endoparasites, while in the larval state, they parasitize mollusks. In the wild, capybaras are frequently infected by trematodes, since they share the aquatic habitat of snails, the intermediate hosts. However, in captivity, the parasite's life cycle is interrupted by the elimination of the intermediate host. Trematodes have not been detected in Argentina's captive breeding facilities, where water provisioning is controlled (Santa Cruz et al. 2005; Sarmiento et al. 2005), and are only usually recorded immediately after new animals have been brought in from the wild (Allekotte 2003).

Cestodes also exhibit a complex life cycle, usually with an intermediate host. However, their intermediate hosts tend to be mites (Acari) or insects, which are present and sometimes abundant on capybara farms, particularly where sanitary practice is poor. The presence of their intermediate host means that cestodes can complete their life cycle even in captivity. The most common Cestode in capybaras belongs to the genus *Monoecocestus*, mostly *Monoecocestus hagmani* (Fig. 9.2) and *Monoecocestus hydrochoeri*, which are found throughout the range of capybaras, both in captivity (Allekotte 2003; Sinkoc et al. 2004) and in the wild (Casas et al. 1995b; Bonuti et al. 2002; Salas and Herrera 2004).

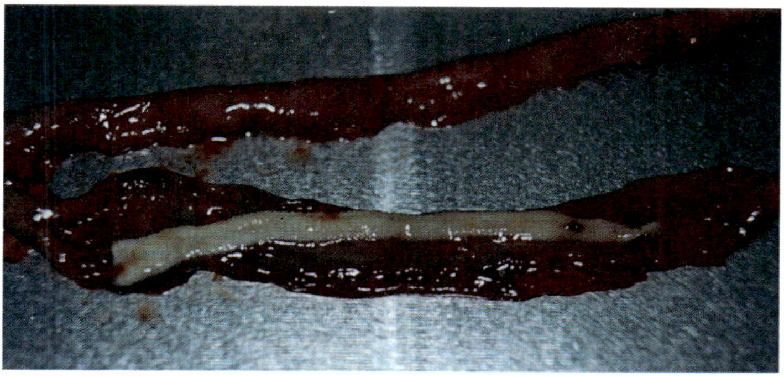

Fig. 9.2 Specimen of *Monoecocestus hagmani* in the small intestine of capybara (Photo by G. R. Cueto)

Nematodes have a direct life cycle, without intermediate hosts. Capybaras are infected by nematodes via the ingestion of infected eggs or larvae of the parasite. The eggs of these parasites are then excreted in the feces and, after a short period of maturation (approximately 10 days) outside the body of the host, can infect new individuals upon ingestion. Nematodes are the most abundant of all parasites of capybaras and can cause great damage in intensive breeding farms. In order to control transmission in pens, fecal material must be swiftly removed. In the wild, Mayaudon (1980) suggests that capybaras gathering around the dwindling pools of the dry season increase transmission of these parasites, whereas during the wet season, natural rates of infection tend to be lower amongst the widely dispersed animals.

The main signs of parasite infection in a capybara are coarse and pointed hair, inactivity, weight loss, anemia, a distended abdomen, dragging the hindquarters as if to scratch their anus, and larvae in feces (Fuerbringer 1974). In order to diagnose the presence of endoparasites, to identify the species present, and to quantify parasite load, fecal material is microscopically examined. Termed coproparasitology, these analyses are not expensive and it is recommended that they are performed regularly to determine the degree of infestation in the breeding grounds (Allekotte 2003; Nogueira and da Cruz 2007). Whenever parasites are detected in feces, a dose of 1 mL/kg of subcutaneous Ivermectin or oral Mebendazol (benzimidazol) in 2 doses every other day should be applied. Alho (1986) recommends anti-helminth treatment every 6 months, with a polyvalent product used for cattle or horses, which may be added to foodstuff or mixed with mineral salt.

9.4.2 Filariae

Filaria is the common name for nematodes of the superfamily Filaroidea, which are parasites of vertebrates, including humans, transmitted by insect bite. Ojasti (1973) reported the presence of macroscopic endoparasites (Nematoda: Filaroidea) in adult capybaras, which were found in the lungs and kidneys of 55.4% of examined specimens.

Eberhard et al. (1976) classified the species, calling it *Cruorifilaria tuberocaudata*. Campo-Aasen (1977) also described the presence of filariae and microfilariae in the muscle tissue, kidney, and dermis of capybaras and described the histological alterations (intense renal panarteritis and severe inflammatory process) suffered by affected tissues. These alterations generally lead to important secondary phenomena including arterial thrombosis, ischemia (restricted blood supply), and tissue degeneration, especially in kidney tissue (Campo-Aasen 1977; Morales et al. 1978).

Five filarial species have been described for capybaras: *Dirofilaria acutiuscula*, *Cruorifilaria tuberocaudata*, *Yatesia hydrochoerus*, *Mansonella longicapita*, and *Mansonella rotundicapita*; all have been recorded in tropical countries such as Venezuela, Colombia, and Brazil (Nascimento et al. 2000), whereas no cases were reported in capybaras from more temperate regions. In a study of wild animals in the state of Mato Grosso do Sul in Brazil (Fig. 9.1), Nascimento et al. (2000) recorded the presence of *Y. hydrochoerus* in 47% of examined capybaras. Although the lesions produced by this filaria were not serious, their presence made the frozen meat unsuitable for sale (Nascimento et al. 2000).

9.4.3 *Protozoa*

9.4.3.1 Coccidia

The genera *Eimeria* and *Cryptosporidium* are among the most important parasitic protozoa worldwide. Coccidia cause enteritis in a wide variety of animal species, and can cause substantial losses among domestic animals (Fitzgerald 1980). Most *Eimeria* infections are asymptomatic or subclinical, but some species in this genus are associated with diarrhea and stunted growth (Newman et al. 2001; Geurden et al. 2005). Several *Eimeria* species are associated with capybaras throughout their distribution. According to the review by Albuquerque et al. (2008), *EImeria trinidadensis, EImeria ichiloensis*, and *EImeria boliviensis* have been reported for Venezuela, Bolivia (Casas et al. 1995a), and Southern Brazil (Fig. 9.1; Gurgel et al. 2007) where the presence of *EImeria araside* was also reported. Carini (1937) recorded the presence of *EImeria capibarae* and *EImeria hydrochoeri* in the state of São Paulo (Brazil) while Albuquerque et al. (2008) found *E. trinidadensis*, and *E. ichiloensis* in the northeast of Brazil (Fig. 9.1). In Argentina, the presence of *E. boliviensis*, *E. hydrochoeri*, *E. ichiloensis*, and *E. trinidadensis* was recorded in commercial capybara farms (Sarmiento et al. 2005; Gonzalez et al. 2007).

Meireles et al. (2007) recorded the presence of *Cryptosporidium parvum* in capybaras in Brazil. The presence of *Cryptosporidium* has zoonotic implications since humans are among the numerous species of mammals it infects (Webster and Macdonald 1995). Cryptosporidiosis is a common cause of (mild to severe) diarrhea in humans and other animals throughout the world. Most people are immunocompetent to control infection without medication; in immunodeficient patients, infection can be persistent (Ramirez et al. 2004).

Coccidiosis, in captive capybaras, is closely linked with farm management and is most common in situations of close confinement, lack of hygiene and/or high humidity (Allekotte 2003). Coccidiosis can be diagnosed by fecal material analysis. Using a simple flotation method, oocytes can be easily identified under an optical microscope (Nogueira and da Cruz 2007). Captive-bred capybaras are frequently parasitized with coccidians but usually with a low parasite load. As long as loads are relatively low, it is recommended that no treatment be applied, because the parasite may become immune to regular control doses. Nevertheless, if there is an acute infection, shown by severe and bloody diarrhea, animals must be treated with sulfonamides, provided in their drinking water for 10 days (Allekotte 2003).

9.4.3.2 Trypanosoma

Trypanosoma evansi is a hemoflagellated protozoan of great veterinary importance and worldwide distribution. It has spread throughout South and Central America since it was first introduced from Spain, probably in the sixteenth century (Canelón and Meléndez 2003). *T. evansi* is the causal agent of the most common and widely distributed trypanosome-generated disease. It can infect most mammals, with camels and horses being the main hosts and suffering the greatest economic losses. Signs of illness caused by infection with *T. evansi* include anemia, edema, paralysis of the hind limbs, infertility, and death. Trypanosomiasis caused by *T. evansi* has a number of local names, such as "surra" in Hindi meaning thin and rotten; "mal de caderas" (hip disease), "quebra-bunda" (broken rump), or "murrina" in Argentina and Brazil; and "derrengadera" (meaning having the hindquarters paralyzed) or "peste boba" (silly plague) in Venezuela, particularly in the Llanos or savannas (Fig. 9.1; Canelón and Meléndez 2003). Infection is mechanically transmitted by blood-sucking insects such as *Tabanus, Stomoxys, Atylotus*, and *Lyperosia* (Brun et al. 1998). Although it is not usually considered a zoonosis, a single case of infection by *T. evansi* in a human (a farmer who probably contracted the disease through contact with an infected animal) was reported in India (Joshi et al. 2005).

In Colombia, capybaras can be asymptomatic carriers of *T. evansi* and may act as a wildlife reservoir for the disease in domestic animals (Morales et al. 1976). Arias et al. (1997) recorded high prevalence of this protozoan in natural populations of capybaras, which showed no visible signs of infection, corroborating their role as reservoir. However, acute symptoms of trypanosomiasis have been detected in capybaras in Argentina, Brazil, and Paraguay (González Jiménez 1995).

Trypanosomiasis can cause perhaps the greatest damage to a captive breeding program, due to the high mortality rates observed when outbreaks of the disease occur in wild populations of endemic zones (Allekotte 2003). Diagnosis is by examination of blood smear, or by serological techniques, which detect antibodies or antigens in the animal blood plasma or by molecular methods (Fernández et al. 2009). Polymerase chain reaction (PCR), which detects parasite nucleic acids, is a very sensitive method that can detect 1 trypanosome/mL of blood. This makes PCR a very useful method for the presymptomatic period and chronic phase of the disease (Fernández et al. 2009).

9.5 Viral Diseases

9.5.1 *Rabies*

Rabies is considered one of the most important viral zoonoses in the world because of the high mortality rate among infected people. People who have rabies suffer irritation of the central nervous system followed by paralysis and death. Capybaras can contract rabies (Bello et al. 1984), generally via infected blood-feeding bats that may bite capybaras when feeding. According to Nogueira and da Cruz (2007), rabies is rarely diagnosed in capybaras but the steady increase in the number of rabies cases in wildlife and the development of intensive capybara farming practices point to the importance of preventive measures to control the spread of this disease.

9.5.2 *Foot-and-Mouth Disease*

Foot-and-mouth disease (FMD) is an important viral zoonosis worldwide, because of high economic losses in animal production and possible danger to human health (Saraiva 2004). There has been only one report of infection of a wild capybara with the FMD virus – in Colombia, in 1976, on a farm where there was an outbreak of foot-and-mouth disease among cattle and pigs (Rocha et al. 1981). Following that incident, it was shown experimentally that the disease can be transmitted to capybaras by infected cattle (Gomes and Rosenberg 1984), by other infected capybaras (Rosenberg and Gomes 1977; Rocha et al. 1981), and infected capybaras can transmit the disease to cattle (Rosenberg and Gomes 1977; Rocha et al. 1981), although FMD virus can be eliminated from capybaras just 23 days after being infected, thanks to a high immune response (Gomes and Rosenberg 1984), the possible role of the species in the dissemination of the disease and as a natural reservoir should be better studied.

9.6 Bacterial Diseases

9.6.1 *Brucellosis*

Brucellosis is an infectious disease caused by intracellular bacteria of the genus *Brucella*. This disease mainly affects cattle and causes abortion, retained placenta, orchitis, epididymitis, infertility, and serious economic damage due to loss of calves and reduced milk production (Acha and Szyfres 2001). Humans are infected by the conjunctival route, skin, or through mucous membranes. Rural workers and veterinarians can spread brucellosis by handling infected animals. Slaughterhouse workers are also exposed, as are people that consume milk or milk products from infected animals.

Brucella abortus is recognized as the principal cause of abortion in cattle but it can also infect other animals, both farmed and wild. Nine biotypes of *B. abortus* are recognized worldwide, each with distinctive epidemiologic features.

Several authors reported the presence of *Brucella* spp. in capybaras. Plata Garcia (1973) presented the first report of antibodies to *Brucella* spp. in capybaras. Bello et al. (1976) isolated *B. abortus* biotype 6 in two wild capybaras and Lord et al. (1981) isolated *B. abortus* biotype 1 in capybaras captured in Venezuela. Meanwhile, Lord and Flores (1983) assessed the role of capybaras as reservoirs of these bacteria from a bacteriological and serological study of 201 wild capybaras captured in the Venezuelan plains. These authors reported a seroprevalence of 58% and among 23 isolates, 8 were identified as *B. abortus* biotypes 2, 3, 4, and 5, and 15 as *Brucella suis*. The results suggest that wild capybaras can be considered an important host of brucellosis in Venezuela. Nevertheless, no antibodies against brucellosis were found in the serological profile of wild capybaras caught in the state of Minas Gerais, in Brazil (Fig. 9.1; Nishiyama et al. 2002; Milagres 2004).

9.6.2 Leptospirosis

Leptospirosis is a bacterial infection caused by pathogenic spirochetes of genus *Leptospira* characterized by enlarged spleen, jaundice, and nephritis, which can cause death. Leptospirosis is thought to be the most widespread zoonotic disease in the world (Webster et al. 1995; Meites et al. 2004). This disease has been recognized as an important emerging global public health problem because of its epidemic proportions and increasing incidence in both developing and developed countries (Vijayachari et al. 2008). The infection is transmitted to humans and animals through contact with contaminated soil and water with urine and secretions from infected animals or through ingestion of contaminated food (Acha and Szyfres 2001). Rats are the universal reservoir of this zoonosis, but almost all known species of mammals can carry and excrete leptospires.

Marvulo et al. (2009) demonstrated that capybaras experimentally infected with a virulent strain of *Leptospira interrogansthe* experience the classic leptospiremic and leptospiruric phases, similar to those previously described in cattle, pigs, dogs, and other animals. These authors also suggest that capybaras can shed *Leptospira* in urine and may serve as a source of infection for other animals.

Studies on the prevalence of leptospirosis in wild capybaras captured in Venezuela (Jelambi 1976) showed that 63% of the analyzed samples ($n=178$) were positive for different serotypes of leptospires with a predominance of *Leptospira canicola*, *Leptospira ballum*, *Leptospira hardjo*, *Leptospira hendomadis*, and *Leptospira wolffi*. Studied animals showed no signs of disease during the sampling period. Studies on seroprevalence in wild capybaras in various regions of Brazil showed rates between 30% and 60% (Silva et al. 2009), registering a seroprevalence of *Leptospira* of 27% (6/22) in capybaras from a slaughterhouse in the state of Rio Grande do Sul, Brazil (Fig. 9.1). These authors consider that capybaras in captivity may act as a reservoir for pathogenic leptospires, emphasizing the occupational risk for people who work on farms or in abattoirs.

9.7 Diseases Associated with Intensive Breeding Systems

Most of the pathologies observed on capybara farms stem from inadequate facilities or poor management practices (Allekotte 2003). For this reason, sanitation in capybara breeding facilities cannot be considered in isolation from other aspects of the system such as infrastructure, feeding, management, etc.

There is a close link between animal welfare and the animal's predisposition to contract disease, and the environmental conditions in which the animals are kept play an important role in the development of several pathologies. On the capybara farm at the Estación Experimental Delta del Paraná, for example, during a period of intensive rains (a consequence of "El Niño" of 1998), a lack of dry places to give birth caused high mortality among the females, who contracted diseases through mud in contact with the genital canal (Allekotte 2003).

Good sanitary management must start with providing the capybaras with suitable conditions for their development. For a pathology to appear, the environment acts directly on the causal agent, which in turn affects the animal.

There are two noninfectious pathologies which have an important impact on captive capybara populations: stress and scurvy. These are the two main causes of mortality or diminished reproduction in captive breeding programs and are at the root of most failed commercial capybara production facilities.

9.7.1 Stress

Stress is a physiological and behavioral response to a perilous situation. Acute stress is thus a reaction which helps the animal escape from potentially dangerous situations. Chronic stress, on the other hand, represents a long-term threat to the health and welfare of captive animals.

The response to stress-generating factors involves increased autonomic activity and shifts in the metabolic profile of an organism as it adjusts to some perceived threat from its surroundings. In the short term, stress responses are associated with behavioral changes such as a permanent vigilant state and increased signs of alertness. Physiological changes can also occur, such as increased heart and breathing rates, an increase in glucose metabolism and glucocorticoids (GCs), which in turn modify the metabolism by increasing energy consumption (Morgan and Tromborg 2007). Long-lasting circulation of elevated GCs in blood make them self-preserving, as they damage areas in the brain in charge of terminating the response to stressful situations (Morgan and Tromborg 2007). Chronic long-term stress produces other behavioral changes such as reduced reproductive and exploratory behavior, an increase in "abnormal" behaviors, and alert and vigilance behaviors (see review by Morgan and Tromborg 2007). Stress may also be manifested by an increase in aggressive behavior and a reduction in the behavioral repertoire.

Stress in capybaras is a serious problem when attempting to introduce wild animals into a farm to create a stock of breeders. Adult animals appear to be highly stressed during the first few days of captivity and adult females appear to be particularly sensitive. Up to 80% mortality can occur among recently captured adult females (Allekotte 2003). Stress, in its early stages, can be detected by a general appearance of despondency in the animal, loss of appetite, and diarrhea. Death, as a result of stress, is less common among males, although general health may deteriorate and death may occur in some cases. These clinical manifestations of stress do not occur when animals weighing less than 20 kg (still classified as juveniles) are introduced to captivity. The response in adults, however, does not appear to lessen over time: adult capybaras tend not to adapt to confinement even after 2 years in captivity (Nogueira et al. 2004).

Although captive-born capybaras do not react to human presence (Nogueira et al. 2004), they can be highly stressed by changes in their environment. Maintaining a daily routine is important to minimize stress in captive situations. Handling the animals should be done only when strictly necessary since this can be a cause of intense stress; medication with a long-lasting effect should be used whenever possible to minimize the number of times an animal needs to be handled. Animals under an extensive handling regime are more likely to lose weight as the consequence of an array of behavioral and physiological responses to stress. It is, therefore, clearly important to understand the "cost" of stressful handling procedures (Gelling et al. 2009). For a captive breeding program to be successful, it is important to use a management plan that minimizes the stress to which the animals may be subjected.

9.7.2 Scurvy

Ascorbic acid (vitamin C) is a cofactor of the proline hydroxydase enzyme, which is essential for the synthesis of collagen (the main protein of connective tissue). Vitamin C deficiency hinders collagen synthesis, and lost collagen fibers cannot be replaced. This causes a generalized degeneration of connective tissue (also called conjunctive tissue), especially in those places where collagen regeneration is most needed, such as in the ligaments that bind teeth to alveoli (tooth sockets).

In most mammals, glucose is converted to ascorbic acid by the hepatic enzyme L-gulonolactone oxidase. However, this enzyme is absent in the apes (suborder Anthropoidea) and in cavies such as guinea pigs (*Cavia porcellus*), so these animals must acquire all vitamin C from their food. It is well established, in these species, that a lack of vitamin C in the diet causes all the known symptoms of scurvy, such as listlessness, bleeding gums and nose, loss of teeth (Fig. 9.3), deterioration of articulations, brittle bones, and, eventually, death (Davies et al. 1976; Kipp et al. 1996; Weinstein et al. 2001).

Cueto et al. (2000) showed that capybaras also require exogenous vitamin C for survival and that the level of this enzyme cofactor strongly affects the reproductive success of this species in captivity. These authors recorded signs of scurvy in all

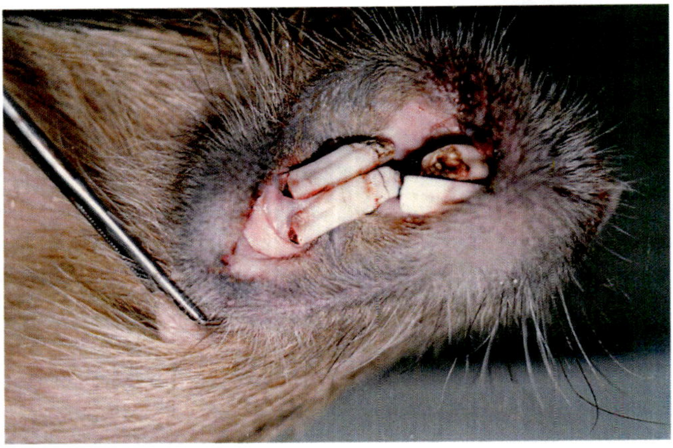

Fig. 9.3 Symptom of advanced stage of scurvy on capybara: loss of all upper and lower incisors (Photo by G.R. Cueto)

capybaras kept on a vitamin C–free diet, while none was observed in a control group (provided with vitamin C). Symptoms of scurvy were observed within 25 days on the vitamin C–free diet, including fracture of the top and bottom incisors, gingivitis, loss of all lower incisors, and death in one animal, and nonlethal symptoms were totally reverted when vitamin C was added to the diet of experimental animals. When reproductively active animals were kept on a diet low in vitamin C (10% of the amount normally offered), only 1 in 40 females (2.5%) managed to complete gestation successfully, although all had copulated. As vitamin C was reincorporated into the diet (300 mg per day for each adult animal), 78% of the females eventually became pregnant and successfully gave birth.

9.8 Final Considerations

Captive capybara breeding is a recent development, so no established sanitary management protocol exists nor are there standard procedures for treating disease. Vaccines, for example, have yet to be used in capybaras. When working with capybaras, it must be kept in mind that they are a wild, nondomesticated species, even if individuals have been born in captivity. Domestic species have lived alongside people for thousands of years, so most pathologies have been described and the emergence of new diseases is a rare occurrence. For capybaras, on the other hand, the literature on pathologies is scarce compared to that available for domestic species, so further diseases apart from the ones described here are likely to be discovered in the future.

As we have seen in this chapter, diseases have different effects on wild and captive capybara populations. Most diseases seem not to play a major role in the regulation of wild populations. In captivity, however, sanitary management is a key to the success of a productive enterprise. A good sanitary management plan should minimize the spread of direct cycle diseases and reduce stress among captive animals. Further research is still needed to improve our knowledge of capybara diseases.

References

Acha PN, Szyfres B (2001) Zoonosis y enfermedades transmisibles al hombre y a los animales, vol 1, 3rd edn, Bacteriosis y Micosis. Organización Panamericana de la Salud, Washington, DC

Albuquerque GR, Berto BP, Catenacci LS, Cunha Nogueira SS, Nogueira-Filho SL, Lopes CW (2008) Eimerid coccidia from capybaras (*Hydrochoerus hydrochaeris*) in southern Bahia, Brazil. Pesquisa Veterinária Brasileira 28:323–328

Alho CJR (1986) Criação e manejo de capivaras em pequenas propriedades rurais, vol 13. EMBRAPA-DDT, Brasília, pp 24–25

Allekotte R (2003) La cría del carpincho. INTA, Buenos Aires

Arias JF, Garcia F, Rivera M, Lopez R (1997) *Trypanosoma evansi* in capybara from Venezuela. J Wildl Dis 33:359–361

Bello AN, Mogollón P, Villegas M, de Laserna R, Gómez G, Regueiro C (1976) La brucelosis en los animales salvajes: I. el chigüire. (*Hydrochoerus hydrochaeris*). Vet Trop 1:117–128

Bello NA, Lord V, Laserna R (1984) Enfermedades infecciosas que afectan el chiguire (*Hydrochaeris hydrochaeris*) en Venezuela. Rev Vet Venez 278:32–46

Bonuti MR, Nascimento AAD, Mapelli EB, Arantes IG (2002) Helmintos gastrintestinais de capivaras (*Hydrochoerus hydrochaeris*) na sub-região de Paiaguás, Pantanal do Mato Grosso do Sul, Brasil. Semina: Ciências Agrárias 23:57–62

Brun R, Hecker H, Lun Z (1998) *Trypanosoma evansi* and *T. equiperdum*: distribution, biology, treatment and phylogenetic relationship (a review). Vet Parasitol 79:95–107

Campo-Aasen I (1977) Microfilariae in the upper middle dermis of capybara (Chigüire). Acta Cient Venez 28:165–166

Canelón JE, Meléndez RD (2003) Posible origen del *Trypanosoma evansi* en Venezuela. Vet Trop 22:155–167

Carini A (1937) Sur deux nouvelles Eimeria *d'Hydrochoerus capibara*. Ann Parasitol Hum Comp 15:367–369

Casas MC, Duszynski DW, Zalles LM (1995a) Three new eimerians in capybara (*Hydrochaeris hydrochaeris*) populations from Eastern Bolivia and Southern Venezuela. J Parasitol 81:247–251

Casas MC, Zalles LM, Patrick MJ, Dailey M (1995b) Intestinal helminths of capybara (*Hydrochaeris [Hydrochoerus] hydrochaeris*) from Bolivia. J Helminthol Soc Wash 62:87–88

Caughley G, Sinclair A (1994) Wildlife ecology and management. Blackwell, Oxford, UK

Cueto GR, Allekotte R, Kravetz FO (2000) Scurvy in capybaras bred in captivity in Argentina. J Wildl Dis 36:97–101

Davies JE, Wilson HK, Hughes RE (1976) Food intake, dietary supplements and survival time of scorbutic guinea pigs. Nutr Metab 20:62–66

Eberhard ML, Morales GA, Orihel TC (1976) *Cruorifilaria tuberocauda* Gen. Et. Sp. N (Nemátoda: Filarioidea) from capybara, *Hydrochoerus hydrochaeris*, in Colombia. J Parasitol 4:604–607

Fernández D, González-Baradat B, Eleizalde M, González-Marcano E, Perrone T, Mendoza M (2009) *Trypanosoma evansi*: a comparison of PCR and parasitological diagnostic tests in experimentally infected mice. Exp Parasitol 121:1–7

Fitzgerald PR (1980) The economic impact of coccidiosis in domestic animals. Adv Vet Sci Comp Med 24:121–143

Fuerbringer J (1974) The capybara: a practical manual for raising them in captivity in Colombia. Temas de Orint Agropecuaria 99:5–59

Gelling M, Mclaren GW, Mathews F, Mian R, Macdonald DW (2009) Impact of trapping and handling on leukocyte coping capacity in bank voles (*Clethrionomys glareolus*) and wood mice (*Apodemus sylvaticus*). Anim Welf 18:1–7

Geurden T, Claerebout E, Vercruysse J (2005) Protozoan infection causes diarrhea in calves. Tijdschr Diergeneeskd 130:734–737

Giraldo D, Ramírez Perilla JA (2001) Guía para el manejo, cría y aprovechamiento sostenible del Chigüiro, Chigüire o Capibara: *Hydrochoerus Hydrochaeris* Linneo. Convenio Andrés Bello, Bogotá

Gomes I, Rosenberg FJ (1984) A possible role of capybaras (*Hydrochoerus hydrochoeris hydrochoeris*) in foot-and-mouth disease (FMD) endemicity. Prev Vet Med 3:197–205

González Jiménez E (1995) El capibara (*Hydrochoerus hydrochaerus*) Estado actual de su producción. Organización de las Naciones Unidas para la Agricultura y la Alimentación, Roma, p 112

Gonzalez AO, Gonzalez JA, Santa Cruz AM, Roux JP, Comolli J, Repetto C, Sarmiento NF (2007) Prevalencia de *Eimeria* sp. en un criadero de carpincho (*Hhydrochaeris hydrochaeris* linnaeus, 1766), en la provincia del Chaco, Argentina. Rev Med Vet 88:3–5

Gurgel ACF, Sartori AS, Araújo FAP (2007) Eimeriosis in capybaras (*Hydrochaeris hydrochaeris*) in the state of Rio Grande do Sul, Brazil. Parasitología latinoamericana 62:76–78

Heijden KMVD, Szabó MPJ, Matushima ER, Veiga MLD, Santos AA, Egami MI (2003) Valores hematológicos e identificação morfo-citoquímica de células sangüíneas de capivaras (*Hydrochoerus hydrochoeris*) parasitadas por carrapatos e capivaras livres de infestação. Acta Sci Anim Sci 25:143–150

Jelambi F (1976) Leptospirosis en Chigüires. CENIAP/FONAIAP, Maracay

Joshi PP, Shegokar VR, Powar RM, Herder S, Katti R, Salkar HR, Dani VS, Bhargava A, Jannin J, Truc P (2005) Human trypanosomiasis caused by *Trypanosoma evansi* in India: the first case report. Am J Trop Med Hyg 73:491–495

Kipp DE, Mcelvain M, Kimmel DB, Akhter MP, Robinson RG, Lukert BP (1996) Scurvy results in decreased collagen synthesis and bone density in the guinea pig animal model. Bone 18:281–288

Labruna MB (2012) Brazilian spotted fever: the role of capybaras. In: Moreira JR, Ferraz KMPMB, Herrera EA, Macdonald DW (eds) Capybara: biology, use and conservation of an exceptional neotropical species. Springer, New York, pp 371–383

Lemos ER, Melles HH, Colombo S, Machado RD, Coura JR, Guimarães MA, Sanseverino SR, Moura A (1996) Primary isolation of spotted fever group rickettsiae from *Amblyomma cooperi* collected from *Hydrochaeris hydrochaeris* in Brazil. Mem Ins Oswaldo Cruz 91:273–275

Lemos ER, Alvarenga FB, Cintra ML, Ramos MC, Paddock CD, Ferebee TL, Zaki SR, Ferreira FC, Ravagnani RC, Machado RD, Guimarães MA, Coura JR (2001) Spotted fever in Brazil: a seroepidemiological study and description of clinical cases in an endemic area in the state of São Paulo. Am J Trop Med Hyg 65:329–334

Lord VR, Flores R (1983) *Brucella* spp. from the capybara (*Hydrochaeris hydrochaeris*) in Venezuela: serologic studies and metabolic characterization of isolates. J Wildl Dis 19:308–314

Lord VR, Bello AN, Moreno J, Rubiela DL, de Ramos J (1981) Aislamientos de *Brucella abortus* biotipo 1 en chiguires (*Hydrochoerus hydrochaeris*) (Estado Apure, Venezuela). Vet Trop 6:45–57

Losson BA, Lonneux JF (1993) Field efficacy of injectable moxidectin in cattle naturally infested with *Chorioptes bovis* and *Sarcoptes scabiei*. Vet Parasitol 51:113–121

Marvulo MFV, Silva JCR, Ferreira PM, De Morais ZM, Moreno AM, Doto DS, Paixão R, Baccaro MR, Vasconcellos SA, Neto JSF (2009) Experimental leptospirosis in capybaras (*Hydrochaeris hydrochaeris*) infected with *Leptospira interrogans* Serovar Pomona. J Zoo Wildl Med 40:726–730

Mayaudon H (1980) Sobre una colección de helmintos del chigüire (*Hydrochoerus hydrochaeris*) en Venezuela. Revista de la Facultad de Ciencias Veterinarias UCV 28:19–47

Meireles MV, Soares RM, Bonello F, Gennari SM (2007) Natural infection with zoonotic subtype of *Cryptosporidium parvum* in capybara (*Hydrochoerus hydrochaeris*) from Brazil. Vet Parasitol 147:166–170

Meites E, Jay MT, Deresinski S, Shieh WJ, Zaki SR, Tompkins L, Smith DS (2004) Reemerging leptospirosis, California. Emerg Infect Dis 10:406–412
Milagres BS (2004) Perfil sorológico de algumas infecções em capivara (*Hydrochaeris hydrochaeris*) capturadas nos estados de São Paulo e Minas Gerais, Brasil. Universidade Federal de Viçosa, Viçosa
Mones A, Martinez S (1982) Estudios sobre la familia Hydrochoeridae (Rodentia) XIII. Parasitosis y patologías de *Hydrochoerus* Brisson, 1972. Revista de la facultad de humanidades y Ciencias, Serie Ciencias Biológicas 1: 297–329
Morales GA, Wells EA, Angel D (1976) The capybara (*Hydrochoerus hydrochaeris*) as a reservoir host for *Trypanosoma evansi*. J Wildl Dis 12:572–574
Morales GA, Guzman VH, Angel D (1978) Vascular damage caused by *Cruorifilaria tuberocauda* in the capybara (*Hydrochoerus hydrochaeris*). J Wildl Dis 14:15–21
Morgan KN, Tromborg CT (2007) Sources of stress in captivity. Appl Anim Behav Sci 102:262–302
Nascimento AAD, Bonuti MR, Tebaldi JH, Mapeli EB, Arantes IG (2000) Natural infections with filarioidea nematodes in *Hydrochaerus hydrochaeris* in the floodplain of Mato Grosso do Sul, Brazil. Braz J Vet Res Anim Sci 37:105–108
Newman C, Macdonald DW, Anwar MA (2001) Coccidiosis in the European badger, *Meles meles* in Wytham Woods: infection and consequences for growth and survival. Parasitology 123:133–142
Nishiyama SM, Moraes MP, Figueiredo AS, Pompermayer LG, Paula TAR (2002) Perfil sorológico de capivaras (*Hydrochaeris hydrochaeris*) mantidas em cativeiro. In: Proc. Assoc. Bras. Vet. Anim. Selv., p 26
Nogueira-Filho SLG, Pinheiro MS, Nogueira SSC (2012) Confined and semi-confined production systems for capybaras. In: Moreira JR, Ferraz KMPMB, Herrera EA, Macdonald DW (eds) Capybara: biology, use and conservation of an exceptional neotropical species. Springer, New York, pp 229–241
Nogueira MF, Da Cruz TF (2007) Doenças da capivara. Embrapa Pantanal, Corumbá
Nogueira SSC, Bernardi LG, Nogueira-Filho SLG (2004) A note on comparative enclosure facility usage by wild and captive-born capybaras (*Hydrochoerus hydrochaeris*). Appl Anim Behav Sci 89:139–143
Nogueira-Filho SL (1996) Manual de criação da capivara. Viçosa – MG, Centro de Produçoes Técnicas – CPT, Brazil
Ojasti J (1973) Estudio biológico del chigüire o capibara. Ediciones del Fondo Nacional de Investigaciones Agropecuarias. Editorial Sucre, Caracas
Pinheiro MS (2007) Criação de Capivara em Sistema Intensivo. Empresa Brasileira de Pesquisa Agropecuária, Pelotas, p 43
Plata Garcia V (1973) Muestreo Serológico en Chigüires (*Hydrochoerus hydrochaeris*). Estado Apure, Informe anual en revisión. Proyecto Conicit DF. DF. 030 SI, 6pp
Ramirez NE, Ward LA, Sreevatsan S (2004) A review of the biology and epidemiology of cryptosporidiosis in humans and animals. Microbes Infect 6:773–785
Rivera MA (1983) Sarna sarcoptica en Chigüires (*Hydrochoerus hydrochaeris*, Linneo). Revista de la Facultad de Ciencias Veterinarias UCV 30:99–115
Rocha RJR, Rueda AF, del Barrera VJC, Jiménez TJM (1981) Transmission experimental de la fiebre aftosa de chiguiros a varias especies animales. Revista ICA 16:193–198
Rosenberg FJ, Gomes I (1977) Suceptibilidad del carpincho o capibara (*Hydrochoerus hydrochaeris*) al virus de la fiebre aftosa. Boletín Centro Panamericano de Fiebre Aftosa 27–28:35–41 and 43–48
Salas V, Herrera EA (2004) Intestinal helminths of capybaras, *Hydrochoerus hydrochaeris*, from Venezuela. Mem Ins Oswaldo Cruz 99:563–566
Santa Cruz AC, Sarmiento NF, González JA, Comolli JA, Roux JP (2005) Parásitos gastrointestinales de carpincho (*Hydrochaeris hydrochaeris*) del criadero "Marchi-E", Baradero, provincia de Buenos Aires, Argentina. Comunicaciones Científicas y Tecnológicas, Universidad Nacional del Nordeste V-038
Saraiva V (2004) Foot-and-mouth disease in the Americas epidemiology and ecologic changes affecting distribution. Ann N Y Acad Sci 1026:73–78

Sarmiento NF, Santa Cruz AC, González AO, González JA, Roux JP (2005) Diagnóstico de endoparásitos de carpincho (*Hydrochaeris hydrochaeris* Linnaeus, 1766) del criadero "Ayuí" de Santo Tomé, provincia de Corrientes, Argentina. Comunicaciones Científicas y Tecnológicas, UNIVERSIDAD NACIONAL DEL NORDESTE V-037
Silva EF, Seyffert N, Jouglard SD, Athanazio DA, Dellagostin OA, Brod CS (2009) Soroprevalência da infecção leptospiral em capibaras (*Hydrochoerus hydrochaeris*) abatidas em um frigorífico do Rio Grande do Sul. Pesquisa Vet Brasil 29:174–176
Sinkoc AL, Brum FA, Muller G, Brum JGW (2004) Helmintos parasitos de capivara (*Hydrochoerus hydrochaeris* L. 1766) na região de Araçatuba, São Paulo, Brasil. Arquivo do Instituto de Biologia, São Paulo 71(3):329–333
Souza CE, Moraes-Filho J, Ogrzewalska M, Uchoa FC, Horta MC, Souza SSL, Borba RCM, Labruna MB (2009) Experimental infection of capybaras *Hydrochoerus hydrochaeris* by *Rickettsia rickettsii* and evaluation of the transmission of the infection to ticks *Amblyomma cajennense*. Vet Parasitol 161:116–121
Szabó MPJ, Castagnolli KC, Santana DÁ, Botelho De Castro M, Romano MA (2004) *Amblyomma cajennense*; ticks induce immediate hypersensitivity in horses and donkeys. Exp Appl Acarol 33:109–117
Vijayachari P, Sugunan AP, Shriram AN (2008) Leptospirosis: an emerging global public health problem. J Biosci 33:557–569
Webster JP, Macdonald DW (1995) Cryptosporidiosis reservoir in wild brown rats (*Rattus norvegicus*) in the UK. Epidemiol Infect 115:207–209
Webster JP, Ellis WA, Macdonald DW (1995) Prevalence of Leptospira and other zoonoses in wild brown rats on UK farms. Mammalia 59:615–622
Weinstein M, Babyn P, Zlotkin S (2001) An orange a day keeps the doctor away: scurvy in the year 2000. Pediatrics 108:e55

Chapter 10
Capybara Scent Glands and Scent-Marking Behavior

David W. Macdonald and Emilio A. Herrera

10.1 Scent Marking in Mammals

Scent marking in mammals can convey a wide range of information (e.g., Brown 1979; Müller-Schwarze 1983; Brown and Macdonald 1985a, b; Arakawa et al. 2008), sometimes linked to agonistic behavior in ritualized contests over resources (e.g., Gosling 1990). Scent marks are used by some mammals to delineate territorial boundaries, as in Ethiopian wolves (Sillero-Zubiri and Macdonald 1998). They can also indicate group membership, as in matrilines of cats, *Felis sylvestris catus* (Passanisi and Macdonald 1990), individual identity, as in dwarf mongooses, *Helogale undulate* (Rasa 1973) or spotted hyenas, *Crocuta crocuta* (Drea et al. 2002) or social and sexual status as in giant otters, *Pteronura brasiliensis* (Leuchtenberger and Mourão 2008). Frequently, scent-marking behavior and the chemistry of the secretion are related to social dominance (e.g., Huck and Banks 1982; Novotny et al. 1990; Ryon and Brown 1990). The latter is especially true in rodents, where status signaling appears to be the most common function of scent marking (Roberts 2007). Scent glands are commonly sexually dimorphic. Capybaras are unusual among caviomorph rodents in having not only anal glands but also a nasal gland, both of which are sexually dimorphic (Macdonald et al. 1984; Macdonald 1985).

D.W. Macdonald (✉)
Wildlife Conservation Research Unit Zoology Department, University of Oxford
The Recanati-Kaplan Centre Oxford, Tubney House, Abingdon Road,
Tubney, Abingdon OX13 5QL, United Kingdom
e-mail: david.macdonald@zoo.ox.ac.uk

E.A. Herrera
Departamento de Estudios Ambientales, Universidad Simón Bolívar, Apartado 89000,
Caracas 1080-A, Venezuela
e-mail: eherre@usb.ve

J.R. Moreira et al. (eds.), *Capybara: Biology, Use and Conservation of an Exceptional Neotropical Species*, DOI 10.1007/978-1-4614-4000-0_10,

10.2 Morphology and Histology of the Scent Glands

10.2.1 Nasal Gland

Capybaras have a sexually dimorphic gland above the snout first described by Rewell (1949), which is very bulbous in the male but often barely visible in females (although some females have small-sized glands; Macdonald et al. 1984). This gland is known as the morrillo (after the Spanish morro, describing a small oblong hillock) and is an oval-shaped, shiny, blackish protuberance (Fig. 10.1). The morrillos of male capybaras are naked except for very sparse short hairs (*c*. 6.0 mm in length) and have a long axis of 6–7 cm and a width of 4–5 cm; they stand proud of the surrounding skin to a height of 1–3 cm (Macdonald et al. 1984). The oily surface glistens in the sun and is punctuated by distended pores out of which drops of creamy white, highly viscous secretion ooze (Macdonald et al. 1984). The size of the morrillo increases with age up to 25 months but thereafter any increase in size is not necessarily associated with age (Costa and Paula 2006). Larger, more dominant males have larger morrillos compared to subordinate males (Herrera and Macdonald 1994). Males with larger testes and higher levels of testosterone have larger morrillos (Herrera 1992; Costa and Paula 2006; López et al. 2008).

The morrillos are covered with a thick epidermal layer (*c*. 0.1–0.2 mm deep) beneath which the glandular layer is variously developed, averaging 8.2 mm (S.D. = 4.7, n = 44) in depth (Macdonald et al. 1984). The maximum depth of the glandular lozenge is different for males and females, at 10.5 mm (S.D. = 4.6, n = 25)

Fig. 10.1 Male capybara showing the morrillo, a thick scent gland on top of the snout (Photo by E.A. Herrera)

and 4.3 mm (S.D. = 1.1, n = 18), respectively, but for each sex the depth of glandular tissue increases with increasing body size class (Macdonald et al. 1984).

The glandular tissue is spongy in appearance, due to the presence of large lacunae or ampullae, measuring up to 1.5 mm in width and 6 mm in depth (Macdonald et al. 1984). The lacunae are filled with secretion and drain into pores leading directly to the surface. Around the hair follicles and the ampullae, there is ample evidence of secretory activity: disintegrating sebaceous cells and distorted nuclei border each lacuna and adjoin alveolar masses of sebaceous acini. One or several acini may drain into either a hair follicle or a lacuna leading directly to a pore. Connective tissue intrudes into the glandular mass, but less so for larger individuals of both sexes (Macdonald et al. 1984). Males have a greater area of sebaceous acini and lacunae per unit area than females and the larger males have more glandular tissue than smaller males, which takes the place of connective tissue (Macdonald et al. 1984).

10.2.2 Anal Glands

In both males and females, the anal glands are located beside and below the anus, lying within a chamber which contains the urogenital and anal pocket openings, all largely covered by the surrounding skin at rest (Macdonald et al. 1984). The morphology of the anal glands differs between males and females. In females, the anal pockets are relatively deep chambers which open through a constricted neck (depth 1.5 cm, chamber internal diameter 1.0–1.5 cm, neck *c.* 3 mm; Macdonald et al. 1984). Within the chamber the skin and hairs are coated with a smear of grayish, greasy material which when abundant may cause the hairs to be matted together (Macdonald et al. 1984). There is a large, solid mass of glandular tissue underlying the chamber (Macdonald et al. 1984).

The male anal pocket differs in being more an open pouch (approximately 4 × 3 cm) that can be easily pulled wide open and is not backed by the knot of tissue found in the glands of females (Macdonald et al. 1984). The pocket also contains hairs, but they are coated with a brittle gray-black deposit, giving them the shape of a club or truncheon (Fig. 10.2). The extent of the coating varies between hairs from a thin coating to a maximum cross-sectional diameter of 2.5 mm and length of 8.0 mm, with larger accumulations associated with longer (and therefore perhaps older) hairs (Macdonald et al. 1984). Hairs from a male's everted anal pocket are easily detached from the skin, and in general larger hairs are more easily detached than smaller ones (Macdonald et al. 1984). The hairs of the male anal pocket are narrower than the more bristle-like hairs of the female anal pocket, but both types are flimsier than body hairs such as those found on the back and inner thigh.

The tissues lining the anal pocket of males are highly glandular. They are, however, shallower and have less sebaceous cell development and sebaceous activity than female anal pockets (Macdonald et al. 1984). They are also different in tissue appearance, with ducts of sebaceous acini opening either into hair follicles or

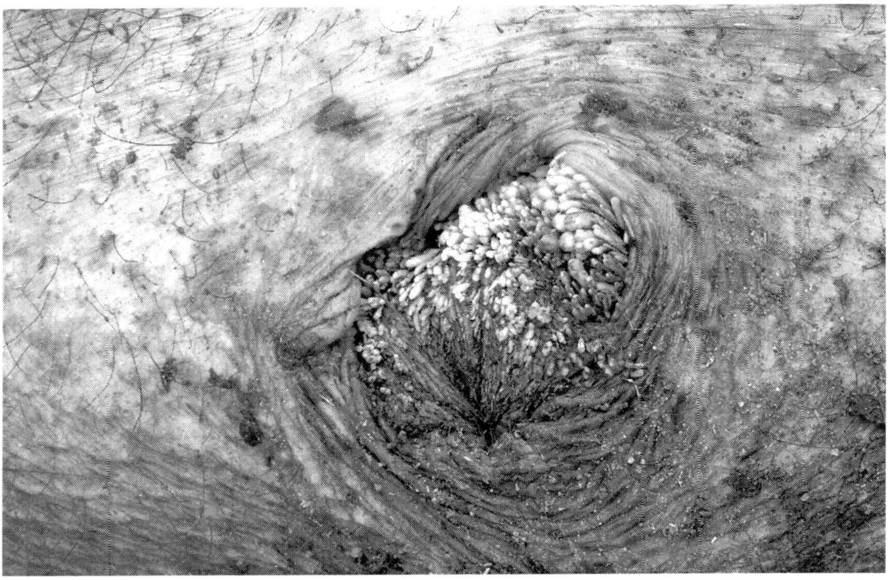

Fig. 10.2 Male anal pocket with hairs coated with a gray-black deposit in the shape of a club (Photo by J.R. Moreira)

directly through the epidermis along empty hair follicles (Macdonald et al. 1984). Within the gland, lacunae up to 2 mm in length and 0.25 mm in width are formed in the acini from cellular debris and secretions (Macdonald et al. 1984). Generally in males, larger individuals have more sebaceous cell development and sebaceous activity than do smaller individuals; however, a small sample of juvenile males had highly active, dense sebaceous tissue comparable to that of adult females (Macdonald et al. 1984). The hairs in the anal pocket of males are coated by an amorphous solid substance which in turn is encased by layers of crystalline material (up to *c*. 20 distinct layers; Macdonald et al. 1984). The annular structure is likely to arise from the drying of successive coatings of secretion, adhesion being aided by the overlapping scale pattern on the surface of the hairs (Macdonald et al. 1984). Bacteria from Streptococci group D, gram negatives, and Clostridia have been found within the hair coating, but it is not known whether these bacteria play a role in developing odors (as has been shown in other mammals; Albone et al. 1978; Lanyon et al. 2007) or were simply contaminants from the intestinal flora (Macdonald et al. 1984).

The swollen knot of tissue under the female anal pocket has highly dense, active sebaceous cells (Macdonald et al. 1984). While the largest size class of females has greater sebaceous cell development, there are no overall differences between size classes. The activity of sebaceous tissue is greater in larger females (Macdonald et al. 1984).

10.3 Chemical Composition of the Secretions

10.3.1 Nasal Gland

Secretions from the male nasal glands are a complex mixture of sterols and/or terpenes, lipids, and amino acids with up to 54 compounds found within a sample from a single individual (Macdonald et al. 1984). Lipids are the main component of the secretion, consisting of a mixture of esters of long chain fatty acids, and the most volatile substance found was a hydrocarbon, $C_{30}H_{50}$ (Macdonald et al. 1984). Each male had most of the compounds present but in significantly different proportions, possibly to aid individual recognition (Macdonald et al. 1984).

10.3.2 Male Anal Gland

The crystal deposits attached to the hairs within the male anal gland are principally composed of a calcium salt with some magnesium and trace amounts of silicon, phosphorous, aluminum, sulfur, and potassium (Macdonald et al. 1984).

10.3.3 Female Anal Gland

The greasy secretion from the female anal gland consists of sterols, terpenes, lipids, and amino acids, with up to 30 compounds found in the secretion from a single individual (Macdonald et al. 1984). The presence and amount of each compound varies between individuals, both in terms of relative and absolute concentrations (Macdonald et al. 1984).

10.4 Capybara Scent Marking Behavior

Scent marking in capybaras is much more common in males than females (Herrera and Macdonald 1994), but during courtship, males and females mark with equal frequency and use both glands (Schaller and Crawshaw 1981). A typical marking sequence for males involves rubbing the morrillo against a shrub or twig (Fig. 10.3), then straddling the plant, pressing the anal pocket onto it (Fig. 10.4) and, at least sometimes, simultaneously urinating on the plant (Azcaráte 1980; Macdonald 1985; Macdonald et al. 1984). During this process, hairs from the anal pocket are detached (Macdonald et al. 1984). Suitable plants for marking are often scarce within capybaras' habitat and so are marked by many individuals from the same group within any single day (Herrera and Macdonald 1994).

Fig. 10.3 Male capybara rubbing his nasal gland on a post (Photo by R. Barreto)

Fig. 10.4 Female capybara rubbing her anal glands on a twig (Photo by E.A. Herrera)

Dominant males have higher marking rates than do subordinate males for both types of gland, and subordinate males are more likely to sniff at a plant before marking it (Herrera and Macdonald 1994). The differences associated with status are most notable for the morrillo, for both scent-marking rates and the size of the gland (Herrera and Macdonald 1994). Marking occurs commonly without any social provocation, but will sometimes follow aggression, in which case the victor or both parties carry out marking, but it is rare for the vanquished male alone to do this (Herrera and Macdonald 1994). It is also very frequent during courtship, where the male overmarks, in particular, female anal marks. On rare occasions, males rub with their morrillos on the necks of females or subordinate males (Macdonald 1985) or females rub their morrillos on the necks of males (Herrera and Macdonald 1994). Males often sniff at the rear of females, particularly during courtship as they herd the female toward water to mate (Macdonald et al. 1984).

The morrillo and anal glands may also be used in separate scent-marking patterns, and females more often tend to mark only with their anal glands than do males; subordinate males do this more commonly than dominant males (Herrera and Macdonald 1994).

Larger groups have lower scent-marking rates than smaller groups, both as a unit and individually (Herrera and Macdonald 1994). Since larger groups defend larger territories (Herrera and Macdonald 1989), it is likely that the home ranges of large groups are less thoroughly covered with scent than those of smaller groups.

10.5 Capybara Scent Gland Marking: Possible Functions

Scent marking is the most common type of social interaction among capybaras (Herrera and Macdonald 1994). Although capybaras do use vocalizations (Azcaráte 1980), chemical communication may be especially effective in an animal that is active during much of the night (Macdonald 1981; Herrera 1986). The differences between individuals in the chemical composition of secretions may facilitate individual recognition from the scent marks (Macdonald et al. 1984; Macdonald 1985; Roberts 2007). One of the main functions of scent marking (particularly using the morrillo) in capybaras is thought to be the maintenance of the strict social hierarchy in males (Herrera and Macdonald 1993; Salas 1999), due to different scent-marking behavior in males of lower status, as outlined above (Herrera and Macdonald 1994). This is consistent with the general pattern observed among rodents (Roberts 2007). In capybaras, since the dominance hierarchy, and especially the dominant position, are maintained year-round and for several years (Herrera and Macdonald 1993), the role of scent marking in the maintenance of social status cannot be overestimated. This is further corroborated and emphasized by the large investment in testosterone-producing tissue at the expense of sperm-producing tissue in testes of capybaras (Moreira et al. 1997; Costa and Paula 2006; see also López et al. 2008; Paula and Walker 2012), leading to a correlation between testosterone concentration in blood and size of the morrillo (Costa and Paula 2006). Moreira et al. (1997) have also suggested that the morrillo may be a visual signal of dominance.

Scent marking is also commonly used in mammals to demarcate territory (Gosling 1990); as capybaras defend territory it is possible that scent marking is also used for this purpose (Herrera and Macdonald 1994). However, as the secretion is always deposited on a plant, there are structural limitations to the locations of scent marks of capybaras. The distribution of bushes and shrubs in capybaras' habitat is patchy and irregular (Herrera and Macdonald 1989) and, therefore, so are the scent marks. For this reason, the territory of a group of capybaras, although precisely limited in space (Herrera and Macdonald 1989), cannot be systematically marked on the borders. As the patterns of space utilization are probably closely linked to activity patterns, it is possible that capybaras do not require an immediate deterrent at territorial borders (Herrera and Macdonald 1994), but marking of territory may provide a mechanism for social cohesion as well as for the defense of limited resources.

It is possible that scent marking also functions in identification of group membership, particularly in the patterns shown by females and subordinate males using the anal gland (Herrera and Macdonald 1994). Females often mark just after the dominant male, possibly to indicate their association with him as well as possibly to demonstrate their group membership (Herrera and Macdonald 1994).

References

Albone ES, Gosden PE, Ware GC, Macdonald DW, Hough NG (1978) Bacterial action and chemical signaling in the red fox (*Vulpes vulpes*) and other mammals. In: Bullard RW (ed) Flavour chemistry of animal foods, vol 67, ACS symposium series. American Chemical Society, Washington, USA, pp 78–91

Arakawa H, Blanchard DC, Arakawa K, Dunlap C, Blanchard RJ (2008) Scent marking behavior as an odorant communication in mice. Neurosci Biobehav Rev 32:1236–1248

Azcaráte T (1980) Sociobiologia y manejo del capybara. Doñana Acta Vertebrata 7–6 (special number):1–228

Brown RE (1979) Mammalian social odors: a critical review. In: Rosenblatt JH (ed) Advances in the study of behavior, vol 10. Academic, New York, pp 103–162, 1–310

Brown RE, Macdonald DW (1985a/b) Social odours in mammals, vols 1 and 2. Oxford University Press, Oxford

Costa DS, Paula TAR (2006) Testosterone level, nasal gland volume and Leydig cell morphometry in capybaras (*Hydrochoerus hydrochaeris*). Arq Bras Med Vet Zoo 58:1086–1091

Drea CM, Vignieri SN, Kim S, Weldele ML, Glickman SE (2002) Responses to olfactory stimuli in spotted hyenas (*Crocuta crocuta*): II. Discrimination of conspecific scent. J Comp Psychol 116:342–349

Gosling LM (1990) Scent marking by resource holders: alternative mechanisms for advertising the costs of competition. In: Macdonald DW, Müller-Schwarze D, Natynczuk SE (eds) Chemical signals in vertebrates 5. Oxford University Press, New York, pp 315–328, 659

Herrera EA (1986) The behavioural ecology of the capybara *Hydrochoerus hydrochaeris*. D.Phil. thesis, University of Oxford, Oxford, p 256

Herrera EA (1992) Size of testes and scent glands in capybaras, *Hydrochoerus hydrochaeris* (Rodentia: Caviomorpha). J Mammal 73:871–875

Herrera EA, Macdonald DW (1989) Resource utilization and territoriality in group-living capybaras (*Hydrochoerus hydrochaeris*). J Anim Ecol 58:667–679

Herrera EA, Macdonald DW (1993) Aggression, dominance, and mating success among capybara males (*Hydrochoerus hydrochaeris*). Behav Ecol 4:114–119

Herrera EA, Macdonald DW (1994) Social significance of scent marking in capybaras. J Mammal 75:410–415

Huck UW, Banks EM (1982) Male-dominance status, female choice and mating success in the brown lemming, *Lemmus Trimucronatus*. Anim Behav 30:665–675

Lanyon CV, Rushton SP, O'Donnell AG, Goodfellow M, Ward AC, Petrie M, Jensen SP, Gosling LM, Penn DJ (2007) Murine scent mark microbial communities are genetically determined. FEMS Microbiol Ecol 59:576–583

Leuchtenberger C, Mourão G (2008) Scent marking of giant otter in the Southern Pantanal, Brazil. Ethology 115:210–216

López M, Muñoz MG, Herrera EA (2008) Reproductive morphology of male capybaras: no evidence for sperm competition? Mamm Biol 73:241–244

Macdonald DW (1981) Dwindling resources and the social behavior of capybaras, (*Hydrochoerus hydrochaeris*) (Mammalia). J Zool 194:371–391

Macdonald DW (1985) The rodents IV: suborder Hystricomorpha. In: Brown RE, Macdonald DW (eds) Social odours in mammals, vol 1. Clarendon, Oxford, pp 480–506, 1–506

Macdonald DW, Krantz K, Aplin RT (1984) Behavioral, anatomical and chemical aspects of scent marking amongst capybaras (*Hydrochoerus hydrochaeris*) (Rodentia, Caviomorpha). J Zool 202:341–360

Moreira JR, Clarke JR, Macdonald DW (1997) The testis of capybaras (*Hydrochoerus hydrochaeris*). J Mammal 78:1096–1100

Müller-Schwarze D (1983) Scent glands in mammals and their function. In: Eisenberg JF, Kleiman DG (eds) Advances in the study of mammalian behaviour, vol 7. Special Publication, The American Society of Mammalogists, Shippensburg, pp 150–197, 1–753

Novotny M, Jemiolo B, Harvey S (1990) Chemistry of rodent pheromones: molecular insights into chemical signalling in mammals. In: Macdonald DW, Müller-Schwarze D, Natynczuk SE (eds) Chemical signals in vertebrates 5. Oxford University Press, New York, pp 1–22, 659

Passanisi WC, Macdonald DW (1990) Group discrimination on the basis of urine in a farm cat colony. In: Macdonald DW, Müller-Schwarze D, Natynczuk SE (eds) Chemical signals in vertebrates 5. Oxford University Press, New York, pp 336–345, 659

Paula TAR, Walker NJ (2012) Reproductive morphology and physiology of the male capybara. In: Moreira JR, Ferraz KMPMB, Herrera EA, Macdonald DW (eds) Capybara: biology, use and conservation of an exceptional neotropical species. Springer, New York, pp 107–129

Rasa OAE (1973) Marking behaviour and its significance in the African dwarf mongoose, *Helogale undulate rufula*. Z Tierpsychol 32:449–488

Rewell RE (1949) Hypertrophy of sebaceous glands on the snout as a secondary male sexual character in the capybara, *Hydrochoerus hydrochaeris*. Proc Zool Soc Lond 119:817–819

Roberts CS (2007) Scent marking. In: Wolff JO, Sherman PW (eds) Rodent societies. An ecological and evolutionary perspective. Chicago University Press, Chicago, pp 255–266, 610

Ryon J, Brown RE (1990) Urine marking in female wolves (Canis lupus): an indicator of dominance status and reproductive state. In: Macdonald DW, Müller-Schwarze D, Natynczuk SE (eds) Chemical signals in vertebrates 5. Oxford University Press, New York, pp 346–351, 659

Salas V (1999) Social organisation in capybaras, *Hydrochoerus hydrochaeris*, in Venezuela. Ph.D. thesis, Cambridge University, Cambridge

Schaller GB, Crawshaw PG (1981) Social organisation in a capybara population. Saugetierkundliche Mitteilungen 29:3–16

Sillero-Zubiri C, Macdonald DW (1998) Scent-marking and territorial behaviour of Ethiopian wolves *Canis simensis*. J Zool 245:351–361

Chapter 11
Capybara Social Behavior and Use of Space: Patterns and Processes

Emilio A. Herrera

11.1 Introduction

Among rodents, the group of caviomorphs (South American Hystricognaths; Vucetich et al. 2012) is usually considered atypical because of the peculiar adaptations of many species in this group, which contrast with features that come to mind when we think about "typical" rodents, such as rats, mice, or squirrels. Among the characteristics that make caviomorph rodents special is of course their large size: from pacas (*Cuniculus paca*, 7–12 kg) to coypus (*Myocastor coypus* 5–9 kg) and capybaras (*Hydrochoerus hydrochaeris*, 50 kg), caviomorphs include the largest of all rodents. Additionally, caviomorphs show a number of unique (among rodents) adaptations and ecological niches, in some cases exhibiting striking convergences with ungulates from other continents (Eisenberg and Mckay 1974; Kleiman 1974), including almost all forms of social behavior and mating systems. Thus, for instance, there are monogamous caviomorphs such as the Patagonian maras (*Dolichotis patagonum*; Taber and Macdonald 1992) while cavies are clearly promiscuous (*Cavia* sp.; Rood 1972; Schwarz-Weig and Sachser 1996). There are also highly social species such as the subterranean social tuco-tucos (*Ctenomys sociabilis*; Zenuto et al. 1999) and the capybaras.

Capybaras are indeed fundamentally social: virtually everywhere they have been studied, they form groups (Azcárate 1980; Macdonald 1981a; Schaller and Crawshaw 1981; Alho et al. 1987; Herrera and Macdonald 1987; Quintana and Rabinovich 1993; Salas 1999; and the review by Herrera et al. 2011); the only – albeit important – exception being the capybaras from the Amazonian rainforest, which live in very small family groups formed simply by the adult pair (or, less often, a trio of one male and two females) and offspring from one or two litters (Soini and Soini 1992),

E.A. Herrera (✉)
Departamento de Estudios Ambientales, Universidad Simón Bolívar,
Apartado 89000, Caracas 1080-A, Venezuela
e-mail: eherre@usb.ve

J.R. Moreira et al. (eds.), *Capybara: Biology, Use and Conservation of an Exceptional Neotropical Species*, DOI 10.1007/978-1-4614-4000-0_11,

demonstrating great behavioral plasticity. Capybara groups exhibit a characteristic social structure that is stable over time (described below). However, within this general trend several studies have clearly shown that there are certain features that are rigid, while others are flexible (Herrera et al. 2011). In this chapter, I review the main features of capybara social behavior and their geographic variations, as well as the species' use of time and space, including its territorial behavior and communication by scent marking. The patterns are described and interpreted within the context of capybaras' adaptation to their ecological circumstances.

11.2 Use of Space and Time

Like many other grazers, capybaras exhibit well-defined patterns for using their space and time. Typically, a groups of capybaras rest in the morning; spend most of the early afternoon (1300–1600) in the water, presumably thermoregulating; and set out to graze from late afternoon into the night, when they alternately feed and rest until dawn (Ojasti 1973; Macdonald 1981a; Herrera 1985). During the morning resting hours, capybaras practice cecotrophy (Herrera 1985, 2012; Borges et al. 1996), an adaptation to their herbivorous diet (Clauss and Hummel 2005), equivalent to rumination in cervids and bovids (Artiodactyla). This general pattern shows seasonal variation. Thus, in the dry season mornings, rest is frequently interrupted for cecotrophy (up to three times an hour), while in the wet season mornings capybaras often get up from resting to graze (Herrera 1985; Barreto and Quintana 2012).

Use of space also follows predictable daily patterns: resting occurs in a small patch very close to the water, while thermoregulation is carried out in a specific section of the pool, and grazing follows a more variable route throughout the home range (Herrera and Macdonald 1989). In fact, some evidence suggests that movement patterns shift as animals cover different parts of their home range on successive days (Fig. 11.1; Barreto and Herrera 1998; Barreto and Quintana 2012), an apparent adaptation to allow grass to recover from grazing.

One important feature of capybara groups (Fig. 11.2), closely linked to the fact that groups have permanent members (see next section), is that they are territorial: the pattern of space use as described by Herrera and Macdonald (1989) clearly demonstrates exclusive use of a particular area (Fig. 11.3). Overlapping sections in Fig. 11.3 are places where animals from each neighboring group would enter the other's territory only to be evicted. Others are sections of a territory that a group from an adjacent territory would cross to reach a patch of their own territory. The area covered by the territory is quite stable, with few changes in shape and size for 3 years and possibly more. Capybara group members actively reject intruders, to the point where several members of a group – females and subordinate males included – will chase members from a neighboring group if for any reason (fleeing from a predator, for instance) they come into close contact (Herrera and Macdonald 1989).

Home range, which may be considered equivalent to the territory (except perhaps for areas visited during the night), covers from 6 to 16 ha (Herrera and Macdonald 1989), although home ranges up to 56 ha (in Colombia; Perea and Ruiz 1977)

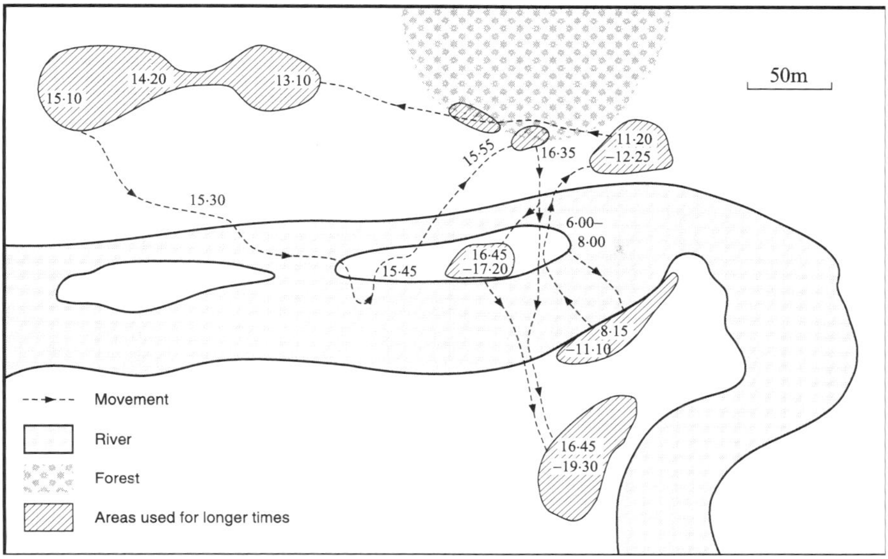

Fig. 11.1 Typical movements of a group of capybaras throughout a day. Large *striped* areas are feeding patches where the animals spend several hours grazing (From Macdonald (1981a))

Fig. 11.2 A group of capybaras at Hato El Cedral, Apure, Venezuela (Photo by E. Congdon)

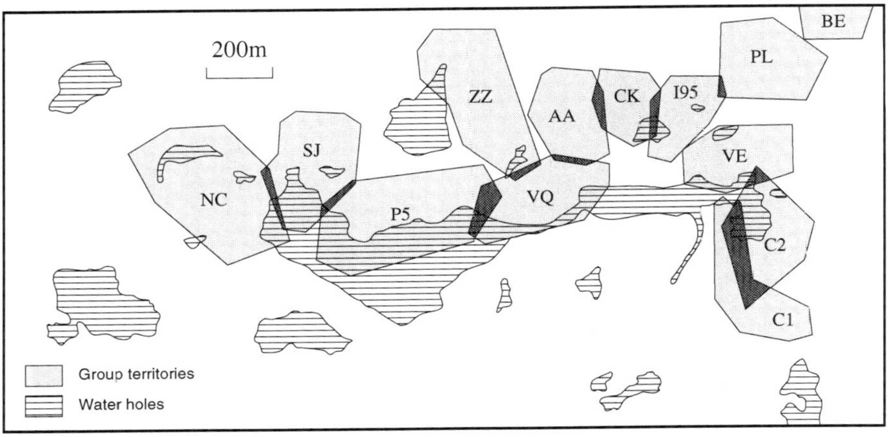

Fig. 11.3 Distribution of capybara group territories (*shaded polygons* with two-letter codes) around a water hole (areas filled with *broken lines*) at Hato El Frío, Apure, Venezuela (From Herrera and Macdonald (1989))

or even 200 ha (in Brazil; Schaller and Crawshaw 1981) in size have been recorded. The latter probably included excursions not normally assumed to be part of the home range (Herrera et al. 2011). Although grass appears to be a widespread, non-defendable commodity in the Llanos, a patch of grass next to a water hole large and deep enough to retain water through the dry season is a scarce resource worth defending (Herrera and Macdonald 1989). In the Peruvian Amazon, capybara groups defend a small patch of a very large home range, for example, 17–22 ha for a pair with their offspring (Soini and Soini 1992). The highly dispersed food patches in this ecosystem seem to determine home range size, as predicted by the Resource Dispersion Hypothesis (Macdonald 1983).

As will be seen in the next section, groups are stable social units which share not only a territory but also less tangible assets such as a social structure in which males avoid overt acts of aggression and females cooperate in the nursing of the group (Nogueira et al. 2000), while all members contribute to group defense by being vigilant and giving alarm barks. Thus, maintaining group membership and rejecting intruders appear to be additional benefits of territorial behavior.

11.3 Social Structure

11.3.1 Group Size and Composition

The salient feature of capybara behavior is undoubtedly their gregariousness, since capybaras are rarely seen in isolation (Fig. 11.2; but see Soini and Soini 1992). Almost immediately after birth, infants join crèches, which are even more cohesive than adult associations. After a wide study of many aspects of capybara biology, ecology, and population dynamics in the Llanos of Venezuela, Ojasti (1973) described capybaras as living in groups of both sexes and all ages, and the herd as a "closed society with a hierarchical organization" (my translation), an insight corroborated by later studies (Herrera and Macdonald 1987, 1993). In a review of capybara social behavior and its variation, Herrera et al. (2011) give a range of group sizes of 5–100, although the more stable units are not more numerous than 20–30 animals. It should be noted that to determine social group size, it is necessary to watch marked animals for several days or even weeks, since many come in and out of sight, and vegetation and water may hide the animals, causing errors in the estimation of group size. For this reason Herrera et al. (2011) excluded casual reports from their review. There is, however, an important phenomenon that creates additional temporary variation in group size in capybaras: in the dry season, capybaras come together to the dwindling pools, causing associations of 100 animals or more to form (Ojasti 1973). Herrera and Macdonald (1987) stress that these associations are ephemeral and do not destroy the social structure of the coalescing groups. In some areas of the Brazilian Pantanal, there is a curious, apparently inverse, effect as animals concentrate on the higher, dry patches at the peak of the wet season (Alho and Rondon 1987), although groups also coalesce in the dry season as they do in the Venezuelan and Colombian Llanos (Jorgenson 1986).

Herrera and Macdonald (1987) and Salas (1999) concluded that groups are effectively closed societies, composed of both adult males and females, with a bias toward females. Group members do not tolerate interlopers, making size and composition of groups relatively stable. Male to female sex ratio in social groups varies from 1:1.7 (Herrera and Macdonald 1989) to 1:3 (Alho and Rondon 1987). Male capybaras are permanent members of the groups, in contrast with other mammalian societies in which polygyny is more marked and a proportion of males live in solitary or bachelor groups (e.g., stag groups in *Cervus elaphus*; Clutton-Brock et al. 1982). Salas (1999), however, found that 40% of all animals she observed on a ranch in Venezuela appeared to be unaffiliated to any particular group, while Alho and Rondon (1987) in Brazil reported that 8% of capybaras were "satellite". In both cases most of these floaters were males.

Herrera and Macdonald (1987) do not give a figure for the percentage of floater males in their study, but it is clearly lower than the 40% reported by Salas (1999) on a ranch just some 50 km away. The two ranches, one where Herrera and Macdonald (1987, 1989, 1993) carried out their study and the other where Salas (1999) performed hers, are apparently very similar since they are located in the same general habitat – the seasonally flooded savannas of the Venezuelan Llanos – but they exhibit important differences in resource availability and distribution for capybaras. At Salas' (1999) study site (Hato El Cedral), a water management system consisting of well-planned dykes with more technical water-level control contrasts with a makeshift, loosely managed system at Hato El Frio, where Herrera and Macdonald (1987, 1993) worked. These differences make resources (water and grass) more stable both in time and space at El Cedral than at El Frio (Herrera et al. 2011). As a consequence, there is a greater density of capybaras at El Cedral, with a concomitant increase in various features of their social structure, such as the formation of larger groups, more floating males and a sex ratio more biased toward females (Salas 1999; Herrera et al. 2011).

11.3.2 Advantages of Group Living

Group size was found to correlate with reproductive success (Herrera and Macdonald 1989), but in the higher density and larger groups of Salas (1999) the relationship was not significant. This may be because, in the latter study site, optimal group size had been exceeded. Groups appear to confer protection against predators, both passively (the "dilution effect"; Krebs and Davies 1993) and actively, as adults protect the young by rounding them up, facing out, especially in the water (Macdonald 1981a). Within a group, females reproduce quite synchronously, with most females giving birth within a period of 2 weeks at the end of the wet season (Herrera 1986). This allows the young to benefit from grouping and the communal nursing of their mothers (Macdonald 1981a). Vigilance behavior is closely associated with group living, as members of larger groups scan less frequently than those in smaller groups, while providing the group with more heads up per hour (Fig. 11.4; Yáber and Herrera 1994).

Females appear to benefit most from group living. They occur significantly more often near the center of the group than males (the dominant male excepted), thus taking full advantage of the group's protection against predators. They participate in territorial behavior and will evict conspecific intruders (Herrera and Macdonald 1989), and they share in the nursing of their young (Macdonald 1981a; Herrera and Macdonald 1987). These phenomena of apparent collaboration suggest a role for kin selection in their evolution (Hamilton 1963), implying that females are related. Herrera et al. (in prep.) did in fact find that females in groups appeared to be significantly more related than the population at large, while males were less so (Herrera, Salas and Carreño, unpublished data).

11.4 The Dominance Hierarchy of Males

The most obvious feature of capybara society is the dominance hierarchy among the males. Originally suggested by Ojasti (1973), it has been corroborated repeatedly in later studies (Herrera and Macdonald 1993; Salas 1999; Bedoya 2007). The dominance hierarchy stems from ritualized agonistic behavior patterns consisting mainly of simple chases. Occasionally a chased male will turn around and attack his aggressor, which results in a brief all-out fight, both males standing on their hind feet trying to bite each other. In most cases of aggression, a loser and a winner are quite obvious, allowing the observer to score the result of the interaction and construct the hierarchy. As can be seen in Table 11.1, such a dominance hierarchy is extremely clear cut with most dyads scoring n-0, and very few reversals (Herrera and Macdonald 1993).

Table 11.1 Dominance hierarchy among capybara males in capybara groups studied by Herrera and Macdonald (1993). Three-digit-letter codes identify the animals. The numbers are the number of times the animal in the left-hand column beat the animal in the top row (a and b are different groups in different years)

(a)

	R15	WTE	WNA	1DG
R15	–	15	9	3
WTE	0	–	8	0
WNA	0	0	–	1
1DG	0	0	0	–

(b)

	I65	G14	WW7	W14
I65	–	4	5	4
G14	0	–	1	4
WW7	0	0	–	3
W14	0	0	0	–

The top-ranking position in the hierarchy may last up to 3 years, while others are also relatively stable. Occasionally, the dominant male is ousted by an intruder or may lose his position for unknown reasons (Herrera and Macdonald 1993). When the top male disappears, the rest of the animals climb one position in the ladder, suggesting the existence of a queuing system as proposed by Kokko and Johnstone (1999). A correlation between age and rank found by Salas (1999) is consistent with this view. Interestingly, during the agonistic encounters, the dominant male is unable to oust the subordinates altogether. Rather, the subordinate describes a wide arc as he walks away from his opponent, which brings him back into the group, avoiding eviction. The costs of all-out fights and the benefits of sharing a territory – and its defense – appear to promote this relatively tolerant behavior on the part of the dominant male.

The main advantage of a top-ranking male is access to receptive females: 75% and 81% of copulations observed by Herrera and Macdonald (1993) and Salas (1999), respectively, were performed by the dominant male of the group. Assuming that mating success correlates with reproductive success, there is an obvious evolutionary advantage for dominant males. Subordinate males do have access to mating, which led to the suspicion that sperm competition may occur between capybara males. However, capybaras do not show the typical anatomical correlates of sperm competition, such as large testes (Birkhead and Pizzari 2002). In fact, the opposite is true: capybara testes are slightly smaller than expected for an animal their size (Herrera 1992a), and their gonads have a large proportion of testosterone-producing tissue, rather than sperm-producing tissue (Moreira et al. 1997). Still, two lines of evidence suggest that a certain degree of sperm competition may occur in capybaras. First, a vaginal plug was described by Ojasti (1973), which is usually interpreted as a mechanism in the competition for fertilization of the female. And second, López et al. (2008) found that subordinate males have a greater proportion of sperm-producing tissue in their testes, apparently to compensate for their smaller gonads (Herrera 1992a). The short estrus of females (López-Barbella 1982) seems to be the factor allowing dominant males to guard their females successfully against other males (Bedoya 2007), thereby reducing the opportunities for sperm competition. Additionally, dominant males often interrupt courtship by subordinates (Herrera and Macdonald 1993; Salas 1999). There is also evidence of female mate choice: when courted by a dominant male, her behavior is less agonistic toward the courting male than when courted by a subordinate, and the route she walks is more complex (more turns and more going in and out of the water), perhaps to allow the dominant male to interfere (Bedoya 2007).

There does not appear to be any other significant advantage of higher ranking for males, in terms of, for instance, access to resources such as water or grass (Herrera and Macdonald 1993). The only possible exceptions to this are water holes used for wallowing, especially in the dry season, to which dominant males have priority of access; and the tick-removing behavior by yellow-headed caracaras (*Milvago chimachima*, Aves: falconiformes; Macdonald 1981b) to which dominants also have preferential access (Herrera and Macdonald 1993).

Subordinate males contribute disproportionately to vigilance: although the mean individual rate of vigilance diminishes with group size (Fig. 11.4), this effect is mainly due to females, since subordinate males maintain a constant rate of scanning even as group size increases (Yáber and Herrera 1994). Subordinates also perform more alarm calls than other group members (Herrera and Macdonald 1993). The tolerance exhibited by dominants toward subordinates seems to be compensated by three forms of "payment" (sensu Kokko and Johnstone 1999): territorial defense, vigilance, and alarm calling.

Females do not appear to have a particular social structure among them, except in captivity where a dominance hierarchy is evident (Ojasti and Sosa-Burgos 1985; Ferraz et al. 2012). Curiously, females in captivity can commit infanticide, but only when unfamiliar females are grouped (Nogueira et al. 1999). Infanticide by males has been very occasionally observed in the wild (Salas 1999), but its rare occurrence precludes interpretation.

11.5 Communication

Living in a closed social unit with a complex social structure means that communication is important in capybaras. Several communication mechanisms have evolved in this species, most notably chemical communication, via two glands: a snout gland, called the "morrillo," which tends to be more developed in males (so much so that it can be used, although not infallibly, to identify the sexes), and an anal gland located on both sides of the anus developed in both sexes (Macdonald and Herrera 2012). Anal glands differ between the sexes: males have dry crystalline hairs in the glands, while females have a more typical moist, greasy gland (Macdonald et al. 1984; Macdonald and Herrera 2012). Possibly, marks with the anal gland play a role in group membership identification while marks with the morrillo have to do with signaling hierarchical status by males.

Visual communication does not appear as important as chemical signaling, but males do sometimes chase each other in response to subtle gestures that are

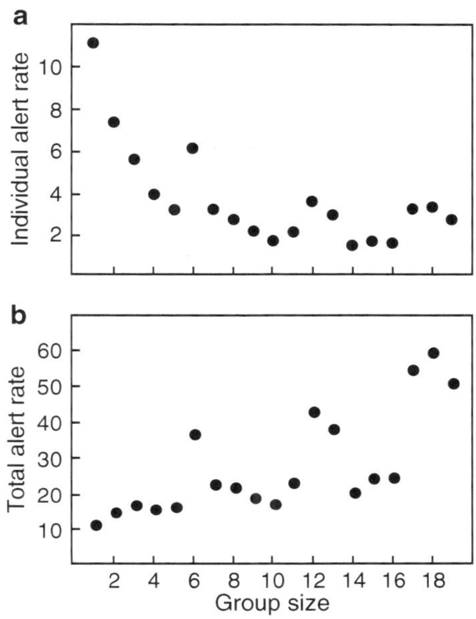

Fig. 11.4 Vigilance in capybaras (**a**) mean number of times per hour (scanning rate) that individual capybaras raise their heads as a function of group size and (**b**) mean group scanning rate as a function of group size (From Yáber and Herrera (1994))

apparently visual, and Moreira et al. (1997) suggest that the morrillo may also function as a visual signal in addition to its more immediate role as a scent gland.

Vocal communication among capybaras is very common but little understood. It is composed of at least seven different sounds that appear to be group specific (Barros et al. 2010), perhaps contributing to group cohesion by membership recognition. Young capybaras emit a characteristic whining or whistling sound very frequently throughout the day. This seems to play a role in maintaining contact both among themselves (young capybaras move around and rest together most of the time) and with their mother or other females. Abandoned young are exposed to predators and may also die of exposure (personal observation), so keeping in touch with the rest of the group is literally a matter of life and death. Other life-saving means of auditory communication are vigilance and associated alarm calling (Herrera and Macdonald 1993; Yáber and Herrera 1994). Scanning is quite subtle; animals simply raise their heads just above a normal standing position. The frequency of scanning is related both to social status (subordinate males do it more often) and to group size (animals in larger groups scan less frequently but are also protected by a greater group scanning rate; Fig. 11.4; Yáber and Herrera 1994). Whenever a group member detects a threat, he or she will emit an alarm call consisting of a loud, low-pitched bark. The immediate reaction of all other group members is to stand alert. Usually, the threat does not materialize and capybaras return to grazing or resting. However, if the threat continues, more animals will start calling and eventually all will run en masse to the nearest water hole where they will remain until the threat is gone. In this situation, the adults may make a circle around the young, facing out (Macdonald 1981a). Subordinate males give more alarm calls than other group members (Herrera and Macdonald 1993).

11.6 Dispersal

Dispersal is an important process in the ecology of any organism as it affects mortality rates, species' distribution patterns and the genetic structuring of populations (Aars and Ims 2000). Using mark-recapture methods, Herrera (1992b) found that in capybaras both sexes dispersed equally and that they appeared to do it in groups, both of which are rare occurrences in mammals (Greenwood 1980). That they dispersed in groups was further supported by observations of groups of juveniles wandering away from their parental group, apparently accompanied by a subordinate male (Herrera and Macdonald 1987). The median dispersal distance recorded by Herrera (1992b) was 3.4 km while the maximum was 5.6 km. In that study, neither philopatry nor dispersal was different between sexes. A similar situation was found in another rodent from northern Africa, the gundi (*Ctenodactylus gundi*; Nutt 2005) where both sexes are equally philopatric while dispersal is male-biased. Dispersing in mixed-sex groups might have a similar effect to philopatry by both sexes, leading to inbreeding or at least strong genetic structuring (Dobson 2007).

In a location not far from Herrera's study, and several years later, Salas (1999) found females to be philopatric, while all males appeared to disperse. More recently, Congdon (2007) found similar results. This location had a more abundant and homogenous resource base (Hato El Cedral, see above) than Herrera's (1992b) study site. Variation in patterns of dispersal in relation to ecology has been observed in, for instance, red deer (*Cervus elaphus*; Pérez-González and Carranza 2009). It should be noted, nonetheless, that Salas (1999) observed one case of a male and a female dispersing together for a distance of 3 km (about six home range widths) and Congdon (2007) saw a group of three juveniles (two males and one female) apparently dispersing together. These observations show that group dispersal cannot be ruled out, even in the different conditions of Salas' and Congdon's studies. Clearly, however, capybara dispersal patterns are affected by ecological circumstances, with greater density (as in Salas' and Congdon's studies) leading to male-biased dispersal. In the latter case, females within groups could be more related than males, by virtue of their philopatry, and this is supported by data from Herrera et al. (in prep.) using DNA samples from Salas' (1999) animals. As stated above, relatedness among females might be associated with their apparent cooperation.

11.7 Summary and Final Remarks

Initially, capybara groups may have formed to escape predators, in the "selfish herd" manner (Hamilton 1971). Several currently observed behavior patterns suggest that this is so: fleeing behavior, permanent closeness to water, vigilance, and other such behaviors. Gregariousness, once established, leads to the evolution of specific behaviors associated with group living, such as the formation of a dominance hierarchy, communal territorial defense, vigilance and the like.

The social structure of capybaras can be briefly described as having closed social units with members of both sexes, a rigid dominance hierarchy among the males, and group territoriality. Additionally, the sharing of nursing among females, vigilance behavior, alarm calling, and group territoriality are all facets of cooperation. Several features of capybara social behavior appear to be quite fixed, such as the stability of the groups and the dominance hierarchy. Others, such as group size and the proportion of floater males, are more variable (Herrera et al. 2011) and seem to be related to ecological factors: in an area with more predictable, abundant, and homogeneous (in time and space) resources, groups were larger and their composition more female-biased, there were more floaters, and females were more philopatric (Herrera et al. 2011). In the rainforest, capybaras live along the rivers in pairs or trios (Soini and Soini 1992), demonstrating an even greater adaptability and behavioral plasticity.

Territoriality seems to be an adaptation of individuals in groups to gain exclusive access to a limited and widely distributed commodity such as grassy patch next to a near-permanent water hole (Herrera and Macdonald 1989). Additionally, the presence of patches of higher ground in which to avoid flooding at the height of the wet season is also a requirement of territories (Salas 1999). The observations by Soini and Soini (1992) that pairs of capybaras lived in home ranges as large as those of 10 to 16-strong groups supports the resource dispersion prediction that resource dispersal affects home range size, while other factors affect group size.

Dispersal patterns were also variable in relation to contrasting ecological circumstances: both sexes dispersing in groups in the lower density, less homogeneous terrain, and female philopatry in the more predictable habitat. The latter would imply genetic structuring such that related females tend to remain in their groups, which may then explain the evolution of cooperative behavior via kin selection (Hamilton 1963), and this is what has been largely found (Herrera et al. in prep.).

References

Aars J, Ims RA (2000) Population dynamic and genetic consequences of spatial density-dependent dispersal in patchy populations. Am Nat 155:252–265

Alho CJR, Rondon NL (1987) Habitat, population densities, and social structure of capybaras, (*Hydrochaeris hydrochaeris*, Rodentia) in the Pantanal, Brazil. Rev Bras Zool 4:139–149

Alho CJR, Campos ZMS, Gonçalves HC (1987) Ecologia da capivara, (*Hydrochaeris hydrochaeris*, Rodentia) do Pantanal: I. Habitats, densidades e tamanho de grupo. Rev Bras Biol 47:87–97

Azcárate T (1980) Sociobiología del capibara (*Hydrochoerus hydrochaeris*). Doñana Acta Vertebrat 7–6:1–228

Barreto GR, Herrera EA (1998) Foraging patterns of capybaras in a seasonally flooded savanna in Venezuela. J Trop Ecol 14:87–98

Barreto GR, Quintana RD (2012) Foraging strategies and feeding habits of capybaras. In: Moreira JR, Ferraz KMPMB, Herrera EA, Macdonald DW (eds) Capybara: biology, use and conservation of an exceptional neotropical species. Springer, New York, pp 83–96

Barros KS, Tokumaru RS, Pedroza JP, Nogueira SSC (2010) Vocal repertoire of captive capybara (*Hydrochoerus hydrochaeris*): structure, context and function. Ethology 116:1–11

Bedoya M (2007) Elección de pareja en hembras de *Hydrochoerus hydrochaeris*. Graduate dissertation, Universidad Simón Bolívar, Caracas

Birkhead TR, Pizzari T (2002) Postcopulatory sexual selection. Nat Rev Genet 3:262–273

Borges PA, Domínguez MG, Herrera EA (1996) Digestive ecophysiology of capybaras. J Comp Physiol B 166:55–60

Clauss M, Hummel J (2005) The digestive performance of mammalian herbivores: why big may not be that much better. Mammal Rev 35:174–187

Clutton-Brock TH, Guinness FE, Albon SD (1982) Red deer. Chicago University Press, Chicago

Congdon ER (2007) Natal dispersal and new group formation in capybaras (*Hydrochoerus hydrochaeris*) in a seasonally flooded savanna of Venezuela. Ph.D. thesis, University of Missouri at Saint Louis, Saint Louis

Dobson FS (2007) Gene dynamics and social behavior. In: Wolff JO, Sherman PW (eds) Rodent societies, an ecological and evolutionary approach. Chicago University Press, Chicago, pp 163–172

Eisenberg JF, Mckay GM (1974) Comparison of ungulate adaptations in the new world and the old world tropical forests with special reference to Ceylon and the rainforests of Central America. In: Geist V, Walther F (eds) The behavior of ungulates and its relation to management. IUCN NS, Morges, pp 585–602

Ferraz KMPMB, Izar P, Sato T, Nishida SM (2012) Social and spatial relationships of capybaras in a semi-confined production system. In: Moreira JR, Ferraz KMPMB, Herrera EA, Macdonald DW (eds) Capybara: biology, use and conservation of an exceptional neotropical species. Springer, New York, pp 243–260

Greenwood PJ (1980) Mating systems, philopatry and dispersal in birds and mammals. Anim Behav 28:1140–1162

Hamilton WD (1963) The evolution of altruistic behavior. Am Nat 97:354–356

Hamilton WD (1971) Geometry for the selfish herd. J Theor Biol 31:295–311

Herrera EA (1985) Coprophagy in the capybara, *Hydrochoerus hydrochaeris*. J Zool (Lond) 207A:616–619

Herrera EA (1986) The behavioural ecology of the capybara. D.Phil. thesis, University of Oxford, Oxford

Herrera EA (1992a) Size of testes and scent glands in capybaras, *Hydrochaeris hydrochaeris* (Rodentia: Caviomorpha). J Mammal 73:871–875

Herrera EA (1992b) Growth and dispersal of capybaras, *Hydrochaeris hydrochaeris*, in the llanos of Venezuela. J Zool (Lond) 228:307–316

Herrera EA (2012) Capybara digestive adaptations. In: Moreira JR, Ferraz KMPMB, Herrera EA, Macdonald DW (eds) Capybara: biology, use and conservation of an exceptional neotropical species. Springer, New York, pp 97–106

Herrera EA, Macdonald DW (1987) Group stability and the structure of a capybara population. Symp Zool Soc Lond 58:115–130

Herrera EA, Macdonald DW (1989) Resource utilization and territoriality in group-living capybaras. J Anim Ecol 58:667–679

Herrera EA, Macdonald DW (1993) Aggression, dominance and mating success among capybara males. Behav Ecol 4:114–119

Herrera EA, Salas V, Congdon ERC, Corriale MJ, Tang-Martínez Z (2011) Capybara social structure and dispersal patterns: variations on a theme. J Mammal 92(1):12–20

Jorgenson JP (1986) Notes on the ecology and behavior of capybaras in northeastern Colombia. Vida Silvestre Neotrop 1:31–40

Kleiman D (1974) Patterns of behaviour in hystricomorph rodents. Symp Zool Soc Lond 34:171–309

Kokko H, Johnstone R (1999) Social queuing in animal societies: a dynamic model of reproductive skew. Proc Roy Soc Lond Ser B 266:571–578

Krebs JR, Davies NB (1993) Introduction to behavioural ecology. Blackwells, Oxford

López M, Muñoz MG, Herrera EA (2008) Reproductive morphology of capybaras, *Hydrochoerus hydrochaeris* (Rodentia: Hystricognathi): no evidence for sperm competition? Mamm Biol 73:241–244

López-Barbella S (1982) Determinación del ciclo estral en chigüires, *Hydrochoerus hydrochaeris*. Acta Cient Venez 33:497–501

Macdonald DW (1981a) Dwindling resources and the social behaviour of capybaras, *Hydrochoerus hydrochaeris*. J Zool (Lond) 194:371–391

Macdonald DW (1981b) Feeding associations between capybaras, *Hydrochoerus hydrochaeris*, and some bird species. Ibis 123:364–366

Macdonald DW (1983) The ecology of carnivore social behaviour. Nature 301:379–384

Macdonald DW, Herrera EA (2012) Capybara scent glands and scent-marking behavior. In: Moreira JR, Ferraz KMPMB, Herrera EA, Macdonald DW (eds) Capybara: biology, use and conservation of an exceptional neotropical species. Springer, New York, pp 185–193

Macdonald DW, Krantz K, Aplin RT (1984) Behavioral, anatomical and chemical aspects of scent marking amongst capybaras (*Hydrochoerus hydrochaeris*) (Rodentia, Caviomorpha). J Zool 202:341–360

Moreira JR, Macdonald DW, Clarke JR (1997) Correlates of testis mass in capybaras *(Hydrochoerus hydrochaeris)*, dominance assurance or sperm production? J Zool (Lond) 241:457–463

Nogueira SSC, Nogueira-Filho SLG, Otta E, Dias CTS, Carvalho A (1999) Determination of infanticide causes in capybara (*Hydrochoerus hydrochaeris*) groups in captivity. Appl Anim Behav Sci 62:351–357

Nogueira SSC, Otta E, Dos Santos Dias CT, Nogueira-Filho SLG (2000) Alloparental behavior in the capybara (*Hydrochoerus hydrochaeris*). Rev Etologia 2:17–21

Nutt KJ (2005) Philopatry of both sexes leads to the formation of multimale, multifemale groups in *Ctenodactylus gundi* (Rodentia: Ctenodatylidae). J Mammal 86:961–968

Ojasti J (1973) Estudio Biológico del chigüire o capibara. FONAIAP, Caracas, p 270. Reprinted 2011, Editorial Equinoccio, USB, Caracas

Ojasti J, Sosa Burgos L (1985) Density regulation in populations of capybara. Acta Zool Fennica 173:81–83

Perea J, Ruíz S (1977) Organización social y hábitos territoriales del cacó. Graduate Dissertation, Biología Universidad Nacional, Santa Fé de Bogotá, p 109

Pérez-González J, Carranza J (2009) Female-biased dispersal under conditions of low male competition in a polygynous mammal. Mol Ecol 18:4617–4630

Quintana RD, Rabinovich JE (1993) Assessment of capybara (*Hydrochoerus hydrochaeris*) populations in the wetlands of Corrientes, Argentina. Wetlands Ecol Manag 2:223–230

Rood JP (1972) Ecological and behavioral comparisons of three genera of Argentine cavies. Anim Behav Monogr 5:1–83

Salas V (1999) Social organisation of capybaras in the Venezuelan Llanos. Ph.D. thesis, University of Cambridge, Cambridge

Schaller GB, Crawshaw PG (1981) Social organization in a capybara population. Säugetierkd Mitt 29:3–16

Schwarz-Weig E, Sachser N (1996) Social behaviour, mating system and testes size in cuis (*Galea musteloides*). Z Säugetierkd 61:25–38

Soini P, Soini M (1992) Ecologia del ronsoco o capibara (*Hydrochoerus hydrochaeris*) en la reserva nacional Pacaya-Samiria, Peru. Folia Amazonica 4:135–150

Taber AB, Macdonald DW (1992) Spatial organization and monogamy in the mara, *Dolichotis patagonum*. J Zool (Lond) 227:417–438

Vucetich MG, Deschamps CM, Pérez ME (2012) Paleontology, evolution and systematics of capybara. In: Moreira JR, Ferraz KMPMB, Herrera EA, Macdonald DW (eds) Capybara: biology, use and conservation of an exceptional neotropical species. Springer, New York, pp 39–59

Yáber MC, Herrera EA (1994) Vigilance, group size and social status in capybaras. Anim Behav 48:1301–1307

Zenuto RR, Lacey EA, Busch C (1999) DNA fingerprinting reveals polygyny in the subterranean rodent *Ctenomys talarum*. Mol Ecol 8:1529–1532

Part II
Production

Chapter 12
Products and Uses of Capybaras

Max S. Pinheiro and José Roberto Moreira

12.1 Introduction

The capybara is used by people from South America for a variety of purposes across its range. Its main products are meat, leather and fat, but it is also the focus of leisure pursuits and craftwork, as well as providing ingredients for alleged medicinal use and featuring in local folklore. Unfortunately, because wildlife production is usually carried out illegally, very few official records or data are available about this use. Even the legal trade of capybara meat and leather has no official records in some South American countries like Brazil and Argentina.

In this chapter, we describe the different uses of capybaras and their products, in some cases providing an evaluation of the quality of the latter. Although some of the descriptions are anecdotal, usually recounted to the authors by hunters, farmers, and traditional populations, we think they clearly illustrate the scope of the species' importance to the South American population.

12.2 Capybara Meat

Here we describe some of the relevant characteristics of capybara meat, including its yield, presentation (retail cuts), composition, quality, flavor and aroma, and potential by-products.

M.S. Pinheiro (✉)
Embrapa Clima Temperado, Rodovia BR392 km78, Caixa Postal 403, 96010-971
Pelotas, Rio Grande do Sul, Brazil
e-mail: max.pinheiro@embrapa.br

J.R. Moreira
Embrapa Recursos Genéticos e Biotecnologia, Parque Estação Biológica,
Final W5 Norte, 70770-900 Brasília, Distrito Federal, Brazil
e-mail: alencar_moreira@yahoo.com.br

J.R. Moreira et al. (eds.), *Capybara: Biology, Use and Conservation of an Exceptional Neotropical Species*, DOI 10.1007/978-1-4614-4000-0_12,

Table 12.1 Carcass yield of capybaras observed by different authors in animals slaughtered at different live weights and under different management systems

Live weight (kg)	Production system	Carcass yield (%)	Reference
40	Intensive	57.5	Allekotte (2003)
34.4–35.1	Intensive	49.8–50.7	Albuquerque (1993)
24–27	Intensive	61.64	Andrade (1996)
35–40	Semi-intensive	58.26	Pinheiro et al. (2007)
40	Semi-intensive	47	Pinto et al. (2004)
–	–	54	Hosken (1999), Paiva (1992)
20	–	54.4	Cardozo et al. (2005)
63.8	–	51.6	Bressan et al. (2002), Miguel (2001)
44.2	Free-living	52	Ojasti and Medina-Padilla (1972)
40	Free-living	51.5	Godoy and Gómez (1976)
–	Free-living	49.89–64.7	Assaf et al. (1976a, b)
42.2 and 38.2	Free-living	Females (45.7) Males (44.3)	González-Jiménez and Parra (1972)

12.2.1 Carcass Yield

Few studies have been undertaken to determine the yield that can be obtained from a capybara carcass. The meat yield is probably between 44.3% and 67.4% of carcass weight (Table 12.1). Variability in yields is probably due to differences in the production system used, the type of food provided, and the age and weight of the animal, among other factors. Carcass yields for wild capybaras are generally about 50% (Ojasti and Medina-Padilla 1972), and thus about 15% lower than those of farmed individuals.

The percentage of muscle in the carcass also varies among studies, from 61.5% to 75.6% of the live weight (Albuquerque 1993; Andrade 1996). Older animals tend to have a higher proportion of fat (usually around 30%; Miguel 2001; Bressan et al. 2002), most of which is subcutaneous, the remainder being intermuscular fat and connective tissues (including skin).

12.2.2 Meat Cuts

Capybaras have a thick layer of subcutaneous fat (Fig. 12.1; Table 12.2), although in the wild the thickness of the fat varies seasonally and possibly also in relation to habitat (Ojasti 1973; Emilio A. Herrera personal communication). Therefore, the way in which the hide is removed affects the meat yield and its composition (fat content). The animal can be skinned by using a knife or by pulling the hide off with a chain. The latter method leaves a greater deal of fat attached to the skin, so that the carcass and the resultant meat cuts are leaner. In areas where there is no market for

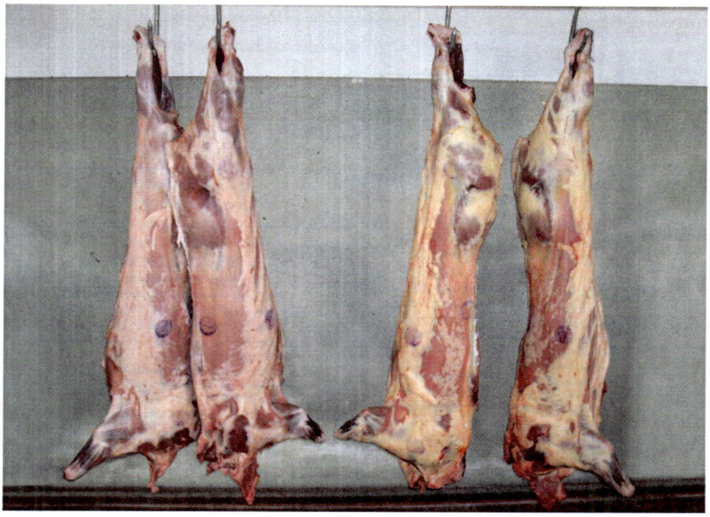

Fig. 12.1 Capybara carcasses in a semi-confined system (Photo by M.S. Pinheiro)

Table 12.2 Characterization of the carcass of female and intact male capybaras reared in semi-confined conditions (Pinheiro et al. 2007)

Parameter	Females	Males
Cold carcass (g)	22,805.00	22,520.00
Cold carcass yield (%)	58.65	57.73
Carcass finishing (1–5)[a]	3.13	3.00
Carcass conformation (1–5)[b]	3.03	2.96
Carcass length (cm)	57.63	55.41
Carcass depth (cm)	23.62	22.68
Leg length (cm)	31.23	30.18
Width of leg (cm)	10.37	9.98
Depth of leg (cm)	13.50	12.48
Backfat at 12th rib (mm)[c]	7.26	6.79
Backfat at 1st rib (mm)	21.61	20.66
Backfat at flank steak (mm)	12.82	11.62
Loin eye area (cm^2)	25.88	25.44
Muscle in the carcass (g)	15,690.79	16,117.09
Muscle in the carcass (%)	67.86	69.92
Fat in the carcass (g)	4,584.58	4,069.41
Fat in the carcass (%)	19.81	17.52
Bone in the carcass (g)	2,837.05	2,887.86
Bone in the carcass (%)	12.33	12.55
Muscle: Bone ratio	5.56	5.62
Edible portion (musc. + fat; g)	20,275.37	20,186.50

[a]1 = Excessively thin; 3 = normal; 5 = excessively fat
[b]1 = Very poor; 3 = good; 5 = excellent
[c]Thickness of subcutaneous fat measured in the loin, at the last rib

Table 12.3 Characterization of the prime cuts of meat from female and intact male capybaras at slaughter weight, reared in a semi-confined system (Pinheiro et al. 2007)

Cut/characterization	Females	Males
Ham		
Weight (g)	3,680.02	3,641.30
Carcass proportion (%)	31.20	31.65
Muscle (g)	2,760.92	2,787.12
Bone (g)	515.83	512.72
Fat (g)	403.28	341.45
Loin		
Weight (g)	920.65	1,048.53
Carcass proportion (%)	7.87	9.14
Muscle (g)	710.53	804.05
Bone (g)	122.99	138.59
Fat (g)	87.13	105.88
Rack		
Weight (g)	1,216.70	1,157.32
Carcass proportion (%)	10.43	10.32
Muscle (g)	826.12	791.28
Bone (g)	223.81	229.61
Fat (g)	166.77	136.42
Palette		
Weight (g)	1,997.97	2,067.37
Carcass proportion (%)	16.76	17.97
Muscle (g)	1,306.28	1,409.30
Bone (g)	263.17	268.13
Fat (g)	428.52	389.93
Rib		
Weight (g)	1,518.12	1,455.07
Carcass proportion (%)	12.96	12.70
Muscle (g)	902.63	909.35
Bone (g)	179.23	168.72
Fat (g)	436.26	377.00

capybara leather, the skin can be left on the carcass and the animal simply shaved, as is done with pigs. The resulting meat yield is high (around 70%), but the meat has a high fat content (Pinheiro et al. 2007).

In general, for the retail market, the carcass of a capybara in Brazil is subdivided into the following cuts (Table 12.3): gammon, palette, ribs, loin, and rack (Pinheiro et al. 2007). The cuts can be sold individually or as a half-carcass for the wholesale market. The loin and ribs are most suitable for barbecuing, a popular way of eating meat in South America. In Venezuela, capybara is sold as one piece of dried meat, only once a year, in Lent (Ojasti 1973, 1991).

Table 12.4 Chemical analysis and cholesterol in muscle tissue of capybaras

Protein (%)	Fat (%)	Energy cal/100g	Cholesterol mg/100g	Reference
20.2–21.4	1.81–4.74	–	45.7–52.1	Girardi et al. (2005)
21.29–22.62	0.36–1.49	–	17.68–33.61	Oda (2002), Oda et al. (2004a)
22.1–22.8	0.3–0.4	–	–	Pinto et al. (2007)
21.17	0.82	–	44	Jardim (2001), Jardim et al. (2003)
20.49	1.4–2.0	–	27–51	Saldanha (2000), Saldanha et al.(2002)
23.1	4.2	–	–	Salgado et al. (1999)
20.04	0.91	–	–	Roça et al. (1999)
24.41–24.54	0.12–0.14	–	–	Lavorenti (1989)
22.1	4.5	135	–	Torres-Gaona (1987)
21.64	0.34	–	–	Godoy and Gómez (1976)

12.2.3 Meat Quality

Capybara muscle tissue has little intramuscular fat and, as a result, an extremely low fat content (0.12–4.5%; Table 12.4). Capybara meat, therefore, has little marbling, a low saturated fat content, high amounts of polyunsaturated fats, and an appropriate balance between ω3 and ω6 fat types, compared to traditional red meats (Oda 2002). The protein content of the meat is high (up to 24%; Oda et al. 2004a).

The relatively low levels of fat/cholesterol and high levels of unsaturated fatty acids (Table 12.5) mean that capybara meat is of higher nutritional quality for humans than other red meats. However, the high polyunsaturated fatty acid content also means that capybara meat begins to decompose, and become rancid, faster than other types of red meat (Oda et al. 2004a, b), although when it is properly packaged and frozen, it can have a shelf life of up to 1 year (Sarkis 2002).

The quality of the meat is also affected by the slaughter method used (Oda et al. 2004b). Slaughtering the animals by shooting resulted in a meat with a higher shear force (5.04 kgf – kilogram force), indicating less tenderness relative to the humane method of slaughtering. The humane method (which consists of stunning the animals before bleeding them) produced meat with a shear force of 3.97 kgf.

12.2.4 Flavor and Aroma

Capybara meat is regarded as having a good flavor, making it suitable for fresh consumption and for sale to restaurants, steakhouses, and supermarkets. However, sometimes capybara meat is recognized as having a strong aroma and flavor that is generally considered off-putting. The cause of this less palatable taste is unknown, despite speculation regarding the possible role of the origin of the

Table 12.5 Fatty acid profile in muscle of farmed animals and in oil of free-living capybaras

	Muscle				Oil	
	Girardi et al. (2005)	Bressan et al. (2004)	Oda (2002)	Saldanha et al. (2002)	Santos et al. (1986)	Fukushima et al. (1997)
Fatty acid	(%)	(%)	(%)	(%)	(%)	(%)
C4:0 or <				25–89	0.92–3.19	
C6:0				0.64	1.03–1.68	
C8:0					0.83–0.94	
C10:0					0.30–0.52	
C12:0					0.29–0.44	
C14:0	1.70–2.93	1.66–3.04	3.64	0.88–1.29	3.99–4.37	3.0
C14i					1.31–1.72	
C15:0				0.70–1.45		
C15:2n-5	0.13–0.19					
C16:0	22.0–24.9	27.64–42.02	29.56	2.66–13.68	20.56–24.32	22.7
C16ri	1.12–2.23				10.01–12.17	
C16:1n-7	1.47–2.60	0.37–2.56	1.90			5.0
C16:1n-9	0.44–0.62			0.57–1.56		
C17:0	1.44–2.61			0.85–1.4		
C17:1n-9	1.13–2.02					
C18:0	3.16–8.16	6.08–9.52	6.57	4.98–6.82	4.16–5.29	4.3
C18:1					35.64–39.88	
C18:1n-6.11c				0.91–1.99		
C18:1n-7	1.03–2.49					
C18:1n-9	23.7–32.0	17.97–35.74	26.9			26.5
C18:1n-9.9tr				0.92		
C18:1n-9.9c				1.19–20.05		
C18:2					4.37–11.83	
C18:2n-6	23.7–28.6	12.10–23.21	19.18	1.28–15.11		19.6

C18:2n-9		3.61–6.01	3.31			
C18:3					0.73–1.95	
C18:3n-3	2.00–3.89	1.00–1.27	4.91	1.73–18.3		17.9
C18:3n-6		0.20–0.32	0.18			
C20					2.58–7.85	
C20:1n-9	0.42–0.70	0.18–0.29	0.20			
C20:1n-11				0.89		
C20:2n-6				0.47–1.45		
C20:3n-3				0.51		
C20:3n-9	0.08–0.40					
C20:4n-6		5.79–12.81	3.45	1.04–1.7		0.7
C20:5n-3		0.11–0.59	0.49	0.79		
C21:0	0.36–2.67					
C21:5n-3	0.06–0.25					
C22:1n-1	0.11–0.13					
C22:4n-6		0.35–0.89	0.74			
C22:6n-3		0.10–0.24	0.19			0.3
Saturated (S)	30.2–42.3	38.36–51.90		50.58–94.87		
Monounsaturated(M)	28.9–39.1	18.6–38.50		1.19–24.05		
Polyunsaturated (P)	27.4–32.7	24.89–67.51		3.48–36.41		
P: M ratio	0.68–1.01			0.036–0.75		0.7
P: S ratio			0.82			
Total n-6		18.48–37.11	23.55	2.86–17.32		
Total n-3		1.44–2.00	5.59	1.73–19.60		
n-6/n-3 ratio			4.21	0.88–3.60		1.12

animal, the rearing methods used, the concentrate level in the food provided, and the age and sex of the animal.

Some authors claim that capybara meat from farmed animals, which receive concentrate-rich feedstuffs, has the best flavor (Cueto and Allekotte 2002; Allekotte 2003; Martinez 2007). Consumers in Brazil rated highly capybara meat from animals raised in a confined system, and slaughtered at a weight of 40 kg, and found no differences in quality in terms of flavor, aroma, texture, and color compared to pork or beef cuts (Nogueira-Filho 1996).

Other researchers associate the strong flavor of the meat with the age of the animal (Montes 1983; José R. Moreira personal observation; Maria C. Bressan personal communication) and suggest that even the meat of an animal fed with concentrates may have a strong flavor if the animal is old. The reason farmed animals usually taste better than wild ones may simply be that confined animals are usually killed at a younger age than wild animals of the same body weight. Hosken (2007) is more specific and suggests that the strong flavor of capybara meat is caused by an increase in carcass fat, and occurs in all farmed animals over the age of 18 months.

Capybara farmers speculate that there is a link between strongly flavored capybara meat and the animal's consumption of grass (G. Castilhos and G. Duarte personal communications; Max S. Pinheiro personal observation). They claim that this occurs when farmed animals graze more extensively, and that this characteristic also occurs in meat from wild animals. Another possible explanation is that the strong smell and taste of some capybara meat is due to the presence of high levels of certain short-chain fatty acids. High levels of butyric acid, for example, were observed in capybara meat from animals fed on pasture in a semi-confined system (Table 12.5; Saldanha et al. 2002). However, there is little information available on the profile of fatty acids in the muscle and fat of capybaras fed on different diets and under different systems with which to make comparisons or to identify particular fatty acids that may be undesirable. Unfortunately no palatability tests with professional tasters have compared the flavor of meat from animals of different ages and fed on different diets to elucidate this important information.

12.2.5 Forms of Meat Processing

Capybara meat has more, and shorter, muscle fibers than beef, which makes it possible to process the meat in diverse ways (Mackey et al. 1976). A very tasty sausage can be made from capybara meat. Fresh sausages, pepperoni, or smoked sausages made adding pork fat are also excellent. Capybara meat is also suitable for the manufacture of frankfurters, mortadella, patés, hams, salamis, cold cuts, hamburgers, and other similar products (González-Jiménez 1995; Marín and Arias 2002; Pinto et al. 2007). Nevertheless, although technically feasible to produce, these products are not available on the market.

In Venezuela, capybara meat is sold in the form of jerky. Salting and drying the meat is the method used for the preservation of meat harvested in remote areas, without ready access to freezers and meat stores. The yield from jerky, however, is only 17% of the animal's live weight since it is dry meat. But it regains weight when rehydrated in cooking.

12.2.6 Head, Viscera, Blood and Bones

As with most domestic animals, farmers aim to use as much of the carcass as possible. From capybaras, the edible viscera (liver, heart, spleen, kidney, and lung), the tongue, and the blood can be used in the manufacture of patés and blood sausages (Marín and Arias 2002; Pinto et al. 2007). The liver can also be used in the form of steaks or strips. The head can be consumed domestically, and both its flesh and the brains are used for stuffing sausage. Bones can be ground for use as a source of calcium in animal feed.

12.3 Fat and Oil

Oil can be extracted from the capybara's subcutaneous and intermuscular fat in a bain-marie, by exposing it to the sun, or by frying. A capybara of commercial slaughter weight (40 kg) provides about 2 kg of fat, from which can be extracted 0.5 L of oil (Allekotte 2003). The price of capybara oil in Argentina is US $100 a liter. Crackling can be produced from the fat remaining after the oil has been extracted.

As with other red meats, capybara fat has a high concentration of palmitic acid (Table 12.5), a major precursor of cholesterol. Therefore, consumption of capybara fat should be avoided by people who have high cholesterol levels; as yet, there have been no studies on the possible effects on human health of eating capybara meat. However, capybara fat is richer in ω3 fatty acids (Table 12.5) than are traditional red meats. The lipids in capybara meat are about 5% ω3 acids (Oda 2002). The high ω3 fatty acid content might be related to the reputed healing properties of capybara oil, as told in traditional folk beliefs (below).

Another quality of capybara fat is its possible use as a natural nutraceutical product to avoid malnutrition. Rats fed capybara oil achieved greater weight gains than rats fed sardine (49.8% more) or horse oils (40% more). Furthermore, rats fed with high cholesterol diets achieved a 60% reduction in blood cholesterol when they consumed the oil extracted from capybara fat for 3 weeks (Fukushima et al. 1997).

12.4 Leather

Leather is also a commercially important capybara product. In Argentina, capybara leather is the most important product, with a far greater value than the meat. An animal of 40–45 kg produces a piece of leather 1.25 m long × 0.75 m wide.

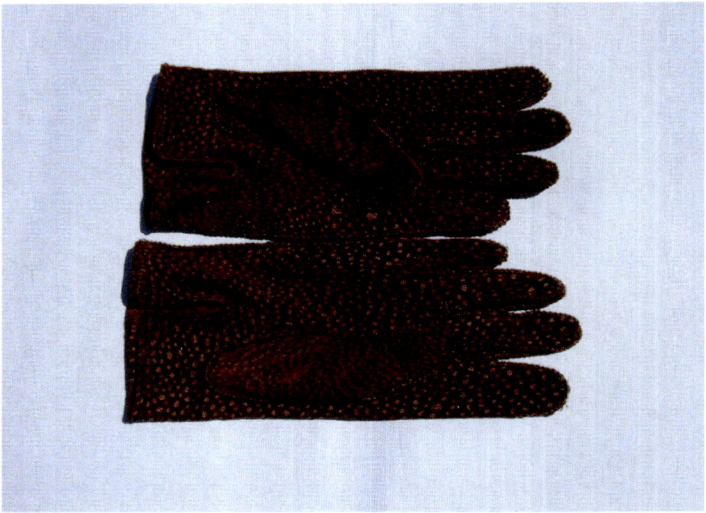

Fig. 12.2 Capybara leather gloves produced in Argentina (Photo by M.S. Pinheiro)

The weight of the leather of an adult capybara is 5.3 kg and it is 0.5 cm thick. A single pelt *in natura* is worth US $ 16–20 (in the year 2003). However, if the length of the leather, including the head, is shorter than 1 m, the tannery pays only half this price (Allekotte 2003).

Salting and drying is the most common method for the preservation of capybara hides. It is critical that the skin is as fat-free as possible when preserving with salt, because fat does not let the salt penetrate, causing stains in these places. Thus, after extraction, the leather should be washed with detergent to remove as much fat as possible.

After tanning, capybara leather is of excellent quality and quite valuable. In the tanning process, hides are subjected to acid and chromium and are dyed tobacco-color, to match the original color of the species. The cost of tanning is approximately US $16 per m^2 but once tanned, the skins are worth between US $50 and 110 a piece (Allekotte 2003). Chrome-tanned leather is extremely strong, smooth, soft, and flexible, like chamois. The thickness of the outer portion can be reduced by removing layers, and its elasticity can be decreased by stretching. In Argentina, it is mostly used for making local crafts. Capybara leather has unique characteristics due to the disposition of the groups of pores that appear with the removal of hair bundles (Fig. 12.2).

Before the restrictions on trading wild animals were imposed in 1967, Brazil was the largest exporter of capybara hides – about 150,000 pelts per year over the period from 1960 to 1969 (Caça 1963–1970). Other exporters were Colombia (25,000 pelts in 1970), Argentina (11,200 pelts per year over the period from 1972 to 1978), and Peru (7,680 pelts from 1962 to 1972). According to Allekotte (2003), the leather from wild capybaras is inferior to that of farmed animals, because of the scars and cuts which reduce the size of the usable skin. Nevertheless, usually only about 70% of the hides obtained from captive animals are used, mainly because of difficulty in cutting large usable pieces.

12.5 Other Uses

Before the arrival of domestic animals in South America, brought by Europeans in the sixteenth century, all the needs of the local people were fulfilled by local wildlife. Capybaras had, and still have, many uses, for economic, ecological, aesthetic, and strategic purposes (Moreira 2001). Today, with the new demands of modern life, new uses are emerging, and some are outlined below.

12.5.1 Organ Donor

A recent innovative use for capybaras is as organ donors for xenotransplantation – that is the use of animal organs in human transplant surgery. The use of capybaras for xenotransplantation of Islets of Langerhan, to replace lost insulin-producing tissues in patients with diabetes, has been suggested due to its large body size (Gray et al. 1998). However, the difficulty of isolating the capybara's Islets of Langerhan from the pancreas means this use is far from practical (Gray et al. 1998).

12.5.2 Leisure

Capybaras have many qualities that make them particularly suitable for ecotourism, especially compared with some of the other neotropical mammals such as the paca (*Cuniculus paca*) that are nocturnal, solitary, and elusive. Capybaras are large, attractive, and interestingly social, denizen of open grasslands, where it is easy to see them by day, especially as they move only slowly. A number of farms in the Pantanal Matogrossense (Brazil; Fig. 12.3), in Venezuela, and other parts of the capybara's range now supplement their income with wildlife tours, in which capybaras feature strongly. Tourists who visit these farms are virtually guaranteed good sightings.

Because capybaras are relatively easy to breed in captivity, they are also often exhibited in public parks, tourist resorts, and zoos around the world. Among the people living on the shores of Lake Paranoá in Brasilia (Brazil; Fig. 12.3), a survey revealed that more take pleasure in the presence of wild capybaras in their gardens than feel uncomfortable with it (Moreira et al. 2001). Capybaras are also kept as pets. Historical accounts report indigenous people keeping them as pets (Anchieta 1997) or even as guard animals because of their alarm call.

Fig. 12.3 Map of Brazil with the locations of the places referred to in the text

Surprisingly, capybaras are also a popular species for sport hunting, despite the minimal challenge involved and the harmless aspect of the trophy. In Argentina, where sport hunting is legalized, there are companies specializing in capybara safaris. People come from abroad just to hunt capybaras.

12.5.3 Handicrafts

Capybara leather is widely used for clothing and in the traditional handicrafts of the South American pampas. Clothes made from capybara leather include waistcoats, coats, jackets, skirts, everyday and sports gloves (Fig. 12.2), hats, caps, belts, money

Fig. 12.4 Advertisement published in Almanac Capivarol® 1955, "Reborn for life," praising the qualities of the tonic once manufactured from capybara oil

belts, gaucho belts, moccasins, boots, gaiters, sandals, and slippers. Wallets, bags and purses, cups for *mate* tea, bracelets, and key rings are also produced. A "carpincho leather" purse costs around US $100 in Argentina (price in 2011). The hides are used in upholstery and in the manufacture of saddles. Bones and teeth, especially the incisors, are popular in handicrafts.

12.5.4 Medicinal Use

According to traditional folk knowledge, capybara oil can be used as a cure for asthma, rheumatism, allergies (Moreira and Macdonald 1996), bronchitis, pneumonia, malnutrition, and to heal wounds. These alleged healing properties have not been proven scientifically. However, capybara oil is used to combat malnutrition and was a component of the once widely used calcium-based tonic Capivarol ® (Fig. 12.4).

For the Krahô indigenous tribes in Brazil (Fig. 12.3), capybara bones are also regarded as having medicinal values (Campos 1995). According to these indigenous people, it is useful as a tonic, being administered in powdered form after being roasted and milled.

12.5.5 Manure

Capybara droppings can be used in compost, in vermiculture (Krolow et al. 2004) and as a fertilizer. The material is usually collected during the cleaning of the enclosure in capybara farms. Where capybaras are kept in semi-confined production systems (Nogueira-Filho et al. 2012), their habit of defecating in water is useful for fertilizing ponds, and thus allows fish rearing in the tanks provided for capybaras without the need for additional resources.

12.6 Folklore and Vocabulary

Because the capybara is widely known and inhabits almost all biomes throughout much of South America, it is often present in the folklore of local people. The "capybara ceremony" is one of the rituals of the indigenous Matis population that live on the Javari River Valley in Brazilian Amazonia (on the border with Peru; Fig. 12.3). In this cacophonous ceremony, the Matis cover their bodies with mud and produce sounds, imitating the vocalizations of the capybara. Being an animist people, they believe that the spirits of animals control their luck in hunting and even their health and prosperity (Pantone 2004).

In the vocabulary of many South American peoples, the word used for the species can have different pejorative meanings. As an example, in Brazil it can mean a foolish person who is trying to look clever, or a woman who wants to look elegant, but dresses vulgarly. It may even mean a woman of loose morals. In chess, it is the name for an inexperienced, inefficient player. In many regions of Brazil the people living along river banks are referred to as "capivaras" (Houaiss et al. 2004). In Argentina, "carpincho" (the species' name in Argentina) is given to the short, spiky haircut worn by some young people (Martin R. Alvarez personal communication).

12.7 Final Remarks

Capybaras are a neotropical species with great potential for economic use. Their main commercial products are meat, leather, and oil. Unfortunately, where meat is produced the leather is wasted, and vice versa; the fat is also rarely made use of. This is a great loss of valuable resources. There is certainly a market for capybaras, but most of the production is illegal. We should also point out that the meat's strong flavor can restrict its use in some areas, especially in the southern part of its distribution. Comparative experiments should be designed to find out what causes the strong flavor, capybaras' consumption of grass or their age.

Another valuable use of capybaras which has recently increased in importance is ecotourism or watching them in public areas. Unfortunately, in some of the areas where there was much potential for such use, there have been fears about Brazilian spotted fever, discouraging its development.

Acknowledgments The present study was kindly reviewed by Maria Cristina Bressan, Roberto de Oliveira Roça, Sandra Oda, Katia Maria Ferraz, and José Augusto de Alencar Moreira, who provided helpful suggestions. MSP is grateful to Embrapa, who provided and financed the research on capybaras, and to partners at the Universidade Federal de Pelotas, whose cooperation helped this study to be carried out. JRM thanks the Lobato family from Belém (Brazil) for their assistance and hospitality and is especially grateful to the employees of Fazenda Eco-Búfalos, on Marajo Island (Brazil). JRM also acknowledges the support of S. Casement, M. Alvarez, and the Fundação de Apoio à Pesquisa Família Alencar (Brazil).

References

Albuquerque NI (1993) Ganho de peso em capivaras (*Hydrochoerus hydrochaeris hydrochaeris*, L. 1766) na fase inicial de crescimento em 3 categorias: machos inteiros, machos castrados e fêmeas. M.Sc. Dissertation, Escola Superior de Agricultura "Luis de Queiroz" – Universidade de São Paulo, Piracicaba

Allekotte R (2003) La cria del carpincho. Ediciones INTA, Buenos Aires

Anchieta J de (1997) Cartas de São Vicente 1560. Conselho Nacional da Reserva da Biosfera da Mata Atlântica, São Paulo. (Série Documentos Históricos, No. 7)

Andrade PCM (1996) Níveis de proteína e energia em rações e manejo de capivaras (*Hydrochoerus hydrochaeris hydrochaeris*, L. 1766) em crescimento. M.Sc. Dissertation, Escola Superior de Agricultura "Luis de Queiroz"/USP, Piracicaba

Assaf A, Cruz Marcano O, Aguero Herrera A, González-Jiménez J (1976a) Estudio de la utilizacion industrial de la carne de chigüire – Fase I. Informe anual del CIEPE, San Felipe, 74 pp

Assaf A, Cruz Marcano O, Aguero Herrera A, González-Jiménez J (1976b) Estudio sobre la capacidad de retencion de agua y poder de emulsificacion de carne de chigüire con relacion a la de res, cerdo y pollo. In: Resumenes del Seminário sobre chiguires (*hydrochoerus hydrochaeris*) y las babas (*Caiman crocodilus*). Universidad Central de Venezuela/Facultad de Agronomia, Maracay

Bressan MC, Miguel GZ, Faria PB, Vieira JO, Oda SHI (2002) Rendimento de carcaça e de cortes comerciais de capivaras (*Hydrochaeris hydrochaeris*, L. 1766). Ciência Agrotécnica Edição Especial:1588–1593

Bressan MC et al (2004) Composição de ácidos graxos dos cortes comerciais de capivara (*Hydrochaeris hydrochaeris* L. 1766). Ciênc Agrotéc 28:1352–1359

Caça (1963–1970) Produção de peles e couros de alguns animais silvestres. Anuário Estatístico do IBGE 24:48–51; 25:58–61; 26:80–83; 27:90–93; 28:76–78; 29:118–120; 30:128–130; 31:102–104

Campos E (1995) Dos trácios aos encourados vaqueiros da caatinga: magia e participação da fauna no quotidiano. In: Paiva MP, Campos E (eds) Fauna do Nordeste do Brasil: conhecimento científico e popular. Parte II Conhecimento Popular. Chap.3. Banco do Nordeste do Brasil, Fortaleza

Cardozo RM et al (2005) Crescimento e avaliação de carcaça em capivaras (*Hydrochaeris hydrochaeris* L. 1766) semiconfinadas, submetidas a manejo com e sem água para banho. Hora Vet 144:66–77

Cueto G, Allekotte R (2002) Cría intensiva de carpinchos. Información para extensión, Série Diversificación No 15. INTA Delta del Paraná, Campana, p 27

Fukushima M, Takayama Y, Habaguchi T, Nakano M (1997) Comparative hypocholesterolemic effects of capybara (*Hydrochoerus hydrochaeris dabbenei*) oil, horse oil, and sardine oil in cholesterol-fed rats. Lipids 32:391–395

Girardi F et al (2005) Proximate composition and fatty acid profile of semi confined young capybara (*Hydrochaeris hydrochaeris hydrochaeris* L. 1766) meat. J Food Comp Anal 18: 647–654

Godoy JF, Gómez EA (1976) Industrializacion de la carne de chiguire. In: Resumenes del Seminário sobre chiguires (*hydrochoerus hydrochaeris*) y las babas (*Caiman crocodilus*). Universidad Central de Venezuela/Facultad de Agronomia, Maracay

González-Jiménez E (1995) El capibara: Estado atual de su producción. Série estudio FAO, Produccion y Sanidade Animal, 122, Roma, p 112
González-Jiménez E, Parra RR (1972) Estudios sobre el chigüire (*Hydrochoerus hydrochaeris*). 1. Peso de diferentes organos y partes del cuerpo. Acta Cient Venez 23:30
Gray DWR, Cahill A, Asakura T, Morris PJ (1998) Investigation of a large, commercially farmed rodent (capybara) as a potential donor for pancreatic islet xenotransplantation. Transplant Proc 30:460–462
Hosken FM (1999) Criação de capivaras. SEBRAE-MT, Cuiabá
Hosken FM (2007) Capivara. Globo Rural 256. Available at: http://revistagloborural.globo.com/GloboRural/0,6993,EEC1445322-4530,00.html. Accessed 28 October 2010
Houaiss A, Villar MS, Franco FMM (2004) Dicionário Houaiss da Língua Portuguesa. Objetiva, Rio de Janeiro
Jardim NS (2001) Sexo e diferentes pesos ao abate na qualidade da carne de capivara (*Hydrochaeris hydrochaeris* L. 1766). M.Sc. Dissertation, Universidade de Lavras, Lavras
Jardim NS, Bressan MC, Lemos AL da S, Thomazini M, Ferreira MW (2003) Teor lipídico e perfil de ácidos graxos da carne de capivara (*Hydrochaeris hydrochaeris*). Ciênc Agrotéc 27:651–657
Krolow IRC et al (2004) Viabilidade da utilização do esterco de capivara (*Hydrochaeris hydrochaeris*) como substrato para criação de minhocas (*Eisenia foetida*). In: Anais do Congresso de iniciação científica da Ufpel. UFPEL, Pelotas
Lavorenti A (1989) Domestication and potential for genetic improvement of capybara. Rev Bras Genét 12:137–144
Mackey A, Flores I, Sosa EM (1976) Utilización de la carne fresca del chigüire. In: Resumenes del Seminário sobre chiguires (*hydrochoerus hydrochaeris*) y las babas (*Caiman crocodilus*). Universidad Central de Venezuela/Facultad de Agronomia, Maracay
Marín AM, Arias R (2002) Mejoras en el sacrificio de los chigüires. Available at: http://www.rlc.fao.org/prior/segalim/pdf/chiguire/1.pdf. Accessed 26 July 2007
Martinez MT (2007) El año del carpincho. El Federal, July, p 24–32
Miguel GZ (2001) Caracterização da carcaça e da carne de capivara (*Hydrochaeris hydrochaeris*, L. 1766) em idade adulta. M.Sc. Dissertation, Universidade de Lavras, Lavras
Montes G (1983) El carpincho. Centro Editor de América Latina. Buenos Aires. (Série Fauna Argentina, 2). p 32
Moreira JR (2001) Os valores e usos dos recursos genéticos faunísticos. In: Anais do 3° Simpósio de Recursos Genéticos para a América Latina e Caribe. IAPAR, Londrina, pp 138–140
Moreira JR, Macdonald DW (1996) Capybara use and conservation in South America. In: Taylor VJ, Dunstone N (eds) The exploitation of mammal populations. Chapman & Hall, London, pp 88–101
Moreira JR, Pinha PRS, Cunha HJ (2001) Capivaras do Lago Paranoá. In: Fonseca FO (ed) Olhares sobre o Lago Paranoá. Secretaria de Meio Ambiente e Recursos Hídricos, Brasília, pp 141–147
Nogueira-Filho SLG (1996) Manual de criação de capivara. CPT, Viçosa
Nogueira-Filho SLG, Pinheiro MS, Nogueira SSC (2012) Confined and semi-confined production systems for capybaras. In: Moreira JR, Ferraz KMPMB, Herrera EA, Macdonald DW (eds) Capybara: biology, use and conservation of an exceptional neotropical species. Springer, New York, pp 229–241
Oda SAI (2002) Diferentes métodos de abate e sexo na qualidade da carne de capivara. 2002. M.Sc. Dissertation, Universidade Federal de Lavras, Lavras
Oda SAI et al (2004a) Efeito do método de abate e do sexo na composição centesimal, perfil de ácidos graxos e colesterol da carne de capivara. Ciênc Tecnol Alimentos 24:236–242
Oda SAI et al (2004b) Efeito do método de abate e do sexo sobre a qualidade da carne de capivara (*Hydrochaeris hydrochaeris*). Ciênc Tecnol Alimentos 24:341–346
Ojasti J (1973) Estudio biologico del chigüire o capibara. FONAIAP, Caracas
Ojasti J (1991) Human exploitation of capybara. In: Robinson JG, Redford KH (eds) Neotropical wildlife use and conservation. University of Chicago Press, Chicago, pp 236–252
Ojasti J, Medina-Padilla G (1972) The management of capybara in Venezuela. Trans North Am Wildl Nat Resour Conf 37:268–277
Paiva R (1992) Capivara: bicho novo no pasto. Globo Rural 7:44–47

Pantone DJ (2004) Matis Indians: the jaguar people. [online] Amazon-indians. Available at: http://www.amazon-indians.org/page01.html. Accessed 21 Aug 2007

Pinheiro MS, Pouey JLOF, Dewantier LR, Garcia CAN, Rodrigues RC, da Silva JJC (2007) Avaliação de carcaça de fêmeas e machos inteiros de capivara (*Hydrochaeris hydrochaeris*) criados em sistema semi-intensivo. Arch Latinoam Producción Anim 15:52–57

Pinto AT, Gomes RM, Greve AG (2004) Tecnologia de abate de capivaras. Higiene Aliment 18:57–59

Pinto MF, Ponsano EHG , Almeida AP da S, Heinemann RJB, Souza WM de (2007) Características e potencial tecnológico da carne da capivara. Ciência Rural 37:868–873

Roça RO de, Veiga N, Silva Neto PB da, Cervi RC (1999) Características sensoriais de carne defumada de capivara. Pesqui Agropecu Bras 34:487–492

Saldanha T (2000) Determinação da composição centesimal nos diferentes cortes da carne de capivara (*Hydrochoerus hydrochaeris*). M.Sc. Dissertation, Universidade Federal Rural do Rio de Janeiro, Rio de Janeiro

Saldanha T, Santana DMN, Gaspar A (2002) Total lipids, fatty acid composition and cholesterol content of capybara (*Hydrochoerus hydrochaeris*) meat. Braz J Food Technol 5:245–250

Salgado JM, Galvão MTEL, Storer M de L, Canniatti-Brazaca SG (1999) Utilização da carne de capivara na elaboração de salsicha e fiambre. Boletim do CEPPA 17:83–92

Santos EC dos, Rodrigues R, Tavares WA, Ribeiro RMP (1986) Ácidos graxos da gordura da capivara (*Hydrochoerus hydrochaeris*). Arq Bras Med Vet Zoo 38:773–778

Sarkis F (2002) Avaliação das condições microbiológicas das carnes de animais silvestres no Estado de São Paulo. M.Sc. Dissertation, Escola Superior de Agricultura "Luis de Queiroz"/ USP, Piracicaba

Torres-Gaona JLT (1987) La carne de chigüiro como alimento. Temas Orientacion Agropecu 9:69–75

Chapter 13
Confined and Semi-confined Production Systems for Capybaras

Sérgio L.G. Nogueira-Filho, Max S. Pinheiro, and Selene S.C. Nogueira

13.1 Introduction

In Latin-American countries capybaras are traditionally hunted for subsistence needs and/or the hide trade (Nogueira-Neto 1973; Ojasti 1973; Robinson and Redford 1991). Environmental law in Brazil and other neotropical countries, such as Peru, Panama, and Paraguay, forbids commercial hunting (Ojasti 1996), but capybaras may be captured for use in captive breeding programs to produce meat and leather. Despite the strict legal framework, the species' high reproductive potential together with its behavioral characteristics – grass-eating, sedentarism, docility, and sociability – have encouraged breeding in captivity by research centers in several South American countries. Increased human demand for animal protein in Latin America (Nogueira-Neto 1973; Ojasti 1973; Lavorenti 1989; González-Jiménez 1995), and for game meat, specifically in urban centers (Nogueira-Filho and Nogueira 2004), has led to the spread of capybara farming to provide a new food resource.

Initially, capybaras were produced within a confined system. A semi-confined system was later developed, after knowledge of capybara behavior improved

S.L.G. Nogueira-Filho (✉)
Departamento de Ciências Agrárias e Ambientais, Universidade Estadual de Santa Cruz, Rodovia Ilhéus Itabuna km 16, 45662-900 Ilhéus, Bahia, Brazil
e-mail: slgnogue@uesc.br

M.S. Pinheiro
Embrapa Clima Temperado, Rodovia BR392 km78, Caixa Postal 403, 96010-971 Pelotas, Rio Grande do Sul, Brazil
e-mail: max.pinheiro@embrapa.br

S.S.C. Nogueira
Departamento de Ciências Biológicas, Laboratório de Etologia Aplicada, Universidade Estadual de Santa Cruz, Rodovia Ilhéus Itabuna km 16, 45662-900 Ilhéus, Bahia, Brazil
e-mail: selene@uesc.br

J.R. Moreira et al. (eds.), *Capybara: Biology, Use and Conservation of an Exceptional Neotropical Species*, DOI 10.1007/978-1-4614-4000-0_13,

(Nogueira et al. 1999), and with the aim of reducing the high production costs associated with confined systems. The semi-confined production system was first described by Ojasti (1991) and later modified by Nogueira-Filho (1996) and Silva-Neto et al. (1996). This chapter describes facility characteristics, husbandry and management practices of both confined and semi-confined capybara farms, as well as a comparison of the costs (investment and production expenses) of the two systems.

13.2 Facilities and Reproductive Husbandry Practices for Confined Production Systems

The facilities required for capybara farming in a confined production system clearly reflect the reproductive husbandry regime adopted. Therefore, these two aspects of capybara farming are considered together. Traditionally, pre-parturition females were isolated until the young were weaned to avoid infanticide (Alho 1986; Lavorenti 1989; González-Jiménez 1995; Cueto 1999). However, with this approach it is difficult to determine the appropriate time to isolate pregnant females and, more importantly, when females return to their original group after weaning, they are not always accepted (Nogueira 1997). To overcome these problems, reduce labor, avoid sanitation problems, and decrease costs (Alvarez and Kravetz 2006), alternative facilities and husbandry procedures were developed.

13.2.1 Traditional Confined System

Capybara farming was originally based on the confined system model used for domestic pigs. This system requires separate facilities for the maintenance of animals at each of the following stages: reproduction, parturition and weaning, and growth and fattening (Fig. 13.1). Reproductive groups (usually comprised of one male and three to eight females) are housed in enclosures ranging from 30 to 120 m^2, provided with sheltered areas of 20–24 m^2 and water tanks, and surrounded by a 1.2–1.5 m-high wire-mesh fence. Pre-parturition females (identified by their low bellies and swollen teats) are isolated in individual pens of 6–20 m^2, equipped with water tanks, for 30–60 days. After weaning, the adult females are returned to their original housing and the young placed in growth paddocks (Alho 1986; Lavorenti 1989; González-Jiménez 1995; Cueto 1999; Alvarez and Kravetz 2006) in mixed-sex groups of 5–20 individuals (Alho 1986; Lavorenti 1989; González-Jiménez 1995; Cueto 1999).

The isolation of pregnant females was introduced on the basis of field observations of wild capybaras, which revealed that free-ranging pregnant capybaras would seek isolation for birth and early nursing, probably to avoid infanticide (Ojasti 1973;

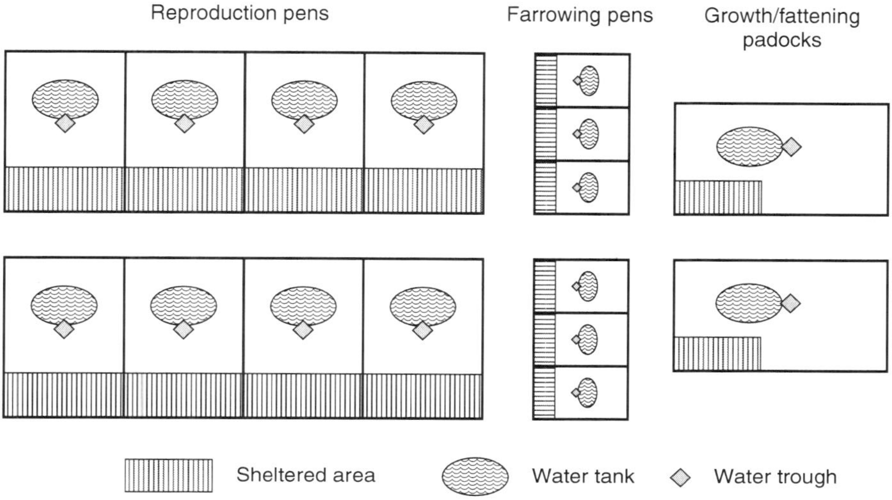

Fig. 13.1 Layout of the traditional confined production system for capybaras. Pen sizes: reproduction pens 30–120 m^2, birthing pens 6.0–20 m^2, growth/fattening paddocks 120–400 m^2

Fuerbringer 1974). However, dominant males were never observed pursuing young from their groups in the wild (Lord 1994). In contrast, group females and the dominant male were observed protecting the young (Azcarate 1980; Macdonald 1981). Moreover, there are no recorded deaths of young due to male attacks in a confined system (Lavorenti et al. 1989). In these conditions, only unfamiliar females killed infants (Nogueira et al. 1999). Therefore, the isolation of pregnant females is unnecessary if capybara groups are composed solely of familiarized females (Nogueira et al. 1999). Isolation of pregnant females also carries significant production costs because of the exclusive facilities and labor-intensive husbandry needed (building and maintaining a number of separate enclosures and carrying out daily observations to assess impending parturition).

13.2.2 Improved Confined System

The improved confined system is based on housing groups (composed of one male and six to nine females) in 400–600 m^2 breeding paddocks, containing shady trees and a water tank, with a space allowance of 40 m^2 per adult animal (Fig. 13.2). As long as the females are familiar with one another, they can give birth and suckle in these enclosures (Nogueira et al. 1999). Weaning occurs when the young are 60 days old, at which point the young of both sexes are transferred to growth paddocks (similar to breeding paddocks) with a space allowance of just 20 m^2 per animal (Nogueira-Filho 1996).

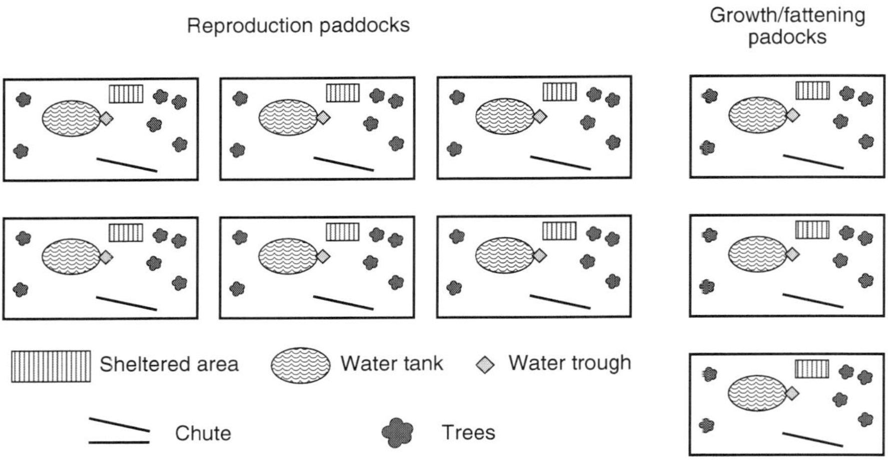

Fig. 13.2 Improved confined production system. Reproduction and growth/fattening areas are 400 m² (From Nogueira-Filho (1996))

The young of mothers from different reproductive groups can be housed in the same growth paddock, making it possible to establish new reproductive groups (Nogueira 1996; Nogueira et al. 1999) and maintain mixed-sex groups until animals are 6 months old. To avoid infanticides in the future, these new reproductive groups must be established at weaning. There appears to be a tolerance level at weaning of approximately 10% for differences in age and live weight (Nogueira-Filho 1996; Nogueira et al. 2003). Young capybaras can reach 20 kg (live body weight) and can be slaughtered at this age if fed a combination of grass and protein/energy supplements (Nogueira-Filho 1996).

To obtain higher live weights at slaughter, 6-month-old males of homogenous live weights are transferred to fattening paddocks (with the same dimensions, characteristics, and stocking density as the growth paddocks). Isolating males of this age prevents male disputes over females, which can lead to injuries and weight losses (Nogueira-Filho 1996).

All the paddocks must be designed for optimum animal performance with full consideration for animal welfare. The provision of a water tank, which should be 20 m² (4 m × 5 m) and at least 0.6 m deep for a 400–600 m² paddock (Fig. 13.2), is particularly important in the breeding paddocks. These tanks are used by capybaras for bathing, exercise, courtship, and copulation. At least one of the tank's sides must have an access ramp to allow the young to access the tank easily and for females to turn and position themselves during copulation (Nogueira-Filho 1996). There should also be at least one water trough and one or two sheltered feeders. Artificial shelter, covering at least 10% of the enclosure, must be provided if there is no natural shade from trees to protect the animals during the hottest hours of the day. The trees in all the paddocks must be protected with a wire-mesh fence (at least 1.0 m high) to prevent capybaras from gnawing the trunks (Nogueira-Filho 1996). Water tanks and sheltered areas are also important for the growing and fattening of capybaras (Nogueira-Filho 1996). Young capybaras in paddocks without water tanks achieve lower weight gains and worse feed-conversion ratios (efficiency in converting feed mass into increased body mass; Silva-Neto 1989), and their meat has a higher saturated fatty acid content (Girardia et al. 2005).

13.3 Facilities and Reproductive Husbandry Practices for Semi-confined Production Systems

Capybaras can live on poor nutritional quality food and thrive in conditions that are difficult for other livestock species, because of their local weather tolerance and resistance to local diseases and parasites (Nogueira-Filho and Nogueira 2004). Therefore, marginal land, which may be inappropriate for traditional agricultural/livestock production (of exotic species) because of soil and/or topography, can be used for capybara production (Lavorenti 1989; Ojasti 1991; Nogueira-Filho and Nogueira 2004).

Lowland flood plains or areas with natural or artificial lakes can be used for semi-confined capybara production. In this production system, capybaras are kept in fenced enclosures of approximately 0.5–3.0 ha (Ojasti 1991; Nogueira-Filho 1996; Silva-Neto et al. 1996). Enclosures should contain trees and/or bushy vegetation or artificial shelters and should resemble natural capybara habitat by also including a water body and dry areas (Ojasti 1973; Alho and Rondon 1987).

Grazing and trampling by capybaras within the enclosure will reduce the amount of available forage and deplete the productivity of the pastures after only a few months (Ojasti 1991; Nogueira-Filho 1996). So it is more economically feasible to plant grasses, such as elephant grass (*Pennisetum purpureum*), outside the breeding paddocks, as recommended by Nogueira-Filho (1996; Mendes and Nogueira-Filho 2012), than to fence unnecessarily large areas which would allow direct grazing, as suggested elsewhere (Ojasti 1991; Silva-Neto et al. 1996).

There are no definitive recommendations for the optimal space requirements, or even the optimal relative proportions of dry and wet areas, for breeding capybaras in semi-confined systems. In a preliminary study, Lopes (2007) was not able to detect a difference in aggression levels between two groups maintained at densities of 300–450 m^2 per adult. Empirical evidence from farmers who used higher stocking densities also indicated no negative impacts. Therefore, stocking densities of approximately 300 m^2 per adult animal could be recommended (Nogueira et al. 1999).

In the semi-confined production system, all phases of the capybara life cycle occur in these enclosures: females give birth and suckle their young, and the young grow up in groups. A male to female sex ratio of 1:7 is recommended within enclosures. The daily weight gain depends on diet. When protein/energy supplements are added to the diet, in addition to fresh grasses, the young can weigh between 6.0 and

8.0 kg at 2 months of age (birth weight: 2 kg) and will reach 20 kg at 6 months, as occurs in the confined system. To satisfy market demands, young capybaras can be slaughtered at this weight/age because at this time they achieve their highest daily weight gains and lowest feed-conversion ratios (Nogueira-Filho 1996).

Heavier slaughter weights can be reached by transferring young males at 6 months to a separate enclosure (a fattening paddock, similar to that described above for confined systems). Young males within the enclosures where they were born may be pursued by dominant males and may avoid approaching the feeder areas as a consequence (Nogueira-Filho 1996), resulting in injuries, reduced daily weight gain, and possible weight loss. Within a separate enclosure (with other young males of a similar size), animals can reach 30–40 kg at 12–18 months of age when fed grass-concentrate diets (Albuquerque 1993; Hosken 1999; Silva-Neto et al. 1996). Groups of more than 60 young are not recommended (Hosken 1999; Hosken and Silveira 2002). No economic advantages appear to result from castrating male capybaras (Albuquerque 1993).

Within a semi-confined system, isolation of 6-month-old females is not necessary. Young females are not aggressively attacked by the males and can replace old females as reproductive adults (Nogueira-Filho 1996). Nevertheless, young females may experience some aggression from adult females upon reaching breeding age. Therefore, higher ranking females over 7 years old and showing reduced reproductive success should be simultaneously culled (Nogueira 1997; Nogueira-Filho and Nogueira 2012). To avoid inbreeding, the dominant male must be replaced every 5 years. It should be kept in mind that all the females in the group should be related to each other or familiar to one another since an early age to avoid conflict and infanticide (Nogueira-Filho 1996).

13.4 Fence Characteristics and Husbandry Structures

Fencing comprises most of the investment required in both confined and semi-confined capybara breeding systems. Wire-mesh fences are the most effective; barbed-wire, electric fences, or hedges are easily breached. In both systems, fences should be 1.5 m high, composed of 0.8–1.0 m of wire-mesh with barbed wire or normal wire spaced in rows 0.15 m apart at the top. The mesh should have from 6.35 to 7.62 cm (2.5″ to 3.0″) inner space between knots and be made of 2.77- or 2.11-mm-thick wire. It is essential to use 6.35 cm mesh for the fences between adjacent enclosures in confined systems, to prevent intergroup fighting between males and potential injuries.

Fences will function better if several recommendations are followed. Wire and wire-mesh should be attached to the paddock side of the posts, spaced 3.0 m apart, to prevent animals injuring themselves on protruding posts when they run along the fence line. When semi-confined systems are established in sandy soil, the fences must have underpinnings in which the mesh-wire is cast. Underpinnings are also necessary if there are free-ranging dogs in the area. Alternatives to underpinning are the use of galvanized steel wire, to add resistance to the fence base, or electric wires.

Fig. 13.3 Corral-trap to capture capybaras in the semi-confined system. The access door is divided into lower (0.45 m high) and upper (1.20 m high) sections that allow animals to be selected by size (From Nogueira-Filho (1996))

Inside semi-confined enclosures, there should be one or more corral-traps constructed for easy capture and handling of the capybaras (Nogueira-Filho 1996; Silva-Neto et al. 1996). These traps should allow the capture of multiple animals to reduce trapping and handling costs. The corral-traps can range from 10 to 100 m^2 and need to be 1.8 m high, fenced with knotted wire-mesh to prevent young from escaping (Fig. 13.3). There are fewer injuries to the animals when the corral-traps are made of wood. At least one of the corral-trap doors must be of the guillotine type. One of the access doors could be divided into two sections; the lower one should be 0.45 m high and the upper 1.20 m high to allow animals to be selected by size (Fig. 13.3). The usual bait to attract capybaras is suspended bundles of grass, or salt; corn grains; and other concentrates in sheltered wooden feeders placed inside and around the corral-traps. In a confined system, handling and everyday husbandry procedures are made easier when the enclosures are designed with chutes in the form of a short-walled corridor (Fig. 13.2), approximately the width of a capybara (0.6 m), to direct animals to transport or restraining cages (Nogueira-Filho 1996). The chutes should have a 1.4 m high wall and are less likely to cause injuries to animals when they are made of wood.

Wooden restraint cages (1.2 m long, 0.6 m wide, 0.6 m high) should be fitted with guillotine or chute doors and can be used for both immobilization and transport of capybaras. The walls should be designed to move in or out so that they can be adjusted to immobilize the animal for handling procedures, such as ear notching, sex determination, and medical procedures. At up to 6 months, capybaras can easily be captured with a net on a 1.5 m-long tubular handle. Only two handlers are needed to handle a group of capybaras when the chute and the restraining cage are well designed and used.

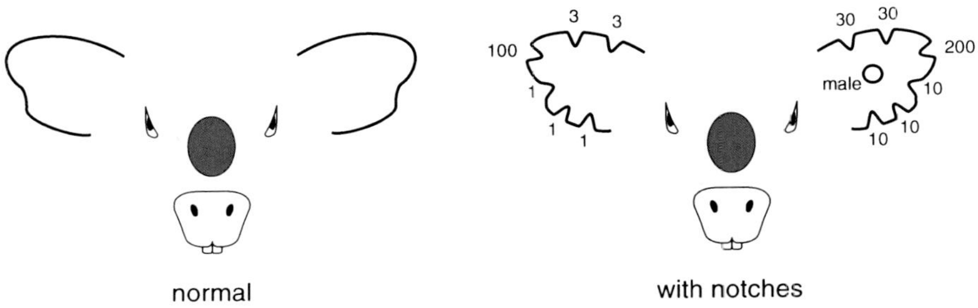

Fig. 13.4 A capybara without ear notches (*left*) and another with all ear notches (*right*). The capybara has a natural indentation at the bottom of the ear that should be emphasized when marked. The animal on the right is a male, number 399

13.5 Sex Determination, Growth and Identification Systems

Capybaras are not sexually dimorphic, and when young it is impossible to determine sex on the basis of morphological characteristics. When adult, dominant males have prominent supra-nasal glands while subordinates have much smaller ones. However, in captivity, some dominant females may also have quite enlarged supra-nasal glands. Therefore, the only reliable method of determining the sex of a capybara is to restrain the animal and protrude its sexual organ (Paula and Walker 2012). However, when determining the sex of young, it is recommended that it should be at least 2 months old to avoid unnecessary stress (Nogueira-Filho 1996).

Young can be individually identified using the Australian method of ear notching for pigs (Fig. 13.4). Being permanent, this method will avoid recapturing for identifications, reducing stress. Other potential marking methods have been proposed (ear tattoos, ear tags, hot or freeze branding, electronic microchips) but they are more expensive, more stressful, and less efficient than ear notching.

13.6 Prophylactic Management Practices

Capybaras are very robust and experience few sanitary problems (Cueto 2012). Nevertheless, to promote good health, some physical and social preventive measures must be adopted. A few simple preventive measures can increase productivity and decrease medical expenses.

The first step is to establish a cohesive reproductive group (Nogueira 1996; Nogueira et al. 1999). Management techniques for group formation and to avoid conflicts should be followed when acquiring the animals. All animals must be captured in the wild on the same day, thus ensuring they are from the same original group (Nogueira 1996). If the capybaras were purchased from another breeder, it is essential that the group's individuals have been living together since weaning (Nogueira et al. 1999). To avoid fights or infanticides, no additional animals should be introduced into the group. Any increase in group numbers should only be through group reproduction (Nogueira 1996).

The second step to increase productivity and decrease medical expenses is to provide appropriate enclosures. Labor expenses in the confined system may represent 12% of the total production cost (Table 13.2; Nogueira-Filho and Nogueira 2004) and involve daily cleaning of enclosures and weekly maintenance of the water tanks. These practices are crucial; otherwise there may be high mortality rates in young due to helminth (*Strongyloides* spp.) infestations. Capybaras, including those in captivity, defecate in water, contaminating their water tank/water source. Observations indicate that they prefer to drink clean water from a different water trough, so they must have more than one water trough inside the enclosure.

Capybaras in both confined and semi-confined systems are vulnerable to some predators. The black vulture (*Coragyps atratus*), for example, can cause huge losses among newborn capybaras. The mothers, and other group members, are normally able to protect the young from attacks by single birds, but are ineffective against mass attacks from 20 or 30 birds. In smaller enclosures, it is possible to limit predator attacks by covering the enclosure with wire or cords attached to the fence posts. In larger enclosures, fireworks can be used to disperse vultures, but they may become used to the noise.

Finally, to ensure optimal capybara health, the stockpeople's experience and training are crucial. Stockpeople must be calm when interacting with the capybaras, avoiding loud noises and unnecessary chasing, so that animals get used to their presence. They also need to constantly observe, record, and report any behavioral changes or anomalies. In addition to the stockperson, a veterinarian is required to establish an appropriate control program for endo- and ectoparasites (Cueto 2012) in both confined and semi-confined production systems.

13.7 Costs of Confined and Semi-confined Production Systems

Commercial production under confined systems is appropriate for rural proprieties near medium to large cities that have a potential market for capybara meat. It is helpful to have abattoir facilities for pigs, sheep, or goats nearby, and there are advantages if the capybara production systems are established on existing dairy or beef cattle farms. On cattle farms, the presence of corn and elephant grass fields, as well as the equipment necessary for hay or silage production, gives farmers an advantage. A group of 10 adult capybaras eats approximately 50 kg of elephant grass per day, the same quantity consumed by just one dairy cow. Thus, there are no significant additional daily expenses for cutting and carrying grass to the capybaras. These favorable conditions will compensate for the extra investments necessary to set up capybara systems (Table 13.1) and will offset the higher production expenses (Table 13.2) in the confined systems when compared to semi-confined ones. Initial

Table 13.1 Comparative investment costs for the maintenance of adult capybaras (24 females and 3 males) in confined and semi-confined production systems in Brazil

Investment (US $)	Confined	Semi-confined
Consultant[a]	500.00	500.00
Registration tax[b]	100.00	100.00
Purchase of animals[c]	1,350.00	–
Materials and labor for captive breeding facilities	6,820.00	4,500.00
Equipment	200.00	200.00
Total	8,970.00	5,300.00

[a]The farmers need to contract a professional consultant to write the project (under Brazilian law)
[b]The farmer must pay an initial charge to IBAMA (Brazilian Environmental Agency) to legalize the project
[c]In the semi-confined system, it is possible to use animals already living on the farm (Nogueira-Filho and Nogueira 2004)

Table 13.2 Comparative expenses per kilogram of live animal weight in confined and semi-confined production systems for capybaras in Brazil

	Confined		Semi-confined	
Expense by item	(US$)	%	(US$)	%
Feeding[a]	1.19	64.3	1.17	75.0
Labor	0.22	11.9	0.10	6.4
Capital investment[b]	0.25	13.5	0.14	9.0
Other expenses[c]	0.19	10.3	0.15	9.6
Total expense per kilogram of live animal weight	1.85	100	1.56	100

[a]US $7.00/t of *Pennisetum purpureum* grass crushed and placed in the feeder (labor included) and US $0.15 per kg of supplementary growth concentrate – both used in confined and semi-confined systems
[b]Includes interest (6% per year) on total initial investments, fuel, medicine, facilities and equipment depreciation, animal acquisition expenses, and fee to consultant
[c]Includes expenses for fuel, electricity, veterinarian and medicines, IBAMA annual charge, and interest on operational capital (6% per year), etc. (Nogueira-Filho and Nogueira 2004)

investments are lower than those required for domestic pig breeding systems (US $700.00 per sow in confined systems; Dalla Costa et al. 1995; US $490.00 per sow in semi-confined systems; Leite et al. 2001).

Given the higher investment and production expenses and the market rule (higher production, lower prices), the confined production system is best suited to producing capybaras for sale as breeders. The price received for healthy breeding stock (captive-reared groups composed of familiarized individuals) can be twice that received for their carcasses, making breeders worth the higher production costs in confined systems.

The confined system is economically viable for farmers with direct access to selling meat and other capybara products, cutting out expensive transport and intermediaries, and it is also appropriate for smallholders producing capybara meat for subsistence. The same conditions recommended for confined systems are desirable in semi-confined ones, such as a nearby abattoir and already existing installations. However, the investment and the production expenses in semi-confinement are lower, and the economic returns higher (Tables 13.1 and 13.2). The use of marginal land and of already available labor will increase the profitability of such a system (Nogueira-Filho and Nogueira 2004).

Only a few big producers can sell their products directly to consumers, mainly restaurants, and so obtain the highest earnings. In Brazil (a country where capybaras can only be produced in captivity), most capybara farmers currently sell their products to intermediaries who pay only US $2.00 per kg of live weight, while the retail price of capybara meat is US $18.00 per kg. Farmers therefore need to become organized to sell their products jointly and directly to consumers. They may obtain even higher profits through the industrialization of capybara leather and fat (Nogueira-Filho and Nogueira 2004).

Acknowledgments We thank Dr. W. Mollineau, Dr. J.R. Moreira, Dr. Sabine Pompéia, Dr. David Macdonald, Dr. Emilio A. Herrera, and anonymous reviewers for valuable comments on this manuscript. Preparation of this chapter was aided by grants from CNPq to Sérgio Nogueira-Filho and to Selene Nogueira (Proc. # 300587/2009-3 and 306154/2010-2, respectively).

References

Albuquerque NI (1993) Ganho de peso na fase final de crescimento e sistematização da avaliação de três categorias de capivaras (*Hydrochoerus hydrochaeris hydrochaeris* L. 1766): machos inteiros, machos castrados e fêmeas. M.Sc. Dissertation, Escola Superior de Agricultura "Luiz de Queiroz", Universidade de São Paulo, Piracicaba

Alho CJR (1986) Criação e manejo de capivaras em pequenas propriedades rurais. Empresa Brasileira de Pesquisa Agropecuária EMBRAPA-DDT Série Documentos, 13, Brasília

Alho CJR, Rondon NL (1987) Habitats, population densities, and social structure of capybaras (*Hydrochaeris hydrochaeris* Rodentia) in the Pantanal, Brazil. Rev Bras Zoo 4:139–149

Alvarez MR, Kravetz FO (2006) Reproductive performance of capybaras (*Hydrochoerus hydrochaeris*) in captivity under different management systems in Argentina. Anim Res 55:153–164

Azcarate T (1980) Sociobiologia y manejo del capibara (*Hydrochoerus hydrochaeris*). Donana Acta Vertebrat 7:1–228

Cueto GR (1999) Biología reproductiva y crecimiento del carpincho (*Hydrochoerus hydrochaeris*) en cautiverio: una interpretación de las estrategias poblacionales. Ph.D. thesis. Universidad de Buenos Aires, Buenos Aires

Cueto GR (2012) Diseases of capybara. In: Moreira JR, Ferraz KMPMB, Herrera EA, Macdonald DW (eds) Capybara: biology, use and conservation of an exceptional neotropical species. Springer, New York, pp 169–184

Dalla Costa OA, Girotto AF, Girotto AD (1995) Análise econômica do sistema intensivo de suínos criados ao ar livre (SISCAL) e confinados (SISCON) nas fases de gestação e lactação. Rev Soc Bras Zootecn 24:615–622

Fuerbringer J (1974) El chigüiro, su cría explotacíon racional. Orientacíon Agropecu 99:5–59
Girardia F et al (2005) Proximate composition and fatty acid profile of semi confined young capybara (*Hydrochoerus hydrochaeris hydrochaeris* L. 1766) meat. J Food Compos Anal 18:647–654
González-Jiménez E (1995) El capybara (*Hydrochoerus hydrochaeris*) – Estado actual de su producción. FAO, (Série Estudio FAO). Producion y Sanidad Animal, Rome, p 122
Hosken FM (1999) Criação e manejo de capivaras. Edição Sebrae, Cuiabá
Hosken FM, Silveira AC (2002) Criação de Capivaras. Aprenda Fácil UFV, Viçosa
Lavorenti A (1989) Domestication and potencial for genetic improvement of capybara. Rev Bras Genét 12:137–144
Lavorenti A, Silva-Neto PB, Marchi AP, Souza MC (1989) Desempenho reprodutivo da capivara (*Hydrochoerus hydrochaeris*) criada em cativeiro. Anais da 26ª. Reunião Anual da Sociedade Brasileira de Zootecnia, Porto Alegre
Leite DMG, Dalla Costa OA, Vargas GA (2001) Análise econômica do sistema intensivo de suínos criados ao ar livre. Rev Bras Zootecn 30:482–486
Lopes PCC (2007) Estudo da estrutura social e das relações materno-filiais de capivaras (*Hydrochoerus hydrochaeris* Linnaeus. 1766) em cativeiro. M.Sc. Dissertation, Universidade Estadual de Santa Cruz, Ilhéus
Lord RD (1994) A descriptive account of capybara behaviour. Stud Neotrop Fauna Environ 29:11–22
Macdonald DW (1981) Dwindling resources and the social behavior of capybara, (*Hydrochoerus hydrochaeris*) (Mammalia). J Zool 194:371–391
Mendes A, Nogueira-Filho SLG (2012) Feeds and nutrition of farmed capybaras. In: Moreira JR, Ferraz KMPMB, Herrera EA, Macdonald DW (eds) Capybara: biology, use and conservation of an exceptional neotropical species. Springer, New York, pp 261–274
Nogueira SSC (1996) Comportamento da capivara. In: Nogueira-Filho Org. SLG, (ed) Criação de Capivaras. Centro de Produções Técnicas (CPT), Viçosa, pp 11–13
Nogueira SSC (1997) Manejo reprodutivo da capivara (*Hydrochoerus hydrochaeris hydrochaeris*) em sistema intensivo de criação. Ph.D. thesis, Instituto de Psicologia – Departamento de Psicologia Experimental of the Universidade de São Paulo, São Paulo
Nogueira-Filho SLG, Nogueira SSC (2012) The impact of management practices on female capybara reproductive parameters in captivity. In: Moreira JR, Ferraz KMPMB, Herrera EA, Macdonald DW (eds) Capybara: biology, use and conservation of an exceptional neotropical species. Springer, New York, pp 275–282
Nogueira-Neto P (1973) Ordem Rodentia. In: A criação de animais indígenas vertebrados: peixes, anfíbios, répteis, aves e mamíferos. Edições Tecnapis, São Paulo, pp 257–273
Nogueira SSC, Nogueira-Filho SLG, Otta E, Dias CTS, Carvalho A (1999) Determination of infanticide causes in capybara (*Hydrochoerus hydrochaeris*) groups in captivity. Appl Anim Behav Sci 62:351–357
Nogueira SSC, Lopes APS, Nogueira-Filho SLG (2003) Como evitar infanticidios e peleas mortales en criaderos de capivaras (*Hydrochoerus hydrochaeris*)? Bol Bur Exch Distrib Inf Minilivestock 12:16–17
Nogueira-Filho SLG (1996) Criação de Capivaras. Centro de Produções Técnicas (CPT), Viçosa
Nogueira-Filho SLG, Nogueira SSC (2004) Captive breeding program as an alternative for wildlife conservation in Brazil. In: Kirsten S, Fragoso JMV, Bodmer R (eds) People in nature: wildlife manegement and conservation in Latin America. Columbia University Press, New York, pp 171–190
Ojasti J (1973) Estudio del chigüire o capibara. Fondo Nacional de Investigaciones Agropecuarias, Caracas
Ojasti J (1991) Human explotation of capybara. In: Robinson JG, Redford KH (eds) Neotropical wild life use and conservation. The University of Chicago Press, Chicago, pp 236–252
Ojasti J (1996) Wildlife utilization in Latin America: current situation and prospects for sustainable management, vol 25, FAO Conservation Guide. Food and Agriculture Organization of the United Nations, Rome

Paula TAR, Walker NJ (2012) Reproductive morphology and physiology of the male capybara. In: Moreira JR, Ferraz KMPMB, Herrera EA, Macdonald DW (eds) Capybara: biology, use and conservation of an exceptional neotropical species. Springer, New York, pp 107–129

Robinson JG, Redford KH (1991) Sustainable harvest of neotropical forest mammals. In: Robinson JG, Redford KH (eds) Neotropical wildlife use and conservation. The University of Chicago Press, Chicago, pp 415–429

Silva-Neto PB (1989) Alimentação e manejo de capivaras (*Hydrochoerus hydrochaeris hydrochaeris*) em cativeiro. M.Sc. Dissertation, Escola Superior de Agricultura "Luiz de Queiroz", Universidade de São Paulo, Piracicaba

Silva-Neto PB, Bortolanza D, Andrade M (1996) Como criar capivara. Agrodata Datavídeo, Curitiba. 1 video tape cassette, 50 min., color, VDS NTSC

Chapter 14
Social and Spatial Relationships of Capybaras in a Semi-confined Production System

Katia Maria P.M.B. Ferraz, Patrícia Izar, Takechi Sato, and Sílvia M. Nishida

14.1 Introduction

Social and spatial relationships between individuals have been used to describe the social structure of animal species (Hinde 1983; Lee 1994), which can be defined as the organizational pattern of social relationships among group members (Lusseau et al. 2008). Understanding the relationship of an individual with conspecifics can be important in determining its social position in the dominance hierarchy (Beacham 2003). As most definitions refer to dominance as an attribute of a pattern of repeated agonistic interactions, i.e., aggression (Bernstein 1981; Drews 1993), dominance hierarchies have been extensively analyzed by quantifying dyadic agonistic interactions (Gauthreaux 1978; de Vries and Appleby 2000).

The capybara (*Hydrochoerus hydrochaeris*) has a rich repertoire of agonistic and affiliative social interactions, both most often seen in confined conditions due to the limitation of resources. Quantifying social interactions allows group social structure to be described, which is fundamental for both effective management and conservation.

K.M.P.M.B. Ferraz (✉)
Departamento de Ciências Florestais, Escola Superior de Agricultura "Luiz de Queiroz", Universidade de São Paulo, Av. Pádua Dias 11, 13418-900, Piracicaba, São Paulo, Brazil
e-mail: katia.ferraz@usp.br

P. Izar • T. Sato
Departamento de Psicologia Experimental, Instituto de Psicologia, Universidade de São Paulo, Av.Prof.Mello Moraes 1721, Cidade Universitaria, 05508-030, São Paulo, Brazil
e-mail: patrizar@usp.br; takesato@usp.br

S.M. Nishida
Departamento de Fisiologia, Instituto de Biociências, Universidade Estadual Paulista "Júlio de Mesquita Filho", Campus de Botucatu Distrito de Rubião Jr., s/n°, 18618-970 Botucatu, São Paulo, Brazil
e-mail: nishida@ibb.unesp.br

J.R. Moreira et al. (eds.), *Capybara: Biology, Use and Conservation of an Exceptional Neotropical Species*, DOI 10.1007/978-1-4614-4000-0_14,

This chapter describes the social interactions of capybaras evaluated in a semi-confined system. Capybaras were kept in a fenced enclosure (about 8 ha) containing an artificial lake, bushy vegetation, open area of grassland, and a small corral with two feeding troughs (10 × 3 m) inside it. Two different groups had been captured in the wild and introduced at different moments. The capybara groups were not submitted to any intervention in this area apart from feeding. Data were quantified from members of only one group (3 males and 12 females), all of which were individually marked. Data were registered for 8 months and analyzed by graph theory methods (Box 14.1) to reveal the strongest dyadic relationships established among the members of a social group (Busacker and Saaty 1965; Sato 1996; Izar et al. 2005, 2006).

Box 14.1 Graph Theory in the Analysis of Social Structures

Graph theory has been used for the analysis of natural and social phenomena. Methods from the theory, such as the Minimum Spanning Tree, Directed Tree, and Dominance Tree allow analysis of the social structure of groups through the graphical representation of dyadic relationships established among the strongest members of the group (Izar et al. 2005). A graph G is a set V of vertices and a set A of arcs, denoted as G = (V, A). Vertices can be individuals in a social group, behaviors, things, groups, positions in a hierarchy, cities, etc. Thus, vertices correspond to elements of set theory. Vertices can be connected by arcs that represent relationships such as "like to," "hit," "together with," "communicate with," "hate," etc. The arcs of a graph correspond to ordered pairs of a relation of set theory (Sato 1996). Thus, the social structure is defined by a set of vertices that are the individuals of the group and a set of arcs that are the dyadic relationships established between them. By this definition, the social structure is a graph.

The Minimum Spanning Tree (MST; Fig. 14.1) is a connected, undirected graph with n vertices and (n − 1) arcs or links, i.e., it has no circuits. It is a

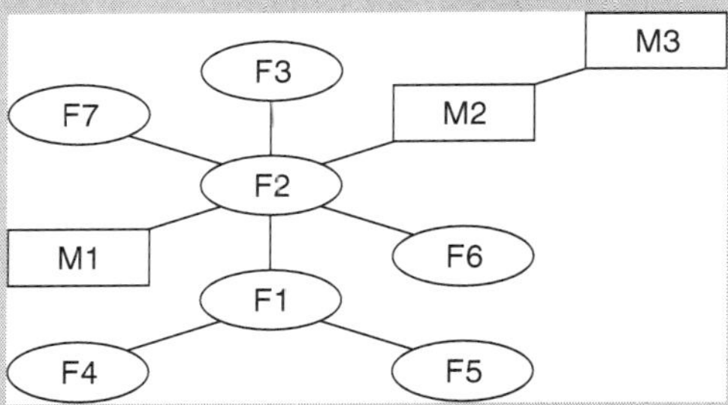

Fig. 14.1 Example of a Minimum Spanning Tree. *Squares* represent males and *circles* represent females

(continued)

Box 14.1 (continued)

symmetric graph, where arcs represent nondirectional relationships between the vertices (e.g., F1 and F2 are in contact). The graphic structure of minimum spanning trees tends to represent the strongest relationships in the matrix of relationships between group members and allows the identification, for example, of subgroups characterized by gender and/or age. Or, assuming that certain group members form small distinct subgroups, the members of a subgroup will appear on branches near the tree, linked to another subgroup through a larger branch (Izar et al. 2005).

The Shortest Directed Tree (ditree; Fig. 14.2) is a partial subgraph of a directed graph, i.e., relations have direction; for example, A moves B. In ditree, each vertex is reached by only one arc and the path between the root and a vertex is always the shortest. A path is a sequence of arcs beginning at a vertex and can take several other vertices. The length of a path is the sum of arcs from the initial to the terminal vertices. The root may be the source or destination. The paths of the shortest directed tree must satisfy the condition $L[r, y] \leq L[r, x] + a(x, y)$ (Japyassú et al. 2006). Therefore, as in a MST, the graphical structure represents the strongest relationships between pairs of group members.

The Dominance Tree (Fig. 14.3) is a method developed to detect and distinguish different ways of organizing dominance relations with partial hierarchies. The method assumes that a hierarchical structure obeys the principles of transitivity (if M1 dominates F1, and F1 dominates M2, M1 must dominate M2), and the transitive matrix (used to represent the social hierarchy) is

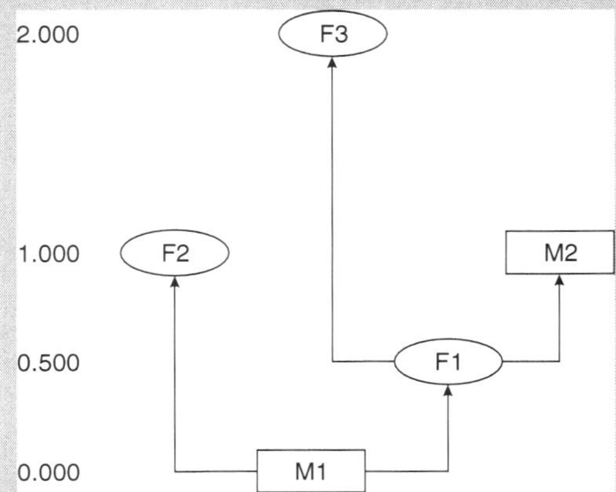

Fig. 14.2 Example of a Shortest Direct Tree. The direction of *arrows* indicates the order of dominance relationships. *Squares* represent males and *circles* represent females

(continued)

Box 14.1 (continued)

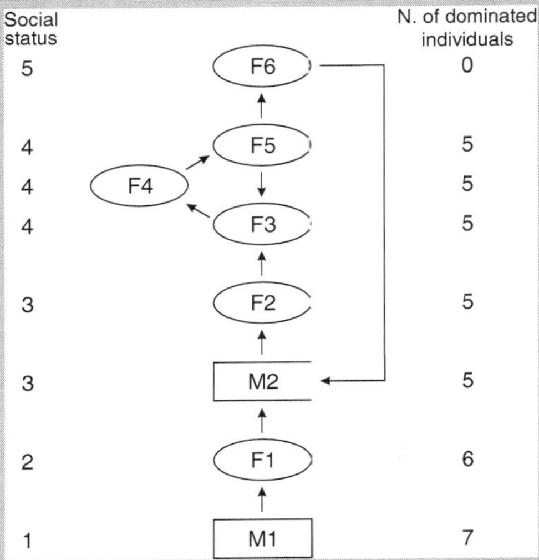

Fig. 14.3 Example of a Dominance Tree. The direction of *arrows* indicates the order of dominance relationships. *Squares* represent males and *circles* represent females

both irreflexive and asymmetric. Unlike conventional methods, dominance trees do not distort the dominance structure to calculate a single index of linearity. It is also possible to identify subgroups of individuals that obey or disobey the principles of linearity. As with the MST and ditree, there is also a graphical representation that allows clear differentiation of the types of hierarchy in a social group, for example, substructures of a partial hierarchy, where more than one line can start from a common dominant, or several lineages converge to a common subordinate (Izar et al. 2006).

14.2 Agonistic Interactions

Agonistic behaviors of capybaras in captivity include biting, chasing, and fighting (described in Box 14.2). Some of these behaviors have been described for capybaras in the wild by Azcárate (1980) and Alho et al. (1989).

There are differences in the frequency of agonistic behavior between male and female capybaras in the wild (Schaller and Crawshaw 1981; Herrera and Macdonald

1993; Salas 1999). Males interact more with each other (Schaller and Crawshaw 1981), whereas female-male and female-female aggressive interactions are rarely observed (Herrera and Macdonald 1993; Lord 1994). Total aggression rates correlate with the number of males in the group (Macdonald 1981; Herrera and Macdonald 1993). According to Azcárate (1980), aggressive dominance among the classes flows in the direction of adult male – adult female – young and infants, representing a gradient of decreasing overall aggressiveness and increasing group cohesiveness.

Agonistic interactions can be intense and frequent among group individuals or between individuals of different groups when confined in the same space. These interactions are typically associated with a limitation of essential resources such as food, shelter, and/or space. In addition to the agonistic interactions between adult males, there is a frequent occurrence of aggression between adult females and between males and females, behaviors rarely observed in natural environments (Herrera and Macdonald 1993; Lord 1994).

In our semi-confined group, we registered 2,181 agonistic interactions by ad libitum sampling, but quantified only 1,846 interactions from the identified individuals. The most commonly recorded agonistic interactions were charging the opponent (58.6%), followed by biting (14.7%) and approaching (13.1%), usually causing the opponent to retreat. The interactions occurred more frequently among females (80.9%). Only 3.9% of the interactions occurred between males and 15.3% among males and females, in this case usually initiated by males.

The frequency of agonistic interactions varied depending on the activity of individuals in this group, becoming more intense during feeding, whether natural (foraging = 4.8%) or artificial (at feeding troughs = 68.0%). The highest frequency of interactions in the enclosure probably occurred due to the concentration of food in the trough and the physical limitation of space.

Box 14.2 Description of Capybara Social Interactions in a Semi-confined Production System: Agonistic Interactions

The term agonistic behavior was coined by Scott and Fredericson (1951) to describe any activity of fighting, encompassing components of attack, defense, and/or escape. The mechanisms underlying the aggressive and defensive components may be distinct, representing two extremes of a continuum or two alternate and interactive dimensions of behavior (Nelson and Chiavegatto 2000).

(continued)

Box 14.2 (continued)

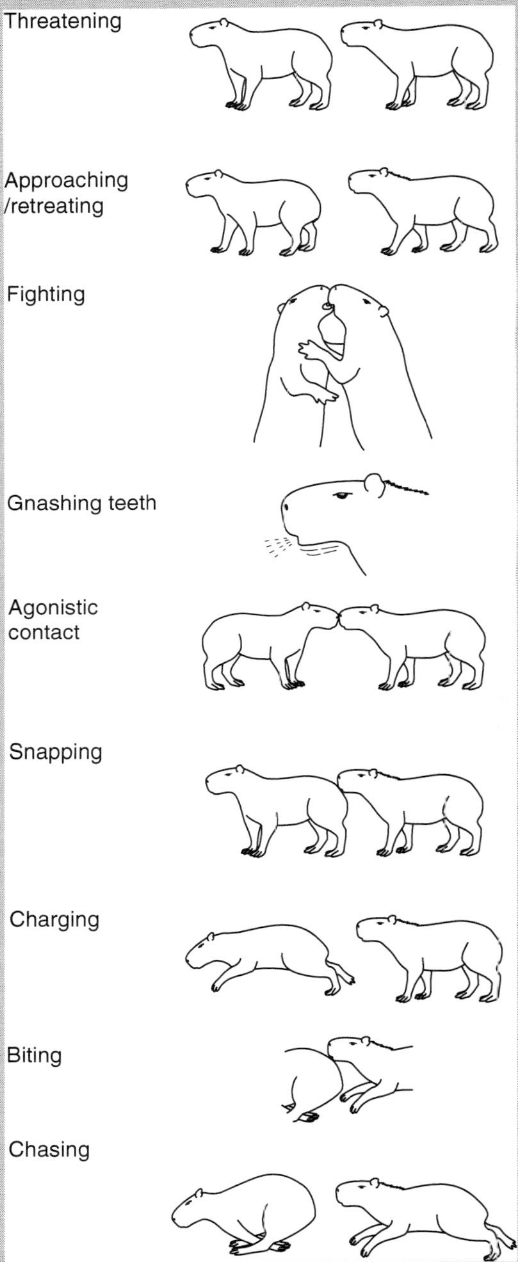

(continued)

Box 14.2 (continued)

Threatening	A standing or sitting individual interrupts its current activity and orientates the head or body toward a cospecific, with or without physical contact with the muzzle. The behavior starts spontaneously or in response to an aggressor, and may be accompanied by vocalization
Approaching/retreating	An individual walks toward an opponent, usually with the head up and hair spiked on the dorsal region and neck, and causes its immediate retreat
Fighting	Two individuals approach one another, rear up on hind legs, and become locked together, supporting front paws on each other. In this position, they alternate in biting the facial region of the opponent, then return to the starting position. This behavior may be accompanied by a chase. In this case, the animal that gives up the fight flees, chased by the opponent
Gnashing teeth	An individual gnashes its teeth toward an opponent, resulting in its retreat
Agonistic contact	An individual approaches and faces an opponent, positioning itself firmly with head held high. Muzzles touch and the position is maintained for several seconds. Then one individual walks away. May also be associated with other agonistic behaviors, including teeth gnashing, chasing, or biting
Snapping	An individual snaps at a cospecific which may or may not cause its removal. This purely agonistic interaction may occur due to the approach of an aggressor and possibly be accompanied by gnashing teeth. It is almost exclusively associated with trough feeding activity, caused by the tense social situation and competition for resources
Charging	The attacker finds the opponent, approaching slowly with its head projecting slightly forward, causing immediate retreat. The attacker may then start a chase
Biting	An individual bites an opponent. May occur as a result of a chase or a threat
Chasing	An individual runs after an opponent. During the chase, the attacker may expose the incisor teeth to the opponent or gnash teeth in its direction. The chase finishes when the attacker gives up or the opponent is displaced. When the chase finishes, the attacker returns to its previous activity, while the opponent remains displaced

14.3 Dominance Structure

The dominance structure of this group of capybaras, including all agonistic interactions recorded in all contexts, was nonlinear (Fig. 14.4). Although the structure could be described as a single lineage, with one dominant and one subordinate female, the principle of transitivity (i.e., if A dominates B, and B dominates C, then A dominates C) was violated at two positions in the hierarchy. In general, this result suggests a flexible hierarchical structure, at least when the group is maintained in a semi-confined production system.

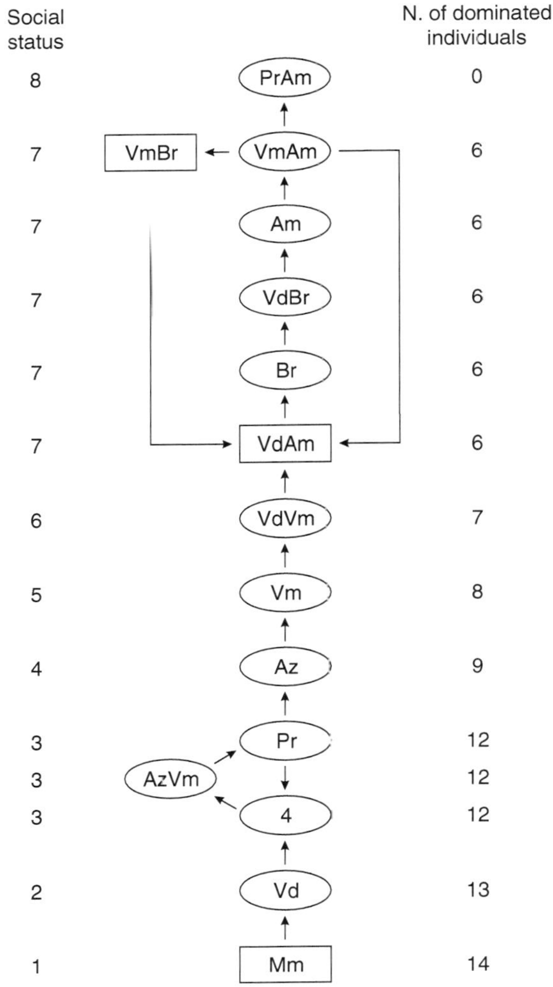

Fig. 14.4 Reduced graph of the social dominance structure of a group of 15 capybaras maintained in a semi-confined production system. The direction of *arrows* indicates the order of dominance relationships. *Squares* represent adult males and *circles* represent adult females

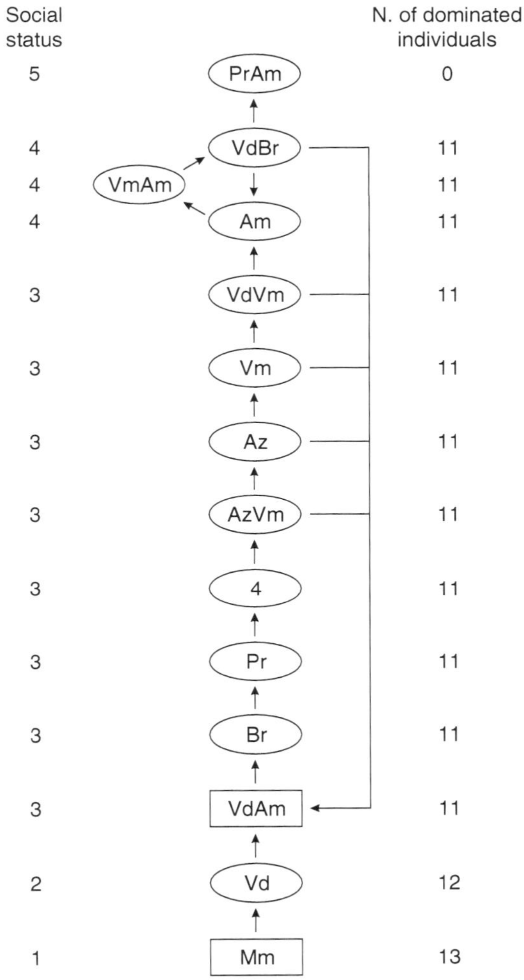

Fig. 14.5 Reduced graph of the social dominance structure of a group of 15 capybaras in a semi-confined production system, during feeding at a trough. The direction of *arrows* indicates the order of dominance relationships. *Squares* represent adult males and *circles* represent adult females

During trough feeding, agonistic relationships did not follow the hierarchy, except in relation to the two most dominant individuals (male Mm and female Vd), and the subordinate individual (PrAm). The other group members behaved as if they had equal ranks (Fig. 14.5). The large increase in the frequency of agonistic interactions in situations of competition for limited resources, and the many apparently circular relationships where an individual had no clear hierarchical position might be explained by the fact that the dominance hierarchy was not sharply defined.

In comparison with the agonistic behavior at the trough, the group's dominance structure during other activities (resting, foraging, and displacement; Fig. 14.6) was clearly more hierarchical. Most individuals occupied different social positions in

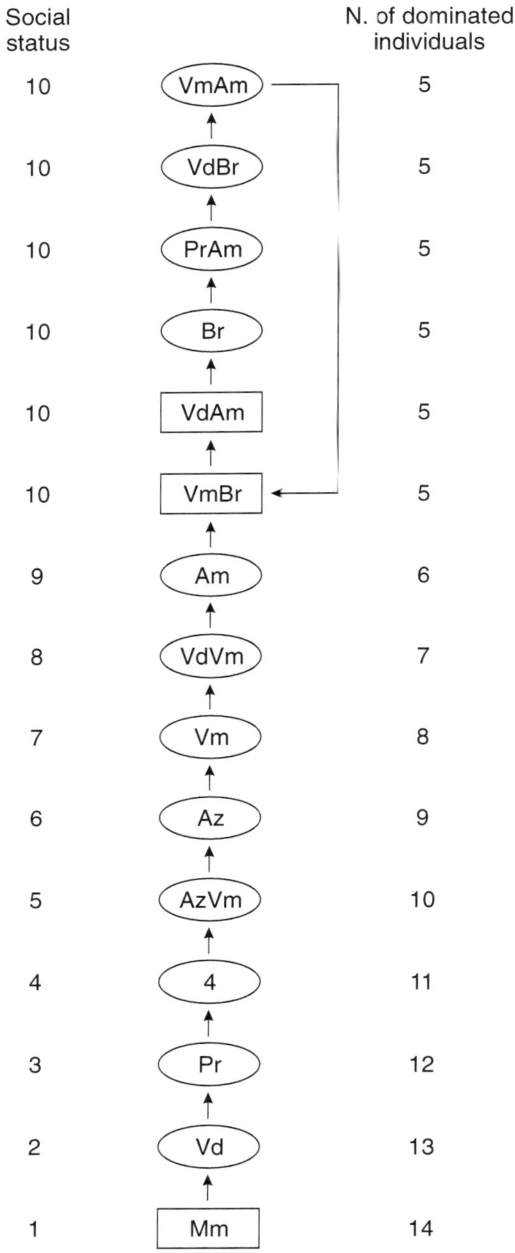

Fig. 14.6 Reduced graph of the social dominance structure of a group of 15 capybaras in a semi-confined production system, during activities other than trough feeding. The direction of *arrows* indicates the order of dominance relationships. *Squares* represent adult males and *circles* represent adult females

the dominance hierarchy and circular relationships occurred only between individuals which held subordinate positions. Côté (2000) observed similar results when comparing the dominance of mountain goats (*Oreamnos americanus*) in a natural situation and during trough feeding. Agonistic interactions were more frequent and

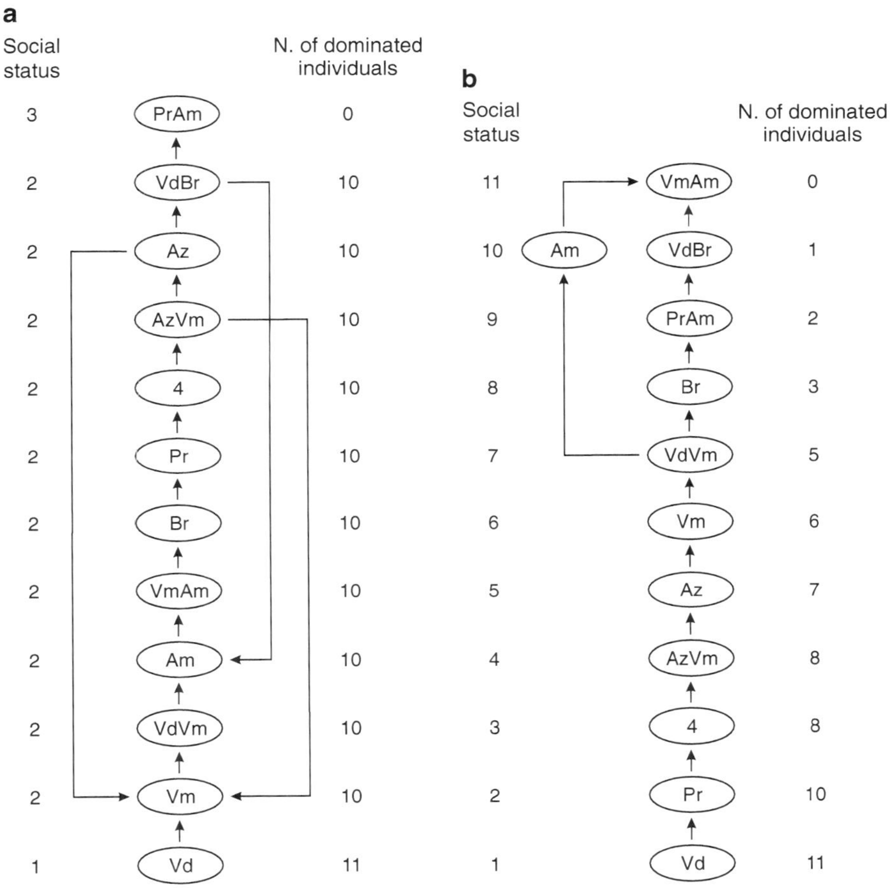

Fig. 14.7 Reduced graphs of the social dominance structure of adult female capybaras in a semi-confined production system during: (**a**) trough feeding and (**b**) other activities. The direction of *arrows* indicates the order of dominance relationships

the dominance order was broken during feeding. Côté (2000) suggested that inconsistencies in the hierarchy arose because individual recognition was hampered by the spatial concentration at the trough and the increased agonistic events.

We suggest, instead, that the linear capybara hierarchy "bends" during feeding time in captivity due to competition for food resources, the benefits of which may offset the potential costs associated with fighting with an individual of slightly higher status. Note that individuals with the highest social status were not challenged, unlike those of intermediate status. Probably, without the benefit of obtaining immediate food resources, subordinate individuals have less incentive to run the risk of injury in an agonistic interaction with higher ranked individuals.

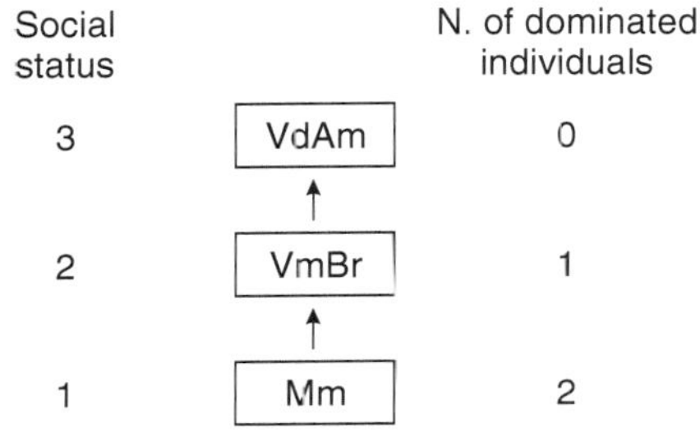

Fig. 14.8 Reduced graph of the social dominance structure of adult male capybaras in a semi-confined production system during trough feeding and other activities. The direction of *arrows* indicates the order of dominance relationships

14.3.1 Dominance Relationships Among Females

Female social structure, including either all agonistic interactions or with the removal of males, proved to be nonlinear, with some females occupying the same social position, shown by circular relationships (Fig. 14.7). Considering that agonistic interactions between females occurred only during trough feeding, dominance does not appear clearly defined, since almost all females occupied equivalent positions in the social hierarchy (regarding the number of individuals dominated by each one). The exception was the dominant female and one subordinate. The dominance relationships of females in activities other than feeding at the trough were more organized and hierarchical.

14.3.2 Dominance Relationships Among Males

The male social structure, including all agonistic interactions, was strictly linear (Fig. 14.8), confirming the results of Herrera and Macdonald (1993) for capybaras in the wild. The outcomes of agonistic interactions between males were invariably won by the dominant individual, ensuring priority at feeding troughs. However, the dominant male behaved differently toward the two subordinates, one of which (VmBr) was excluded from the group, while the other (VdAm) was not. Dominance relationships were the same whether at the feeding trough or under other circumstances.

14.4 Capybara Affiliations

We defined as affiliative interactions the following behavior patterns: sniffing, nibbling, amicable contact, and embracing (described in Box 14.3). Although most of these behaviors have been rarely reported, and partially in the wild, by Azcárate (1980), Alho et al. (1989), and Salas (1999), they occur frequently, both in nature and captivity.

Box 14.3 Description of Capybara Social Interactions in a Semi-confined Production System: Affiliative Interactions

Affiliative or non-agonistic social interactions are nonaggressive behaviors between two individuals with the primary purpose of promoting group cohesion.

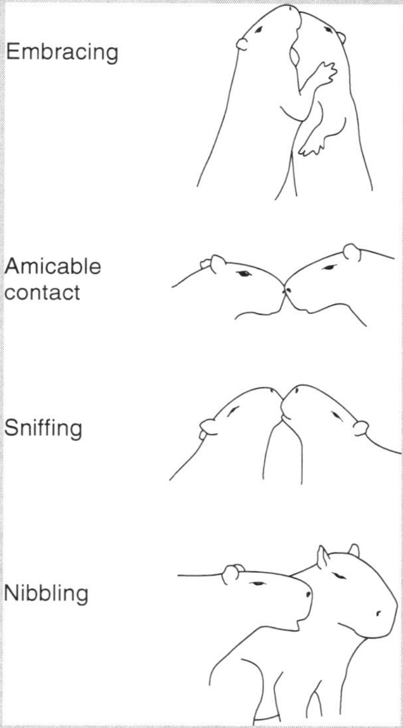

Embracing	Two individuals approach one other, touch muzzles, rise up on their hind feet, becoming upright and supporting each other. They entwine their front legs, remaining in this position and repeating the same sequence several times
Amicable contact	Consists of physical contacts made with the muzzle, usually on the posterior or lateral sides of the muzzle of the recipient
Sniffing	Two individuals come together and touch muzzles
Nibbling	An individual nibbles a cospecific, repeating the contact at short intervals. The recipient may respond by moving or remaining indifferent to the contact

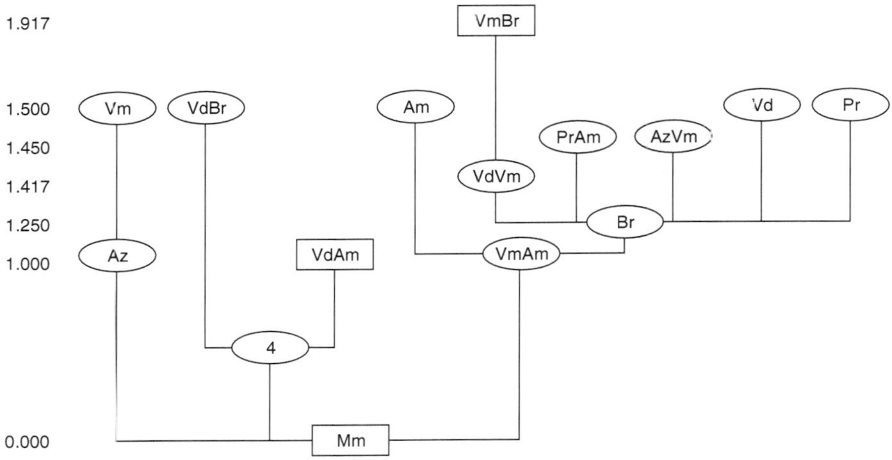

Fig. 14.9 Shortest Directed Tree with the dominant male (Mm) as the source root (agent). Numbers to the *left* represent the arc length (the shorter the arc, the stronger the relationship). *Squares* represent males and *circles* females

We registered 1,016 affiliative interactions by ad libitum sampling (Lehner 1998), but quantified only 307 interactions from identified individuals. Most affiliative interactions were nibbling (51.4%) and amicable contacts (47.6%). The interactions usually occurred between females (83.9%) and to a lesser extent between males and females (15.7%). The vast majority of the interactions between males and females were initiated by females (91.1%). No affiliative interactions between males were recorded.

The Shortest Directed Tree Method (ditree; Box 14.1), based on affiliative interactions, revealed that all individuals in the group participated in the social structure (Fig. 14.9). There was no correlation between the size of the source tree or the destiny tree of each individual as the root and its hierarchical position. The average size of trees with a source root (0.31 ± 0.15) was significantly lower (Wilcoxon $z = -2.78$, $p < 0.001$) than that of trees with a destiny root (0.47 ± 0.06), showing that the animals received interactions less frequently than they gave them. This indicates that the affiliative relationships do not involve reciprocity in the frequency of interactions.

14.5 Social Proximity

The study of social structure through proximity between individuals has grown recently with the advent of social network analyses conducted with many mammal species, including primates, dolphins, hyenas, and elephants (Wittemyer et al. 2005; Smith et al. 2008; Ramos-Fernández et al. 2009; Wiszniewski et al. 2009).

Social proximity can help define the social status of individuals because it results from individual tendencies to approach or avoid conspecifics, thereby allowing the description of a network of social relationships within a group (Tiddi et al. 2011).

During daily activities in captivity, individuals in a group of capybaras are organized in small clusters whose interindividual distance apparently varies depending on the type of activity, on the individuals involved, and on the dyadic social interactions. The composition of small clusters was registered every 5 min, resulting in 3,399 scans. Data were analyzed using Minimum Spanning Trees (Box 14.1).

The spatial structure of capybaras in the group varied according to the situation and activity (Fig. 14.10): pairs of individuals were more frequently observed within the same subgroup during displacement and at the feeding trough, indicating a higher degree of organization in interindividual spatial relationships in these more tense social contexts. The displacements occurred in single file with an apparently organized line of individuals moving along preexisting trails.

Some individuals were always in central locations (e.g., Br and VdVm), whereas others were located in positions peripheral to the group structure (e.g., AzVm, 4 and VdAm). The central position indicates that the animal maintains close relationships with several individuals in the group, which may involve transition between subgroups. This transition can result from different phenomena, depending on the hierarchical position of an individual. It can result from intolerance shown by other individuals, or from the freedom of a dominant animal to visit the various subgroups in search of food resources.

14.5.1 Social Proximity Among Males

Spatial proximity between males was much less frequent than among females. The three males were located at the extremities of the structures in all periods of activity (Fig. 14.10) and in distant locations in all structures except during displacement. This position suggests a role in patrolling females, thereby excluding potential competitors.

Males maintained greater distances between one another than in relation to the group of females in most situations, except in the most rigidly organized activities (feeding at the trough and displacements). In such activities, the proximity was characterized by strictly agonistic behaviors.

14.5.2 Social Proximity Among Females

Spatial relationships varied among females (Fig. 14.10), in terms of the numbers of other females with whom they maintained proximity, and the stability of their

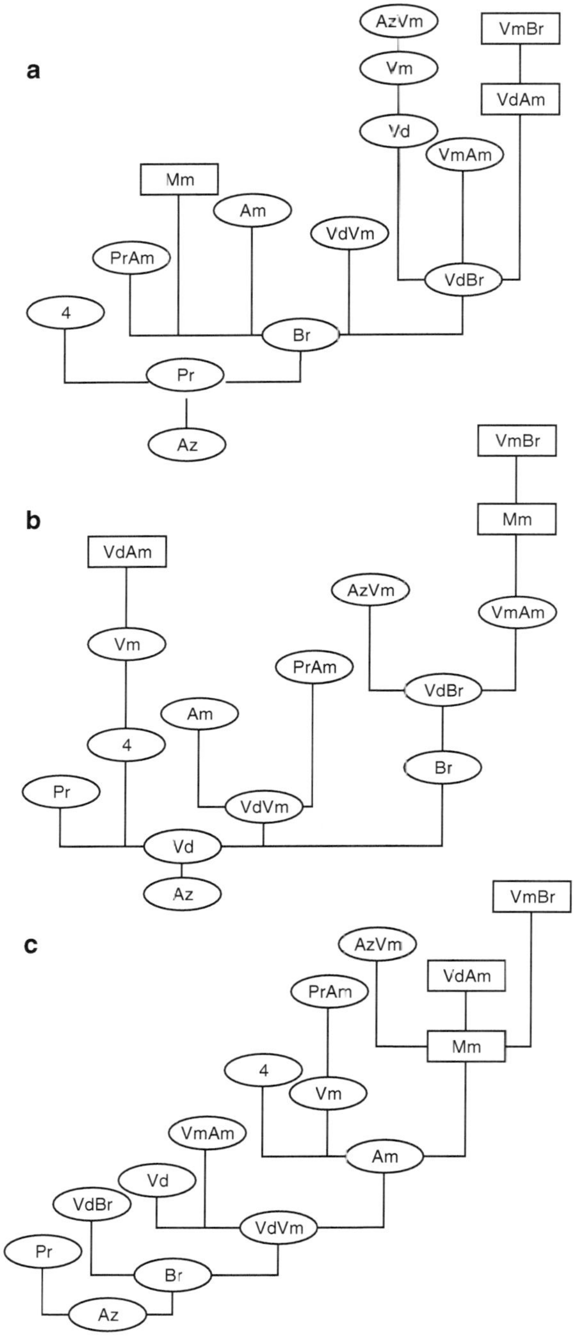

Fig. 14.10 Minimum Spanning Trees during periods of (**a**) resting, (**b**) foraging, and (**c**) trough feeding. *Squares* represent males and *circles* females

affiliations. Some never interacted with other females, and some formed particularly stable affiliations although the biological basis for such behavior was unknown.

14.5.3 Social Proximity Between Males and Females

Males did not maintain any preferential relations with females (Fig. 14.10). Similarly, the intersexual spatial geometry was variable: some females were never close to males; others affiliated with only one male and, if this was the dominant one, they were inevitably distanced from the subordinate.

14.6 Implications for Capybara Management in Captivity

Concentration of resources in captivity may intensify social tensions around water, food, and shelter. Therefore, we recommend that these resources should be provided widely spread. Providing food at various dispersed points within the area and the use of larger troughs may enable individuals to distribute naturally during feeding, lessening the chances of conflict and consequently social tension.

It is important to remember that confinement, however large the containment area, directly interferes with the natural dynamics of groups, altering the age structure and sex ratio over time. Management actions must remove surplus individuals, keeping the original group structure, similar to the process that occurs in natural populations. Moreover, confinement also interferes in hierarchical relationships, since the animals are forced to live together and share the same resources. Thus, any procedure to minimize encounter rates and the frequency of aggressive behavior should be prioritized.

Acknowledgments We thank the Coordenação de Aperfeiçoamento de Pessoal de Nível Superior (CAPES), Brazil, for financial support. We also thank José Roberto Moreira for illustrations of capybara behavior and for his valuable contribution for improving the chapter. We thank Paulo Bezerra for logistic support and the Buana Parque Farm, Guareí, São Paulo state, Brazil.

References

Alho CJR, Campos ZM, Gonçalves HC (1989) Ecology, social behavior and management of the capybara in the Pantanal of Brazil. In: Redford KH, Eisenberg F (eds) Advances in neotropical mammalogy. Sandhill Crane Press, Gainesville, pp 163–194

Azcárate TA (1980) Sociobiología y manejo del capibara (*Hydrochoerus hydrochaeris*). Doñana Acta Vertebrata 7(6):228

Beacham JL (2003) Models of dominance hierarchy formation: effects of prior experience and intrinsic traits. Behaviour 140:1275–1303

Bernstein IS (1981) Dominance: the baby and the bathwater. Behav Brain Sci 3:419–458

Busacker RG, Saaty TL (1965) Finite graphs and networks. McGraw-Hill, New York
Côté SD (2000) Determining social rank in ungulates: a comparison of aggressive interactions recorded at a bait site and under natural conditions. Ethology 106:945–955
De Vries H, Appleby MC (2000) Finding an appropriate order for a hierarchy: a comparison of the I&SI and the BBS methods. Anim Behav 59:239–245
Drews C (1993) The concept and definition of dominance in animal behaviour. Behaviour 125(3–4):283–313
Gauthreaux SA Jr (1978) The ecological significance of behavioral dominance. In: Bateson PPG, Klopfer PH (eds) Perspectives in ethology. Social behaviour. Plenum, New York, pp 17–54
Herrera EA, Macdonald DW (1993) Aggression, dominance, and mating success among capybara males (*Hydrochaeris hydrochaeris*). Behav Ecol 4(2):114–119
Hinde RA (1983) Primate social relationships: an integrated approach. Blackwell, London
Izar P, Japyassú HF, Alberts CC (2005) Métodos da Teoria dos Grafos para Análise Quantitativa do Comportamento. In: Garcia A, Tokumaru RS, Borloti EB (eds) Etologia: uma perspectiva histórica e tendências contemporâneas. Multiplicidade, Vitória, pp 249–258
Izar P, Ferreira RG, Sato T (2006) Describing the organization of dominance relationships by dominance-directed tree method. Am J Primatol 68:189–207
Japyassú HF, Alberts CC, Izar P, Sato T (2006) EthoSeq: a tool for phylogenetic analysis and data mining on behavioural sequences. Behav Res Methods 38(4):549–556
Lee PC (1994) Social structure and evolution. In: Slater PJB, Halliday TR (eds) Behaviour and evolution. Cambridge University Press, Cambridge, MA, pp 266–303
Lehner PN (1998) Handbook of ethological methods. Garland STPM Press, New York
Lord D (1994) A descriptive account of capybara behaviour. Stud Neotrop Fauna Environ 29:11–22
Lusseau D, Whitehead H, Gero S (2008) Incorporating uncertainty into the study of animal social networks. Anim Behav 75(5):1809–1815
Macdonald DW (1981) Dwindling resources and the social behaviour of capybara, (*Hydrochoerus hydrochaeris*) (Mammalia). J Zool Lond 194:371–391
Nelson RJ, Chiavegatto S (2000) Aggression in knockout mice. Mouse Behav Mod Biomed Res 41(3). http://dels.nas.edu:80/ilar_n/ilarjournal/41_3/Aggression.shtml
Ramos-Fernández G, Boyer D, Aureli F, Vick lG (2009) Association networks in spider monkeys (*Ateles geoffroyi*). Behav Ecol Sociobiol 63:999–1013
Salas V (1999) Social organization of capybaras in the Venezuelan Llanos. University of Cambridge, Cambridge, UK
Sato T (1996) Relatório Técnico CNPq
Schaller GS, Crawshaw PG (1981) Social organization in a capybara population. Saugetierkundliche Mitteilungen 29:3–16
Scott JP, Fredericson E (1951) The causes of fighting in mice and rats. Physiol Zool 24:273–309
Smith JE, Kolowski JM, Graham KE, Dawes SE, Holekamp KE (2008) Social and ecological determinants of fission-fusion dynamics in the spotted hyaena. Anim Behav 76:619–636
Tiddi B, Aureli F, Schino G, Voelkl B (2011) Social relationships between adult females and the alpha male in wild tufted capuchin monkeys. Am J Primatol 73:812–820
Wiszniewski J, Allen SJ, Möller lM (2009) Social cohesion in a hierarchically structured embayment population of Indo-Pacific bottlenose dolphins. Anim Behav 77:1449–1457
Wittemyer G, Douglas-Hamilton I, Getz M (2005) The socioecology of elephants: analysis of the processes creating multitiered social structures. Anim Behav 69:1357–1371

Chapter 15
Feeds and Nutrition of Farmed Capybaras

Alcester Mendes and Sérgio L.G. Nogueira-Filho

15.1 Introduction

The capybara (*Hydrochoerus hydrochaeris*) is, potentially, an excellent source of meat and leather (González-Jiménez 1995) due to its fecundity and other characteristics: it is robust, docile, and adapts easily to captive conditions. The main factor usually considered when domesticating any new species for meat production, however, is the animal's diet, both from the economic and the biological point of view (Emmons 1987). Most domestic species that are widely used for meat production feed on grasses and/or grain crops, as well as agricultural by-products that can be mass-produced at low cost. Capybaras consume mainly grass (Ojasti 1973; Alho et al. 1987; Barreto and Quintana 2012), converting it into high-quality animal protein for human consumption (Murphey et al. 1985; Emmons 1987; Frasson and Salgado 1990; González-Jiménez 1995). This is one of the notable advantages to farming capybaras compared to other neotropical wild species, such as peccaries (*Pecari tajacu* and *Tayassu pecari*) and pacas (*Cuniculus paca*). In this chapter we examine the capybara's nutritional requirements and describe the feeds and feeding practices used in scientific and commercial breeding centers.

A. Mendes (✉) • S.L.G. Nogueira-Filho
Departamento de Ciências Agrárias e Ambientais, Universidade Estadual de Santa Cruz, Rodovia Ilhéus Itabuna km 16, 45662-900 Ilhéus, Bahia, Brazil
e-mail: alcester1969@hotmail.com; slgnogue@uesc.br

J.R. Moreira et al. (eds.), *Capybara: Biology, Use and Conservation of an Exceptional Neotropical Species*, DOI 10.1007/978-1-4614-4000-0_15,

15.2 Some Anatomical, Physiological, and Behavioral Features of the Capybara Digestive System

The dentition of the capybara is similar to that of most other rodents with the exception of some specific features. Capybara incisors, sharp teeth around 2.8″ (7.0 cm) in length, allow them to cut grass near the soil surface in natural pastures, while their molariform teeth mash the ingested food into tiny particles (0.001–0.30 mm^2) which results in highly efficient extraction of soluble substances (Ojasti 1973; González-Jiménez 1977a; Parra et al. 1981; Alho et al. 1987; Herrera 2012).

Capybaras are also highly efficient in the digestion of roughage, due to the microbial fermentation that occurs in the cecum. The cecum provides an anaerobic and slightly acidic environment (pH around 6.0) that enables cellulolytic microorganisms (bacteria and protozoa; Herrera 2012) to survive. These digest the structural carbohydrates (cellulose and hemicellulose) that comprise the cell walls of forage consumed (Parra and González-Jiménez 1971).

15.2.1 *Energy and Protein Metabolisms*

The moment at which fermentation occurs in relation to chemical digestion is what distinguishes the microbial action between ruminants and capybaras. Fermentation in bovines and other ruminants occurs before enzymatic digestion, whereas fermentation in capybaras occurs in the cecum after the food has undergone enzymatic digestion in the stomach and nutrients have been absorbed in the small intestine. The undigested compounds, predominantly structural carbohydrates, which arrive in the cecum are decomposed by microbial fermentation (Herrera 2012), a process that provides energy for the microorganisms and by-products such as volatile fatty acids (VFA). The latter are absorbed through the cecum wall, which is thin and abundantly vascularized, transported to and metabolized in the liver of the capybara (Parra et al. 1981; Baldizán et al. 1983).

The concentration of VFAs in the capybara cecum varies according to diet. Under natural conditions, VFA concentration is 107 (± 33) $mM.L^{-1}$, composed of 72.5% acetic acid, 11% propionic acid, and 16.5% butyric acid (Borges et al. 1996). These levels are similar to those found in the rumen of pasture-fed ovines (114 $mM.L^{-1}$ VFA; Pond et al. 2004). In captive capybaras, when the amount of concentrate feedstuffs increases, while fiber decreases (Baldizán et al. 1983), VFA concentration falls, ranging between 41 and 44 $mM.L^{-1}$ (81–83% acetic acid, 14–15% propionic acid, and 2–4% butyric acid). This low VFA concentration nevertheless indicates active microbial fermentation.

The microbes in the cecum also play an important additional role by providing a direct source of protein in a process that is crucial for the protein balance of capybaras. These microorganisms are ingested through cecotrophy (the ingestion of cecal content or cecotrophe; Fig. 15.1), in a similar manner to that of lagomorphs

Fig. 15.1 Cecotrophy behavior sequence in capybaras: (**a**) cecotrophe ingestion; (**b**) shrinkage of the anus (Photo by José Eduardo Moreira)

Fig. 15.2 Two different types of excreta produced by capybaras: standard feces and misshapen material – "cecotrophe" (Photo by A. Mendes)

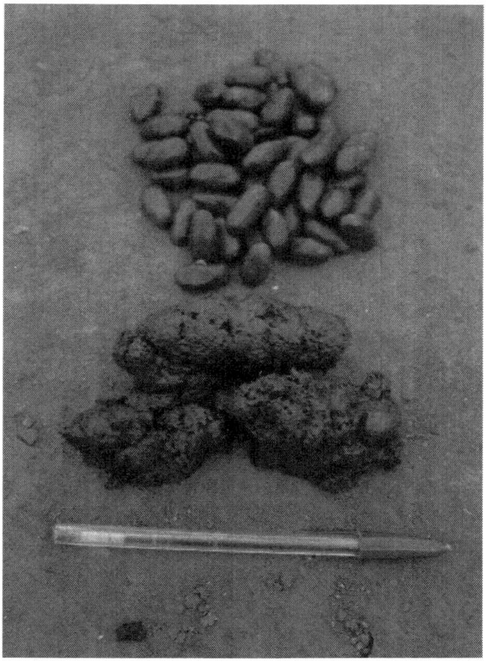

(rabbits and hares) and other rodents (Proto 1984; Herrera 1985; Cheeke 1987; Mendes 1999; Mendes et al. 2000; Hirakawa 2002; Herrera 2012). There are differences, however, in the material ingested by capybaras in natural conditions and in captivity. The cecotrophes of wild capybaras comprise a chain of fecal pellets bound together by yellow mucous (Herrera 1985), whereas in captivity they are a shapeless mass of doughy consistency (Mendes et al. 2000). This material is easily distinguished from the usual oval-shaped feces (Fig. 15.2), and it contains up to

37% more protein and 30% less fibrous material, depending on the diet (Mendes et al. 1996). The differences between these two kinds of excreta are similar to those found between the feces and cecotrophes of rabbits (Cheeke 1987).

The cecotrophes have more protein because they are produced after the cecal content is submitted to bacterial and protozoan fermentation with incorporation of nutrients such as protein and vitamins B and K. The microorganisms themselves are comprised of 40% high biological value proteins (Pond et al. 2004).

15.2.2 Digestibility and Variation in Cecotrophe Production

Cecotrophy in capybaras varies in frequency and it can even stop altogether when food is rich in protein (Herrera 1985; Gonzalez-Jimenez 1977b; Herrera 2012). It is most frequent when the nutritional quality of the diet is low. Diets with a higher proportion of concentrate (containing 14% of crude protein, dry matter basis) produce fewer cecotrophes (Andrade 1996; Mendes et al. 1996; García et al. 2000). In studies of wild populations, there was also a higher occurrence of cecotrophy during the dry season in the Venezuelan llanos, when food was scarce and lacking nutrients (Herrera 1985). However, there were no differences in the occurrence of cecotrophy in captive capybaras fed with diets ranging from 7.1% to 11.5% of crude protein (Carvalho 2010). In addition, when fed twice a day, the frequency of such behavior decreased (Sérgio L. G. Nogueira-Filho personal observation). Clearly, this behavior has implications for farming since the occurrence of cecotrophy can decrease expenditure on protein and vitamin supplementation. Further studies to determine the protein percentage in the diet and the feeding regime that best balances the natural occurrence of cecotrophy, protein gain and reduced costs would be beneficial.

Digestibility coefficients (that is the proportion of food consumed which is digested) of roughage in capybaras (Table 15.1) are similar to those in ovines, and 20% higher than in rabbits (Parra and González-Jiménez 1971; González-Jiménez and Parra 1972; Ojasti 1973; González-Jiménez and Escobar 1975). The rabbits' cecum only ferments the most easily fermentable substrates, such as soluble and short particles; large particles are barely fermented and the effect of cecotrophy on fiber digestibility is low (Cheeke 1987). Capybaras, with their highly efficient mastication and long retention time of undigested compounds in the cecum, can efficiently digest fibrous feedstuffs. Moreover, previous studies suggest a higher digestibility of fiber contents after adding energy and protein supplements in the diet (González-Jiménez and Escobar 1975; Baldizán et al. 1986; Bernardi 1993).

Digestive efficiency is also affected by the conditions in which capybaras are kept. When they live in metabolic cages, cecotrophy is inhibited, hence reducing digestive efficiency and not favoring the use of high fiber diets (Ferreira 2007). The average retention time of roughage in the digestive tract of capybaras is 12 (± 1.9) hours, with a passage rate of 8.4 (± 1.3)% h^{-1}. Thus, the turnover of the contents of a capybara's digestive tract occurs twice a day (Mendes 1999; Mendes et al. 2007). However, no comparative data exist on the passage rate for pelletized and non-pelletized feeds for capybaras.

Table 15.1 Apparent digestibility coefficients (%) for capybara digestion of dry matter (DM), organic matter (OM), crude protein (CP), and crude fiber (CF)

	Digestibility coefficients (%)			
Diets	DM	OM	CP	CF
Paspalum fasciculatum[a]	–	53.5	64.5	54.5
Urochloa mutica=*Brachiaria mutica*[b]	60.7	61.7	60.2	–
Cenchrus ciliaris[c]	44.6	–	–	50.7
Cenchrus ciliaris and 30% Cassava meal	56.6	–	–	52.0
Concentrate feed with 15% of *Cynodon dactylon* hay[d]	–	79.8	–	78.6
Concentrate feed with 43% of *Cynodon dactylon* hay[d]	–	62.4	–	62.1
Pennisetum purpureum[e]	57.9	59.6	63.4	–
Cassava leaf hay[e]	71.8	82.4	84.1	–
Palm oil meal[e]	85.7	86.2	85.4	–

[a]Ojasti (1973)
[b]Parra and González-Juménez (1971)
[c]Baldizán et al. (1986)
[d]Bernardi (1993)
[e]Ferreira (2007)

15.3 Nutritional Requirements for Maintenance, Growth, and Reproduction of Capybaras

To formulate appropriate diets for capybaras in captivity, the breeder will need to know their nutritional requirements, especially their energy and protein requirements. Capybaras need approximately 65.5 kcal/kg of metabolic weight (MW = live weight$^{0.75}$), or about 1,042 kcal/day for an adult capybara weighing 40 kg (Ojasti 1973). Under natural conditions, capybaras obtain energy mainly from the microbial fermentation of grasses in the cecum, as do all mammals that practice cecotrophy (Parra and González-Jiménez 1971; Hirakawa 2001).

Dietary protein content is essential to promote proper biological functioning and growth (Pond et al. 2004). The maintenance requirement of digestible protein (DP) for capybaras is also relatively low: 2.5 g DP/kg per day of MW (González-Jiménez and Escobar 1975). In contrast, the Nelore – one of the breeds of *Bos indicus* cattle most farmed in the neotropics – requires 4.03 g DP/kg of MW per day (Véras et al. 2007). Capybaras need less DP because of cecotrophy. More recently, Carvalho (2010) repeated González-Jiménez and Escobar's (1975) study under captive conditions that favored cecotrophy and found a lower protein requirement for capybaras (1.6 g DP/kg of MW per day). This shows the importance of cecotrophy behavior for the capybara's nitrogen balance (Carvalho 2010). Therefore, it is estimated that a diet containing 2.4% of DP (dry matter basis) would be sufficient to maintain an adult capybara weighing 40 kg and consuming 2.6% of its body weight per day.

Growth and reproduction require higher levels of protein and energy than maintenance alone. Tropical grasses are low in both energy and protein (Pond et al. 2004). Because of this, some concentrate feedstuffs with higher energy and protein contents than grass should also be provided for growing and reproducing capybaras. This will result in higher reproductive performance (Nogueira-Filho and Nogueira 2012), better feed conversion rate (efficiency in converting feed mass into increased body mass), and higher daily weight gain in comparison to a grass-only diet (Silva Neto 1989; Andrade 1996; Table 15.2). A diet for growing or reproducing capybaras must contain at least 12% crude protein and 3,500 kcal/kg gross energy (Silva Neto 1989; Andrade 1996). This can be provided with high-quality roughage and concentrate feedstuffs, such as corn grain and soybean meal.

There is little information available about the capybara's requirements for other nutrients, such as minerals and vitamins. The deficiency of these minerals in the diet of animals in captivity may cause harmful effects on production. Thus, it is recommended to provide ad libitum mineralized salt formulated for cattle in sheltered feeders (Nogueira-Filho 1996). Unlike most other mammals, capybaras cannot synthesize vitamin C from glucose (Cueto 1999; Cueto et al. 2000). Fresh grass is the natural source of this nutrient, and when a continuous supply is not available, it is recommended that a vitamin C supplement (300 mg/day) be added to the diet, to prevent scurvy (Cueto et al. 2000).

15.4 Capybara Feeding in Confined and Semi-confined Systems

The main feed of capybaras, grass, is cheap to produce as a feedstuff. Although selective when food is plentiful, capybaras will eat tree bark, palm seedlings, and bromeliad stalks when resources are scarce (Ojasti 1973; González-Jiménez 1977b; Alho et al. 1987). This somewhat generalist trait makes it easier to formulate their diets: capybara farmers can use several grasses and concentrates such as corn and soybean meal, and even formulations intended for rabbits and horses. In Brazil, capybara husbandry involves feeding with cut grass (e.g., varieties of elephant grass, *Pennisetum purpureum*, and Guinea grass – *Urochloa maxima* = *Panicum maximum*). These grasses are highly productive and usually eaten readily by capybaras (González-Jiménez 1978; Silva Neto 1989; Nogueira-Filho 1996).

15.4.1 Elephant Grass Supply and Alternatives to Its Seasonality

Elephant grass can reach annual production yields of 80 tons of dry matter per hectare, equivalent to 440 tons of green mass per hectare (Santos 1995). In Brazil, the productivity of elephant grass is highest during spring and summer

Table 15.2 Daily feed intake (DFI), daily weight gain (DWG), and feed conversion rate (FCR) of growing capybaras according to animal weight category, diet type, and crude protein (CP) content (%) of the diet

Animal category (live weight in kg)	Diet	CP content (%)	DFI	DWG (g/day)	FCR
14–21	Grass + ration 18% CP (50:50)	–	–	127.2	–
22–27	Grass + ration 14% CP (50:50)	–	–	89.3	–
18–25	Grass + ration 18% CP (56:44)	–	69.8 $g/kg^{0.75}$	110.0	7.5: 1
	Grass + ration 14% CP (43:57)	–	78.4 $g/kg^{0.75}$	91.8	7.4: 1
Above 20	Ration + Grass ad libitum[a]	13.5	1120.7 g	105.3	13.6: 1
	Ration + Grass ad libitum[a]	17.0	1123.4 g	106.3	13.9: 1
	Ration + Grass ad libitum[a]	20.5	1075.0 g	110.9	13.1: 1
14–21	Grass ad libitum	–	669.0 g	39.7	–
	Ration + Grass ad libitum[a]	17.0	588.5 g	100.1	6.7: 1
9.0–24.9	Ration ad libitum	13.5	299.6 g	62.7	5.1: 1
	Ration ad libitum	20.0	285.1 g	77.4	6.2: 1
	Elephant grass ad libitum	8.8	430.9 g	18.8	22.2: 1

Source: Adapted from Andrade (1996)

[a]The total grass intake ranged from 15% to 80% in proportion to concentrate feed comprised of corn grain, soybean meal, coast-cross (*Cynodon dactylon*) hay, mineral salts, and vitamin supplements

(September–March), when the highest rainfall and mean temperatures occur. Three to four cuts can be made in those seasons at intervals of 45–60 days, with production potential per cut between 20 and 40 tons of green mass/ha or 4–8 tons of dry matter/ha (Lima 2006).

This forage should be fed daily to capybaras in hanging bundles, thus avoiding trampling. When grass is seasonally scarce, it can be fed chopped in feeders (which prevents selectivity). The animals should be fed during the late evening, imitating their habit in the wild, where they forage during the cooler hours of the day (Ojasti 1973; Macdonald 1981; Alho et al. 1987). This is also an appropriate time to harvest grasses to avoid wilting and thus prolong nutritional quality. From the practical point of view, it is certainly feasible to harvest and provide grass in the same evening, since just one person can cut enough elephant grass to feed up to 100 adult capybaras in less than 4 h (Sérgio L. G. Nogueira-Filho personal observation).

An adult capybara weighing 40–60 kg should receive 4.0–8.0 kg of fresh forage daily. Younger capybaras with live weights of 2.0–20.0 kg should receive 1.0–3.0 kg daily (Ojasti 1973; González-Jiménez and Escobar 1975; Silva Neto 1989; Andrade 1996; Nogueira-Filho 1996; Cueto 1999; Alvarez and Kravetz 2006). Thus, the mean production of 30 tons of green matter/ha of elephant grass (cut at intervals of approx 60 days) is sufficient to feed about 80 adult capybaras, with an average weight of 50 kg, during the six rainiest months.

During the dry season (June-August in Brazil, January-March inVenezuela and Colombia), there may be little or no production of elephant grass. One solution is to expand the cropped area so that zones can be reserved for harvesting during months of low productivity. Nevertheless, this harvest will be less nutritious due to greater lignification, and so the diet should be supplemented with concentrate to meet capybaras' nutritional requirements. It is also possible to feed capybaras with sugar-cane when elephant grass is scarce. However, sugar-cane should be supplied only on alternate days, whole or chopped, or comprising up to 25% of the diet, because more than this can result in diarrhea. Another alternative for periods of forage shortage is corn and cob meal (whole ear ground with corn and straw) mixed with chopped elephant grass and mineral salt supplement (at 1%) to improve palatability (Nogueira-Filho 1996).

Hay tends not to be consumed willingly and is usually only eaten when ground and mixed with corn grain, soybean meal, and mineralized salt. Such formulations need vitamin supplementation (vitamin premix) containing vitamins A, E, and C. Corn and elephant grass silage could be another alternative for forage shortage periods. Capybaras generally avoid silage, due to its strong odor, but if it is crushed and mixed with corn grain, soybean meal, and mineral salt supplements, it becomes acceptable (Nogueira-Filho 1996).

15.4.2 Concentrate Supplements

Capybaras show their best growth performance on a diet of 60–80% grass supplemented with 20–40% of protein and energy concentrates (González-Jiménez 1977b;

Parra and Escobar 1978; Lavorenti 1989; Nogueira-Filho 1996; Andrade et al. 1998). Commercial livestock feedstuffs (for rabbits or horses) can be used (with about 12–14% of crude protein and 3,500–4,000 kcal/kg of gross energy) if the market price allows.

A suitable concentrate formula can be prepared on the farm, comprising a mixture of triturated corn grain and corn cob, cassava (*Manihot esculenta*) roots, cassava leaf hay, palm oil (*Elaeis guineensis*) meal, soybean meal, wheat meal, mineral supplements and common salt (Table 15.3). There is no need to provide the concentrate in pelletized form since capybaras readily eat ground concentrates dispensed from appropriate feeders. All feedstuffs and supplements should be provided with the cut grass in a single meal to stimulate the occurrence of cecotrophy (Sérgio L. G. Nogueira-Filho personal observation).

In temperate regions such as southern Brazil and Argentina, alternative feed types have been tested. These include: natural pastures (*Scirpus californicus, Echinochloa polystachya, Alternanthera philoxeroides, Cynodon dactylon, Paspalum* sp., *Lolium multiflorum, Bromus catharticus = Bromus unioloides*), cultivated pastures of oats and sorghum (*Sorghum bicolor*), alfalfa (*Medicago sativa*) hay, dehydrated cubes of alfalfa, cereal and oleaginous grains (oats, wheat, corn, sunflower), and carrots. Only sorghum was rejected by the animals, perhaps due to the high tannin content of its seeds (Cueto 1999; Max S. Pinheiro personal communication). Hay is readily available in these regions, but it must be of good quality to avoid problems with aflatoxin contamination (Cueto 1999).

15.4.3 Feeding Females and Young Capybaras

The costs associated with the gestation period are relatively high for capybaras in captivity. The weight of the whole litter at birth ranges from 3.8 to 7.5 kg and represents from 7.4% to 16.5% of the mother's live weight before gestation (Cueto 1999). Therefore, it is important that a supply of high-caloric and high-quality feeds is offered to maximize the nutrients consumed. Corn silage is a good source of energy if animals have previously become used to eating it. Other options are elephant grass mixed with concentrate or commercial livestock feeds.

The supply of protein/energy supplements should be limited to females during late gestation to avoid dystocia – prolonged or difficult labor (Cueto 1999; Alvarez and Kravetz 2006) – and should be fed in a proportion not exceeding 20% of voluntary intake, approximately 250 g for a capybara weighing 50 kg. Capybaras lose up to 25% of their weight after parturition (Sérgio L. G. Nogueira-Filho personal observation). Therefore, the farmer must resume pre-gestation feeding levels immediately after birth to ensure their recovery and successful milk production.

Special care should also be taken in feeding the young before weaning. Newborns lose weight during the first days of life (Cueto 1999), probably because of the low quantity of milk produced, and its low protein and energy levels. However, further studies are needed on the composition of capybara milk in order to establish appropriate practices for artificial feeding when necessary.

Table 15.3 Nutrient digestibility coefficients of some conventional and non-conventional feedstuffs for capybaras, rabbits, cattle, and buffalo

	Digestibility coefficients				
Feeds	DM	CP	NDF	GE	Reference
Capybaras					
Elephant grass	58	63	72	84	Ferreira (2007)
Corn grain	88	97	83	92	Ferreira (2007)
Cassava leaf[a]	72	84	43	86	Ferreira (2007)
Soybean meal	95	98	90	94	Ferreira (2007)
Rabbits (*Oryctolagus cuniculus*)					
Elephant grass	46	65	42	45	Cheeke (1987)
Corn grain	–	85	–	87	Scapinello et al. (1995)
Cassava leaf[a]	–	52	–	49	Herrera (2003)
Soybean meal	–	88	–	85	Scapinello et al. (1995)
Crossbred steers (*Bos taurus taurus* x *Bos taurus indicus*)					
Elephant grass	57	58	58	59	Ribeiro et al. (2008)
Cassava leaf[a]	52	40	35	–	Euclides et al. (1979)
Buffalos (*Bubalus bubalis*)					
Elephant grass	56	64	52	56	Grant et al. (1974)

[a]*Cassava leaf* Cassava leaf hay, *DM* dry matter, *OM* organic matter, *CP* crude protein, *GE* gross energy

This initial weight loss stabilizes in the second week of life when the young consume more solid feeds (grass and concentrate diet) than milk and start gaining weight (Cueto 1999). Therefore, the newborn must have access to fresh forage and protein/energy supplements similar to those provided for adults, from the first days of their life, in the proportions of 20–40% of the total feed consumed.

Early weaned young receiving diets with too much grain are subject to an accumulation of gas in the intestine and, consequently, a distended abdomen (timpanism) because their cecum is not completely developed yet (Cueto 1999). Therefore, farmed capybaras are best weaned at between 45 and 60 days old (Nogueira-Filho 1996).

15.5 Final Remarks

The total expense per kg of live animal weight production is US $1.85 and $1.56 in confined and semi-confined systems, respectively (prices in 2011). Feed expenses account for 64–75% of these production costs (Nogueira-Filho and Nogueira 2004). To achieve greater economic gains, farmers should take advantage of the natural anatomical/physiological and behavioral characteristics of this species. To stimulate cecotrophy, good-quality roughage should be provided with protein, energy, mineral, and vitamin supplements, which will lead to higher weaning, slaughter, and carcass weights.

Acknowledgments We thank Dr. W. Mollineau, Dr. J.R. Moreira, Dr. M. Pinheiro, and anonymous reviewers for valuable comments on this manuscript. Alcester Mendes was supported by CAPES (Proc. 03010/2009-0) and Sérgio Luiz Gama Nogueira-Filho was supported by CNPq (Proc. 300587/2009-3).

References

Alho CJR, Campos ZMS, Gonçalves HC (1987) Ecologia de capivara (*Hydrochaeris hydrochaeris*, Rodentia) no Pantanal: I habitats, densidades e tamanho de grupo. Rev Bras Biol 47(1/2):87–97

Alvarez MR, Kravetz FO (2006) Reproductive performance of capybaras (*Hydrochoerus hydrochaeris*) in captivity under different management systems in Argentina. Anim Res 55:153–164

Andrade PCM (1996) Níveis de proteína e energia em rações e manejo de capivaras (*Hydrochoerus hydrochaeris hydrochaeris*, L. 1766) em crescimento. M.Sc. Dissertation Escola Superior de Agricultura “Luiz de Queiroz”, Universidade de São Paulo, Piracicaba

Andrade PCM, Lavorenti A, Nogueira-Filho SLG (1998) Efeitos do tamanho de área, da dieta e da idade inicial de confinamento sobre capivaras (*Hydrochoerus hydrochaeris hydrochaeris* L. 1766) em crescimento. Rev Bras Zootec 27(2):292–299

Baldizán A, Dixon RM, Parra R (1983) Digestion in the capybara (*Hydrochoerus hydrochaeris*). S Afr J Anim Sci 1(13):27–28

Baldizán A et al (1986) Efecto de la adicion de harina de yuca a una dieta de heno (*Cenchrus ciliaris*) sobre la digestibildad de la materia seca y pared celular en ovejas y chiguire. Informe Anual IPA:60–61

Barreto GR, Quintana RD (2012) Foraging strategies and feeding habits of capybaras. In: Moreira JR, Ferraz KMPMB, Herrera EA, Macdonald DW (eds) Capybara: biology, use and conservation of an exceptional neotropical species. Springer, New York, pp 83–96

Bernardi LG (1993) Efeito dos níveis crescentes de volumoso sobre a digestibilidade de nutrientes de rações para capivaras (*Hydrochoerus hydrochaeris hydrochaeris*, L. 1766). M.Sc. Dissertation, Escola Superior de Agricultura "Luis de Queiroz", Universidade de São Paulo, Piracicaba

Borges PA, Dominguez-Bello MG, Herrera EA (1996) Digestive physiology of wild capybara. J Comp Physiol B 166:55–60

Carvalho MAG (2010) Determinação da exigência protéica para manutenção de capivaras (*Hydrochoerus hydrochaeris*) em cativeiro. M.Sc. Dissertation, Universidade Estadual de Santa Cruz, Ilhéus

Cheeke PR (1987) Rabbit feeding and nutrition, vol 3, Animal feeding and nutrition. Academic, Orlando, pp 15–33, cap

Cueto GR (1999) Biología reproductiva y crecimiento del carpincho (*Hydrochoerus hydrochaeris*) en cautiverio: una interpretación de las estrategias poblacionales. Ph.D. thesis, Universidad de Buenos Aires, Buenos Aires

Cueto GR, Allekotte R, Kravetz FO (2000) Scurvy in capybaras bred in captivity in Argentine. J Wildlife Dis 36:97–101

Emmons LH (1987) Ecological considerations on the farming of game animals: capybaras yes, pacas no. Vida Silvestre Neotrop 1:54–55

Euclides VPB, Silva JM, O'Donovan PB (1979) Efeito da suplementação com feno da parte aérea da mandioca sobre o consumo e digestibilidade da palha de arroz. Comunicado Técnico (September, 1979), EMBRAPA Gado de corte, Cuiabá 1:1–3

Ferreira DN (2007) Estudo sobre a digestibilidade *in vivo* e in vitro de alimentos volumosos e concentrados fornecidos às capivaras *Hydrochoerus hydrochaeris* Linnaeus, 1766 (Rodentia: Caviidae) em cativeiro. M.Sc. Dissertation, Universidade Estadual de Santa Cruz, Brazil

Frasson C, Salgado JM (1990) Capivara – Uma opção contra a fome e a deficiência de proteína animal. In: Anais Interface Nutrição X Agricultura (2o Simpósio)

García A et al (2000) La crianza del carpincho (*Hydrochoerus hydrochaeris*) _ Parámetros técnicos y recomendaciones en alimentación y reproducción. In: Berhouet D et al Org. Producciones alternativas: ranas, martineta y carpincho. (Série FPTA-INIA; 03) INIA. Montevideo, Uruguay, p 84

González-Jiménez E (1977a) The capybara an indiginous source of meat in tropical America. World Anim Rev 21:24–30

González-Jiménez E (1977b) Digestive physiology and feeding of capybaras. In: Rechcigl M (ed) Handbook series in nutrition and food. CRC Press, Cleveland, pp 167–177

González-Jiménez E (1978) The capybara: an indigenous source of meat in tropical America. World Anim Rev 21:24–30

González-Jiménez E (1995) El capybara (*Hydrochoerus hydrochaeris*) – Estado actual de su producción, vol 122, Producion y sanidad animal. FAO, (Série Estudio FAO), Rome, Italy, pp 1–112

González-Jiménez E, Escobar A (1975) Digestibilidad comparada entre chiguire (*Hydrochoerus hydrochaeris*), coejos y ovinos con raciones de diferentes proporciones de forrajes y concentrado. Agron Trop 25:283–290

González-Jiménez E, Parra R (1972) Fisiologia digestiva del chigüire: 2. Cinética de los nutrientes en el tracto diestivo. In: Informe al Proyecto CONICIT, DF 030-S1. IPA, Un. Central de Venezuela, Maracay, p 7

Grant RJ, Van Soast PJ, MacDowell RE, Peres CB Jr (1974) Intake, digestibility and metabolic loss of napier grass by cattle and buffaloes when fed wilted, chopped and whole. J Anim Sci 39:423–434

Herrera EA (1985) Coprophagy in capybara, *Hydrochoerus hydrochaeris*. J Zool Ser A 207:616–619

Herrera APN (2003) Eficiencia produtiva e avaliação nutricional de dietas simplificadas a base de forragens para coelhos em crescimento. Ph.D. thesis, University of Oxford, Oxford

Herrera EA (2012) Capybara digestive adaptations. In: Moreira JR, Ferraz KMPMB, Herrera EA, Macdonald DW (eds) Capybara: biology, use and conservation of an exceptional neotropical species. Springer, New York, pp 97–106

Hirakawa H (2001) Coprophagy in leporids and other mammalian herbivores. Mammal 31(1):61–80, Printed in Great Britain

Hirakawa H (2002) Supplement: coprophagy in leporids and other mammalian herbivores. Mammal 32(2):150–152, Great Britain

Lavorenti A (1989) Domestication and potencial for genetic improvement of capybara. Rev Bras Genét 12(3):137–144

Lima GFC (2006) Reservas estratégicas de forragem: uma alternativa para melhorar a convivência dos rebanhos familiares com a seca, vol 1, Série Circuito de Tecnologia para a Agricultura Familiar. Empresa de Pesquisa Agropecuária do Rio Grande do Norte – EMPARN, Natal, Brazil, pp 1–80

Macdonald DW (1981) Dwindling resources and the social behavior of capybara. (*Hydrochoerus hydrochaeris*) (Mammalia). J Zool 194:371–391

Mendes A (1999) Determinação da ocorrência de cecotrofia em capivaras (*Hydrochoerus hydrochaeris hydrochaeris* L. 1766). M.Sc. Dissertation Escola Superior de Agricultura "Luiz de Queiroz", Universidade de São Paulo, Piracicaba

Mendes A, Nogueira SSC, Nogueira-Filho SLG (1996) Cecotrofagia em capivara (*Hydrochoerus hydrochaeris*). In: Congresso Brasileiro de Zoológicos, Anais, p 20

Mendes A, Nogueira SSC, Lavorenti A, Nogueira-Filho SLG (2000) A note on the cecotrophy behavior in capybara (*Hydrochaeris hydrochaeris*). Appl Anim Behav Sci 66:161–167

Mendes A, Nogueira-Filho SLG, Lavorenti A (2007) Comparação da taxa de passagem da digesta em capivaras (*Hydrochaeris hydrochaeris*). In: 13° Encontro Científico dos Pós-Graduandos no CENA/USP, Piracicaba

Murphey RM et al (1985) Behavioral observations in a capybara colony (*Hydrohchoerus hydrochaeris*). Appl Anim Behav Sci 14:89–98

Nogueira-Filho SLG, Nogueira SSC (2012) The impact of management practices on female capybara reproductive parameters in captivity. In: Moreira JR, Ferraz KMPMB, Herrera EA, Macdonald DW (eds) Capybara: biology, use and conservation of an exceptional neotropical species. Springer, New York, pp 275–282

Nogueira-Filho SLG (1996) Manual de criação de capivaras. Centro de Produções Técnicas – CPT, Viçosa, pp 1–50

Nogueira-Filho SLG, Nogueira SSC (2004) Captive breeding program as an alternative for wildlife conservation in Brazil. In: Kirsten S, Fragoso JMV, Bodmer R (eds) People in nature: wildlife management and conservation in Latin America. Columbia Univerty Press, New York, pp 171–190

Ojasti J (1973) Estudio biológico del chigüire o capibara. In: Fondo Nacional Investigación Agropecuária (ed), Sucre, Caracas, pp 1–275

Parra R, Escobar A (1978) El chiguire su potencial biologico y su cria en confinamento. In: Instituto de Producción Animal, Informe anual – 1978, Maracay, pp 83–94

Parra R, González-Jiménez E (1971) Fisiologia digestiva del chiguire (*Hydrochoerus hydrochaeris*). 1-Capacidad de los diferentes compartimientos del tracto digestivo. In: Informe sobre proyecto CONICIT DF 030-S1, IPA, Un. Central de Venezuela, Maracay, pp 1–14

Parra R, Escobar A, González-Jiménez E, Baldizán A, Szabuniewicz M (1981) Fisiología digestiva comparada. In: Instituto de Producción Animal, Informe anual, Maracay, pp 52–53

Pond WG, Church DC, Pond KR, Shoknecht PA (2004) Basic animal nutrition and feeding, 5th edn. Wiley, Hobeken, p 580

Proto V (1984) Aspetti applicativi della ciecotrofia nell'allevamento del coniglio. Rivista de Coniglicoltura 11:27–35

Ribeiro EG, Fontes CAA, Palieraqui CEM, Martins CE, Cóser AC, Sant'Ana NF (2008) Influiência da irrigação durante as épocas secas e chuvosas na taxa de lotação, no consumo e no desempenho de novilhos em pastagens de capim-elefante e capim-monbaça. Rev Bras Zoot 37(9):1546–1554

Santos FAP (1995) Manejo de pastagens de capim elefante. In: Peixoto AM et al (eds) Volumosos para bovinos, 2nd edn. Fealq, Piracicaba, São Paulo, pp 1–20

Scapinello C, Tafuri ML, Rostagno HS, Furlan AC (1995) Valor nutritivo do milho, do farelo de soja e do feno de aveia para coelhos em crescimento. Rev Bras Zoot 24(6):1001–1007

Silva Neto PB (1989) Alimentação e manejo de capivaras (*Hydrochoerus hydrochaeris hydrochaeris*) em cativeiro. Dissertação (Mestrado), ESALQ – Universidade de São Paulo, Piracicaba

Véras RML, Valadares Filho SC, Valadares RD (2007) Balanço de compostos nitrogenados e estimativa das exigências de proteína de mantença de bovinos Nelore de três condições sexuais. Rev Bras Zoot 36:1212–1217

Chapter 16
The Impact of Management Practices on Female Capybara Reproductive Parameters in Captivity

Sérgio L.G. Nogueira-Filho and Selene S.C. Nogueira

16.1 Introduction

The high reproductive potential of capybaras is extremely relevant to their management. In the wild they can produce up to two litters per year, of one to seven young, with a mean of four (Ojasti 1973), and this while subject to seasonal variations in food availability and quality (Moreira et al. 2012). Thus it might be expected that animals reared in captivity, with a constant supply of high-quality food, would produce larger litters within lower birth intervals, and could breed throughout the year. However, the reproductive parameters recorded so far in captivity are similar to those observed in the wild. A number of studies have shown that different husbandry practices impact reproduction differently, some suggesting significant potential improvement in reproductive indices in captivity. In this chapter, therefore, we evaluate the impact of management practices on capybara reproductive parameters, such as litter size and sex ratio, birth intervals, breeding age, and newborn mortality in confined farming systems and zoos. The captive breeding data presented here were obtained from research stations in Piracicaba, State of São Paulo, Brazil (Nogueira 1997), and in Buenos Aires, Argentina (Cueto 1999; Alvarez and Kravetz 2006), and from zoos all around the world (Chapman 1991).

S.L.G. Nogueira-Filho (✉)
Departamento de Ciências Agrárias e Ambientais, Universidade Estadual de Santa Cruz, Rodovia Ilhéus Itabuna km 16, 45662-900 Ilhéus, Bahia, Brazil
e-mail: slgnogue@uesc.br

S.S.C. Nogueira
Departamento de Ciências Biológicas, Laboratório de Etologia Aplicada, Universidade Estadual de Santa Cruz, Rodovia Ilhéus Itabuna km 16, 45662-900 Ilhéus, Bahia, Brazil
e-mail: selene@uesc.br

J.R. Moreira et al. (eds.), *Capybara: Biology, Use and Conservation of an Exceptional Neotropical Species*, DOI 10.1007/978-1-4614-4000-0_16,

16.2 Breeding Seasons

Captive experiments have shown that capybaras may breed throughout the year in the tropics, while they present two breeding peaks in temperate regions. Capybaras in subtropical Piracicaba gave birth in all months of the year except July (Fig. 16.1). However, 75% of the births occurred between October and January, the wettest months locally. On the other hand, in Buenos Aires there were two birth peaks (Cueto 1999). The first occurred during the early rainy season between October and November in this temperate region, resulting in 37.5% of all births. The other birth peak (32.5%) occurred at the end of the rainy season between the months of April and May.

It could be suggested that the lower number of births during the dry winter months in Piracicaba and Buenos Aires may be related to seasonal variations in fresh forage availability and quality. In both research stations, the animals received a balanced diet comprised of a concentrate and triturated grass hay (as a source of nutritional fiber) during the periods of reduced forage. However, no vitamin C supplement was added to this diet. Only recently was it confirmed that, unlike most other mammals, capybaras cannot synthesize vitamin C from glucose (Cueto 1999; Cueto et al. 2000). Fresh grass is the natural source of this nutrient, so the use of hay probably led to vitamin C deficiency, the vitamin being lost during the sun-drying process (Mendes and Nogueira-Filho 2012). The addition of vitamin C supplements to the capybara's diet (300 mg/day) in May resulted in 77.5% pregnancy of the females during the dry winter (August) in Argentina, where a lack of vitamin C supplementation led to low pregnancy rate (Cueto 1999).

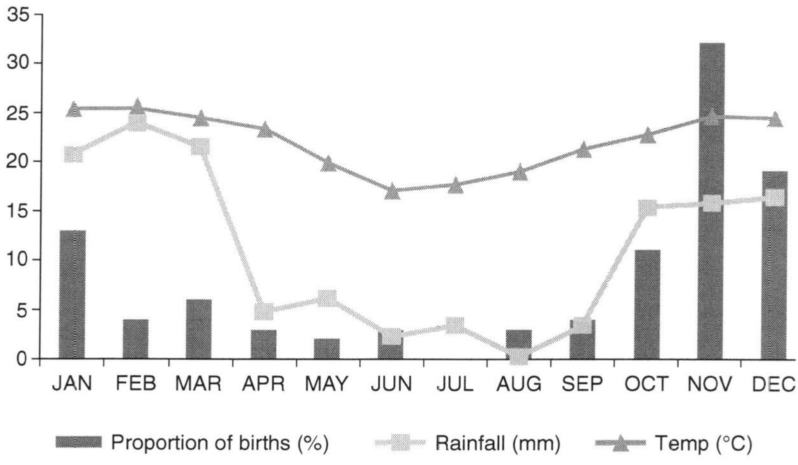

Fig. 16.1 Proportion of monthly births (N = 80) during the year related to monthly mean rainfall (mm) and temperature (°C) for captive capybaras in Piracicaba, State of São Paulo, Brazil

16.3 Litter Size, Sex Ratio, and Birth Weights

The research station in Piracicaba recorded mean litter size of 3.3±1.5 (N=80; Nogueira 1997), while in the Buenos Aires breeding center, litters ranged from 3.1±2.1 (N=12) to 3.8±1.6 (N=61; Alvarez and Kravetz 2006). In this evaluation, in Argentina, the litter size per parturition was independent (P=0.23) of the husbandry practices adopted (Alvarez and Kravetz 2006). On the other hand, the addition of vitamin C supplements to the diet increased the average litter sizes from 2.7±1.7 (N=24) to 3.7±1.7 (N=62; Cueto 1999). Such litter sizes in captivity are similar to those in the wild: Ojasti (1973) reported a mean litter size of 3.5±1.5 (N=93) for wild capybaras in Venezuela (Table 16.1).

In captivity, litter sizes appear to be affected by maternal age, with a sharp decrease occurring by the fourth birth (Fig. 16.2), when captive capybaras are approximately 7 years old (Nogueira 1997). The oldest capybaras, between the seventh and tenth years of life, produce just one or two young per birth, probably because they are reaching the climacteric period of age-related reproductive decline (Nogueira 1997). On the other hand, Chapman (1991) did not find any relationship between age and litter size among capybaras kept in zoos.

The birth sex ratio is approximately 1:1 for captive capybaras (Nogueira 1997). However, Cueto (1999) found that the addition of vitamin C supplements to pregnant females' diet increased (P=0.002) the proportion of males in the litters from 0.42 to 0.55. The same author also recorded that, when vitamin C supplementation was added to the pregnant females' diet, newborn males were heavier (2.2±0.5 kg, N=21) than newborn females (1.7±0.4 kg, N=23). This corroborates a previous finding in the wild: capybaras giving birth during the dry season, when there is less vitamin C in the dry forage, showed a slight tendency to have female-biased litters (Herrera 1998; Macdonald et al. 2007). These results are consistent with Trivers and Willard's (1973) hypotheses of sex allocation. The prediction of this hypothesis is

Table 16.1 Reproductive parameters of captive capybaras under different breeding systems in Argentina compared with parameters obtained from wild capybaras in the *llanos* of Venezuela

Breeding system	Litter size	Birth intervals (days)	Newborn mortality (%)
Confined IP[a,b]	3.5±1.5 (N=22)	197±10 (N=7)	41.3
Confined IF[a,c]	3.1±2.1 (N=12)	238±68 (N=6)	41.2
Confined MP[a, d]	3.8±1.6 (N=61)	193±23 (N=39)	24.7
Wild (Venezuela)[e]	3.5±1.5 (N=93)	1.5 (1.2–1.8) farrows per year	33.0

[a]Alvarez and Kravetz (2006)
[b]Confined Isolated Pens (IP): the traditional confined production system composed of three different types of enclosures for reproduction, parturition and growth
[c]Confined Isolated Females (IF): formed by a central enclosure (reproduction pen), attached to 12 parturition/growing pens
[d]Confined Mixed Pens (MP): an enclosure of 35°×10°m divided into a central area for reproduction and adjacent parturition pens
[e]Ojasti (1973)

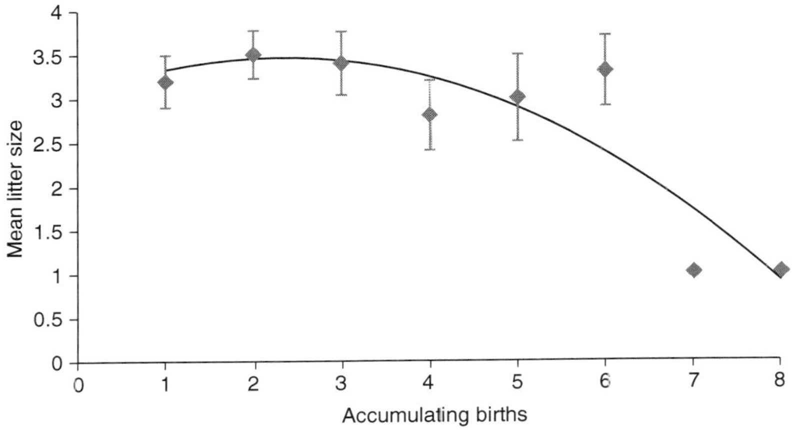

Fig. 16.2 Relationship between accumulating number of births (AB) and mean (±SE) litter size ($LS = -0.08*AB^2 + 0.35^*AB + 3.03$, $R^2 = 0.84$, $P = 0.01$) for captive capybaras in Piracicaba, State of São Paulo, Brazil (From Nogueira (1997))

that parents in the healthiest condition should invest more in male offspring and those in poor condition in female offspring. If the species is polygynous, as capybaras are, a successful son might father many more descendents for these females in good condition.

16.4 Birth Intervals

Capybaras have a gestation period of 5 months (López-Barbella 1984). Since they are able to conceive during lactation, (Nogueira 1997; Cueto 1999), capybaras have the potential to produce up to two litters per year (Ojasti 1973). Nevertheless, there has been a high variation in the birth interval found in different institutions, ranging from 193 ± 23 days (N = 39) in Buenos Aires (Table 16.1, Alvarez and Kravetz 2006) to 514.6 ± 348.0 days (N = 50) in Piracicaba (Nogueira 1997). The different facilities and husbandry procedures tested in Argentina did not interfere (P = 0.74) in the birth interval (Alvarez and Kravetz 2006).

The main causes for variation in birth interval may have been the females' age and other factors, such as seasonal variation in the nutritional quality of feedstuff, postpartum uterine infections, and reproductive suppression by high ranking females. Nogueira (1997) reported that females up to 6 years old showed regular birth intervals, but between the seventh and tenth years of life there was a sharp increase in this parameter (Fig. 16.3). In Buenos Aires, 17% of females died during parturition or failed to reproduce again due to postpartum uterine infections (Cueto 1999). The high content of dietary energy supplied to these capybaras may have resulted in

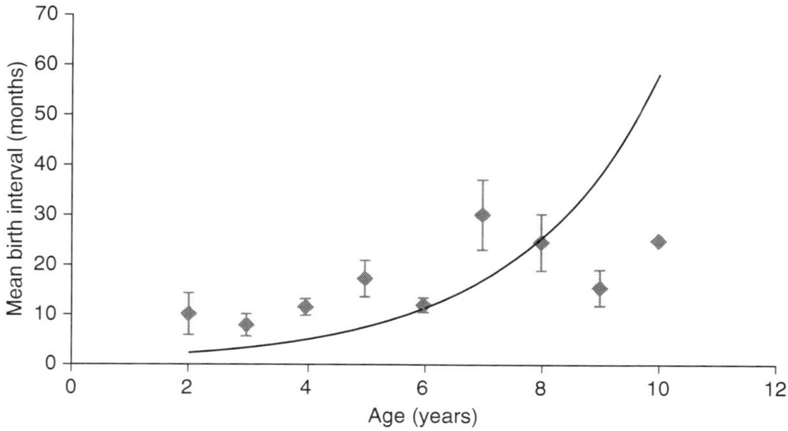

Fig. 16.3 Relationship between age (years) and mean (±SE) birth intervals in Piracicaba, State of São Paulo, Brazil. Birth interval (months) = $5.03^{*}Age^{0.67}$, $R^2=0.60$, $P=0.01$ (From Nogueira (1997))

dystocia (prolonged or difficult labor), which could have caused postpartum uterine infections and deaths.

The variation in birth intervals found among captive capybaras might be explained by a mechanism of suppression as suggested by Maldonado-Chaparro and Blumstein (2008), since even in the wild not all females are reproductively active in a certain period of time (Ojasti 1973). Cueto (1999) observed that in captivity the most subordinate females only had the opportunity to copulate after the dominant females were removed from the group for parturition.

Shorter birth intervals were reported for capybaras kept in zoos (average of 8.4 months; Chapman 1991). The shortest birth interval recorded was 4.2 months, which is even shorter than capybaras' normal 5-month gestation period (Cueto 1999; López-Barbella 1984). Chapman (1991) did not identify any relationship between birth interval and the age of female capybaras kept in zoos.

16.5 Breeding Age

The age at which capybaras in the Piracicaba research station first undergo successful parturition is hugely variable, ranging from 16.2 to 95 months (48.3 ± 24.8 months, N = 30; Nogueira 1997). Capybaras born in zoos had a mean first parturition age of 31.0 ± 2.5 months (N = 19), ranging from 12 to 53 months (Chapman 1991). Histological analysis determined that female capybaras could potentially reach breeding age at 6 months with an average body weight of 25 kg (Altermann and Leal-Zanchet 2002). However, in the wild, the youngest age recorded for a female at first parturition was 19 months, with a body weight of 34 kg (Ojasti 1973).

The large variation in age of first successful parturition in captivity may, in part, be due to the management procedures adopted. Current information suggests that young males become sexually active between 15 and 24 months of age (Ojasti 1973). Therefore, male capybaras reach sexual maturity later than females (López-Barbella 1993; Altermann and Leal-Zanchet 2002). This may result in a delay in the female's breeding age (Nogueira 1997), but may also explain the wide range in ages at first parturition recorded in captivity: young females paired with immature males will reproduce later.

16.6 Newborn and Young Mortality

One of the major challenges faced by farmers attempting to breed capybaras is to reduce the high mortality rates among newborns and young. At the captive breeding center in Piracicaba, 67 (27.6%) out of 243 newborns were killed immediately after birth by strange females in the group (Nogueira-Filho et al. 2012). High newborn mortality was also found in Buenos Aires, ranging from 24.7% to 41.3% (Table 16.1). In zoos, young mortality reached 33% (Chapman 1991). These deaths occurred when a female gave birth in the presence of an unfamiliar female, which then killed the entire litter (Nogueira et al. 1999). This happened because the stockman failed to detect impending parturition and to transfer the females to separate parturition facilities (Nogueira et al. 1999) or to ensure that breeding groups include only related or familiarized females (Nogueira et al. 2003; Nogueira-Filho et al. 2012).

Twenty-five newborns and young out of 243 (10.3%) died due to other causes in the experiments in Piracicaba, including dystocia, predation by black vultures (*Coragyps atratus*), and endoparasites (*Strongyloides* spp. and *Eimeria* spp.; Lavorenti et al. 1989; Nogueira-Filho 1996; Nogueira 1997).

16.7 Final Remarks

Capybaras easily adapt and reproduce in captivity. Newborn and young mortality is the most limiting factor of capybara production in captivity, but mortality rates can be reduced by employing specific procedures for group compositions. The breeding group should be composed only of females related to or familiar with each other (Nogueira et al. 1999; Nogueira et al. 2003; Nogueira-Filho et al. 2012). This procedure must be adopted together with predator protection measures and appropriate sanitary management.

Specific studies are required to establish balanced diets that maximize reproductive potential. These diets should enable captive capybaras to produce two litters per year and are likely to boost litter size. At the same time, breeders must avoid over-feeding pregnant females, because this can result in dystocia (Mendes and Nogueira-

Filho 2012). It is essential to monitor females after birth for possible postpartum uterine infections, to cull females over 7 years old and create husbandry practices that give low-ranking females a chance to mate.

Acknowledgments We thank Dr. W. Mollineau, Dr. J.R. Moreira, Dr. M. Pinheiro, Dr. Sabine Pompéia, Dr. Emilio A. Herrera, Dr. David Macdonald, and Dr. Lauren Harrington for valuable comments on this manuscript. We also thank all stockholders and students from the former CIZBAS, with special gratitude to Dr. Abel Lavorenti, its former coordinator. The studies made at CIZBAS had the support of FINEP, FAPESP, CNPq, and CAPES. Preparation of this chapter was aided by grants from CNPq to Sérgio Luiz Gama Nogueira-Filho (Proc. # 300587/2009-3) and to Selene Siqueira da Cunha Nogueira (Proc. # 306154/2010-2).

References

Altermann GTM, Leal-zanchet AM (2002) Puberdade em fêmeas de capivaras *Hydrochaeris hydrochaeris* (Mammalia, Rodentia), mantidas em sistema de criação intensiva no Rio Grande do Sul. Acta Biol Leopoldensia 24:185–197

Alvarez MR, Kravetz FO (2006) Reproductive performance of capybaras (*Hydrochoerus hydrochaeris*) in captivity under different management systems in Argentina. Anim Res 55: 153–164

Chapman CA (1991) Reproductive biology of captive capybaras. J Mammal 72:206–208

Cueto GR (1999) Biología reproductiva y crecimiento del carpincho (*Hydrochoerus hydrochaeris*) en cautiverio: una interpretación de las estrategias poblacionales. Ph.D. thesis, Universidad de Buenos Aires, Buenos Aires

Cueto GR, Allekotte R, Kravetz FO (2000) Scurvy in capybaras bred in captivity in Argentine. J Wildlife Dis 36:97–101

Herrera EA (1998) Reproductive strategies of female capybaras: dry season gestations. Symp Zool Soc Lond 71:381–392

Lavorenti A, Silva-Neto PB, Marchi AP, Souza MC (1989) Desempenho reprodutivo da capivara (*Hydrochoerus hydrochaeris*) criada em cativeiro. In: Anais da 26ª Reunião Anual da Sociedade Brasileira de Zootecnia. Porto Alegre, p 271

López-Barbella S (1984) Una contribuicion al conocimento de la reproduccion del chigüire (*Hydrochoerus hydrochaeris*). In: Informe Anual de Instituto de Producion Animal (IPA), Maracay, pp 109–117

López-Barbella S (1993) Pubertad en hembras chigüires (*Hydrochoerus hydrochaeris*). Revista da Facultad de Agronomia 19:121–127

Macdonald DW, Herrera EA, Taber AB, Moreira JR (2007) Social organization and resource use in capybaras and maras. In: Societies R, Wolff JO, Sherman PW (eds) Rodent societies, an ecological and evolutionary perspective. University of Chicago Press, Chicago, pp 393–402

Maldonado-Chaparro A, Blumstein DT (2008) Management implications of capybara (Hydrochoerus hydrochaeris) social behavior. Biol Conserv 141:1945–1952

Mendes A, Nogueira-Filho SLG (2012) Feeds and nutrition of farmed capybaras. In: Moreira JR, Ferraz KMPMB, Herrera EA, Macdonald DW (eds) Capybara: biology, use and conservation of an exceptional neotropical species. Springer, New York, pp 261–274

Moreira JR, Wiederhecker H, Ferraz KMPMB, Aldana-Domínguez J, Verdade LM, Macdonald DW (2012) Capybara demographic traits. In: Moreira JR, Ferraz KMPMB, Herrera EA, Macdonald DW (eds) Capybara: biology, use and conservation of an exceptional neotropical species. Springer, New York, pp 147–167

Nogueira SSC (1997) Manejo reprodutivo da capivara (*Hydrochoerus hydrochaeris hydrochaeris*) em sistema intensivo de criação. Ph.D. thesis Universidade de São Paulo, São Paulo

Nogueira-Filho SLG, Pinheiro MS, Nogueira SSC (2012) Confined and semi-confined production systems for capybaras. In: Moreira JR, Ferraz KMPMB, Herrera EA, Macdonald DW (eds) Capybara: biology, use and conservation of an exceptional neotropical species. Springer, New York, pp 229–241

Nogueira SSC, Nogueira-Filho SLG, Otta E, Dias CTS, Carvalho A (1999) Determination of infanticide causes in capybara (*Hydrochoerus hydrochaeris*) groups in captivity. ApplAnim Behav Sci 62:351–357

Nogueira SSC, Lopes APS, Nogueira-Filho SLG (2003) Como evitar infanticidios e peleas mortales en criaderos de capivaras (*Hydrochoerus hydrochaeris*)? Bol BEDIM 12:16–17

Nogueira-Filho SLG (1996) Criação de Capivaras. Centro de Produções Técnicas (CPT), Viçosa

Ojasti J (1973) Estudio del chigüire o capibara. Fondo Nacional de Investigaciones Agropecuarias, Caracas

Trivers RL, Willard DE (1973) Natural selection of parental ability to vary the sex ratio of offspring. Science 179:90–92

Chapter 17
The Sustainable Management of Capybaras

José Roberto Moreira, Luciano M. Verdade, Katia Maria P.M.B. Ferraz, and David W. Macdonald

17.1 Introduction

The capybara (*Hydrochoerus hydrochaeris*) has always been an important item in the diet of South American natives and subject to intensive hunting in Venezuela, Brazil, and Argentina since the nineteenth century. Commercial exploitation began in early twentieth century, especially in the Llanos of Venezuela (Fig. 17.1). Studies of the capybara across its range have shown differences in population structure and dynamics that vary principally according to the type of habitat occupied and also depend on the seasonality of environmental resources (Ojasti 1973; Aldana-Domínguez et al. 2002). Understanding factors that influence or change aspects of population structure and dynamics is crucial for defining appropriate strategies for sustainable exploitation and conservation of a species. In this chapter, we look at

J.R. Moreira (✉)
Embrapa Recursos Genéticos e Biotecnologia, Parque Estação Biológica, Final W5 Norte, 70770-900 Brasília, Distrito Federal, Brazil
e-mail: alencar_moreira@yahoo.com.br

L.M. Verdade
Laboratório de Ecologia Isotópica, CENA, Universidade de São Paulo, Caixa Postal 96, 13416-000 Piracicaba, São Paulo, Brazil
e-mail: lmverdade@usp.br

K.M.P.M.B. Ferraz
Departamento de Ciências Florestais, Escola Superior de Agricultura "Luiz de Queiroz", Universidade de São Paulo, Av. Pádua Dias 11, 13418-900 Piracicaba, São Paulo, Brazil
e-mail: katia.ferraz@usp.br

D.W. Macdonald
Wildlife Conservation Research Unit Zoology Department, University of Oxford The Recanati-Kaplan Centre Oxford, Tubney House, Abingdon Road Tubney, Abingdon OX13 5QL, United Kingdom
e-mail: david.macdonald@zoo.ox.ac.uk

J.R. Moreira et al. (eds.), *Capybara: Biology, Use and Conservation of an Exceptional Neotropical Species*, DOI 10.1007/978-1-4614-4000-0_17,

Fig. 17.1 Map of South America with the locations of the places referred to in the text

inferences made by modeling capybara harvest on the effect on the sustainability of the population. We also present what is known about the effects of capybara harvest on populations and the productive potential of the species in the neotropical region.

17.2 Modeling the Sustainable Harvest of Capybaras

Here we present the results of several studies that sought to model capybara harvest. To enable comparison between different models, when harvest rates were presented in absolute number of animals to be removed, they have been transformed into the proportion of the population. In some cases we used the data collected by Moreira and Macdonald (1995) on Marajó Island, to model the effect of different harvest schemes on the sustainability of capybara populations.

17.2.1 Maximum Sustainable Yield

The example of Venezuela, where capybaras have been exploited commercially for more than half a century, illustrates the importance of reliable estimates of a sustainable rate of harvest. The first attempt at modeling the population dynamics and exploitation of capybaras was undertaken by Bone (1977), an undergraduate student of Juhani Ojasti at the Universidad Central de Venezuela. Bone (1977) used a deterministic seasonal age-structured Leslie model (1945), but without incorporating density dependence to estimate population growth. With his model, Bone (1977) calculated that an annual rate of harvest of 31.5% was sustainable for capybaras under the conditions of Hato El Frio, State of Apure, Venezuela (Fig. 17.1).

Reproductive and population data from capybaras on Marajó Island (State of Pará – Brazil; Fig. 17.1) were used to model the exploitation of the species (Moreira and Macdonald 1996). A deterministic seasonal age-structured Leslie model (1945) was used to assess the exploitation limits for capybaras in flooded neotropical savannas. This model assumed that the mortality of young, but not adults, was density dependent. The harvest rate limit calculated by the model was 17.2%, when using the exploitation system adopted in Venezuela (Moreira and Macdonald 1996). The model estimated that the local population in Venezuela would become extinct within 27 years if the harvest rate seen at the time (30%) continued.

Population data from the Pantanal (State of Mato Grosso do Sul – Brazil; Fig. 17.1) were also used for modeling the exploitation of the capybara (Paglia 1997), using the software Vortex 7.0 (Lacy et al. 1995). Vortex simulates the demographic stochasticity of populations, modeling population processes as sequential and individual events, with probabilistic outcomes determined by a random number generator. When population growth was not density dependent, the annual harvest found to be sustainable for capybaras in the Pantanal of Mato Grosso was 12.4% (Paglia 1997). When density dependence was incorporated into the model, the sustainable harvest decreased to only 7.8% (calculated as a proportion of K).

Here, we test the impact of including demographic stochasticity into modeled populations, using Vortex 8.0 software (Miller and Lacy 1999), to simulate the exploitation of capybara populations using data from Marajó Island (Moreira et al. 2012). It was assumed that the mortality of capybaras was not density dependent. Setting the condition that the risk of population extinction should not exceed 5%, the model suggested that the highest sustainable rate of harvest was 30% (Fig. 17.1). The average population growth rate (r) was 0.2016 ± 0.0007. The observed heterozygosity after 100 years of exploitation was 0.9156 ± 0.0023 with 21.94 ± 0.27 alleles.

17.2.2 Effects of the Frequency and Seasonality of Harvest

It is important to evaluate different harvest strategies prior to implementing management to maximize production and minimize negative effects. Subsistence hunting, by definition, seeks to extract from the population a number of individuals that meet the needs of the hunters. Commercial exploitation in Venezuela and Colombia depends on only one harvest per year (to provide capybara meat for consumption during Lent). Which management strategy would best sustain this market? Federico and Canziani (2005) compared continuous harvest with seasonal extraction of capybaras, using a deterministic seasonal age-structured Leslie model (1945) without incorporating density dependence for population growth. As they had no data on capybara survival or population growth, they used an iterative method testing different parameters. Their model indicated that capybara production was greater where the harvest was conducted once a year rather than continuously. The extent of this increase depended on the age class removed. However, this model evaluated production in terms of the number of animals slaughtered rather than their biomass, so it was not apparent whether a smaller harvest taking only adult capybaras could yield more meat than what was produced when all age classes were harvested.

On some properties in Venezuela, capybara harvests are implemented only every second year (Bone 1977). In this situation, production (number of animals extracted) can be 6.25–7.85% higher than that achieved with annual harvests (Bone 1977; Paglia 1997). However, Paglia (1997) reported that this increase in production when harvests were modeled to occur every second year was canceled out when density-dependent survival was included in the model.

We compared harvest at different stages of the capybara life cycle, assuming seasonal breeding and using data from Marajó Island (Moreira et al. 2012). When harvest occurred after mating, we assumed that there was an increase in the mortality of young (four times the hunting pressure). When harvest occurred after mating or before breeding, it was assumed that there was an effect on the rate of abortions (1.25 times hunting pressure when after mating or 0.75 times hunting pressure when before breeding). The model predicted that the optimal time for harvest was before mating, allowing higher rates of extraction and higher MSY (measured as the production of both meat and leather), when individuals were killed from each age group at random. When only adults and subadults were slaughtered either before or after mating, the sustainable harvest could be higher (as a proportion of the population). According to the model, 15% harvest of adults and subadults just before breeding produced the highest MSY (US $2,461; Fig. 17.2).

17.2.3 Effects of Selective Harvest by Gender

Hunting techniques and effort, combined with intrinsic behavioral and ecological factors, determine how selective hunting affects the sex ratio and age structure of an

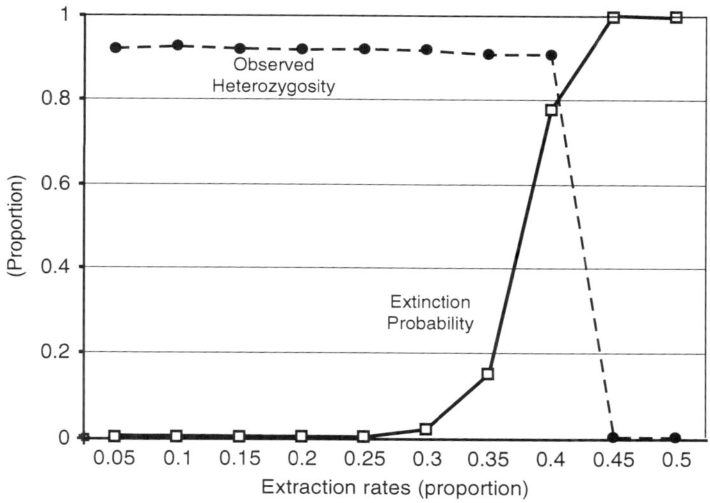

Fig. 17.2 Extinction probability and observed heterozygosity after 100 years of exploitation in capybara populations subjected to different harvest rates of both genders

exploited population (Ginsberg and Milner-Gulland 1994). Such changes affect the social organization and reproductive biology of the population, which in turn affect the recruitment rate and sex ratio (Ginsberg 1991). Examples of the possible positive effects of hunting that result from a greater harvest of males include reduced mortality and reduced aggression between the surviving males and reduced harassment of females. Potentially undesirable effects include: the disruption of the territorial structure, high mortality of young born out of season (Poole and Thomsen 1989), artificial selection for smaller males, an inadequate number of males to ensure insemination of females, and changes to the sex ratio at birth (Ginsberg 1991; Tuyttens and Macdonald 2000). Potentially undesirable effects of hunting resulting from a greater harvest of females include increased rate of harassment by males, increased aggression among males (Ginsberg and Milner-Gulland 1994), increased mortality of young born to inexperienced young females, and instability in the relationships required for normal conception by females (Poole and Thomsen 1989; Tuyttens and Macdonald 2000).

Ojasti (1991) suggested that the productivity of a harvested population of capybaras could be increased by taking proportionally more males to increase the proportion of reproductive females. Paglia (1997) supported Ojasti's predictions when modeling the harvest of capybaras in the Pantanal Matogrossense. A 75% increase in the sustainable harvest occurred when the harvest system removed twice as many males as females from the population, allowing the removal of 21.7% of the population (Paglia 1997).

Recently, Maldonado-Chaparro and Blumstein (2008) modeled the effects of reproductive competition on the dynamics of a capybara group of up to 25 individuals

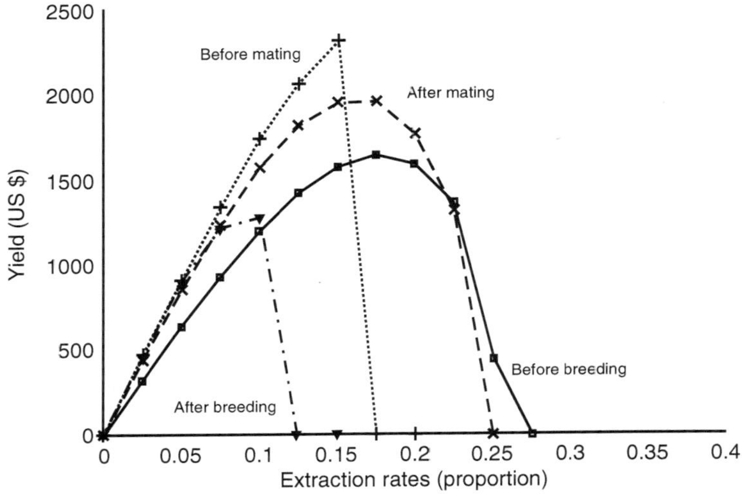

Fig. 17.3 Effects of harvest rate and seasonality of only subadult and adult capybaras on yield (expressed as US $ of meat and leather in the 1990s). The prices considered were US $3.50/kg for meat and US $4.00/ind for leather (Ojasti 1991)

in a closed population. The removal of adult females had a strong effect on group size, when the harvest was greater than approximately 7.5% of the group (presented as the removal of 30% of all adult females). However, removal of only adult males had no effect on the dynamics of the modeled capybara group, even when the harvest was approximately 12.5% of the group (presented as the removal of 50% of all adult males). The modeled group size being harvested ranged from 10 to 11 individuals, and less than half the group members were adult. Therefore it is important to stress that the harvest of 50% of adult males was unlikely to remove more than 1 individual from the group.

The possible consequences of unrestricted access of males to females were noticed by Bone (1977), who found capybara production increased when harvest removed animals at a sex ratio of 0.4 (given as ♂/♀). Despite increased production, Bone (1977) found that after only 3 years of this system no males over 4 years old remained in the modeled population.

Moreira and Macdonald (1995) tested the effects of male-biased harvest when the males from the modeled capybara population had limited access to females. The model assumed that a male could inseminate up to 12 females, which is an overestimate insofar as commercial capybara breeders use one male for every five to eight females (Nogueira-Filho et al. 2012). When a capybara population is exploited at relatively low harvest rates (Fig. 17.3), the sex ratio of the individuals killed has little effect on the resultant population size (Moreira and Macdonald 1995). With an increase in the harvest rate and/or the proportion of males harvested, the sex ratio of the breeding population becomes biased toward females (Fig. 17.4), leading to a

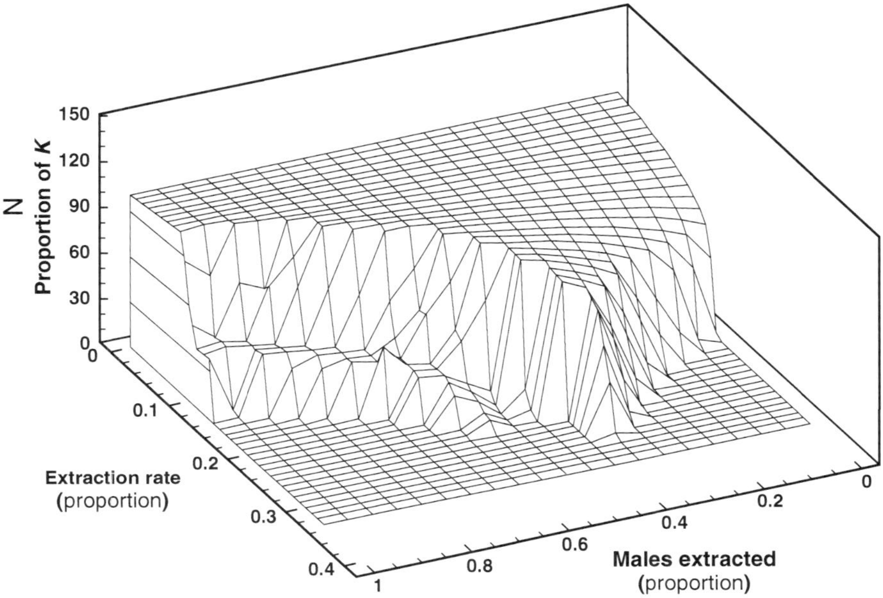

Fig. 17.4 Effects of harvest rate and selective exploitation by gender of only adult capybaras on *N* (population size, expressed as a proportion of *K*). Males with an insemination limit of up to 12 females. *K* is the environmental carrying capacity

rapid population collapse (Fig. 17.3). On the other hand, when a higher proportion of females is taken, the model predicts a decrease in the number of reproductive females in the population, and a gradual reduction in population size until extinction. The MSY (US $3,672; price for the 1990s) was found at a harvest rate of 26% of the population per year, and a sex ratio of 0.61 (given as ♂/♀). The simulations of this model therefore are at odds with Ojasti's (1991) predictions. The reason is that the model assumes (as is the reality) that male and female capybaras become reproductively active at different ages (3 years for males and 2 years for females; Moreira et al. 2012), but are harvested at a similar age. This limits the number of sexually active males for breeding (Moreira and Macdonald 1995).

When an unlimited capacity of capybara males to inseminate females is assumed, males are never the limiting sex (Fairall 1985), with mortality from hunting having a linear effect on population size (Fig. 17.5). When more males than females are slaughtered, harvest rates exceed plausible values. However, the sex ratio of the reproductive population (Fig. 17.6) does not differ much from that found when the ability of males to inseminate females is limited (Fig. 17.4) until it is close to zero (Moreira and Macdonald 1995). This shows an unreal situation where only one male is needed in the population to inseminate all females. A sensitivity analysis showed that unlimited

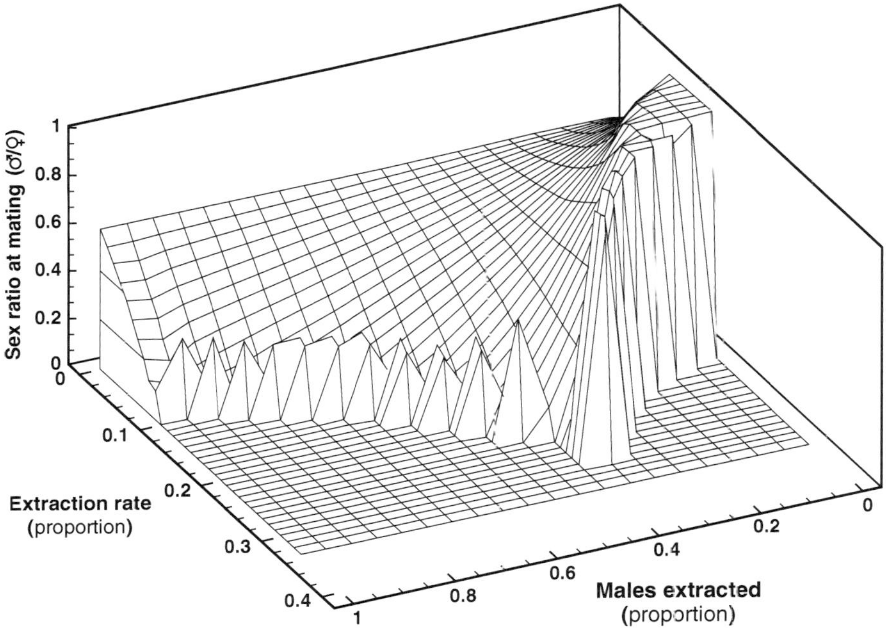

Fig. 17.5 Effects of harvest rate and selective exploitation by gender of only adult capybaras on the sex ratio at mating (given as ♂/♀). Males with an insemination limit of up to 12 females

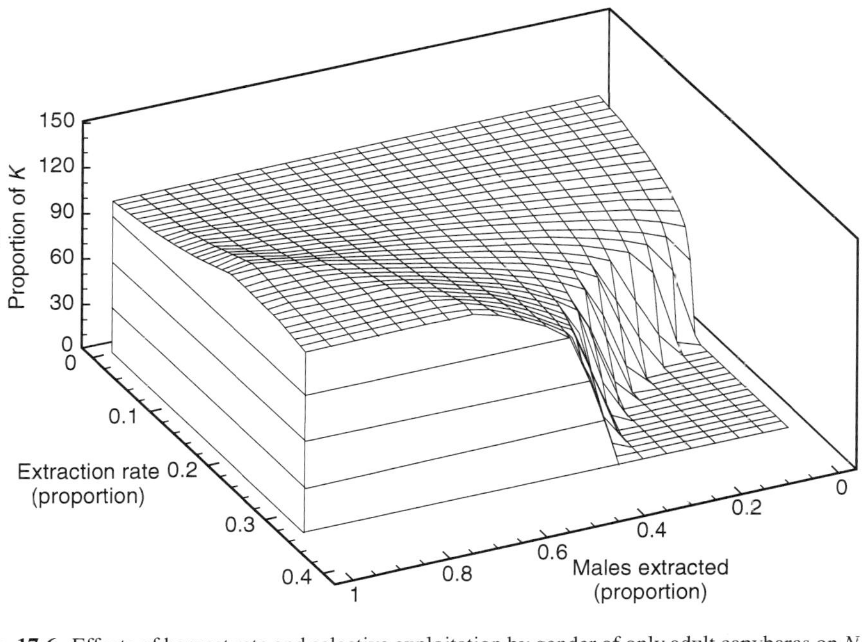

Fig. 17.6 Effects of harvest rate and selective exploitation by gender of only adult capybaras on *N* (population size, expressed as a proportion of *K*). Males without an insemination limit. *K* is the environmental carrying capacity

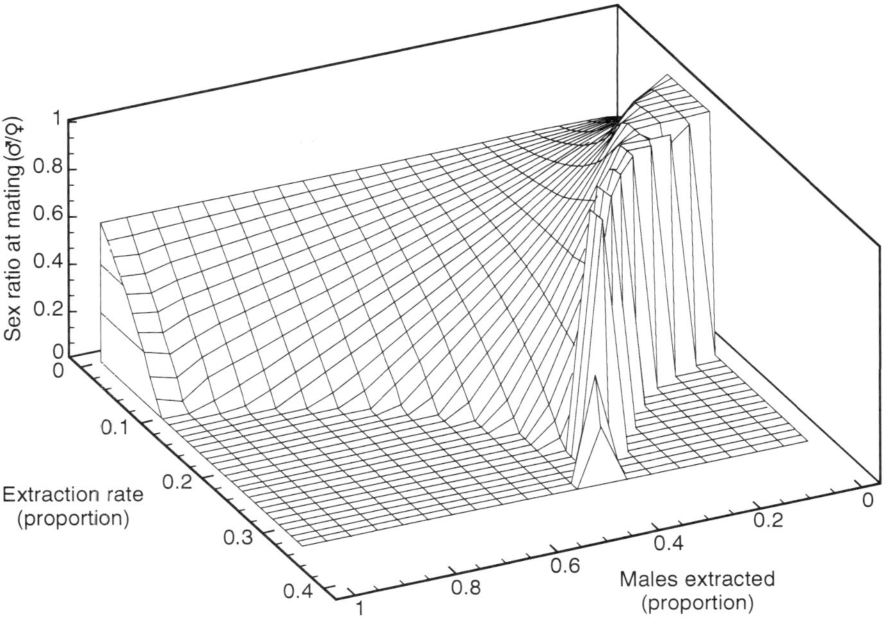

Fig. 17.7 Effects of harvest rate and selective exploitation by gender of only adult capybaras on the sex ratio at mating (given as ♂/♀). Males without an insemination limit

access of males to females allows higher rates of exploitation (Fig. 17.7). However, reducing the limit on the number of females that can be accessed by each male reduces the modeled rate of sustainable harvest (Moreira and Macdonald 1995).

17.2.4 Effects of Selective Harvest by Age Class

The exploitation of all age classes of capybaras indiscriminately allows higher extraction rates than does selective hunting of any single age class (Bone 1977). Using data from Marajó Island (Moreira et al. 2012), we modeled the effects of selective harvest by age class. We found that harvesting only adults reduced the proportion of animals taken from the population by 0.16 in relation to harvesting all age classes (Fig. 17.8). Nevertheless, since the products of capybara exploitation are meat (value of US $3.50/kg; price for the 1990s) and leather (value of US $4.00/ind), harvesting only adults led to a better MSY (Fig. 17.9) in terms of financial value produced (US $3,119 with harvest of only adults, compared to US $2,461 with harvest of sub-adults and adults and US $1,533 with the harvest of all age classes). In conclusion, this model suggested that the system of capybara exploitation which provides the best payoff in terms of production of meat and leather involves hunting only adults at a rate of 15.2% of the population (Fig. 17.10).

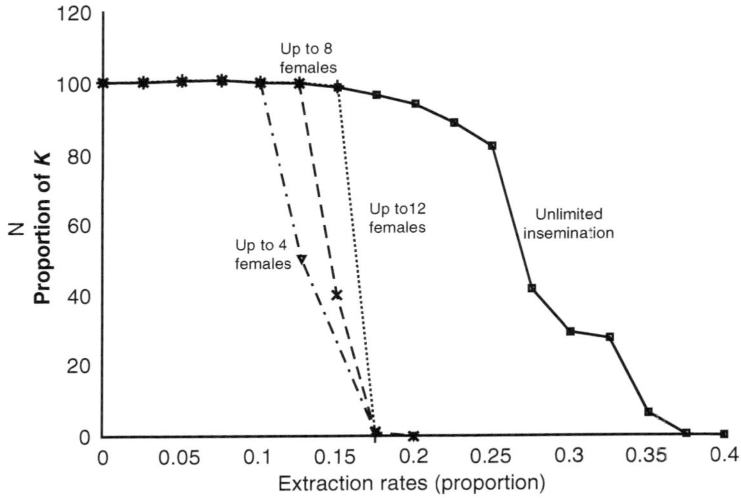

Fig. 17.8 Sensitivity analysis for three different insemination limits of female capybaras by males or with unlimited insemination. The effect of the harvest rate of only adult capybaras on *N* (population size, expressed as a proportion of *K*) is shown

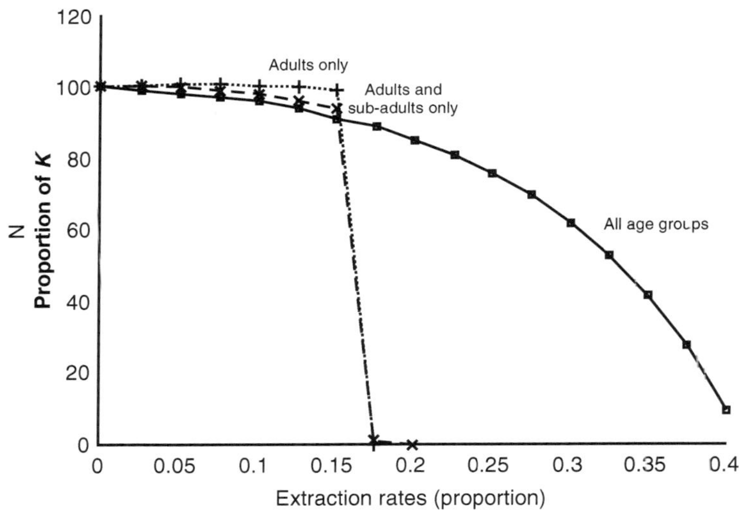

Fig. 17.9 Effects of harvest rate and selective exploitation by age class of capybaras on *N* (population size, expressed as a proportion of *K*). Three removal systems were tested: all age groups, adults and subadults only, and adults only

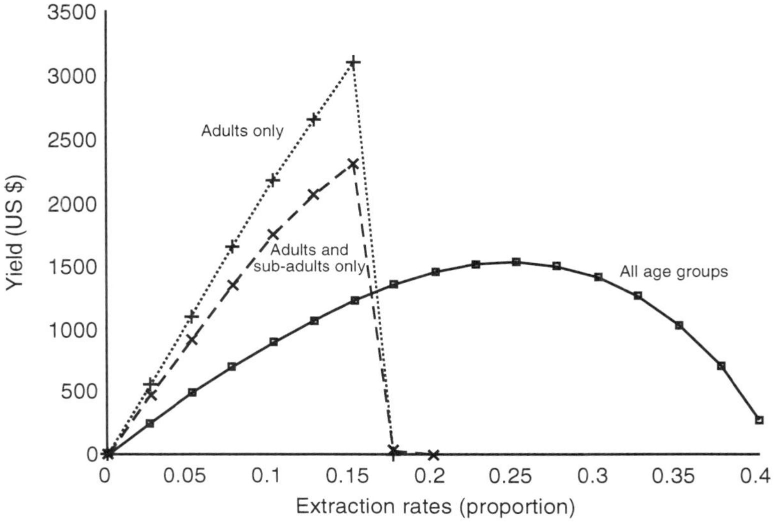

Fig. 17.10 Effects of harvest rate and selective exploitation by age class of capybaras on yield (expressed as US $ of meat and leather). The prices considered were US $3.50/kg for meat and US $4.00/ind for leather (Ojasti 1991). Three removal systems were tested: all age groups, adults and subadults only, and adults only

17.2.5 Effects of Social Behavior

Tuyttens and Macdonald (2000) introduce the concept of perturbation to describe how the behavior of survivors may change when some individuals from a population are removed, thereby changing the social dynamics amongst the survivors. Capybaras live in structured social groups, and the behavior of survivors may change after the removal of group members. Part of the adaptive significance of capybara groups is protection from predators (Macdonald 1981). Nonetheless, females leave the group temporarily when they give birth (Ojasti 1973). Herrera (1986) suggested that infanticide by subordinate males within the group was a possible explanation for this behavior. However, although apparent attempts at infanticide by males have been witnessed (Viviana Salas personal communication; José R. Moreira personal observation), they are at most a rare occurrence. Infanticide by extra-group females has been one of the main problems encountered in rearing captive capybaras (Nogueira et al. 1999).

Reproductive suppression has been observed in various social rodents (Hackländer et al. 2003) and is apparently related to cooperative rearing of young. Amongst capybaras, not all females in a group are reproductively active (Ojasti 1973; Jorgenson 1986), and reproductive suppression of subordinate females in the presence of the dominant female has been observed in captive capybaras (Nogueira-Filho et al. 2012).

Maldonado-Chaparro and Blumstein (2008) assessed the interactions between capybara social behavior and hunting, and their effects on group dynamics. They tested the effects on infanticide when the dominant male was removed, and reproductive suppression when there was a high density of females in the group. Their model suggested that the effects of infanticide by males were limited, whereas those of reproductive suppression were significant. Indeed, this model suggested that reproductive suppression had the potential to be the most important factor in controlling group size, and even more important than infant mortality (assumed to be only 0.029).

17.2.6 What Can Be Inferred from These Models?

Different models have estimated a sustainable harvest for capybaras from different populations ranging from 7.8% to 31.5% (Bone 1977; Moreira and Macdonald 1996; Paglia 1997). Empirically, the harvest rate in Venezuela, which was reduced from 30% to 20% in 1999, appears to be sustainable (Ojasti 1991).

The models suggest that biannual exploitation may be a good exploitation strategy, causing a small increase in production relative to that achieved by a single annual harvest. The models also suggest that harvest is best undertaken before the capybara breeding season, and that a male-biased exploitation strategy can lead to such a drastic reduction in the number of sexually active males that there is a risk of population extinction.

The model for controlling capybara populations considered to be pests is best achieved by harvesting all age classes. In contrast, maximum production of meat and leather is predicted from a harvest of only adults. Finally, if the effects of capybara harvest on the reproductive suppression of females are valid, it is probably better to harvest whole groups instead of just a proportion of the individuals from them. The role of reproductive suppression in capybara population process therefore merits further research.

17.3 The Effects of Harvesting on Capybara Populations

Any management of wildlife is selective, be it in the wild or in captivity. It is rarely straightforward to estimate the effects of hunting pressure in neotropical regions (Ojasti 1991). Very little information exists (Cordero and Ojasti 1981; Herrera 1992; Moreira and Macdonald 1996; Ángel-Escobar and Aldana-Domíngues 2007; Payán 2007) and, in the majority of cases, contains inaccuracies (Box 17.1) or comes from a low sampling effort.

Mortality in the first year of life is a major (if not the main) factor that regulates the size of wild capybara populations (Paglia 1997; Federico and Canziani 2005; Moreira et al. 2012). Therefore, it is essential to estimate this parameter

Box 17.1 Mistakes Commonly Encountered in Studies that Evaluated the Effects of Hunting Pressure on Capybara Populations in the Neotropic

Several studies have attempted to assess the effects of hunting pressure on capybara populations (Cordero and Ojasti 1981; Herrera 1992; Moreira and Macdonald 1996; Ángel-Escobar and Aldana-Domíngues 2007; Payán 2007). However, in most cases, the methodologies used are ambiguous or misleading, featuring at least one of two main mistakes (if not both): (a) confusing survival with mortality and (b) comparing incomplete sets of data. Additionally, most studies do not mention in their analysis whether the three conditions required by "Method 6" of Caughley (1977:92) were met: (1) stable age distribution, (2) knowledge of the rate of population increase, and (3) age distribution calculated from the birth peak. Knowing that the majority of capybaras are killed in Venezuela during the 4 months after the birth peak and that this corresponds to the time that data were collected in the studies used, it is unlikely that condition 3 has been met by researchers in this country (Cordero and Ojasti 1981; Herrera 1992). For populations without a birth peak (as is apparently the case for populations in Colombia; Aldana-Domíngues et al. 2012) the mortality rate during the first 6 months of life should be obtained for the calculation of survival to begin from 6 months of age (Caughley 1977:95). However, in the work of Payán (2007) from the Colombian Llanos, there is no mention of whether or not this condition was met.

Many of these studies, despite their methodological shortcomings and questionable conclusions, reached their goals. They aimed to, and in the end did, provide a timely warning of the negative effects that overexploitation may have on capybaras. However, it is essential to apply greater technical rigor in the development of research into the effects of hunting pressure on populations of neotropical mammals, to produce valid and more robust predictions.

Confounding Mortality with Survival

This confusion stems in no small part from the origin of the data collected – dead animals. Mortality (*dx*) is the measure of the number of animals that die in a population at any given time. Survival (*lx*) is the measure of the number of live animals in a population at this same time. They are two different measures that complement each other.

One way to calculate *dx* is by searching for remains of dead animals in a population. One way to calculate *lx* is by counting the number of live animals in each age class of the population. If we capture an unbiased sample of individuals in the population, the sample represents a proportion of live animals in each age class. The same is true if the sample is of individuals harvested (killed) from the population. The fact that animals were killed to get the

(continued)

Box 17.1 (continued)

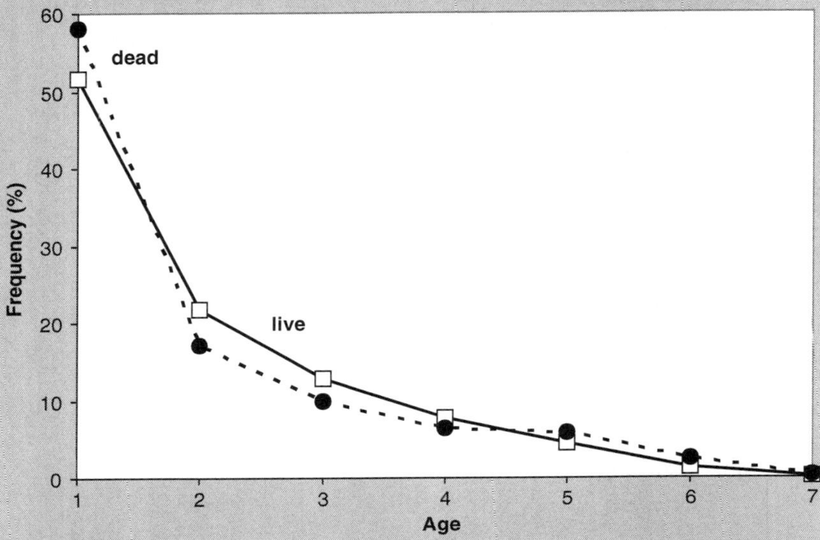

Fig. 17.11 Age distribution of live and dead animals in a hypothetical population. Each age class is plotted as a proportion of the whole population

demographic information is irrelevant (Caughley 1977:93) – the sample is still that of animals that were alive at the moment they were killed and represents the survival of different age classes of the population. The age distribution of live animals differs from the age distribution of deaths of the same population (Fig. 17.11).

There are formulae for calculating survival from mortality data and vice versa (Caughley 1977:87; Moreira et al. 2012).

$$l_x = \sum_{y=x}^{\infty} d_y$$

$$d_x = l_x - l_{x+1}$$

Therefore, mortality and survival are not similar parameters and cannot be grouped (for an exception, see Caughley 1977:94). It is not possible to compare the distribution of live animals from one population with the distribution of deaths from another, as can be found in the capybara literature (Herrera 1992). It is also not possible to group a sample of individuals that died from natural causes with that of another sample containing hunted animals, in the hope that this new set is representative of the age distribution of the population (Payán 2007).

(continued)

Box 17.1 (continued)

Comparing Incomplete Sets of Data

Comparing incomplete sets of data is the most common error in studies that evaluate the effects of hunting pressure on populations of neotropical mammals, especially because they are incomparable (Cordero and Ojasti 1981; Herrera 1992; Angel-Escobar and Aldana-Domíngues 2007; Payán 2007). Newborn animals or those in early developmental stages are small, decompose quickly when they are killed, or are eaten whole when they are predated. They are rarely seen either dead or alive. These difficulties in detection mean that the first age class is often not considered in population surveys. As a rule, only the other age classes are compared in these studies (evaluating "100% of the adult population") based on an assumption that their proportion of the population size did not depend on the size of the first (unsampled) age class. Below is a hypothetical example (Fig. 17.12) where the removal of data from the first age class leads to opposite conclusions to those obtained when using a sample from the entire population. While population B has a low mortality rate in the first year and the highest proportion of individuals of reproductive age (Fig. 17.12a), when age class 1 is removed (Fig. 17.12b), population B transforms to that with the lowest proportion of reproductive individuals. Unless you have samples of all age classes of the populations, these data are in no way comparable.

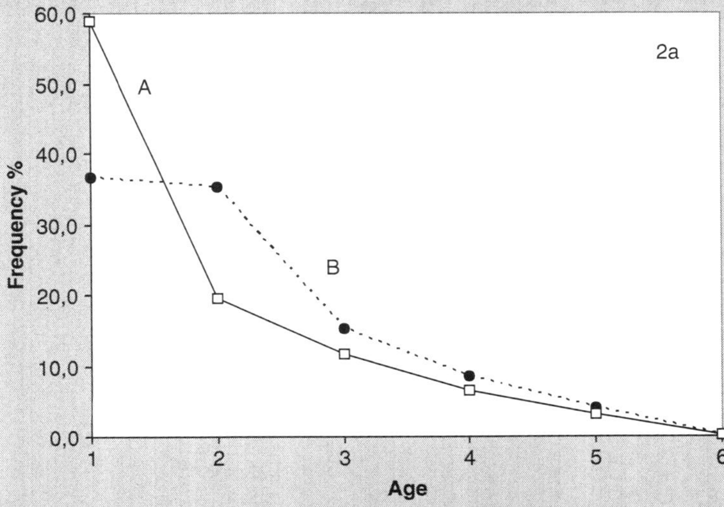

Fig. 17.12 Age distribution of two hypothetical populations with different mortality rates, especially in age class 1. Figure 17.12a shows the complete age distribution for both populations. Figure 17.12b excludes age class 1 and each age class is calculated as a proportion of the total from classes 2 to 6 in both populations. Each age class is plotted as a proportion of the whole population in Fig. 17.12a and as a proportion of the sampled population in Fig. 17.12b

(continued)

Box 17.1 (continued)

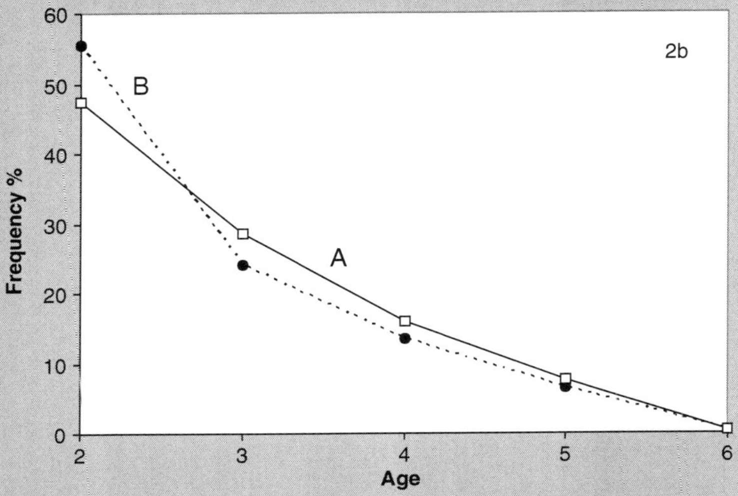

Fig. 17.12 (continued)

accurately before any inference can be made about the effects of hunting on population structure and dynamics. However, this first-year mortality was ignored by the majority of studies that sought to evaluate the effects of hunting on capybaras (Cordero and Ojasti 1981; Herrera 1992; Angel-Escobar and Aldana-Domíngues 2007; Payán 2007). Two other factors known to regulate the size of large ungulate populations, and probably also the capybara, are second-year fecundity and the age at first reproduction (Gaillard et al. 1998). These parameters were also disregarded by most studies. As Caughley (1977:123) notes, "... without prior knowledge of this kind, interpretation of an age distribution is an exercise in clairvoyance."

Although data on the effects of hunting pressure on the capybara are few or flawed, some inferences can be made: intense hunting pressure can modify the behavior of exploited animals (Tuyttens and Macdonald 2000), causing changes in activity patterns (Verdade 1996), alteration of habitat use, increase in time devoted to vigilance, and increase in the minimum flight distance (Ojasti 1973; Moreira and Macdonald 1997). Some of these changes may reduce reproductive success. The change from diurnal to nocturnal habits, observed among capybaras in regions with high hunting pressure (Verdade 1996), can cause changes in trophic strategy, social behavior, predator–prey relationships, and thermoregulation. The increase in time devoted to vigilance can reduce the time dedicated to foraging. The use of more densely forested areas for shelter (Cordero and Ojasti 1981) leads to the use of suboptimal habitat for growth. Alternatively, habitat change can lead to the use of more productive areas and an increase in population size (Verdade and Ferraz 2006; Ferraz et al. 2007).

The selectivity and intensity of hunting may also affect the social and age structure of hunted populations. As they are hunted for meat and/or leather, larger animals are usually killed. Commercial harvests in Venezuela and Colombia often lead to the virtual elimination of the adult animals in the groups (Herrera and Macdonald 1987), with consequential effects on the reproductive rate of the population (Ojasti 1973; Moreira and Macdonald 1993; Moreira et al. 2012). Possible additional effects also include the breakdown of the territorial structure, increased mortality of infants born to inexperienced females and artificial selection of inexperienced and smaller males (Gruver et al. 1984; Poole and Thomsen 1989; Moreira and Macdonald 1996). In general, the selective forces exerted on a hunted population will favor a smaller body size and younger breeding age (Caughley 1977), and this may occur in capybaras (Herrera 1992).

17.4 Productivity of Capybaras in the Neotropics

Previous studies that sought to estimate the productivity of capybaras in the Neotropics used different methods, analyses, and assumptions and differed in their estimates (Kleiman et al. 1979; Ojasti 1991; Robinson and Redford 1991; Moreira and Macdonald 1996; Paglia 1997). Estimates of capybara productivity (Table 17.1) ranged from 143 kg/km^2/year in the Pantanal (Paglia 1997) to 1,200 kg/km^2/year in the Venezuelan Llanos (Ojasti 1991). Robinson and Redford (1991) estimated productivity from capybaras of 74 kg/km^2/year in forest environments, whereas in agricultural landscapes, it was 630 kg/km^2/year, according to Verdade and Ferraz (2006).

Table 17.1 Potential production of capybaras in different localities in the neotropical region, according to different studies

References	Location	kg/km^2/year
Capybara living in savannas		
Kleiman et al. (1979)	–	244
Robinson and Redford (1991) equation	–	834
Ojasti (1991)	Llanos, Apure Department – Venezuela	1,200
Moreira and Macdonald (1996)	Marajó Island, Pará State – Brazil	841
Paglia (1997)	Pantanal, Mato Grosso do Sul State – Brazil	143
Capybara living in forests		
Robinson and Redford (1991) equation	–	74
Capybara living in agricultural landscapes		
Verdade and Ferraz (2006)	Piracicaba River, São Paulo State – Brazil	630

17.5 Future Directions for Modeling the Management of Capybaras

Most capybaras consumed in the Neotropic originate from hunting or sustainable management carried out in Venezuela, Colombia, and Argentina. In Brazil, where the production of wild animals is only permitted in captivity, there is a perceived need to control capybaras as pests. Few simulation models of the exploitation of capybaras have been performed, and, in general, there is little detailed knowledge of the management of wild capybara populations, leaving a considerable knowledge gap regarding the effects of the alternative management systems used. As a consequence, there is also a lack of suitable training for the technicians specializing in capybara management. Thus, a priority is to raise awareness of the sustainable management of the species (and all neotropical wildlife) and to train technicians in this subject.

To date, mathematical models of capybara harvests have had to rely on few data on basic life history and reproductive parameters which may differ between regions. These are fundamental parameters necessary to assess or model management actions. Improving the quality of data collected is another area in need of attention. From these basic data, it will be possible to calculate the intrinsic rate of population increase for the capybara in different regions across its range and for different environments and situations, and under different assumptions. It is also evident from this chapter that the effects of selective harvest by gender and age class should be better evaluated. The effects exerted by peculiar behavioral characteristics on the demography of exploited capybara populations are still very poorly studied and cannot be neglected. The modeling of population trends within this myriad of ecological environments is fundamental for the use and conservation of this valuable neotropical resource.

Acknowledgments Some of the simulation models used by J. R. Moreira and D. W. Macdonald were adapted for the capybara from originals kindly supplied by E. J. Milner-Gulland, to whom we are deeply grateful. J. R. Moreira thanks T. Tsikas for valuable guidance in the use of the program Turbo-Pascal. We also acknowledge Anderson C. Sevilha and two anonymous reviewers for their careful revisions and suggestions to the text.

References

Aldana-Domínguez J, Forero-M J, Betancur J, Cavelier J (2002) Dinámica y estructura de la población de chigüiros (*Hydrochaeris hydrochaeris*: Rodentia, Hydrochaeridae) de Caño Limón, Arauca, Colombia. Caldasia 24(2):445–458

Aldana-Domínguez J, Vieira-Muñoz MI, Bejarano P (2012) Conservation and use of the capybara and the lesser capybara in Colombia. In: Moreira JR, Ferraz KMPMB, Herrera EA, Macdonald DW (eds) Capybara: biology, use and conservation of an exceptional neotropical species. Springer, New York, pp 321–332

Angel-Escobar DC, Aldana-Domíngues J (2007) Estrutura de las poblaciones Del chigüiro (Rodentia: *Hydrochoerus hydrochaeris*) em el departamento Del Casanare, Colombia. In: Aldana-Domíngues J, Vieira-Muñoz MI, Angel-Escobar DC (eds) Estudios sobre la ecología

del chigüiro (*Hydrochoerus hydrochaeris*), enfocados a su manejo y uso sostenible en Colombia. Instituto de Investigación de Recursos Biológicos Alexanser von Humboldt, Bogotá, pp 69–83
Bone TG (1977) Un Modelo de Simulacion para la Explotacion Comercial del Chigüire (*Hydrochoerus hydrochaeris*). Graduation dissertation, Universidad Central de, Caracas, Venezuela
Caughley G (1977) Analysis of vertebrate populations. Wiley, Hobeken
Cordero RGA, Ojasti J (1981) Comparison of the capybara populations of open and forested habitats. J Wildlife Manage 45:267–271
Fairall N (1985) Manipulation of age and sex ratios to optimize production from impala (*Aepyceros melampus*) populations. S Afr J Wildlife Res 15:85–88
Federico P, Canziani GA (2005) Modeling the population dynamics of capybara *Hydrochaeris hydrochaeris*: a first step towards a management plan. Ecol Model 186:111–121
Ferraz KMPMB, Ferraz SFB, Moreira JR, Couto HTZ, Verdade LM (2007) Capybara (*Hydrochoerus hydrochaeris*) distribution in agroecosystems: a crossscale habitat analysis. J Biogeogr 34:223–230
Gaillard J-M, Festa-Bianchet M, Yoccoz NG (1998) Population dynamics of large herbivores: variable recruitment with constant adult survival. Trends Ecol Evol 13:58–63
Ginsberg JR (1991) The consumptive use of wildlife: harvesting, sociobiology, and implications for population dynamics. In: Grootenhuis JG, Njunguna SG, Kat PW (eds) Wildlife research for sustainable development. Agricultural Research Institute, Nairobi, pp 74–75
Ginsberg JR, Milner-Gulland EJ (1994) Sex-biased harvesting and population dynamics in ungulates: implications for conservation and sustainable use. Conserv Biol 8:157–166
Gruver BJ, Guynn DC Jr, Jacobson HA (1984) Simulated effects of harvest strategy on reproduction in white-tailed deer. J Wildlife Manage 48:535–541
Hackländer K, Möstl E, Arnold W (2003) Reproductive suppression in female Alpine marmots, *Marmota marmota*. Anim Behav 65:1133–1140
Herrera EA (1986) The behavioural ecology of the capybara, *Hydrochoerus hydrochaeris*. D.Phil. thesis, University of Oxford, Oxford
Herrera EA (1992) The effect of harvesting on the age structure and body size of a capybara population. Ecotropicos 5:20–25
Herrera EA, Macdonald DW (1987) Group stability and the structure of a capybara population. Symp Zoo Society of London 58:115–130
Jorgenson JP (1986) Notes on the ecology and behaviour of capybaras in Northeastern Colombia. Vida Silvestre Neotropical 1:31–40
Kleiman DG, Eisenberg JF, Maliniak E (1979) Reproductive parameters and productivity of caviomorph rodents. In: Eisenberg JF (ed) Vertebrate ecology in the northern neotropics. Smithsonian Institution Press, Washington, pp 173–183
Lacy RC, Hughes K, Zimmerman R (1995) Vortex, version 7.0. Chicago zoological society, Chicago
Leslie PH (1945) On the use of matrices in certain population mathematics. Biometrika 33: 183–212
Macdonald DW (1981) Dwindling resources and the social behaviour of capybaras, (*Hydrochoerus hydrochaeris*) (Mammalia). J Zool 194:371–391
Maldonado-Chaparro A, Blumstein DT (2008) Management implications of capybara (*Hydrochoerus hydrochaeris*) social behavior: results of a simulation model. Biol Conserv 141(8):1945–1952
Miller PS, Lacy RC (1999) VORTEX Version 8 users manual. A stochastic simulation of the simulation process. IUCN/SSC Conservation breeding specialist group. Apple Valley
Moreira JR, Macdonald DW (1993) The population ecology of capybaras (Hydrochaeris hydrochaeris) and their management for conservation in Brazilian Amazonia. In: Mayo SJ, Zappi DC (eds) Biodiversity and environment; Brazilian themes for the future. The Linnean Society of London/The Royal Botanic Gardens – Kew, London, pp 26–27
Moreira JR, Macdonald DW (1995) Simulação do efeito da seleção de sexo no abate de capivaras (*Hydrochaeris hidrochaeris*) na Ilha de Marajó – Pará. In: Bacellos AO, Mariante AS, Rocha

CMC, Leite GG (eds) Anais da XXXII Reunião Anual da Sociedade Brasileira de Zootecnia. Sociedade Brasileira de Zootecnia, Brasília, pp 399–402

Moreira JR, Macdonald DW (1996) Capybara use and conservation in South America. In: Taylor VJ, Dunstone N (eds) The exploitation of mammal populations. Chapman & Hall, London, pp 88–101

Moreira JR, Macdonald DW (1997) Técnicas de manejo de capivaras e outros grandes roedores na Amazônia. In: Valadares-Pádua C, Bodmer RE (eds) Manejo e Conservação de Vida Silvestre. Conselho Nacional de Desenvolvimento Científico e Tecnológico, Brasília, pp 186–213

Moreira JR, Wiederhecker H, Ferraz KMPMB, Aldana-Domínguez J, Verdade LM, Macdonald DW (2012) Capybara demographic traits. In: Moreira JR, Ferraz KMPMB, Herrera EA, Macdonald DW (eds) Capybara: biology, use and conservation of an exceptional neotropical species. Springer, New York, pp 147–167

Nogueira-Filho SLG, Pinheiro MS, Nogueira SSC (2012) Confined and semi-confined production systems for capybaras. In: Moreira JR, Ferraz KMPMB, Herrera EA, Macdonald DW (eds) Capybara: biology, use and conservation of an exceptional neotropical species. Springer, New York, pp 229–241

Nogueira SSC, Nogueira-Filho SLG, Otta E, Dias CTS, Carvalho A (1999) Determination of infanticide causes in capybara (*Hydrochoerus hydrochaeris*) groups in captivity. Appl Anim Behav Sci 62:351–357

Ojasti J (1973) Estudio Biologico del Chigüire o Capibara. FONAIAP, Caracas

Ojasti J (1991) Human exploitation of capybara. In: Robinson J, Redford KH (eds) Neotropical wildlife use and conservation. The University of Chicago Press, Chicago, pp 236–252

Paglia AP (1997) Ecologia populacional e modelagem da exploração econômica da capivara (*Hydrochaeris hydrochaeris*) no Pantanal da Nhecolândia, Mato Grosso do Sul. M.Sc. Dissertation, Universidade Federal de Minas Gerais, Belo Horizonte

Payán E (2007) Análisis de proporciones de edad y sexo a partir de cráneos resultantes de cosechas de chigüiros en los Llanos Orientales de Colombia, departamento del Casanare, para el uso y conservación de la espécie. In: Aldana-Domíngues J, Vieira-Muñoz MI, Angel-Escobar DC (eds) Estudios sobre la ecología del chigüiro (*Hydrochoerus hydrochaeris*), enfocados a su manejo y uso sostenible en Colombia. Instituto de Investigación de Recursos Biológicos Alexanser von Humboldt, Bogotá, pp 85–103

Poole JH, Thomsen JB (1989) Elephants are not beetles: implications of the ivory trade for the survival of the African elephant. Oryx 23:188–198

Robinson JG, Redford KH (1991) Sustainable harvest of Neotropical forest mammals. In: Robinson JG, Redford KH (eds) Neotropical wildlife use and conservation. University of Chicago Press, Chicago, pp 415–429

Tuyttens FAM, Macdonald DW (2000) Consequences of social perturbation for wildlife management and conservation. In: Gosling LM, Sutherland WJ (eds) Behaviour and conservation. Cambridge University Press, Cambridge, pp 315–329

Verdade LM (1996) The influence of hunting pressure on the social behavior of vertebrates. Rev Bras Biol 56:1–13

Verdade LM, Ferraz KMPMB (2006) Capybaras (*Hydrochoerus hydrochaeris*) on an anthropogenic habitat in Southeastern Brazil. Braz J Biol 66:371–378

Part III
Conservation

Chapter 18
Capybaras as a Source of Protein: Utilization and Management in Venezuela

Emilio A. Herrera and Guillermo R. Barreto

18.1 Introduction

Capybaras are hunted for the direct consumption of their meat throughout their geographical range by local peasants as well as by indigenous peoples (Ojasti 1991). A number of their biological characteristics, the habitats where they are found, as well as cultural elements of the human populations who share the land with capybaras have made this species a potential source of sustainably usable products (Emmons 1987; Herrera 1999). In Argentina, for example, capybaras are managed for their skin (Quintana and Bolkovic 2012), while in Venezuela, meat is the main product sought from them (Ojasti 1991).

Wildlife management is now a full-fledged discipline (Caughley and Sinclair 1997) and, in the case of the exploitation of capybaras in Venezuela, its principles (Moreira et al. 2012a) have been applied almost literally, with little or no human intervention required (except for the kill itself). One of the aims of wildlife management is the commercial exploitation of a given animal species in a sustainable way, and this is the topic of this chapter as it applies to the management of capybara populations in Venezuela. We will briefly discuss the case of subsistence hunting, followed by a historical outline of capybara exploitation in this country, the present management plan and, finally, the future of capybara management in Venezuela. We will not cover more intensive management or farming since that approach is not practiced in this country.

E.A. Herrera (✉)
Departamento de Estudios Ambientales, Universidad Simón Bolívar, Apdo. 89000, Caracas 1080-A, Venezuela
e-mail: eherre@usb.ve

G.R. Barreto
Departamento de Biología de Organismos, Universidad Simón Bolívar, Apdo. 89000, Caracas 1080-A, Venezuela
e-mail: guibarre@usb.ve

J.R. Moreira et al. (eds.), *Capybara: Biology, Use and Conservation of an Exceptional Neotropical Species*, DOI 10.1007/978-1-4614-4000-0_18,

18.2 Subsistence Hunting

Capybaras have been a source of protein for Creole (individuals of European parentage born in the Americas) and indigenous peoples since time immemorial. In Venezuela, for instance, capybara remains were found by Garson (1980) in human settlements in the Western Llanos (plains; Fig. 18.1), part of the "Complejo Caño del Oso," which covers a period from 230 B.C. to 650 A.D., indicating their use by the Guahibo and Achagua peoples among others, many years before the arrival of the Spanish conquistadores.

Capybaras are hunted for their meat with methods ranging from bow and arrow to firearms, including the use of dogs to flush them out of bushes and hunting from boats at night, spotlighting to kill them with a harpoon or a gun. In practice, capybaras are often killed opportunistically (Mondolfi 1977; Ojasti 1991). Although capybaras can produce as much as 20 kg of fresh meat, their use by indigenous peoples as a food source is not as common as that of much smaller species such as brocket deer (*Mazama* sp.), agoutis (*Dasyprocta* sp.), or even monkeys (*Cebus* sp., *Lagotrix* sp., *Alouatta* sp.).

In Venezuela, studies on the use of wildlife by Creole or indigenous peoples are rare, but all agree that capybaras are not an important food source for these people.

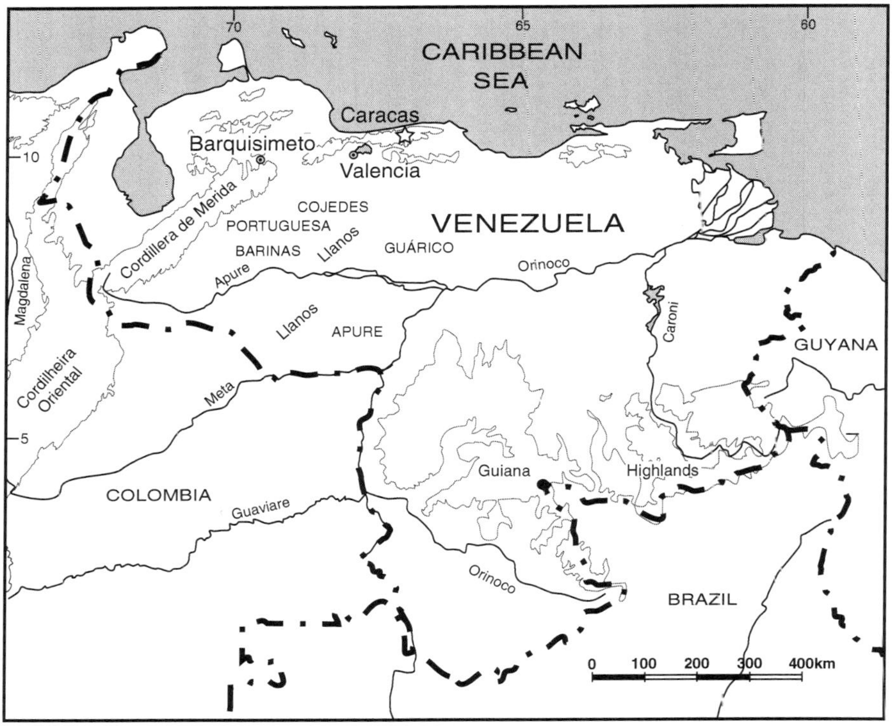

Fig. 18.1 Map of Venezuela with the locations of the places referred to in the text

In the rainforests south of the Orinoco river (Fig. 18.1), Bisbal (1994) ranked the capybara as seventh and eighth in importance for Creoles and indigenous peoples, respectively, and estimated that capybara consumption was just 0.07% of consumed biomass. North of the Orinoco, Cordero (1990) refers to the capybara among hunted species, but does not consider it important, while Bisbal (2000) found in Central Venezuela (around Lake Valencia; Fig. 18.1) that agoutis were more important than capybaras among hunted species.

These studies were based on surveys carried out by researchers and no field verifications were carried out to estimate the number of animals actually hunted, or the precise biomass taken. In the seasonally flooded plains (low Llanos) of Venezuela, where capybara populations reach their highest densities (Ojasti 1973; Mones and Ojasti 1986), illegal hunting is widespread and may in fact represent a greater component than that reported in the studies mentioned (Méndez-Arocha and Medina-Padilla 1982).

Despite an apparent lack of interest in capybara meat by subsistence hunters, its commercial harvesting has been much more important and, in Venezuela, the historical beginnings of commercial harvesting date back more than 200 years (Humboldt 1820). The remainder of this chapter will therefore be devoted to discussing this type of exploitation.

18.3 Commercial Exploitation of Capybaras in Venezuela

18.3.1 Background

Commercial hunting of capybaras on a large scale is carried out mostly in the Llanos of Venezuela (Ojasti 1991). Dried and salted capybara meat is sold in cities in Central Venezuela to satisfy a demand for it as a traditional Lenten dish. Its consumption probably dates to colonial times when capybaras were culled because they were considered livestock competitors and carriers of diseases (Ojasti 1973). The meat produced from the cull was salted and dried in the sun, perhaps to avoid it going to waste, and destined for local consumption. In 1784 and after several attempts at obtaining a Vatican license, a Papal Bull (decree) allowed the consumption of capybara flesh during Lent (López de Ceballos 1974), perhaps due to the semiaquatic habits of this animal but, more likely, because the dried, salted meat was similar in color and shape to dried fish. Hence, this local habit turned into a religious tradition. During the nineteenth century, Codazzi (1841) refers to the trade in capybara meat originating in the Llanos states of Barinas and Apure and ending in cities such as Barquisimeto, Valencia, and Caracas (Fig. 18.1). To this day, consumption is concentrated in these cities.

Capybara hunting, in those days, was carried out without any control or management plan. Demand for capybara meat grew until 1958 when the government was forced to allow it to be imported from neighboring Colombia, due to a decline in capybaras in Venezuela (Méndez-Arocha and Medina-Padilla 1982). In 1962, a 5-year ban on capybara hunting was imposed to allow capybara populations to recover from overexploitation. During that time, the Ministry of Agriculture decided to study capybara biology to provide the basic knowledge required for a sustainable management program; the Finnish biologist Juhani Ojasti was appointed to the job (Box 18.1).

Ojasti's (1973) meticulous and thorough study covered many aspects of capybara biology in the Llanos of Venezuela. One of the most important results, however, was the proposed management program that was immediately put into practice and remains so today (2011), with only minor modifications. Ojasti used his own demographic data (on female reproductive rates, birth and death rates, individual growth rates, age at sexual maturity, etc.) to build a life history table, which he used as a basis to calculate potentially sustainable extraction rates. Ojasti (1973) proposed a 30% extraction rate and undertook an experimental extraction at one ranch to test the sustainability of this rate. His experiment confirmed that it was indeed sustainable; at least during the period that he applied it.

Box 18.1 Juhani Ojasti

Juhani Ojasti was the "founding father" of wildlife management in Venezuela. Having obtained a degree in biology in his native Finland, he arrived in the late 1950s to visit his father who had established himself as a farmer in Eastern Venezuela a few years previously. There he met Volkmar Vareschi, an Austrian botanist, who invited him to work at the natural history museum in Venezuela's Central University (UCV). Soon, he pioneered the teaching of ecology at UCV and became part of the school of biology staff. In the mid-1960s a visionary from the Ministry of Agriculture, Gonzalo Medina, launched a project to study capybaras, a species already traditionally consumed during Lent, with the aim of establishing the scientific bases for their sustainable exploitation. This was a really pioneering initiative since sustainable wildlife exploitation was in its infancy. Ojasti was thus the biologist hired to carry out the study, which sent him to the Llanos with his family for nearly 4 years. In those days (late 1960s) roads were mostly dirt and much traveling was done on horse- or mule-back. By 1970, Ojasti had submitted his final report to the Ministry of Agriculture. A couple of years later, while studying for his Ph.D. in Athens, Georgia (USA), he received a package with a few copies of his report published as a book (Ojasti 1973). The quality of his work was such that Gonzalo Medina decided that it could be made into a book with virtually no changes. That book established in full detail the procedure for the capybara management program which has been in operation in Venezuela, with little modification, until today (2011). Ojasti is now retired in Finland but his legacy remains and his book has just been reissued with revisions by the author. He published many scientific papers during his years at UCV and has been a consultant in all of Latin America and elsewhere for wildlife management programs.

18.3.2 The Program and Its Implementation in the Field

Ojasti's management plan has been in operation in Venezuela since 1969, initially on private ranches ("hatos"), in the states of Apure and Barinas, and latterly in the states of Cojedes, Guárico, and Portuguesa (all within the area known as the Llanos, the plains, or more specifically the seasonally flooded plains; Fig. 18.1). The cull is carried out in February and March, at the height of the dry season. Legislation (Ley de protección a la fauna silvestre 1970 and Resolución 205 of Ministerio de Ambiente 1980), which has been in force with only minor changes since 1970, defines the following procedure. Any ranch owner interested in exploiting capybaras on their property submits a request for a permit to the Wildlife Service of the Ministry of the Environment. The Wildlife Service sends a group of two or three civil servants to the ranch who then carry out a direct and exhaustive count of the capybara population on the entire property. Such a count is possible due to the openness of the habitat, the size of the animal, its gregarious and mostly diurnal habits, and the fact that, at the time of the counting (at the height of the dry season), capybaras tend to be concentrated around the dwindling pools (Macdonald 1981). The dry land allows for easy access to all areas of the ranch in a four-wheel drive vehicle. If the population is of adequate size, a permit allowing the cull of a percentage (see below) of the whole population is granted. Adult animals of all ages and both sexes can be culled. Capybaras reach adult size at 35 kg (Ojasti 1973) and so only animals weighing 35 kg or more may be harvested; body size is estimated by eye but later checked postmortem to allow adjustment of field estimates if necessary. When the law was passed in 1968, a minimum population size of 400 animals was required, and the permitted percentage was 30%, following Ojasti's (1973) guidelines. Now, a minimum breeding stock (*pie de cría*) is required and this varies from ranch to ranch (see below). Once the license is obtained, the ranch contracts additional personnel to carry out the slaughter and purchases whatever supplies are necessary, mostly salt. The whole production is sold during Lent, which is the only time of the year when it is legally hunted.

The procedure in the field is based on the work of a group of men on horseback who herd as many capybaras as possible toward another group on foot (the "crew"); each man is armed with a stick made of extremely tough wood or, more recently, a piece of thick metal tubing. After a slow chase lasting less than 20 min (a distance of 200–500 m), capybaras, following their natural tendency to huddle (Macdonald 1981), stop in a dense aggregation, surrounded by the men on foot (Fig. 18.2). Under ideal conditions, a skilled cowboy hits the chosen animals one by one (avoiding small animals and visibly pregnant females, but see below) with a sharp blow to the top of the head. The animals die almost instantly without emitting a sound, and there is no sign of panic amongst the survivors. Some people regard this as inhumane, but if the cowboy hits the animal squarely on the head, there is little apparent suffering and death is almost instantaneous. "Modern" methods, such as the use of a shotgun or rifle, frequently only wound the animal, which escapes, only to die later, sometimes by drowning or after extensive suffering, out of the workers' sight.

Fig. 18.2 A herd of capybaras surrounded by men with sticks, about to be culled. Hato El Frío, Apure, Venezuela (Photo by Emilio Herrera)

Thus, these more modern methods do not necessarily improve animal welfare and the value of the culled animal can be lost. Further, because several shots are often necessary to kill the animal, the skin is ruined and not usable as a by-product of the cull. The use of firearms adds to the costs of the procedure, introduces a very serious risk for the workers and the noise frightens the animals. However, in more forested habitats (e.g., the eastern part of the state of Apure, close to rivers, and in the state of Barinas), using shotguns may be the only practical way to carry out the cull, since the men on horseback cannot herd the animals toward the crew on foot. Our personal experience has shown that skilled shooters minimize suffering and loss of wounded animals using this method.

Despite the efforts to avoid pregnant females and the fact that this is the time of year when reproduction is at its lowest point, up to 43% of culled females have been found to be pregnant (in this study about 50% of the cull was female but normally it is much less; Herrera 1998). Reproduction during the dry season may, in part, be facilitated by the dykes built on ranches for flooding and drought control, which maintain a certain amount of green grass even at the height of the dry season.

The cull is generally carried out in the mornings, except in more forested ranches where workers prefer to work at night when animals come out of the woods to graze. Usually, animals are eviscerated in the field, immediately after being killed, to minimize the risk of the meat rotting in transit. Carcasses are then taken to a base camp where the meat is separated from the bones in one piece, called the "salón," washed thoroughly, salted with about 2.5 kg salt per

Fig. 18.3 Salted capybara meat drying in the sun. Hato Santa Luisa, Apure, Venezuela (Photo by Emilio Herrera)

animal, and left to dry in the sun (Fig. 18.3). At this time of year, rains are almost nonexistent and it is very windy, which contributes to the quick drying of the meat. Meat is re-salted the next day and it is ready for sale a couple of days later. Lack of rains also helps to reduce the presence of flies and other insects. Although sanitary conditions are far from optimal, cross-specific transmission of disease has not been reported.

The quality of the carcasses is inspected by the Wildlife Service prior to being transported to market (Fig. 18.4). Each carcass is labeled with a unique number to prevent the sale of poached animals. Traditionally, capybara meat in Venezuela is prepared only as a stew of shredded meat (previously boiled to extract the salt). Capybara meat has been experimentally used to make salamis and sausages, as well as smoked ham, and this has been well accepted in taste trials (González-Jiménez 1977), but has never been made commercially. More recently, the Wildlife Service has permitted the experimental culling of capybaras outside the few months presently allowed, with the aim of selling it fresh and to diversify the way it is served (Omar Hernández personal communication).

The convergence of ecological factors (the dry season, when reproduction is at its lowest and animals have little or no subcutaneous fat), biological factors (fast growth rate; high reproductive output: four young per litter), practical factors (animals are easy to capture in the dry season), economic factors (low costs of maintenance of capybara populations and of the cull and processing of carcasses, plus no need for refrigeration), and cultural and religious factors (tradition and a Papal Bull) make the capybara management program a success in Venezuela. The procedure described has been in operation since colonial times; although several attempts have been made to modify and "modernize" the program over the years, they have had little success (Box 18.2).

Fig. 18.4 Carcasses being weighed and identified by personnel from the Venezuelan Wildlife Service. Hato El Frío, 2007 (Photo by Gisselle Perdomo)

Box 18.2 Chronology of the Capybara Management Program in Venezuela

The program as we know it has been in operation since 1969. Previously, the Hunting Law of 1944 defined specific periods when hunting (including of capybaras) was allowed. Resolution #212 by the Ministry of Agriculture (1953) authorized the hunting of capybaras with no limitation except for the payment of a small tax per animal killed. In 1968, after a 5-year ban, Resolution #91 of the Ministry of Agriculture allowed capybara hunting under license, in the states of Apure and Barinas. In 1979, the Law for the Protection of Wildlife was passed, which replaced the Hunting Law. This law established general guidelines for research, protection, and exploitation of wildlife. Open and closed seasons were defined, as well as specific permitted methods, provisions for the mobilization of wildlife products, and sanctions for those breaking the law. In 1976, the Ministry of the Environment was created, and it became the state's agency in charge of the capybara management program. Importing capybara meat was prohibited in 1980, and the percentage allowed

(continued)

Box 18.2 (continued)

to be culled was increased to 35% of the population. In 1994, Resolution #133 by the Ministry of the Environment reduced the permitted percentage cull back to 20% (lower than Ojasti's recommendation of 30%). Since 1999, a management plan and a minimum density of 0.3 capybaras per ha are required for a license to cull capybaras. Additionally, a breeding stock defined as 70% of the population as counted in 1988 is required. Resolution #172 of the Ministry of the Environment (2006) changed the breeding stock requirement to 70% of the population counted in 2005.

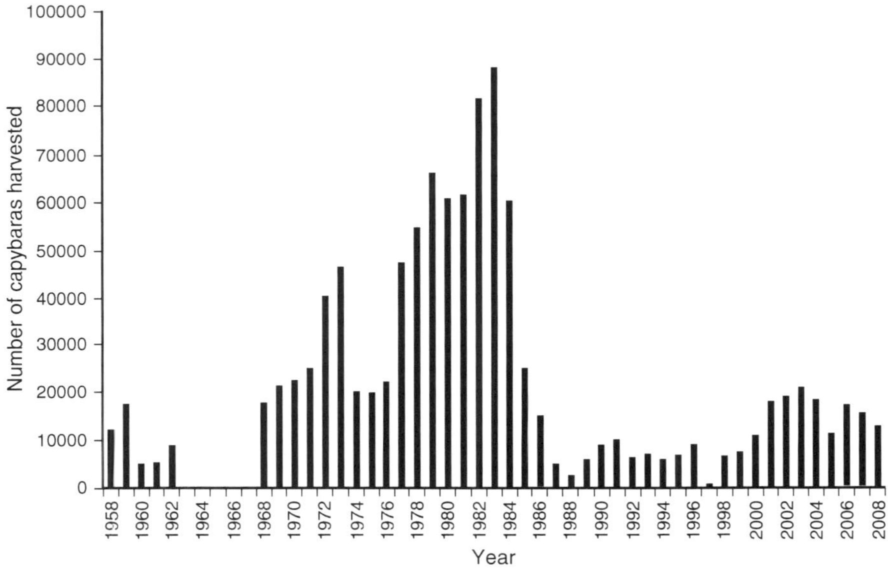

Fig. 18.5 Capybara harvest in Venezuela from 1958 to the present. During the interval 1963–1967, a moratorium on capybara hunting was established to allow the populations to recover and to carry out a study of capybara biology (From Ojasti and Medina-Padilla (1972), Ojasti (1983, 1991)). The actual number of slaughtered animals may be lower since not all authorizations materialized

18.4 The Capybara Cull

There have been considerable fluctuations in the annual harvest of capybaras over the years (Fig. 18.5). There was a substantial increase in the number of animals culled at the end of the 1970s and into the 1980s, doubling and even tripling the cull of the previous decade. During that period, and in particular after 1983, the economic crisis that shook Venezuela (together with most of Latin America) caused an increase in

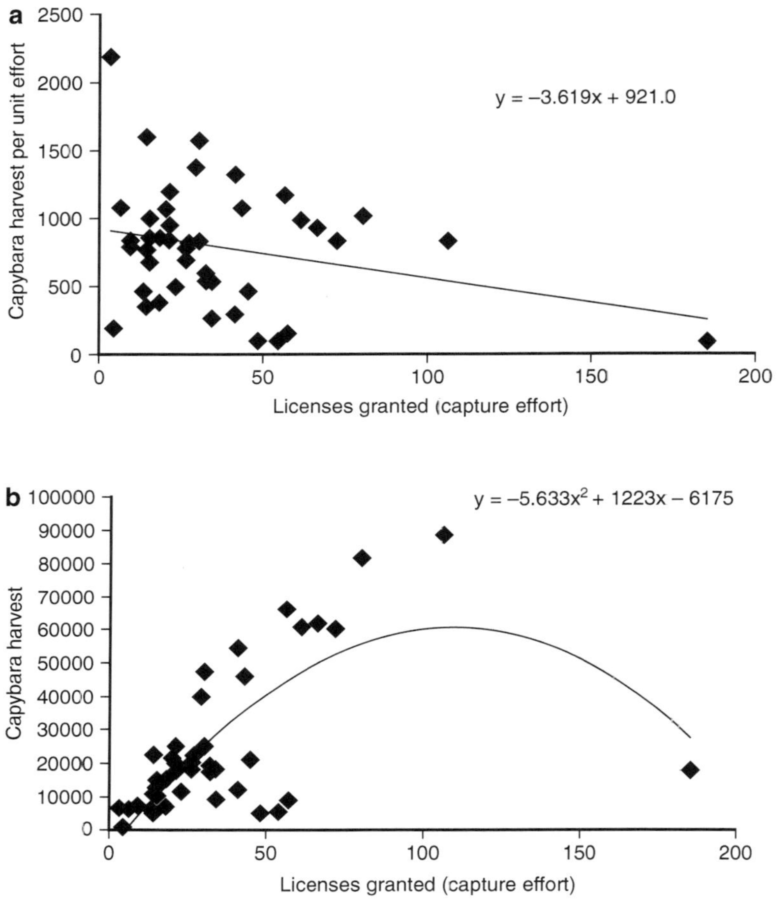

Fig. 18.6 Capybara harvest per unit effort, and the number of licenses granted (the capture effort) from 1958 to 2008, revealing overexploitation (**a**) The effort defining the maximum sustainable yield would be around 100 licenses if the licenses had relatively constant numbers (**b**) (see text)

unemployment and a reduction in personal income, which led many people to resort to intensive exploitation of natural resources (Rodríguez 2000). Indeed, from 1985 on, a sharp decline in the number of harvested animals suggests overexploitation. From 1990, the program appears to have recovered and stabilized, and even to have increased in 2000, albeit well below the levels of the 1960s and 1970s.

Figure 18.6 shows the relationship between the capybara harvest and capture effort as estimated by the number of licenses granted. The number of individuals captured per unit effort decreased as the total effort increased, indicating overexploitation (Fig. 18.6a). The level of effort associated with the maximum sustainable harvest appears to be just under 100 licenses per year (Fig. 18.6b). The number of capybaras per license is variable (mean = 792, SD = 440). The overharvest that

Table 18.1 Mean population growth rate (μ), variance (σ2), and 95% confidence interval estimates for three ranches that have exploited capybaras since 1991. The variables were calculated following the method proposed by Dennis et al. (1991) based on a regression model (The data come from the official censuses carried out by the Ministry of the Environment)

Ranch	μ	σ^2	95% Confidence interval
1 (Apure)	0.0647	0.1361	−0.1301; 0.2597
2 (Apure)	0.0934	0.0673	−0.0457; 0.2325
3 (Barinas)	0.0841	0.0926	−0.0790; 0.2473

occurred in the 1980s was a product of both a large number of licenses and the large quotas per license, which in turn was apparently a consequence of overestimates of population sizes. It is important to point out that in the 1980s, indirect methods of population estimation were used, and these did not have a solid scientific basis (pers. obs.).

Another way to estimate the level of exploitation is to observe the actual yield obtained. Yields ranging from 834 up to 1,200 kg/km^2 have been reported (Ojasti 1991; Robinson and Redford 1991), but a calculation based on the harvests of 1988–1999, for which we know the area under exploitation, showed much lower yields (543, 781, and 600 kg/km^2, respectively), again suggesting overexploitation. In 1988, direct counting resumed, avoiding population estimates based on extrapolated values. Establishing a requirement to maintain breeding stocks and reducing the allowable proportion culled to 20% was aimed at the recovery of certain populations. Although the precise figures adopted were arbitrary and not based on scientific study, they appear to be producing the expected results. In 2008, 15 licenses were granted to 13 ranches in the state of Apure and 2 in Barinas. A total of 64,453 capybaras were counted and the slaughter of 12,890 individuals was allowed.

Analysis of three emblematic ranches in the states of Apure and Barinas (Table 18.1) that have maintained more or less stable extraction rates from 1991 to 2008 shows average growth rates slightly above zero, indicating that their populations are relatively healthy, despite exploitation. Population growth rates are, however, subject to strong variation, responding to underlying fluctuations in population demographics. To allow for such variation, managers should be encouraged to adopt conservative extraction rates and perhaps redefine their breeding stock, taking into account capybara demography (e.g., reproductive patterns and survival rates), to allow regular updates of estimates of the populations under exploitation.

18.5 Impact of the Slaughter on the Populations

The capybara cull could have a deleterious effect on populations (Moreira et al. 2012a). When hunting selects larger animals, as capybara slaughter generally does (personal observation), it can alter the size and age structure of the surviving population. A population biased toward younger animals would doubtless lose productivity (Ojasti 1973), which in turn would reduce population density. Cordero and Ojasti (1981) found that in places where capybaras were exploited there was indeed a bias toward younger animals, and survivors tended to use preferably wooded areas, where the bias was less noticeable. Herrera (1992a) found that animals were older and heavier in areas where the cull had been suspended for 3 years prior to data collection compared with areas with a continuous annual cull. These types of effects could have impacts on the social structure and organization of capybara groups. Younger females have been found to have lower reproductive success (Ojasti 1973) and smaller litters (Moreira 1995; Moreira et al. 2012b), which would negatively affect population growth. Additionally, these factors also have the effect of producing fewer kilograms of meat per animal, with concomitant economic losses.

Moreira (1995) did not find any differences between the survival curves or life expectancy in hunted and non-hunted populations on Marajó Island (Brazil). Herrera (2002) also did not detect any trend over time in a number of capybara characteristics (such as litter size or proportion of pregnant females) in a population that was regularly culled over several years, after a fallow period. There was, however, an initial decrease in weight which thereafter remained stable.

18.6 The Future of the Program and Final Comments

We consider the capybara management program in Venezuela to be a success. It is a success in the sense that it is a functioning application of wildlife management principles to the commercial exploitation of a tropical mammal. It is also a success in the sense that the method has proved to be sustainable: ranches where the program has been applied continuously harbor healthy populations of capybara, despite some overexploitation causing temporary reduction of populations. Importantly, the moment this reduction is noted by officials, a local, temporary ban is applied until the population has recovered to acceptable levels. A third measure of success is the production of additional income for both owners and workers and the establishment of a tradition that respects norms.

Nevertheless, we are aware that the program has been applied in a fragmented fashion, that is, on individual ranches rather than throughout the region. This is because no one controls what happens in ranches that do not apply for a cull license. So, on some of these ranches, capybaras are simply overexploited to virtual local extinction. The other problem – overexploitation in ranches participating in the program – has more to do with not following the regulations, fraud, and corruption than with the principles sustaining the program. In some cases, there has been overexploitation, either by the owners themselves or by their tolerance of poaching on their property. In other cases, lack of official supervision or monitoring of the population count has led to population overestimates and the granting of licenses above the maximum sustainable yield.

The implications of poor management can be illustrated by a case study of a traditionally capybara-producing ranch. The success and failure of the program, in this case, can be explained by changes in management policy and inaccurate estimates of population size. From the 1960s, the owners of this ranch had adopted a conservation and management policy on their land that favored an increase in the size of the capybara population, as well as other wildlife, making the ranch a role model for capybara management in Venezuela. In the 1990s, what appeared to be unwise and possibly corrupt management practices, accompanied by excess harvest apparently caused by an overestimation of population size, caused a dramatic drop and virtual collapse of the capybara population. This in turn led to a rejection of all licenses applied for by the ranch between 1996 and 2003, when no capybaras were culled. The population does not appear to have recovered so far: our analysis of the population trend between 1991 and 2007 indicates a mean growth rate of –0.061 with a variance of 0.079.

Thus, we conclude that the success of the commercial management program of capybaras in Venezuela is based on three factors: (1) the design of the program based on the biology and ecology of the species; (2) the correct implementation of the program as it was designed, with strong and committed monitoring, enforced by the authorities; and (3) a proactive attitude from the management promoting the protection of wildlife, particularly with regard to protection from poaching.

In relation to factor 1, it will become important, in the near future, to reevaluate both the required breeding stocks and the 20% recommended extraction rate. A model by J.R. Moreira, E.J. Mulner-Gulland, and D.W. Macdonald (cited by Moreira and Macdonald 1996; Moreira et al. 2012a), which simulates the viability of capybara populations under varying management plans, showed the maximum extraction rate to be just 17%, not selecting for gender. Above this level, the viability of the population would be compromised. Although the data on which the model was based came from Brazilian Amazonia, and are not necessarily applicable to the Venezuelan seasonally flooded savannas, it is important to highlight the possibility of overexploitation, particularly in light of local examples described above.

Correct implementation of the program, factor 2, depends on strict controls and adequate funding to allow the Wildlife Service personnel to attend the counts, cull, processing, and transportation of the animals. Being vigilant to the sale of illegal capybara meat is also vital. As to factor 3, it is important to remember that the program is designed to work on large properties owned by a company or a family. To evaluate the possibility of a program being managed by the local community is a future challenge for the program's administrators. A community-managed program would allow a wider distribution of revenue. It could also involve the inhabitants of the region directly in the management and conservation of capybaras in the Llanos ecosystem.

The capybara management program in Venezuela is compatible with the practice of extensive cattle ranching and it requires little or no habitat modification, so it has minimal environmental impact, making it possibly the only sustainable wildlife management program in this region. It may even contribute not only to the conservation of capybaras in particular since many capybara-exploiting ranches do protect the species – if only for its economic benefits – but also to Llanos biodiversity in general. This is because measures to protect capybaras – essentially keeping poachers at bay – also contribute to protection of wildlife in general. Additionally, the water management infrastructure of many ranches contribute to water-associated wildlife (wading birds such as herons and plovers, ducks, caiman, capybaras) while providing drinking water for land-dwelling animals such as deer. Additionally, many capybara-managing farms have started ecotourism ventures which, if properly implemented, can also contribute to the conservation of the ecosystem, since well-preserved nature is the attraction. If an intensification of the program or an increase in production is planned, it is important to maintain the key features of the present plan, which are its sustainability and low environmental impact. The capybara is autochthonous to the Llanos, therefore adapted to the ecological conditions and with a great resistance to infections and diseases, even those introduced such as *Trypanosoma evansi* (Reverón 1992), contributing to the low cost and impact of the program since no special procedures to keep the animals healthy are necessary. This, together with its gregarious habits, relative docility, large size, abundance in open habitat, fast growth, and high reproductive rates (Ojasti 1973; Herrera 1992b), makes the capybara an ideal species to exploit under a management program such as the one in Venezuela. Nevertheless, cultural, economic, and ecological factors, perhaps unique to this country and possibly playing an important role in the success of the program, cannot be ignored, so the implementation of a similar program in a different region must take all these factors into account.

Acknowledgments We thank Edis Solórzano, Director in charge of the National Office of Biological Diversity at the Ministry of Popular Power for the Environment, for authorizing the use of census and harvest data, as well as Alexis Rengifo and Roldán de Sola, for providing such data.

References

Bisbal F (1994) Consumo de fauna silvestre en la zona de Imataca, Estado Bolívar, Venezuela. Interciencia 19:1–6

Bisbal F (2000) Consumo de la fauna en el Lago de Valencia, Estados Aragua y Carabobo, Venezuela. Bol Cent Investig Biol 34:362–375

Caughley G, Sinclair A (1997) Wildlife ecology and management. Blackwells, Oxford

Codazzi A (1841) Resumen de la Geografía de Venezuela. Imprenta de H. Fournier et Co., Paris

Cordero G (1990) Aprovechamiento de la fauna silvestre en Barlovento, Estado Miranda, Venezuela. Vida Silvestre Neotrop 2:70–74

Cordero G, Ojasti J (1981) Comparison of capybara populations of open and forested habitats. J Wildl Manage 45:267–271

Dennis B, Munholland PL, Scott JM (1991) Estimation of growth and extinction parameters for endangered species. Ecol Monogr 61:115–143

Emmons LH (1987) Ecological considerations on the farming of game animals: capybaras yes, pacas no. Vida Silvestre Neotrop 1:54–55

Garson AG (1980) Prehistory, settlement and food production in the savanna región of La Calzada de Páez, Venezuela. Ph.D. thesis, University of Yale, Yale

González-Jiménez E (1977) El capibara, una fuente indígena de carne de la América Tropical. Rev Mundial Zootec 21:24–30

Herrera EA (1992a) Effect of the slaughter on the age structure and body size of a capybara population. Ecotropicos 5:20–25

Herrera EA (1992b) Growth and dispersal of capybaras, *Hydrochaeris hydrochaeris*, in the Llanos of Venezuela. J Zool (Lond) 228:307–316

Herrera EA (1998) Reproductive strategies of female capybaras: dry season gestations. Symp Zool Soc Lond 71:281–296
Herrera EA (1999) Comportamiento, conservación y manejo de fauna silvestre: el caso del capibara en Venezuela. Etología 7:41–46
Herrera EA (2002) Ecología, manejo y conservación del chigüire, *Hydrochoerus hydrochaeris*. Report to full professorship, Universidad Simón Bolívar, Caracas
Humboldt A von (1820) Voyages aux régions équinoxiales du Nouveau Continent fait en 1799, 1800, 1801, 1802, 1803 et 1804 par A. Von Humboldt et A. Bompland. VI. Chez N. Maze, Paris, p 318
López de Ceballos E (1974) Fauna de Venezuela y su conservación. Editorial Arte, Caracas
Macdonald DW (1981) Dwindling resources and the social behaviour of capybaras, *(Hydrochoerus hydrochaeris)* (Mammalia). J Zool (Lond) 194:371–391
Méndez-Arocha JL, Medina-Padilla G (1982) Fauna Silvestre de Venezuela. Sistemas Ambientales Venezolanos. Proyecto VEN/79/001. Ministerio del Ambiente y de los Recursos Naturales Renovables, Caracas
Mondolfi E (1977) Fauna silvestre de los bosques húmedos tropicales. In: Hamilton LS, Steyermark J, Veillon JP, Mondolfi E (eds) Conservación de los bosques húmedos de Venezuela. Ministerio del Ambiente y de los Recursos Naturales Renovables, Caracas, pp 113–181
Mones A, Ojasti J (1986) *Hydrochoerus hydrochaeris*. Mamm Species 264:1–7
Moreira JR (1995) The reproduction, demography and management of capibaras (*Hydrochaeris hydrochaeris*) on Marajó Island-Brazil. D.Phil thesis, University of Oxford, Oxford
Moreira JR, Macdonald DW (1996) Capybara use and conservation in South America. In: Taylor VJ, Dunstone N (eds) The exploitation of mammal populations. Chapman & Hall, London, pp 88–101
Moreira JR, Verdade LM, Ferraz KMPMB, Macdonald DW (2012a) The sustainable management of capybaras. In: Moreira JR, Ferraz KMPMB, Herrera EA, Macdonald DW (eds) Capybara: biology, use and conservation of an exceptional neotropical species. Springer, New York, pp 283–302
Moreira JR, Wiederhecker H, Ferraz KMPMB, Aldana-Domínguez J, Verdade LM, Macdonald DW (2012b) Capybara demographic traits. In: Moreira JR, Ferraz KMPMB, Herrera EA, Macdonald DW (eds) Capybara: biology, use and conservation of an exceptional neotropical species. Springer, New York, pp 147–167
Ojasti J (1973) Estudio biológico del chigüire o capibara. Fondo Nacional de Investigaciones Agropecuarias, Caracas, 2nd edn. 2011, editorial Equinoccio, Universidad Simón Bolívar, Caracas
Ojasti J (1983) Ungulates and large rodents of South America. In: Bourlière F (ed) Tropical savannas. Elsevier, Amsterdam, pp 427–439
Ojasti J (1991) Human exploitation of capybara. In: Robinson J, Redford KH (eds) Neotropical wildlife use and conservation. The University of Chicago Press, Chicago, pp 236–252
Ojasti J, Medina-Padilla G (1972) The management of capybara in Venezuela. Trans North Am Wildl Nat Res Conf 37:268–277
Quintana RD, Bolkovic ML (2012) Use of capybaras in Argentina. In: Moreira JR, Ferraz KMPMB, Herrera EA, Macdonald DW (eds) Capybara: biology, use and conservation of an exceptional neotropical species. Springer, New York, pp 345–356
Reverón I (1992) Tripanosomiasis en chigüires, *Hydrochaeris hydrochaeris*. Graduate dissertation, Universidad Simón Bolívar, Caracas
Robinson JG, Redford KH (1991) Sustainable harvest of neotropical forest mammals. In: Robinson JG, Redford KH (eds) Neotropical wildlife use and conservation. The University of Chicago Press, Chicago, pp 415–429
Rodríguez JP (2000) Impact of the Venezuelan economic crisis on wild populations of animals and plants. Biol Conserv 96:151–159

Chapter 19
Conservation and Use of the Capybara and the Lesser Capybara in Colombia

Juanita Aldana-Domínguez, Maria Isabel Vieira-Muñoz, and Patricia Bejarano

19.1 Introduction

In Colombia, there are two capybara species: the lesser capybara (*Hydrochoerus isthmius*) and the capybara (*Hydrochoerus hydrochaeris*). The lesser capybara is found in the Caribbean region, the northern end of the Pacific region and the inter-Andean valleys of the Cauca and Magdalena rivers, while the capybara inhabits the biogeographic regions of Amazonia and Orinoquia (Fig. 19.1; León 1974; Torres and Sanabria 1976). Capybaras from the Orinoquia region were taken to recreational country houses in the Cauca valley, and from there, they escaped into river and wetland ecosystems, to which they successfully adapted (Moreira et al. 2012a). Encounters between two species of the same genus, such as the capybara and the lesser capybara, can lead to hybridization, which may have detrimental effects on hybrid descendants or, if extensive, may result in the local extinction of both parental species.

Capybaras are a relatively well-known species in Colombia, and the authorities and local communities are supportive of their conservation and sustainable use. However, some wild capybara populations are declining, particularly in Arauca (Aldana-Domínguez et al. 2002). Much less is known about the lesser capybara,

J. Aldana-Domínguez (✉)
Independent consultant on biodiversity and landscape, Calle 16, No. 1 F-12, Puerto Colombia, Atlantico, Colombia
e-mail: juanitaldana@gmail.com

M.I. Vieira-Muñoz
Instituto de Investigación de Recursos Biologicos Alexander von Humboldt, Calle 45 No. 66a–28 (apto.402), Bogotá, DC, Colombia
e-mail: mivieira@humboldt.org.co

P. Bejarano
Conservación Internacional Colombia, Cra. 13 No. 71-41, Bogotá, DC, Colombia
e-mail: p.bejarano@conservation.org

J.R. Moreira et al. (eds.), *Capybara: Biology, Use and Conservation of an Exceptional Neotropical Species*, DOI 10.1007/978-1-4614-4000-0_19,

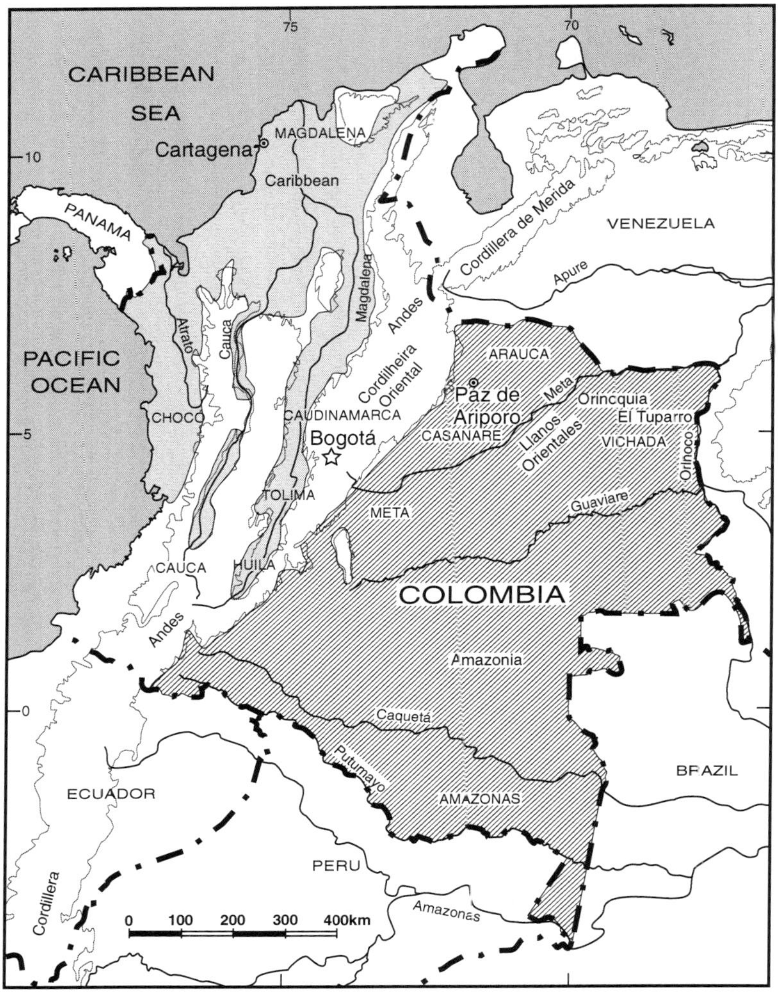

Fig. 19.1 Map of Colombia showing the distribution area of the lesser capybara (*gray area - Hydrochoerus isthmius*) and of the capybara (*dashed area - Hydrochoerus hydrochaeris*) and the places cited in the text

as there are very few, if any, studies of their population status and their current distribution in the country is partially unknown.

In this chapter, we initially present the advances in the knowledge of the capybaras and lesser capybaras, and analyze the studies of their wild populations. Next, we outline the ways in which the capybara and lesser capybara have been used in Colombia and discuss relevant legislative development. Finally, we reflect on opportunities for the conservation and management of both species in Colombia.

19.2 Current Knowledge About the Capybara and the Lesser Capybara in Colombia

To provide a general perspective on the current state of knowledge about the capybara and the lesser capybara in Colombia, a bibliographical survey of 15 of Colombia's libraries was conducted, along with an electronic mail survey of 11 of Colombia's main regional environmental authorities ("Corporaciones Autónomas Regionales," hereafter CARs). The bibliographical survey revealed a total of 150 documents over a 50-year period (Table 19.1). The three thematic areas with the greatest number of contributions were capybara uses, ecology, and captive breeding. The least-studied themes were reproduction, growth, age, and historical analysis (Table 19.1). In Colombia, the great majority of the studies (94%) focused on the capybara, and only a few mentioned the lesser capybara (6%).

Most of the answers (82%) came from authorities working in areas originally inhabited by the lesser capybara. In most regions, the capybara and the lesser capybara are reported as scarce species (Table 19.2). It was only in the Orinoquia region (which includes the subregions of Arauca, Casanare, and Vichada; Fig. 19.1) that the capybara was reported as abundant. Both species are hunted in almost all of the regions (Table 19.2). In the Andean and Amazonia regions, capybaras are hunted for subsistence; while in the Caribbean and Orinoquia regions commercial hunting is also reported. The main product used is its meat, which is mainly exported to Venezuela, with a small amount for local consumption; the current use of capybara leather is almost negligible (Table 19.2). The CARs have supported some studies on captive breeding and on the evaluation of the status of wild populations. In general,

Table 19.1 Thematic classification of the available bibliography on capybaras and lesser capybaras in Colombian libraries

Thematic area	Number of documents/ contributions	Percentage (%)
Historical analysis	1	1
Reproduction, growth and age	1	1
Systematics and genetics	3	3
Diet	7	6
Publicity articles	7	6
Anatomy, histology, and physiology	8	7
Children's stories	9	8
Parasites, diseases, and sanitary aspects	9	8
Behavior and social structure	12	10
General studies	14	12
Captive breeding	22	18
Uses: products, value, processing, and market	28	24
Ecology and management of natural populations	29	24
Total	*150*	

Table 19.2 Results of a survey of Colombia's main regional environmental authorities "Corporaciones Autónomas Regionales" (CARs) on the abundance and use of the capybara and lesser capybara in Colombia

Region	CAR	Species present	Common name	Abundance	Hunting purpose	Product
Amazonia	Corpoamazonia	Capybara	Chigüiro, yulo	Scarce	Subsistence	Meat
Andean	Corantioquia	Capybara and lesser capybara	Chigüiro (capybara), ponche, lancho(lesser capybara)	Medium	Subsistence	Meat
Andean	CVC	Capybara	Chigüiro	Scarce	Subsistence	Meat
Andean	Cornare	Not known	Yulo	Scarce	Not present	–
Andean	CAR	Not known	Chigüiro	Very scarce	Not known	–
Andean	CRQ	Not known	Chigüiro	Scarce	Subsistence	Meat
Andean	Cortolima	Not present	–	–	–	–
Caribbean	Corpomojana	Lesser capybara	Chigüiro, ponche, cacó	Scarce	Subsistence and commercial	Meat
Caribbean	CVS	Lesser capybara	Chigüiro, ponche, cacó	Scarce	Subsistence and commercial	Meat
Caribbean	Corpouraba	Lesser capybara	Lancho, cacó, ponche	–	Subsistence	Meat
Orinoquia	Corporinoquia	Capybara	Chigüiro, chigüire, pompo	Abundant	Commercial	Meat and leather

captive breeding initiatives have been unsuccessful, and the projects associated with them have lacked the continuity needed to identify and implement the required adjustments.

19.3 Studies on Wild Population Densities

Densities reported for the capybara in Colombia range from 0.1 to 7.1 ind/ha (Table 19.3). Such a broad variation could arise from the differences between the methods employed in each study; the extent and quality of the habitat suitable for the species in each region, and the historical hunting pressure.

Surveys carried out in the Arauca subregion reveal high densities of capybaras, but these are local estimates that do not represent the typical situation across the region (Aldana-Domínguez et al. 2002). Illegal hunting has decimated capybara populations in the subregion and, as a result, only small populations are found in private areas where hunting is not allowed.

The situation in the Casanare subregion is divided. In the Hato Corozal municipality, in the north, proximity to the Arauca subregion (Fig. 19.1) generates much illegal hunting, stimulated by the trade of capybara meat to Venezuela. In contrast, in the Paz de Ariporo municipality, where capybara populations abound in most areas, the owners of cattle ranches are interested in exploiting the capybara legally and, to this end, have favored population growth by implementing effective control of illegal hunting and building wells to increase the amount of water available for the capybara (Caro et al. 2005; Aldana-Domínguez and Ángel-Escobar 2007). There are also habitat differences between the two regions that likely contribute to

Table 19.3 Capybara population ecological densities in the Orinoquia region

Subregion	Locality	Ecological densities (individuals/hectare)	Date	Reference
Arauca	Caño Limón	0.3–1.7	1998–1999	Aldana-Domínguez et al. (2002)
Arauca	Laguna Venero	4.06–7.1	1976–1978	Jorgenson (1986)
Casanare	Hato Corozal	1.06	2002–2003	Bejarano et al. (2004)
Casanare	Hato Corozal	0.38–0.39	2004	Caro et al. (2005)
Casanare	Hato Corozal	0.11–0.14	2003	Aldana-Domínguez and Ángel-Escobar (2007)
Casanare	Orocué	1.38	2002–2003	Rodríguez et al. (2003)
Casanare	Paz de Ariporo	1.18	2002–2003	Rodríguez et al. (2003)
Casanare	Paz de Ariporo	2.64–2.86	2004	López et al. (2006)
Casanare	Paz de Ariporo	6.21–6.44	2005	López et al. (2006)
Casanare	Paz de Ariporo	4.80–5.93	2006	López et al. (2006)
Casanare	Paz de Ariporo	2.22–4.11	2003	Aldana-Domínguez and Ángel-Escobar (2007)

the differences in capybara densities (at least in part) (Fig. 19.1; Aldana-Domínguez and Ángel-Escobar 2007).

Very few population evaluations have been made for the lesser capybara. In the Caribbean region (Fig. 19.1), censuses were made along the banks of rivers and channels where the presence of the species had previously been reported (Ballesteros 2001). The resulting study reports densities of 16–0.1 individuals per lineal kilometer of river and average geographical densities of 0.3 ind/ha; however, the lesser capybara was not found in the majority of the areas searched. The decline in the lesser capybara populations of Colombia's Caribbean region has been going on for the past three decades and is the result of commercial and subsistence hunting and of the habitat destruction generated by agricultural encroachment (Ballesteros 2001).

19.4 Hunting of the Capybara and the Lesser Capybara in Colombia

In Colombia, wild populations of capybaras and lesser capybaras have been used by indigenous and peasant communities as food. Indigenous communities of the Amazonas, Chocó, Casanare, Vichada, and Meta subregions include capybara meat in their diet and trade it on a local scale (Fig. 19.1; Rodríguez and van der Hammen 2003; Tafur 2004; Castro and Peñuela 2006; Sánchez 2007).

In Colombia's Caribbean region, particularly in the "Canal del Dique" (near Cartagena; Fig. 19.1), lesser capybara populations have declined, mainly as a result of the disappearance of wetlands and significant hunting pressure (Medrano-Bitar no date). Local hunters consume 30% of their kill and sell the rest. In the 1980s, lesser capybaras were illegally hunted and traded on a massive scale. Smoked and fresh meat is currently sold in ports, marketplaces, and traditional restaurants. Prices range between $3,000 and $5,000 Colombian Pesos per kilogram (equivalent to US $1.5–2.55 in July 2010; Medrano-Bitar n.d.).

In Colombia's "Llanos Orientales" (seasonally flooded plains to the East of the country, bordering with Venezuela), the illegal hunting of capybaras is chiefly a result of the demand for dried and salted meat in Venezuela, the abundance of capybaras and the lack of control exerted by the authorities in Colombia. In addition to the demand from Venezuela, in Colombia there is also an illegal domestic market for capybara meat. In Bogotá alone, there are at least 15 restaurants selling "typical" food from the "Llanos Orientales" region that claim to sell capybara meat. The authorities, despite having full knowledge of this situation, maintain a permissive attitude, and fail to confiscate meat or exert any other control in the matter (López et al. 2002).

Illegal hunting, which takes place during the dry season (January, February, and March), has reduced wild populations and, in some cases, has even caused local extinctions (Hernández et al. 1983). The hunts are locally known as "chigüiranzas" when horsemen round up herds, and selected individuals are killed by a blow to the head, similar to the Venezuelan process (Herrera and Barreto 2012). The meat from the hunt is salted and dried in the sun for sale in the form of dried meat (*salones*), while the viscera, skin, and bones are abandoned in the savanna.

Hunting tends to be selective and adults are generally preferred over young. However, hunters tend not to select males, so adult females may be taken in proportion to their abundance. This may be deleterious to population birth rates and, as a result, render exploitation unsustainable (Payán 2007; but see Moreira et al. 2012b). The trafficking of capybara meat is a difficult problem to solve, given the large areas over which the species is distributed and the limited capacity of the authorities to control intricate trade routes. Currently, illegal hunting is the main threat to capybaras in Colombia.

Captive breeding trials for capybaras in confined production systems have not produced favorable results in Colombia. Mortality rates have been high and the benefits did not outweigh the costs, given the large economic investment that has to be made at the beginning of the project (Ramírez-Perilla 1992). However, there have been some successful experiences of captive breeding, where peasant farmers keep a limited number of animals in pens for local consumption (Coral 2004).

19.5 History of the Legislation for the Conservation and Use of the Species

The first resolution to regulate the exploitation of capybaras in Colombia was issued by the country's Ministry of Agriculture in 1964. Due to the rapid decline in wild capybara populations, hunting was banned throughout the country. Since this ban was ineffective, in 1969, Colombia's National Institute of Natural Resources (INDERENA) issued regulations for hunting capybaras in an attempt to protect the species. Sport hunting was banned, and commercial hunting was only allowed during the months of January, February, and March each year. The following year, the hunting season was closed due to a decline in some populations. In 1974, the National Code of Renewable Natural Resources and Protection of the Environment was introduced. In 1976, commercial hunting was allowed again, but only for 2 months and only in the Arauca and Casanare subregions. So several closed and open seasons for hunting capybaras succeeded each other for some years. At the same time, the government supported research on captive breeding in confined and semi-confined conditions (Ministerio de Agricultura 1980).

Growing interest in commercial exploitation of capybaras led the environmental authorities to formulate alternatives to the extraction of individuals from their natural habitat. To this end, in 1985, Colombia's Ministry of Agriculture and the INDERENA authorized wildlife specimens to be hunted for the establishment of breeding farms. The capybara captive breeding program was organized using both intensive and semi-extensive breeding schemes. The semi-extensive approach was only approved for the Orinoquia region and was not permitted in the Cauca, Magdalena, Cundinamarca, Huila, and Tolima subregions (Fig. 19.1), given the vulnerability of the wild populations in these areas. Under the semi-extensive scheme, commercial

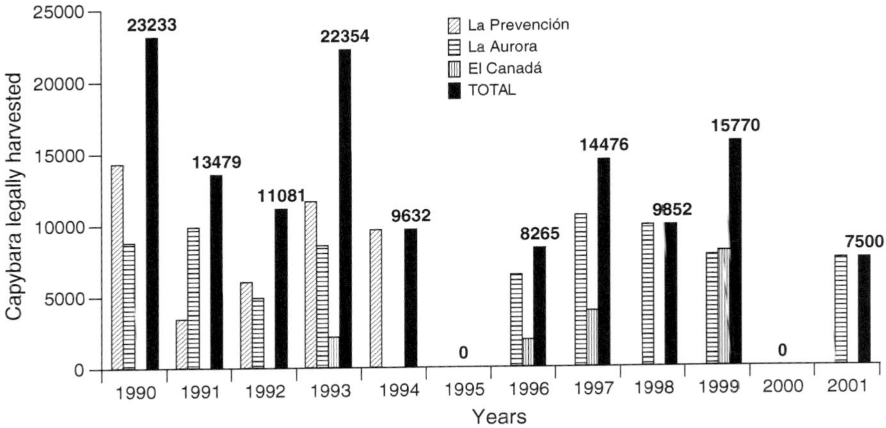

Fig. 19.2 Number of legally hunted animals in the "La Prevención," "La Aurora" (both in Casanare subregion), and "El Canadá" (in Arauca subregion) breeding ranches between 1990 and 2001 (From López et al. (2002))

exploitation of capybaras in their natural environment was also allowed. Between 1990 and 2001, 135,642 individuals were legally hunted and sold. These animals came from the Orinoquia region and, more specifically, from the breeding ranches "El Canada" (in the Arauca subregion), "La Aurora" and "La Prevención" (in the Casanare subregion). Most of the meat was exported to Venezuela (Fig. 19.2).

Breeding permits were issued only for *H. hydrochaeris* and only in the Orinoquia region. The harvest quota was calculated as a percentage (between 20% and 30%) of the total population, and was estimated by the environmental authorities using a visual encounter survey. However, the fact that a standard methodology was never defined made it impossible to assess the sustainability of the exploitation. Additionally, the lawfully established breeding farms served as a façade for "legalizing" meat that was obtained from illegal hunting.

Law 611 of the year 2000 was then issued to provide greater clarity in the management of wildlife. It established that the exploitation of wildlife could be done through direct harvesting in the natural environment or through captive breeding in open and/or closed cycles. In the open breeding approach ("ranching" sensu CITES), individuals of any age were periodically captured and transported to farms, where they were reared until they reached the stage of development appropriate for their final use. In contrast, closed cycle breeding ("farming" sensu CITES) involved obtaining parental stock (from wild populations or any other wildlife management system) and then breeding and raising in captivity the individuals to be exploited.

Despite the efforts made to legalize and control exploitation of capybaras in Colombia, illegal traffic continued and, in March 2000, an unauthorized breeding ranch illegally exported 100,000 kg of dried meat to Venezuela. This meat came from more than 10,000 capybaras from the wild populations of two municipalities in the Casanare department. The incident was reported by local citizens, and Colombia's

Council of State condemned the environmental authorities (Colombia's Ministry of Environment, Housing, and Territorial Development and the CAR from Orinoquia – Corporinoquia) for allowing the slaughter, and demanded that they implement the necessary measures to rectify the situation (Consejo de Estado, fifth session 2001).

Between 2002 and 2007, the authorities promoted several research projects on the ecology of the capybaras to generate a baseline database that could be used to formulate regulations for the conservation and sustainable use of the capybaras in the Orinoquia region (Rodríguez et al. 2003; Aldana-Domínguez et al. 2003, 2004; Caro et al. 2005; López et al. 2006; Montenegro et al. 2006). Some capybara populations increased in density between 2004 and 2005 and it was recommended that such populations be exploited under a legal regulatory framework. In contrast, other populations, which had suffered severe declines, required the implementation of recovery measures.

It was established that anyone wishing to hunt capybaras would have to obtain a license from the respective environmental authority, i.e., the CAR. For this, the CARs would have to define the total harvest quota, calculated using a simulation model and the information gathered from monitoring the wild populations to be exploited. A simulation of capybara population dynamics suggested that it is possible to harvest 20% of the population in a sustainable manner, but that the harvested animals should predominantly be males, with a maximum of 20% being adult females (Montenegro et al. 2006).

Based on the censuses of 2004 and 2005, a total harvest quota of 70,000 individuals was established for the Orinoquia region. However, there is still no agreement between the consumers and the environmental and public health authorities on the public health standards that the meat must fulfill to be considered approved for human consumption and legal export, neither are there agreements on the features that slaughterhouses must have: abattoir operators argue that the current sanitary requirements are impossible to fulfill (Claudia Rodríguez personal communication). Thus, all parties have so far failed to reach consensus on the legal exploitation of the capybara in the Orinoquia region.

19.6 Management Opportunities

So little is known about the lesser capybara, its distribution and abundance, that there is virtually no foundation on which to base mitigation of the impacts of its possible exploitation. The species' habitat, which includes swamps, wetlands, and other water bodies in Colombia's Caribbean region and its inter-Andean valleys, has been considerably transformed to make way for livestock, crops, and human settlements. These have put great pressure on the environment, including hunting, logging for firewood, and the pollution of water bodies (Ballesteros 2001). Additionally, the progress that has been made with the legislative and the consensus-building processes for sustainable exploitation has focused on the capybara, largely neglecting the lesser capybara.

The capybara is more abundant than the lesser capybara and there may be greater scope for its conservation. The populations of the rainforests of the Amazonian region (Fig. 19.1) are smaller than those in the savannas of the Orinoquia region. In the rainforest, capybara populations have declined around areas inhabited by humans and along the larger rivers, used as the main transport routes (Emmons 1997).

Sites where capybaras may be successfully preserved include Protected Natural Areas such as National Natural Parks, private reserves, indigenous reserves, and cattle ranches. In the Orinoquia region, there are few protected areas, encompassing only 2% of the total area. They include only the National Natural Park El Tuparro (548,000 ha) and 23 private reserves, which collectively cover 23,649 ha. The indigenous reserves in Orinoquia and Amazonia encompass 2,230 km^2 and these include remote areas in good conditions for conservation.

In the Colombian Orinoquia, capybaras have disappeared entirely from some cattle ranches, or persist at only low population densities due to loss of habitat and illegal trade. In response, Corporinoquia is currently developing the Regional Program for the Recovery and Conservation of the Capybara in the Colombian Orinoquia, which includes evaluating and monitoring of the populations and their habitats, as well as the sustainable use and recovery of these populations. This program also introduces environmental education and community participation as essential tools for the successful implementation of the initiatives. Finally, institutional management and strengthening will play an important role in defining the long-term continuity of the program's activities (López et al. 2006).

On the other hand, some cattle ranches retain large capybara populations that have the potential for sustainable harvesting. In the Casanare subregion, partnerships have been developing among the owners of large areas of land who are interested in complementing their extensive cattle husbandry with commercial exploitation of capybaras. Currently, there are five associations of capybara ranchers or farmers ("chigüireros"). Although a sector is emerging in the Casanare subregion to exploit capybaras commercially, it lacks competitive strength due to poor logistical and technical capacity in the exploitation and production processes and an inadequate organizational structure. Limited local capacity was also identified in aspects such as business planning, marketing, logistics, and administrative, productive, and executive management. This is mainly the result of the rural profile of the entrepreneurs and the limited investments that they are able to make (Lozada 2007).

Since 2002, in Colombia there has been an energetic development of regulations and institutional agreements to govern capybara harvests on private properties. It is recognized that this species can add value to the seasonally flooded savannas. This economic alternative takes on even greater importance in the face of land-use changes and the new projects that are being developed in the Orinoquia region.

Acknowledgments The authors thank the officers of the Corporaciones Autónomas Regionales (CARs) who participated in the survey: Luisa Fernanda Molina (CVS), Alberto Vivas (CORPOURABA), Derly Carrillo (CORTOLIMA), Dalila Camelo (CAR), Johana Talia Lugo (CORPOAMAZONIA), Mauricio Orozco Zapata (CORNARE), Diego Duque Montoya (CRQ), Arbei Osorio Restrepo and Juan Camilo Restrepo (CORANTIOQUIA), Emiro Cordero (CORPOMOJANA), Saulo Orduz (CORPORINOQUIA), and Joaquín Romero (CVC). They also express their gratitude to Claudia Rodríguez, from Colombia's Ministry of Environment, Housing and territorial development (MAVDT), for supplying valuable information for this chapter.

References

Aldana-Domínguez J, Ángel-Escobar DC (2007) Evaluación del tamaño y densidad de poblaciones silvestres de chigüiros en el departamento del Casanare. In: Aldana-Domínguez J, Vieira-Muñoz MI, Ángel-Escobar DC (eds) Estudios sobre la ecología del chigüiro (*Hydrochoerus hydrochaeris*), enfocados a su manejo y uso sostenible en Colombia. Instituto Alexander von Humboldt, Bogotá, pp 33–48

Aldana-Domínguez J, Forero-M J, Betancur J, Cavelier J (2002) Dinámica y estructura de la población de chigüiros (*Hydrochaeris hydrochaeris*: Rodentia, Hydrochaeridae) de Caño Limón, Arauca, Colombia. Caldasia 24(2):445–458

Aldana-Domínguez J, Ángel-Escobar DC, Vieira-Muñoz MI (2003) Evaluación del estado actual de las poblaciones silvestres de chigüiros (*Hydrochaeris hydrochaeris*) y los hábitat asociados en los municipios de Paz de Ariporo y Hato Corozal-departamento del Casanare. Fase I. Informe final. Instituto Alexander von Humboldt- Ministerio de Ambiente, Vivienda y Desarrollo Territorial, Bogotá

Aldana-Domínguez, J, Ángel-Escobar DC, Vieira-Muñoz MI, Payán CE, Lozada PA, Ramírez G, Castro G (2004) Conservación y uso sostenible del chigüiro (*Hydrochaeris hydrochaeris*) en el departamento del Casanare. Fase II. Informe final. Instituto Alexander von Humboldt – Ministerio de Ambiente, Vivienda y Desarrollo Territorial, Bogotá

Ballesteros J (2001) Estado de conservación del chigüiro o ponche (*Hydrochaeris hydrochaeris isthmius*) en el departamento de Córdoba, Colombia. M.Sc. Dissertation, Universidad Pontificia Javeriana, Bogotá

Bejarano P, Rodríguez M, Vélez N, Torrijos P (2004) Caracterización de las poblaciones silvestres de chigüiro (*H. hydrochaeris*) y sus hábitats en las sabanas anegables del Departamento de Casanare. In: Memorias encuentro latinoamericano sobre investigación y uso sostenible de *Hydrochoerus hydrochaeris* (chigüiro, capibara, carpincho). Yopal

Caro C, López HF, Sánchez P, Mesa E, Atuesta N, Lenis AR, Camargo A, Hernández Y (2005) Fortalecimiento al control y seguimiento a las poblaciones naturales de chigüiro para la implementación de proyectos de uso sostenible en Casanare- Municipios de Paz de Ariporo y Hato Corozal. Fase III. Informe final. UNILLANOS-Ministerio de Ambiente, Vivienda y Desarrollo Territorial, Villavicencio

Castro F, Peñuela L (2006) Caracterización de usos de la biodiversidad e identificación de opciones de manejo de recursos de la biodiversidad en el resguardo indígena Caño Mochuelo, Casanare. Informe de consultoría. Proyecto: Biodiversidad y Desarrollo en Ecorregiones Estratégicas de Colombia-Orinoquia. Fundación Horizonte verde, Instituto de Investigación de Recursos Biológicos Alexander von Humboldt, Bogotá

Coral AJ (2004) Zoocría de chigüiros en el patio. In: OIKOS, Corporinoquia (eds) Memorias del encuentro latinoamericano sobre investigación y aprovechamiento sostenible del chigüiro. Yopal, pp 56–58

de Agricultura M (1980) Proyecto chigüiro: Arauca. División de fauna terrestre, Bogotá, p 18

Emmons HL (1997) Neotropical rainforest mammals. The University of Chicago Press, Chicago, p 307

Hernández J, Rodríguez J, Pachón J (1983) Evaluación de poblaciones de chigüiro (*Hydrochaeris hydrochaeris*) en los hatos "Brasilia", "Guamito", "La Aurora", "La Borra", "El Danubio", "La Veremos" y "Mapurisa", municipio de Hato Corozal. INDERENA, Bogotá

Herrera EA, Barreto GR (2012) Capybaras as a source of protein: utilization and management in Venezuela. In: Moreira JR, Ferraz KMPMB, Herrera EA, Macdonald DW (eds) Capybara: biology, use and conservation of an exceptional neotropical species. Springer, New York, pp 305–319

Jorgenson JP (1986) Notes on the ecology and behavior of capybaras in northeastern Colombia. Vida Silvestre Neotrop 1(1):31–40

León CA (1974) Zoonimia y distribución geográfica del Chigüiro en Colombia. Primer simposio internacional sobre chigüiro y babilla. INDERENA, Santa Fe de Bogotá

López HF, Mesa E, Sánchez JH (2002) Implementación de la fase inicial del programa de manejo de chigüiro (*Hydrochaeris hydrochaeris*) en semicautiverio, como estrategia de conservación de la biodiversidad en la Orinoquia colombiana y acopio de información complementaria. Informe final. Universidad Nacional de Colombia – Ministerio de Ambiente, Vivienda y Desarrollo Territorio, Bogotá

López HF, Sánchez P, Montenegro O, Quiroga G, Mesa E, Atuesta N, Camargo AA, Guzmán AR, Álvarez O, Delgado R, Tovar D, Arenas R (2006) Conservación y uso sostenible del chigüiro (*Hydrochaeris hydrochaeris*) en la Orinoquia Colombiana. Fase IV. Informe final. Universidad Nacional de Colombia – CORPORINOQUIA, El Yopal

Lozada P (2007) Análisis del mercado nacional e internacional para productos derivados del chigüiro. In: Aldana-Domínguez J, Vieira-Muñoz MI, Ángel-Escobar DC (eds) Estudios sobre la ecología del chigüiro (*Hydrochoerus hydrochaeris*), enfocados a su manejo y uso sostenible en Colombia. Instituto Alexander von Humboldt, Bogotá, pp 147–173

Medrano-Bitar S (n. d.) La conservación del humedal o el ocaso del ponche. Corporación Autónoma del Canal del Dique. Cartagena

Montenegro O, Sánchez P, Mesa E, Atuesta N, Maldonado A, Álvarez O, Sarmiento C, Correa C, Bejarano P (2006) Conservación y uso sostenible del chigüiro (*Hydrochaeris hydrochaeris*) en la Orinoquia Colombiana. Fase V. Informe final. Universidad Nacional de Colombia – Ministerio de Ambiente, Vivienda y Desarrollo Territorio, Bogotá

Moreira JR, Alvarez MR, Tarifa T, Pacheco V, Taber A, Tirira DG, Herrera EA, Ferraz KMPMB, Aldana-Domínguez J, Macdonald DW (2012a) Taxonomy, natural history and distribution of the capybara. In: Moreira JR, Ferraz KMPMB, Herrera EA, Macdonald DW (eds) Capybara: biology, use and conservation of an exceptional neotropical species. Springer, New York, pp 3–37

Moreira JR, Verdade LM, Ferraz KMPMB, Macdonald DW (2012b) The sustainable management of capybaras. In: Moreira JR, Ferraz KMPMB, Herrera EA, Macdonald DW (eds) Capybara: biology, use and conservation of an exceptional neotropical species. Springer, New York, pp 283–302

Payán E (2007) Análisis de proporciones de edad y sexo a partir de cráneos resultantes de cosechas de chigüiros en los Llanos Orientales de Colombia, departamento del Casanare, para el uso y conservación de la especie. In: Aldana-Domínguez J, Vieira-Muñoz MI, Angel-Escobar DC (eds) Estudios sobre la ecología del chigüiro (*Hydrochoerus hydrochaeris*), enfocados a su manejo y uso sostenible en Colombia. Instituto Alexander von Humboldt, Bogotá, pp 87–103

Ramírez-Perilla J (1992) El chigüiro (*Hydrochaeris hydrochaeris*, RODENTIA): una alternativa pecuaria adicional para el neotrópico. Trabajo docente. Universidad Nacional de Colombia. Revista: ACOVEZ 11:89–94

Rodríguez CA, Van der Hammen MC (2003) Manejo indígena de la fauna en el medio y bajo río Caquetá (Amazonía colombiana). Tradición, transformaciones y desafíos para su uso sostenible. In: Polanco-Ochoa R (ed) Manejo de fauna silvestre en amazonia y Latinoamérica. Selección de trabajos V Congreso internacional. CITES, Fundación Natura, Bogotá, pp 325–338

Rodríguez M, Vélez N, Bejarano P, Torrijos P, Burgos N, Ladino E, Barrera N, González R (2003) Caracterización de las poblaciones silvestres de chigüiro y sus habitats en las sabanas anegables del departamento de Casanare con miras a formular una propuesta de conservación y uso sostenible. Informe final. Grupo de Estudios Ecológicos OIKOS – Gobernación de Casanare, Bogotá

Sánchez LF (2007) Caracterización de los grupos humanos rurales de la cuenca hidrográfica del Orinoco en Colombia. Instituto de Investigación de Recursos Biológicos Alexander von Humboldt, Bogotá

Tafur MP (2004) Acercamiento a la percepción sobre el chigüiro (*Hydrochaeris hydrochaeris isthmius*), hábitat, cacería, dieta y caracteres morfológicos básicos de esta sub-especie en la comunidad Isla de los Palacios, en el complejo humedal "Los Platillos" Medio Atrato, Antioquia – Chocó, Colombia. Graduate dissertation, Pontificia Universidad Javeriana, Bogotá

Torres E, Sanabria J (1976) Aspectos ecológicos del chigüiro y establecimiento de un zoocriadero. Graduation dissertation, Universidad Nacional de Colombia. Santa Fe de Bogotá

Chapter 20
Capybara Production in Brazil: Captive Breeding or Sustainable Management?

José Roberto Moreira and Max S. Pinheiro

20.1 Introduction

When the Portuguese arrived in Brazil in the year 1500 they were startled by what they found. The land was well populated with healthy indigenous people but without a single specimen of the domestic breeds they were used to eating in Europe. In his famous letter to the King of Portugal, Pero Vaz de Caminha (1999), one of the captains of the fleet that "discovered" Brazil, reported what he had just seen in the new-found land:

> They neither reap nor farm. Nor are there here bulls or cows, goats, sheep or chickens, or any other animal which is used to man's life. ... And with this they walk so wiry and glowing with health, much more than us, no matter how much wheat or legumes we eat.

The indigenous people of Brazil acquired their main source of protein from hunting and fishing; they had domesticated only one species – the muscovy duck (*Cairina moschata*; Moreira et al. 2008).

Before any Europeans arrived in South America, capybaras were eaten and sometimes raised by the indigenous people. Archaeological studies show that capybaras were an important food item in the diet of those from Santa Catarina Island (State of Santa Catarina), in the South of Brazil (Fig. 20.1), in precolonial times (Castilho and Simões-Lopes 2005). In a letter sent to Portugal in 1560 by the Jesuit Father Anchieta José de (1933), he writes that capybaras are "...raised at home like dogs: go out to forage and come back home on their own."

J.R. Moreira (✉)
Embrapa Recursos Genéticos e Biotecnologia, Parque Estação Biológica,
Final W5 Norte, 70770-900 Brasília, Distrito Federal, Brazil
e-mail: alencar_moreira@yahoo.com.br

M.S. Pinheiro
Embrapa Clima Temperado, Rodovia BR392 km78, 403, 96010-971
Pelotas, Rio Grande do Sul, Brazil
e-mail: max.pinheiro@embrapa.br

J.R. Moreira et al. (eds.), *Capybara: Biology, Use and Conservation of an Exceptional Neotropical Species*, DOI 10.1007/978-1-4614-4000-0_20,

Fig. 20.1 Map of Brazil with the locations of the places referred to in the text

Until the 1960s, capybaras were profitably hunted for their leather, meat, and oil throughout Brazil. However, there was no control of capybara hunting and local populations were exploited almost to extinction (Moreira and Macdonald 1996). In 1967, wildlife hunting was banned in Brazil, and the new law permitted only products from captive-bred animals to be traded. Since then, capybara production in Brazil has been the subject of a debate between those who advocate sustainable management, and others who believe the solution lies in captive breeding. The two authors of this chapter belong to opposite sides of this debate, and join here in synthesizing the arguments. The objectives of this chapter are to describe capybara use in Brazil, show the positive and negative sides of the policies and legislation in force, identify what we can learn from the past, and pinpoint the opportunities we foresee for the future of capybara use in this country.

20.2 The History of Capybara Use in Brazil

From the time of the European colonization, capybaras were indiscriminately hunted in Brazil for meat or leather until 1967, when hunting was prohibited. Leather was the main motivation for capybara hunting in the south of Brazil, while in the north it was both meat and leather (Moreira and Macdonald 1996). From 1960 to 1967, the only period when hunting production was recorded, Brazil officially produced 1,546,696 capybara hides (Caça 1963–70), but the real figure is certainly much higher. From 1967 onward, Decree 5197 prohibited the utilization, persecution, destruction, hunting, and capture of wildlife in Brazil (Presidência da República 1967). Nevertheless, commercial captive exploitation of this resource by farmers remained legal. Today, even though hunting is prohibited, salted capybara meat is still found in village markets in Brazilian Amazonia, where it is sometimes called "vaquinha" (little cow; Moreira and Macdonald 1996).

Captive breeding of capybaras begun in Brazil only in response to the ban on exploiting wild populations (Nogueira-Filho 1996). Before the ban, traditional rural affairs magazines like *Chácaras e Quintais*, *Caça e Pesca*, and the Agricultural Supplement of the newspaper *O Estado de São Paulo* (Nogueira-Neto 1973) occasionally included information about rearing capybaras in captivity. Even after the ban in 1967, very few attempts to raise capybaras in captivity were recorded before the 1980s.

Research on capybara captive rearing started in Brazil in the 1980s with the creation of the (no longer extant) Interdepartmental Center for Wildlife Zootechny and Biology (CIZBAS; Fig. 20.2), at the University of São Paulo, in Piracicaba in

Fig. 20.2 Capybaras raised in a confined production system (Photo by K.M.P.M.B. Ferraz)

Fig. 20.3 Capybaras raised in a semi-confined production system (Photo by M.S. Pinheiro)

the State of São Paulo (Fig. 20.1; Lavorenti 1989), and with a semi-confined production experiment in the Pantanal Mato-grossense (State of Mato Grosso do Sul; Fig. 20.1) run by Embrapa, the Brazilian Agricultural Research Corporation (Alho 1986). Later, other captive rearing research centers were created in different regions of the country, but most are no longer active today. With wildlife exploitation in the wild forbidden, captive rearing flourished in Brazil. The reproductive potential of capybaras (Moreira et al. 2012) was widely publicized and curiosity increased demand for capybara meat in large cities, especially in São Paulo, the largest and economically most important city in Brazil. At this time, capybara meat commanded a high price – US $20.00/kg (Silva-Neto et al. 1996) – and many farmers invested in rearing them. At first, capybaras were reared in captivity in a confined system which proved to be both costly and poorly suited to the species' behavior; this was then replaced by a semi-confined system (Fig. 20.3; Pinheiro et al. 2005; Nogueira-Filho et al. 2012). The meat was initially sold in grill-house restaurants, and in specialist game butchers and supermarkets, and the market grew. Farms raising capybaras in captivity spread throughout the country, and by the end of the 1990s there were over 200 commercial capybara breeders registered at IBAMA, the Brazilian Ministry of Environment agency (Nogueira-Filho and Nogueira 2004). Today, the capybara is the native mammal most bred in captivity in Brazil.

Nevertheless, the only farmers who profited from this new market were the early entrants who were well placed to sell valuable breeding stock and to charge for their knowledge. Consultants providing advice on rearing capybaras in captivity, and obtaining the permits required to breed the species, were able to charge over US $4,000 for their services. However, knowledge of capybara captive rearing at the time was limited and the importance of their social system not fully appreciated (Nogueira-Filho et al. 2012); as a result, production was low. Since capybara farmers were predominantly smallholders, they were dependent on middlemen who took a large cut, so only a few producers managed to breed capybaras and make a profit.

By the end of the 1990s, early curiosity about capybara meat had dwindled, leading to a fall in demand. To make matters worse, in 2005, a capybara-related health scare arose, particularly in the state of São Paulo, following six human deaths from Brazilian spotted fever, a disease that capybaras may have a role in spreading (Labruna 2009, 2012). The market for capybara meat shrank although the meat does not transmit the disease. The number of capybara farms has fallen dramatically since the mid 1990s.

Meanwhile, wild capybara populations have increased in some areas of Brazil, partly due to the dramatic expansion of agricultural land since the early 1990s, which has provided a new and very suitable habitat for capybaras: farmers have replaced gallery forest with reservoirs and sugar-cane, soybean, and maize plantations; they have eliminated large predators and curtailed poaching (Nogueira-Filho and Nogueira 2004; Ferraz et al. 2007). Human-capybara conflict is now common (Box 20.1), especially in highly populated areas in the Southeast and South of Brazil, and is exacerbated by fears of Brazilian spotted fever. At the same time, wildlife conservationism has attracted new followers in large cities, with an increase in the number of people against hunting.

Box 20.1 Capybara-Human Conflicts in Brazil

Damage control is an increasingly important issue in some capybara populations, especially in southeastern Brazil, because of expanding human occupation and intensified land-use practices. Concurrent with this growing need to reduce capybara-people conflicts, public attitudes, and environmental regulations are restricting some of the traditional ways to control populations, such as hunting.

Capybaras use agricultural lands, which often results in conflict, with the species being viewed negatively by producers because of crop damage issues. Although we know that capybaras eat corn, sugarcane, rice, and soybean (Bilenca and Kravetz 1995; Ferraz et al. 2003), agroindustry does not consider them as important as insects for pest control (e.g., Gallo et al. 1978).

(continued)

Box 20.1 (continued)

However, in small plantations, capybaras can destroy up to one-fourth of the plants (Ferraz et al. 2003), which can be significant for the smallholders involved. It is noticeable, nevertheless, that even in small plantations, the damage caused by capybaras is concentrated along the edges of the field, especially close to the forest habitat, which suggests that the species actively avoids open areas in such conditions (Ferraz et al. 2003).

Capybaras have recently been considered a hazard for public health in southeastern Brazil, because they host ticks that can transmit the Brazilian spotted fever caused by *Ricketsia ricketsii* (Folha de São Paulo 2002; Labruna et al. 2004; Labruna 2012). Six people died between 2002 and 2006, presumably of Brazilian spotted fever, in the municipality of Piracicaba, in the State of São Paulo. Although the outbreak was geographically restricted, the press took up the story, and capybaras were scapegoated (Gazeta de Piracicaba 2005a, b, c). The result was that today people all over Brazil are afraid that capybaras may be carrying and transmitting the disease. This became a threat to capybaras because people want to restrict their occurrence in areas where they share the space with the species.

With human population increases in urban areas across southeastern Brazil and the growing fragmentation of the environment, management of pest capybaras is bound to increase in the future and spread to other regions. Pest capybaras in built-up areas have important social, economic, and environmental impacts. Capybaras start invading gardens, eat ornamental plants, drown in swimming pools, attack dogs, contaminate the garden with droppings and ticks, cause road accidents (Moreira et al. 2001), and in some rare cases attack people (Rechenberg et al. 2000). This type of conflict in urban areas inevitably affects a large number of people, which makes it difficult to obtain agreement on acceptable forms of pest management.

20.3 History of Legislation on Wildlife Use

Prior to the Brazilian wildlife protection law – Decree n° 5197 – there had been no regulation of wildlife use in Brazil. Before 1967, according to environmental law n° 5894 of 1943 (Câmara dos Deputados 1943), ownership of wildlife did not exist. Hunting and trading wildlife products were not controlled, and anybody could capture a wild animal and keep it in captivity. With Decree n° 5197, wildlife became the property of the State and only animals bred in captivity (or their products) could be sold (Presidência da República 1967).

In the late 1980s and early 1990s, as the number of capybara farms rose and control issues emerged, IBAMA was forced to implement more specific regulations for wildlife production. This came with the publication of Ordinance n° 118

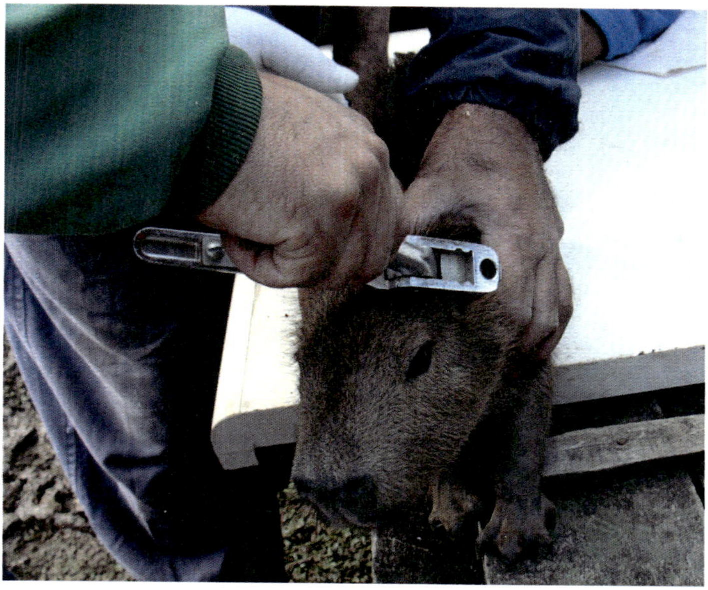

Fig. 20.4 Young capybara being identified by an ear-mark (Photo by J.R. Moreira)

of 1997 (IBAMA 1997b), which regulates commercial exploitation of captive wildlife. Among several other demands, a farmer now has to provide IBAMA with a detailed project plan under the auspices of a technician, all the animals need to be marked for identification (Fig. 20.4), and an annual report of all the farm's activities has to be presented (IBAMA 1997b). Only animals born in captivity can be used commercially.

At the same time, Ordinance n° 117 was published to regulate trade in live Brazilian wildlife or its products (IBAMA 1997a). In addition to farmers, anyone who intends to kill or sell wildlife, or use its parts, has to be registered at IBAMA. One of the requirements of this Ordinance is that wildlife products be identified so that they can be traced back to their producer, thus discouraging illegal trade. The bureaucratic burden and associated taxes in fact discourage farmers, abattoirs, and traders from registering with IBAMA, and this has restricted the uptake of capybara breeding and the expansion of the wild meat market in Brazil.

20.4 Capybara Production in Brazil Now and in the Future

Despite this heavy burden of bureaucracy created as an effort to control illegal trade, IBAMA has no control or oversight of capybara captive production in Brazil. Simple statistics like the number of capybara breeding farms, number of abattoirs which can slaughter capybaras, number of capybara captive individuals, or number of

Fig. 20.5 Capybaras crossing a busy road (Photo by Denilson Pinto)

capybaras commercially killed per year are completely unknown by the environmental authorities (asked in September 2011). This is the only reason why these numbers are unfortunately not shown here.

From talking to capybara farmers, the authors can state that most capybara production in Brazil takes place in a low number of small enterprises (an average of 15 female breeders per farm). Only a few farmers are really producing commercially. Many farmers have just a few animals to ensure they can occasionally invite friends over for a capybara barbecue or to embellish their landscape. Nevertheless, there is a market, albeit small, for capybara meat in Brazil and abroad, but this market is not supported by a network of large distributors (Moreira 2004). The number and size of capybara farms do not meet the commercial requirements of quantity and reliability of supply that would make such a distribution network viable, and this limits the development of individual enterprises. Often, the producer must rely on a fixed market, such as a chain of restaurants, supermarkets, or a specialist game butcher. Additional problems are due to the lack of specialist abattoirs (Moreira 2004).

With the increase in wild capybara populations and the scare of contamination by Brazilian spotted fever, IBAMA has constantly received complaints about the presence of the species on properties. Capybaras are viewed as competing with cattle, devastating agriculture, infesting landscapes with contaminated ticks, destroying gardens, and being dangerous to road traffic (Fig. 20.5) etc. There is a case for controlling capybara populations in these areas. Legally, the animals could be transferred to a capybara farm or the area could be fenced to create a new capybara farm. Obviously, this also raises the question of why the law does not allow culled individuals to be used.

20.4.1 In Defense of Capybara Management

With the diversity of habitats, landscapes, cultures, and economic wealth found in Brazil, capybara production must be adapted to this variety. Captive rearing may be suitable for areas close to large cities, where there is a big market for capybara meat and feedstuffs are cheaper. In natural flooded savannas like the Pantanal Matogrossense, however, the best choice would be sustainable use in the wild. This is also the best choice where capybaras are pests, as a control management.

The captive rearing of wildlife provides farmers with no incentive to protect the natural environment and natural populations of capybaras, and makes no contribution to wildlife conservation. Where capybaras are an agricultural pest and farmers have to pay for their control, they are likely to illegally exterminate them on their properties. On the other hand, if farmers were allowed to control capybara population size (by culling) and were to profit from it, they would better conserve the population and the environment where capybaras live. This would depend, obviously, on proper management regulations, adequate evaluation of the harvest rates, and a strict control of the market.

20.4.2 In Defense of Captive Rearing

With current Brazilian public opinion against hunting it is unlikely that the law will be changed to permit wildlife management. The ban on hunting is necessary so that captive rearing can flourish. But for this to happen, it is also necessary to have governmental support such as technical advice and tax exemption.

Farmed capybaras produce higher quality meat than do wild ones (Nogueira-Filho 1996; Pinheiro et al. 2007), because the slaughter process respects all the appropriate public health standards (Fig. 20.6), and the meat has a better flavor and smell and better quality fat. Leather from farmed capybaras is of better quality and commands a higher price.

Within an integrated farming system, capybaras can be raised in parts of the farm unsuitable for other stock, adding to the potential profitability of capybara farming (Pinheiro et al. 2009). If capybara captive rearing is suitably managed, it can be economically profitable (Pinheiro et al. 2009) and may help the species' conservation. Where capybaras are pests, the area can be transformed into a captive farm or the animals can be transferred to a captive farm, thus controlling the population.

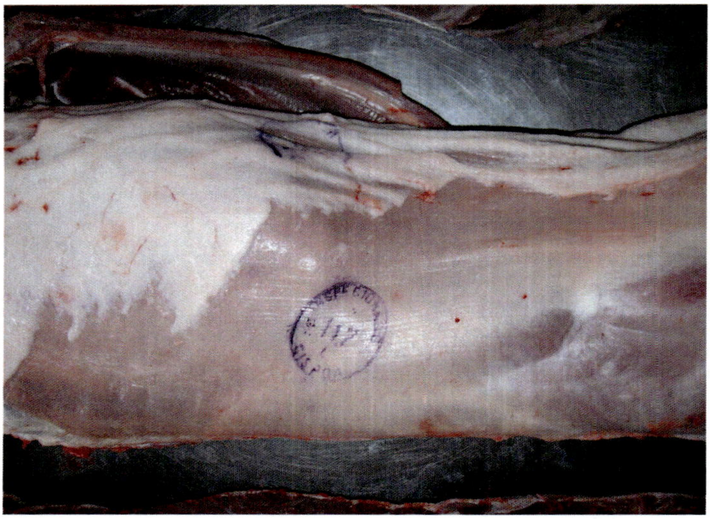

Fig. 20.6 Capybara meat submitted to state sanitary inspection (Photo by M.S. Pinheiro)

20.5 Final Remarks

The Brazilian ban on hunting in 1967 came after strong international pressure for the conservation of wildlife in Brazil. The choice of a total ban, however, arose because the governing body had no adequate system to monitor or regulate the sustainable use of wildlife. Even today, although laws exist to regulate the commercial use of wildlife, control of wildlife farms and traders is minimal. Meanwhile, concern for wildlife conservation and animal welfare has developed, especially in the urban areas of Southeast and South of Brazil. This culminated in the embargo, in 2008, on sport hunting in the State of Rio Grande do Sul (Fig. 20.1). This was the only state in Brazil where sport hunting had been legal and controlled.

A combination of numerous laws, minimal inspection and inadequate enforcement has created conditions that discourage farmers from stocking capybaras. Is this what wildlife needs in Brazil? Are farmers being encouraged to protect natural habitats and the native populations? Are they encouraged to breed them in captivity and sell their products? Conservation should be aimed at populations, species, and ecosystems, rather than at the salvation of individuals. It is time to choose between wildlife conservation policies that are easier and those that are necessary. The future existence and well-being of our natural heritage will depend on our present decisions.

Acknowledgments JRM thanks the Fundação de Apoio à Pesquisa Família Alencar (Brazil) for its support and S. Casement for editorial suggestions. MSP is grateful to Embrapa, who provided and financed his research on capybaras, and to partners at the Universidade Federal de Pelotas. The authors are grateful for the information provided by S.L.G. Nogueira Filho, S.S.C. Nogueira, L.M. Verdade, and K.M.P.M.B. Ferraz and by farmers, hunters, and IBAMA staff.

References

Alho CJR (1986) Criação e manejo de capivaras em pequenas propriedades rurais. Embrapa-DDT, Brasília

Anchieta J de (1933) Cartas, Informações, Fragmentos Históricos e Sermões. Civilização Brasileira, Rio de Janeiro

Bilenca DN, Kravetz FO (1995) Daños a maiz por roedores en al región pampeana (Argentina), y un plan para su control. Vida Silvestre Neotrop 4(1):51–57

Caça (1963–1970) Produção de peles e couros de alguns animais silvestres. Anuário Estatístico do IBGE, 24:48–51; 25:58–61; 26:80–83; 27:90–93; 28:76–78; 29:118–120; 30:128–130; 31:102–104

Câmara Dos Deputados (1943) Decreto-Lei n° 5.894, de 20 de Outubro de 1943. Câmara dos Deputados. Available from http://www2.camara.gov.br/legin/fed/declei/1940-1949/decreto-lei-5894-20-outubro-1943-415862-publicacao-1-pe.html. Accessed 27 Jan 2011

Caminha PV (1999) Carta de Pero Vaz de Caminha. In: Pereira PR (org). Os três únicos testemunhos do descobrimento do Brasil. Nova Aguilar, Rio de Janeiro, pp 31–59

Castilho PV, Simões-Lopes PC (2005) A capivara, *Hidrochoerus hydrochaeris* (Mammalia: Rodentia), no sítio arqueológico SC PRV 02, Ilha de Santa Catarina – Santa Catarina. Biotemas 18(2):203–218

Gazeta de Piracicaba (2005a) Rumo à cidade. Piracicaba, 24 Feb 2005

Gazeta de Piracicaba (2005b) Dossiê Capivara. Piracicaba, 4 Sept 2005

Gazeta de Piracicaba (2005c) Capivara em xeque. Piracicaba, 15 Sept 2005

Folha de São Paulo (2002) Morte por maculosa põe a ESALQ em alerta. São Paulo, 21 Sept 2002

Ferraz KPMB, Lechevalier M-A, Couto HTZ, Verdade LM (2003) Damage caused by capybara (*Hydrochaeris hydrochaeris*) on a cornfield in São Paulo, Brazil. Sci Agr 60(1):191–194

Ferraz KMPMB, Ferraz SFB, Moreira JR, Couto HTZ, Verdade LM (2007) Capybara (*Hydrochoerus hydrochaeris*) distribution in agroecosystems: a cross-scale habitat analysis. J Biogeogr 34:223–230

Gallo D, Nakano O, Silveira-Neto S, Carvalho RPL, Batista GC, Berti-Filho E, Parra JRP, Zucchi RA, Alves SB (1978) Manual de Entomologia Agrícola. Editora Agronômica Ceres, São Paulo

IBAMA (1997a) Portaria n° 117-N, de 15 de Outubro de 1997. Instituto Brasileiro do Meio Ambiente e dos Recursos Naturais Renováveis. Available from http://www.ibama.gov.br/fauna-silvestre/wp-content/files/port_117_97.pdf. Accessed 27 Jan 2011

IBAMA (1997b) Portaria n° 118-N, de 15 de Outubro de 1997. Instituto Brasileiro do Meio Ambiente e dos Recursos Naturais Renováveis. Available from http://www.ibama.gov.br/fauna-silvestre/wp-content/files/port_118_97.pdf. Accessed 27 Jan 2011

Labruna MB (2009) Ecology of *Rickettsia* in South America. Ann N Y Acad Sci 1166:156–166

Labruna MB (2012) Brazilian spotted fever: the role of capybaras. In: Moreira JR, Ferraz KMPMB, Herrera EA, Macdonald DW (eds) Capybara: biology, use and conservation of an exceptional neotropical species. Springer, New York, pp 371–383

Labruna MB, Whitworth T, Horta MC, Bouyer D, Mcbride J, Pinter A, Popoiv V, Gennari SM, Walker DH (2004) Rickettsia species infecting Amblyomma cooperi ticks from an area in the state of São Paulo, Brazil, where Brazilian spotted fever is endemic. J Clin Microbiol 42(10):90–98

Lavorenti A (1989) Domestication and potencial for genetic improvement of capybara. Rev Bras Genét 3(12):137–144

Moreira JR (2004) Avaliação do atual estado de uso e da conservação da capivara no Brasil. In: XLI Reunião Anual da Sociedade Brasileira de Zootecnia. Embrapa Gado de Corte, Campo Grande, pp 517–527

Moreira JR, Macdonald DW (1996) Capybara use and conservation in South America. In: Taylor VJ, Dunstone N (eds) The exploitation of mammal populations. Chapman & Hall, London, pp 88–101

Moreira JR, Pinha PRS, Cunha EHJ (2001) Capivaras do Lago Paranoá. In: Fonseca FO (ed) Olhares sobre o Lago Paranoá. Secretaria de Meio Ambiente e Recursos Hídricos, Brasília, pp 141–147

Moreira JR, de S Aguiar LM, de Camargo AJA (2008) Aptidão da fauna do Cerrado para o uso sustentável. In: Parron LM, de S Aguiar LM, Duboc E, Oliveira-Filho EC, de Camargo AJA, de G Aquino F (eds) Cerrado: desafios e oportunidades para o desenvolvimento sustentável. Embrapa Cerrados, Planaltina, pp 165–192

Moreira JR, Wiederhecker H, Ferraz KMPMB, Aldana-Domínguez J, Verdade LM, Macdonald DW (2012) Capybara demographic traits. In: Moreira JR, Ferraz KMPMB, Herrera EA, Macdonald DW (eds) Capybara: biology, use and conservation of an exceptional neotropical species. Springer, New York, pp 147–167

Nogueira-Filho SLG, Pinheiro MS, Nogueira SSC (2012) Confined and semi-confined production systems for capybaras. In: Moreira JR, Ferraz KMPMB, Herrera EA, Macdonald DW (eds) Capybara: biology, use and conservation of an exceptional neotropical species. Springer, New York, pp 229–241

Nogueira-Neto P (1973) Ordem Rodentia. In: A criação de animais indígenas vertebrados: peixes, anfíbios, répteis, aves e mamíferos. Edições Tecnapis, São Paulo, pp 257–273

Nogueira-Filho SLG (1996) Criação de Capivaras (ed). Centro de Produções Técnicas (CPT). Viçosa

Nogueira-Filho SLG, Nogueira SSC (2004) Captive breeding program as an alternative for wildlife conservation in Brazil. In: Kirsten S, Fragoso JMV, Bodmer R (eds) People in nature: wildlife management and conservation in Latin America. Columbia University Press, New York, pp 171–190

Pinheiro MS, Silva JJCDA, Rodrigues RC (2005) Sistemas de criação de capivara. Embrapa Clima Temperado. Documentos, 152. Available from http://www.cpact.embrapa.br/publicacoes/download/documentos/documento_152.pdf. Accessed 29 Apr 2008

Pinheiro MS, Garcia CAN, Pouey JLOF, Rodrigues RC, Ruas JL, Silva JJCDA, Sinkoc AL, Wendt LW (2007) Desempenho e manejo de capivaras em sistema semi-intensivo na região costeira sul do rio grande do sul. Embrapa Clima Temperado. Boletim de Pesquisa e Desenvolvimento, 47. Available from http://www.cpact.embrapa.br/publicacoes/download/boletins/boletim_47.pdf. Accessed 05 June 2008

Pinheiro MS, Azambuja IHV, Garcia CAN (2009) Resultado preliminar de avaliação econômica de criação de capivaras (*Hydrochaeris hydrochaeris*) em sistema semi-intensivo. Anais do 14° congresso brasileiro de zootecnia. Águas de Lindóia. Available from http://www.abz.org.br/publicacoes-tecnicas/anais-zootec/artigos-cientificos/pequenos-animais-domesticos-silvestres/21018-resultado-preliminar-avaliao-econmica-criao-capivara-hydrochaeris-hydrochaeris-sistema-semi-intensivo.html. Accessed 03 Feb 2011

Presidência da República (1967) Lei n° 5.197, de 3 de Janeiro de 1967. Presidência da República Federativa do Brasil. Available from http://www.planalto.gov.br/ccivil_03/Leis/L5197.htm. Accessed 27 Jan 2011

Rechenberg E, Bertelli PW, Oechsler A, Debatin ET (2000) Ocorrência de três ataques de capivara (*Hydrochaeris hydrochaeris*) à espécie humana no município de Blumenau (SC). In: Resumos do Congresso Brasileiro de Zoologia, vol 23. Universidade Federal de Mato Grosso, Cuiabá, p 639, (Resumo MA228)

Silva-Neto PB, Bortolanza D, Andrade M (1996) Como criar capivara. Agrodata Datavídeo, Curitiba, VHS, 50 min., color, VDS NTSC

Chapter 21
Use of Capybaras in Argentina

Rubén D. Quintana and María Luisa Bolkovic

21.1 Introduction

Capybaras (*Hydrochoerus hydrochaeris*) have been used in Argentina since pre-Columbian times as a source of hide, meat, and fat. Hide is the most valuable capybara by-product and, at present, represents the major hunting pressure for this species in the country. Until recently it has been among the most frequently consumed wild terrestrial vertebrates after the coypu (*Myocastor coypus*) and the tegu lizard (*Tupinambis* spp.), representing an important live natural resource due to the social and economic impact of its use in all provinces with wild populations of this rodent. In particular, products made from the manufacture of leather goods feature within the "typical regional products", which are sought by both local buyers and foreign tourists. Furthermore, due to its high abundance in many localities within its distribution range, this species also has an important ecological role in wetlands associated with large river basins in the northeastern region of the country. This chapter deals with the ecological and socioeconomic importance of capybaras in Argentina, providing an overview of the capybara's status, its historical use, present legislation, and associated trade issues. Finally it analyzes capybara production systems, progress in wild population management, and future needs for administration of this natural resource.

R.D. Quintana (✉)
Instituto de Investigación e Ingeniería Ambiental (3iA), Universidad Nacional de San Martín, Peatonal Belgrano 3563, B1650ANN San Martín, Pcia. de Buenos Aires, Argentina
e-mail: rubenq@ege.fcen.uba.ar; rquintana@unsam.edu.ar

M.L. Bolkovic
Dirección de Fauna Silvestre, Secretaría de Ambiente y Desarrollo Sustentable de la Nación, San Martín 451, C1004AAI Buenos Aires, Argentina
e-mail: mluisa_bolkovic@ambiente.gob.ar

J.R. Moreira et al. (eds.), *Capybara: Biology, Use and Conservation of an Exceptional Neotropical Species*, DOI 10.1007/978-1-4614-4000-0_21,

21.2 Current Status

The southernmost section of the capybara geographical distribution extends from northern to central-eastern Argentina, to the south of the Buenos Aires province (Moreira et al. 2012). As the species is closely associated with water, it is found in the major wetland systems of Argentina within the del Plata Basin, particularly those of large rivers such as the Paraná, Uruguay, Paraguay, Pilcomayo, and Bermejo (Fig. 21.1; Kandus et al. 2008). This predominantly tropical and subtropical species finds its southern distribution limit at about 38°S, probably due to climatic constraints, while its occurrence to the west is restricted by lower temperatures and water shortages in arid and high-altitude regions. On this basis, the potential distribution of capybaras in Argentina was estimated to cover 450,000 km^2, 40% of which qualifies as optimal or good habitat (Fig. 21.1; Adámoli et al. 1988).

In Argentina, changes in land use during recent decades may have affected the present population status of the capybara. Although the agricultural frontier has expanded mainly into native forest areas, it has also affected other ecosystems. Many former cattle grazing areas are currently under crops, mainly soybean which currently enjoys high international prices. As a result, traditional cattle production has been displaced to marginal lands, including wetland habitats where capybara densities have historically been high. Moreover, extensive areas of wetland are being drained for agriculture or replaced by rice crops or forestry. At a national level, these disturbances, together with mining, urbanization, poaching, and organized hunting have led to a generalized biodiversity crisis largely due to the loss of habitat for many wildlife species. Capybaras are still relatively abundant in this country, and they remain very abundant in some areas such as wetlands in the province of Corrientes (Fig. 21.1). In water bodies and wetlands to the south of this province, Quintana and Rabinovich (1993) estimated densities between 1 and 52 capybaras per km of waterside (0.35 capybaras/ha, range 0.01–0.61), depending on habitat and hunting pressure. However, their present status is very different from that in colonial times. Chronicles written in the eighteenth century by naturalists like Father Pauke (UNT 1943) or Félix de Azara (1802) report the occurrence of capybaras over most of the wetlands associated with rivers and shallow lakes of the de la Plata Basin and other freshwater systems. There are still few data on capybara abundance over most of their distributional range in Argentina, but it is clear that some populations have declined dramatically or even disappeared near urbanized areas, especially when leather demand increases.

21.3 Historical and Present Use

Today, capybara hunting represents a traditional activity, usually practiced by residents of towns in the proximity of capybara habitats. In contrast to Venezuela, Colombia and Brazil, where capybaras are killed mostly for their meat (González-Jiménez 1995; Giraldo and Ramírez 2001; Silva Neto 2005), in Argentina this product

Fig. 21.1 Map of Argentina with the locations of the places referred to in the text. Also shown is the potential habitat suitability for the capybara in Argentina (From Adámoli et al. (1988))

is rarely sold in urban centers. However, capybara meat is a valuable resource for the subsistence of poor communities and many ethnic groups who consume it traditionally (Quintana et al. 1992; Barbarán 2000; González 2001; Arenas 2003). The species is occasionally consumed by agricultural landowners and laborers, residents of nearby villages, foreign hunters, or fishermen. Oil extracted from capybara fat is also used locally as an alternative remedy for lung diseases like asthma.

The main use of capybaras in Argentina is as a source of high-quality leather. This country has a strong cattle industry, and technology developed for leather processing and tanning has been adapted for the capybara hide industry, which includes production, manufacturing and sale of footwear, clothing, luggage, handicrafts, and accessories.

Past use of the capybara varied greatly, reflecting the different peoples who have occupied its range. In the anthropological review by Palermo (CEAL 1983), he mentions that the use of wildlife by pre-Columbian indigenous cultures was based on mythical or religious concepts which limited hunting and avoided overexploitation. By contrast, archaeological records from localities of the Paraná River basin show that the capybara was absent or poorly represented in the diet of some ethnic groups, despite being among the most abundant mammals in those areas (Loponte 2008; Acosta et al. 2011; Santini et al. 2011; Sartori and Colasurdo 2011) and one of the highest ranking prey in terms of its potential economic performance (Loponte 2008). This could be related to limits placed on its use or as a result of food taboos (Feulliet Terzaghi 2002; Acosta 2005). On the other hand, for hunter-gatherers of the northern portion of the Paraná basin and for Amazonian horticulturalists of the Lower Delta of the Paraná River, capybaras appear to have had a relatively higher incidence in their diet (Loponte and Acosta 2003–2005; Pérez Jimeno 2007; Acosta et al. 2010).

After European settlement in America, this modest predator-prey relationship was replaced by more intensive use, leading to overexploitation. Initially, hunting was mostly of young capybaras, probably less than 1 year old, for their soft flesh. The use and profitability of the hide and the intrusion of capybaras into crops and pastures, with consequent damage, resulted in increased hunting pressure on bigger animals. The increasing demand for capybara hides prompted the appearance of the *carpinchero* or capybara hunter. During the first decades of the Republic, capybara hides were a valuable natural resource. For example, they were already listed in the 1891 official export figures of livestock products in Entre Ríos province (Fig. 21.1; Comisión para la Exposición Universal de Chicago 1893). Hunting was already high in the 1880s and, according to the official reports, about 12,100 hides were shipped from ports of that province between 1886 and 1889. This figure does not include poaching, smuggling, shipping from other ports, local use, or other losses (CEAL 1983).

At present, it is difficult to determine precisely the harvest volume from official documents because, even though provincial administrations have regulations for its use, much of the commercial trade is still illegal. In the first decade of this century, there was an estimated annual harvest of about 45,000 capybaras from the Iberá Natural Reserve alone (Fig. 21.1; Fraga 2003), most of them illegally. Demand for capybara leather products in the local market rose since foreign tourist affluence increased (up 76.5% between 2003 and 2010). As well as locally sold products, approximately 10,000 hides were exported each year, mainly to the European Union, from 2000 to 2009 (M. Lamarque personal communication), but these figures showed a marked decrease in 2010 with less than 350 units exported. The hide trade involves an organized and complex network of stakeholders from hunting to

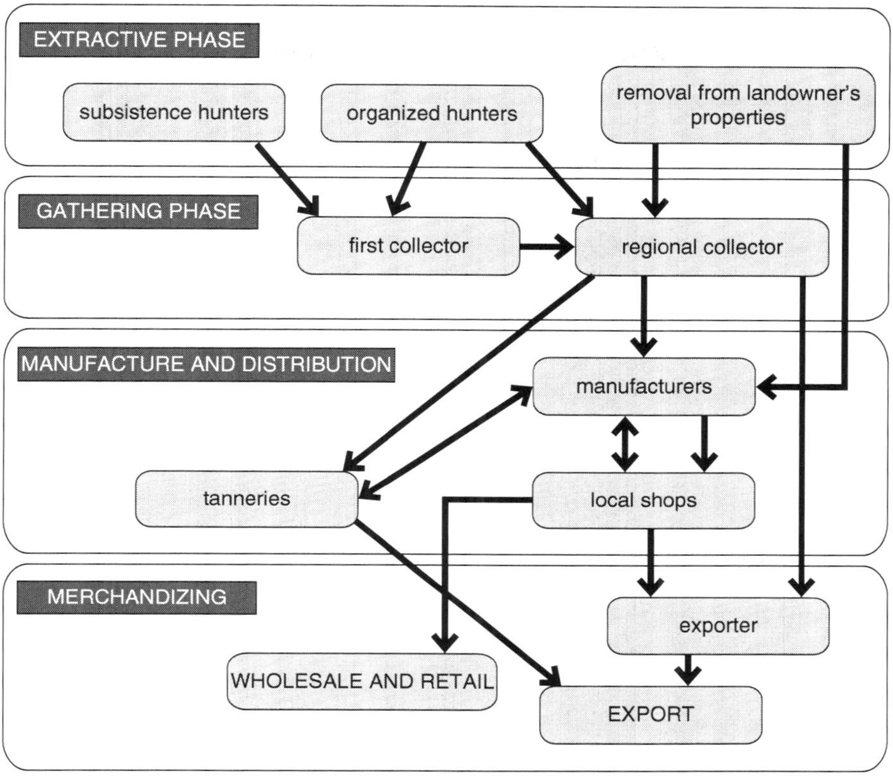

Fig. 21.2 Trade network for capybara hide in Argentina (From Rabinovich et al. (2003))

commercialization (Fig. 21.2). A sustainable management plan for this species is still lacking at a national level to encourage actions when demand increases. Coordinated actions with authorities of the provinces that support capybara populations are needed, in order to ensure this resource is used in a more goal-oriented, sustainable manner (Bolkovic et al. 2006).

21.4 Current Legislation and Trade Issues

At the national level, capybaras are protected under the Law for Wildlife Protection and Conservation (Law 22421, Decree 666/97) and the species is listed as non-threatened (Resolution 1030/04). Nevertheless, the Argentinean Society of Mammals classified capybaras as potentially vulnerable (Díaz and Ojeda 2000). Regarding wildlife use and conservation, in Argentina each Province has its own regulations. Hunting capybaras is totally prohibited in the provinces of Entre Ríos, Buenos

Aires, and Santa Fe. Subsistence hunting, by which poor people can consume capybara meat and trade the hide, is authorized in the provinces of Formosa, Chaco, and Corrientes. In the province of Chaco, commercial and sport hunting are restricted to a hunting season between May and August (winter), with quotas set each year (e.g., in 2008 the quota was 20,000 individuals; Disposition 58/08) and restrictions on the size of hides, setting the minimum length at 80 cm. In the province of Corrientes capybara hunting is allowed (and hide trade authorized) only after the provincial Wildlife Department certifies damage to pastures due to excessively abundant local capybara populations. There, sport hunting has a quota of four capybaras per person for local hunters or one per person for tourists that come from other provinces, during winter.

Differences in tax rates and hunting prohibitions between provinces may explain the illegal trade between them. There is not enough control, and this makes it difficult to set up legal production chains, even though many capybara populations have the potential for sustainable exploitation. The trade in capybara by-products is, indeed, made more difficult to oversee due to the complexity of the trade system (Fig. 21.2) and the wide range of items involved. These find their greatest demand on the local and less regulated market and are sold in numerous retail stores distributed across the country. It is almost impossible to monitor their movements. In contrast, it is easier to control products that are exported (mainly raw and tanned hides followed by manufactured articles) thanks to the lower number of checkpoints, such as airports and other points of exit out of the country (Bolkovic et al. 2006). As for live capybaras, shipment is forbidden (Resolution 62/86), except for those reared in captivity. From a legal standpoint, captive breeding is regulated by Resolution 26/92, and the National Service for Agrifood Safety and Quality (SENASA) is vested by Law 22421 with authority for the surveillance of wildlife diseases.

21.5 Capybara Production Systems in Argentina

In Argentina, traded capybara hides come from wild populations. Although there are still no plans for sustainable use, the species remains abundant in many areas, while overexploitation has caused local extinctions in others (Quintana 1996). As a result, a large proportion of the by-products are lost, especially when hunters flay the animal to obtain the hide and abandon the rest of the carcass, wasting the meat. In addition, the poorly regulated trade minimizes the profits of the lower links in the commercial chain because illegal hides are cheaper than those established under a legal system with quota restrictions. This system still benefits the collectors, who make large profits by buying at a low price and selling at high price (Fig. 21.2; Table 21.1a; Quintana 1996). Illegal activities are sometimes linked to legal subsistence hunting, and become the source of a large number of hides entering the commercial market (Quintana et al. 2005).

A system of intensive farmed capybara production was developed in Argentina at the beginning of the 1990s (Cueto 1999; Álvarez 2002; Allekotte 2003).

Table 21.1 Prices of capybara by-products and manufactured articles in the town of Mercedes (Corrientes province) and in Buenos Aires in August 1988 (Source: Quintana 1996)

(a) Raw materials (in Corrientes Province)	
Legal hide	USD 15.00
Illegal hide	USD 6.25
Meat	USD 0.50[a]/kg
Oil	Equivalent to 5 kg of sugar/l
(b) Manufactured articles	
(b.1) *Sold by craftsman in Mercedes*	
Short leather boots	USD 50.00
Medium leather boots	USD 55.00
Tall leather boots	USD 60.00
Shoes	USD 32.00
(b.2) *Sold in capybara leather shop in Mercedes*	
Short leather boots	USD 75.00
Tall leather boots	USD 100.00
Shoes	USD 47.00
(b.3) *Sold in capybara leather shop in Buenos Aires*	
Tall leather boots	USD 150.00
Shoes	USD 78.00
Jacket	USD 500.00

[a]At present, meat is sold in Paso de los Libres (Corrientes Province) at about USD 2.5/kg (Aparicio personal communication)

The experimental phase, during which the impacts of different population parameters on breeding output were surveyed, was carried out as a joint venture between the University of Buenos Aires, the National Council for Scientific and Technical Research (CONICET) and the Agricultural Experimental Station EEA-Delta of the National Institute of Agricultural Technology (INTA). The program was officially abandoned about 10 years later, partly because it had achieved the proposed research objectives, and partly because of institutional and economic factors. Nevertheless, some breeding farms were implemented under this project and they continued operating in a productive way. They were then grouped in the Capybara Breeders' Association of Argentina. Breeding farms were located in the provinces of Buenos Aires, Corrientes, Entre Ríos, and Santa Fe, and each maintained between 40 and 100 breeding females (ACCA 2007).

The main goal of these thriving farms was meat production, giving value to a by-product that has not previously been regarded as commercially viable in Argentina. The emphasis on meat production resulted from the possibly higher profit margin than hides. These endeavors were only accomplished on a small scale and their economic feasibility was controversial. In fact, from the seven farms that existed in 2007 not one is still running. Raising capybaras to produce only meat has not been successful, probably because it is very difficult to compete in a country that primarily consumes beef. Including hides has not been worth it either, as farmed hides cannot compete in price with those obtained from wild populations.

21.6 The Future Management of Wild Populations in Argentina

In Argentina, the presence of appropriate bio-ecological conditions for sustaining large capybara populations and the existence of high demand for capybara leather articles (Table 21.1) should motivate the development of a sustainable management plan for this species in the wild. In addition, the fact that there are still high densities over a large range of the capybara's distribution suggests that, despite its intensive use, the species has considerable resilience to hunting pressure. However, the mid- and long-term sustainability of capybara harvesting still needs to be assessed in different habitat types. As mentioned, there is demand for hide products on domestic and foreign markets, and the potential market for capybara meat is high because of Argentinean eating habits. In this context, Bolkovic et al. (2006) proposed strategies for use according to habitats and social contexts that exist across the distributional range of capybaras in the country. They include: (a) extensive management with harvest of wild populations on large farms with good availability of open habitats, such as those in Corrientes province; (b) a semi-extensive farming for sites where the species is relatively under protection, but with farms of about 500 ha or less, as is the case of many farms in Entre Ríos province, (c) captive breeding on small farms where suitable habitats are unavailable, and (d) direct hunting under a strict registration of hunters and gatherers in many areas where subsistence hunters rely on this resource for meat with the possibility of selling hides on the legal market, as would be suitable for the Delta region of the Paraná River.

These strategies must be part of a national management program for the sustainable use of the species, incorporating ecological, social, and economic aspects. In this regard, from mid-2002 to mid-2003 the Secretary of Environment and Sustainable Development (SAyDS), through the former Direction of Wild Fauna and Flora (DFyFS), set up the "Capybara Project", which included the participation of the private sector for the supply of funds for research and agreements with wildlife management authorities of the provinces of Entre Ríos, Santa Fe, Chaco, Formosa, and Corrientes in order to support of this project. During this preliminary phase management, biological and socioeconomic aspects were addressed and a number of recommendations were issued (Table 21.2).

The second phase of the project was to start in 2007 and focused on putting these management recommendations into practice, aiming to assess more accurately the current status of capybaras on a regional scale. Unfortunately, economic constraints arose, related to the management of private funds for sustainable management programs, which delayed the implementation of the plan as a whole. Meanwhile, the DFS-SAyDS began population and socioeconomic research in the Paraná River islands in Santa Fe province to assess the importance of this species to local hunters and fishermen. This is one of the provinces that need to regulate capybara use more urgently, in order to reduce illegal transport of hides. The results of this research will soon be discussed with local authorities to design a strategy for incorporating the existing use of this species into the legal framework. Evaluation of a potential

Table 21.2 Recommendations derived from the first phase of the Capybara Project (Adapted from Bolkovic et al. 2006)

Aspects	Recommendations
Management	Implementation of mechanisms for disclosing actual hide stocks per jurisdiction
	Organization of commercial activity under a unified protocol for all provinces
	Setting a provisional hunting quota per province, subject to possible future adjustments according to local extraction methods
	Implementation of a single and obligatory interjurisdictional transit route for the trade of by-products
	Implementation of a control system for identifying the source of hides
	Development of a database system for control enforcement
Biological	Assessment of the current status of capybara populations at the national level
	Survey of potential habitat suitability based on habitat quality, quantity, and availability at a regional scale
	Undertaking of field studies to fill key information gaps
Socioeconomic	Implementation of actions to make full use of the resource (hide, meat, and oil)
	Development of a system for simultaneous exploitation of different wild species, or for management that is complementary to current traditional agricultural practices and compatible with wetland conservation
	Development of protocols for good production practices contributing to conservation of capybaras and other wildlife species
	Development of a population monitoring system at sites where crop fields were damaged to minimize the negative impacts

habitat suitability model was also carried out at two levels. At a local level, a habitat suitability index was created (Quintana and Bolkovic in press), considering the species' requirements and, on a regional scale, a spatial model was made to identify priority areas for field evaluations of capybara populations in Corrientes province (Schivo 2009). Validation for the latter is currently being carried out in this province by researchers and technicians of the national Wildlife Agency (Dirección de Fauna Silvestre) and the Universities of San Martín and Buenos Aires, and the next step is to develop a model for the distribution of the species in the whole country.

As long as demand does not increase, the species seems likely to remain at low risk. However, risk assessment needs to take into account other factors that influence population dynamics, including climatic constraints, human use of ecosystems, the expansion of the agricultural frontier, and a potential but unpredictable increase in demand for capybara by-products. But as it is not clear if demand is falling or if the improvement in monitoring and control is lowering supply at the retail outlet, actions to ensure sustainable use of the species should be carried out in the short or medium term. There is a worldwide consensus on the importance of wetlands as ecosystems providing a wide variety of goods and services to human societies (Stolk et al. 2006; Kandus et al. 2011). A sustainable management plan for the capybara should take on additional importance, given that it will also contribute to conservation of wetlands, supporting the maintenance of their structure and function and favoring other wetland species, while boosting harvest profits to the benefit of local communities.

Acknowledgments The authors were supported by the Dirección de Fauna Silvestre (Secretaría de Ambiente y Desarrollo Sustentable de la Nación) and under the X-481 grant of the UBACyT Program of the University of Buenos Aires.

References

Acosta AA (2005) Zooarqueología de cazadores recolectores del extremo nororiental de la provincia de Buenos Aires (Humedal del Paraná inferior, Región Pampeana). Ph.D. thesis, Facultad de Ciencias Naturales y Museo, Universidad Nacional de La Plata, La Plata

Acosta AA, Escudero S, Feulliet Terzaghi MR, Loponte D, Pérez Jimeno L (2010) Conectando registros: variabilidad aqueológica en la cuenca del Paraná. In: Berón M, Luna L, Bonomo M, Montalvo C, Aranda C, Carrera Aizpitarte M (eds) Mamül Mapu: pasado y presente desde la arqueología pampeana, vol I, Tomo. Libros del Espinillo, Ayacucho, pp 17–28

Acosta AA, Mucciolo L, Musali J, Arrizurieta M (2011) Avances y problemas relacionados con el estudio del registro arqueofaunístico generado por los grupos cazadores-recolectores del extremos sur de la provincia de Entre Ríos (Humedal del Río Paraná inferior). In: Feuillet Terzaghi MR, Colasurdo B, Sartori JI, Escudero S (eds) Avances y perspectivas en la arqueología del nordeste. Municipalidad de Santo Tomé, Santo Tomé, pp 43–58

Adámoli J, Sennhauser E, Astrada E, Agraz JL (1988) Propuesta para la delimitación del área geográfica potencial de los carpinchos en Argentina. Consejo Federal de Inversiones, Buenos Aires, pp 1–88

Allekotte R (2003) La cría del carpincho. Ediciones INTA, Buenos Aires

Álvarez M (2002) Manejo sustentable del carpincho (*Hydrochoerus hydrochaeris,* Linnaeus 1766) en Argentina: un aporte al conocimiento de la biología de la especie desde la cría en cautiverio. Ph.D. thesis, Universidad de Buenos Aires, Buenos Aires

Arenas P (2003) Etnografía y alimentación entre los Toba-Nachilamoleek y Wichí-Lhuku'tas del Chaco Central (Argentina). Latin Gráfica, Buenos Aires

Asociación De Criadores De Carpinchos En Argentina (2007) La cría de carpinchos en la Argentina. http://www.criaderoscarpinchos.com.ar/

Barbarán FR (2000) Recursos alimenticios derivados de la caza, pesca y recolección de los Wichi del Río Pilcomayo (Provincia de Salta, Argentina). In: Cabrera E, Mercolli C, Resquin R (eds) Manejo de Fauna Silvestre en Amazonia y Latinoamérica. CITES Paraguay – Fundación Moisés Bertoni – University of Florida, Asunción, pp 507–527

Bolkovic ML, Quintana RD, Ramadori D, Elisech M, Rabinovich J (2006) Proyecto Carpincho. Propuesta para el uso sustentable del carpincho (*Hydrochaeris hydrochaeris*) en la Argentina. In: Bolkovic ML, Ramadori D (eds) Manejo de fauna silvestre en la Argentina. Programas de uso sustentable. Dirección de Fauna Silvestre, Secretaría de Ambiente y Desarrollo Sustentable de la Nación, Buenos Aires, pp 105–119

CEAL (1983) El carpincho. Serie Fauna Argentina, vol 2. Centro Editor de América Latina, Buenos Aires, pp 1–33

Comisión para la exposición universal de chicago (1893) La Provincia de Entre Ríos. La Velocidad, Paraná

Cueto G (1999) Biología reproductiva y crecimiento del carpincho (*Hydrochaeris hydrochaeris*) en cautiverio: una interpretación de las estrategias poblacionales. Ph.D. thesis, Universidad de Buenos Aires, Buenos Aires

de Azara F (1802) Apuntamientos para la Historia natural de los quadrúpedos del Paraguay y Río de la Plata. Imprenta de la Viuda de Ibarra. España, Madrid

Díaz G, Ojeda R (2000) Libro rojo de mamíferos amenazados de la Argentina. SAREM (Sociedad Argentina para el Estudio de los Mamíferos), Mendoza

Feulliet Terzaghi MR (2002) El registro arqueofaunístico de vertebrados del sitio Playa Mansa I (Arroyo Seco, Dpto. de Rosario, Pcia. de Santa Fe). Aportes y perspectivas. B.S. thesis,

Facultad de Humanidades y Artes, Escuela de Antropología, Departamento de Antropología, Universidad Nacional de Rosario, Santa Fe
Fraga V (2003) Diagnóstico preliminar sobre la situación del furtivismo en la Reserva de Iberá. Informe Fundación Ecos, Mercedes
Giraldo Hernández D, Ramírez Perilla JA (2001) Guía para el manejo, cría y aprovechamiento sustentable del chigüiro, chigüire o capibara *Hydrochaeris hydrochaeris* Linneo, vol 99, Serie Ciencia y Tecnología. Convenio Andrés Bello, Bogotá, pp 1–76
González G (2001) Conociendo lo nuestro. La cultura de los grupos guaraníticos del río Paraná y su construcción del sistema alimentario actual. B.S. thesis, Universidad de Belgrano, Buenos Aires
González Jiménez E (1995) El capibara (*Hydrochoerus hydrochaerus*). Estado actual de su producción, vol 122, Producción y Sanidad Animal. FAO, Roma, pp 1–114
Kandus P, Minotti PG, Malvárez AI (2008) Distribution of wetlands in Argentina estimated from soil charts. Acta Sci 30:403–409
Kandus P, Quintana RD, Minotti PG, Oddi JP, Baigún C, González Trilla G, Ceballos D (2011) Ecosistemas de humedal y una perspectiva hidrogeomórfica como marco para la valoración ecológica de sus bienes y servicios. In: Laterra P, Jobbagy E, Paruelo J (eds) Valoración de servicios ecosistémicos: conceptos, herramientas y aplicaciones para el ordenamiento territorial. Instituto Nacional de Tecnología Agropecuaria (INTA), Buenos Aires, pp 265–290
Loponte DM (2008) Arqueología del humedal del Paraná inferior. Arqueología de la Cuenca del Plata, vol 1, Serie Monográfica. Ediciones Riel, Buenos Aires
Loponte DM, Acosta AA (2003–2005) Nuevas perspectivas para la arqueología "guaraní" en el humedal del Paraná inferior y Río de la Plata. Cuadernos del Instituto Nacional de Antropología y Pensamiento Latinoamericano (INAPL), 20:179–197
Moreira JR, Alvarez MR, Tarifa T, Pacheco V, Taber A, Tirira DG, Herrera EA, Ferraz KMPMB, Aldana-domínguez J, Macdonald DW (2012) Taxonomy, natural history and distribution of the capybara. In: Moreira JR, Ferraz KMPMB, Herrera EA, Macdonald DW (eds) Capybara: biology, use and conservation of an exceptional neotropical species. Springer, New York, pp 3–37
Pérez Jimeno L (2007) Investigaciones arqueológicas en el sector septentrional de la llanura aluvial del Paraná – margen santafesina: La variabilidad del registro arqueológico. Ph.D. thesis, Facultad de Ciencias Naturales y Museo, Universidad de La Plata, La Plata
Quintana RD (1996) Análisis y evaluación del hábitat del carpincho (*Hydrochaeris hydrochaeris*) en relación con la heterogeneidad del paisaje y las interacciones con ganado doméstico. Ph.D. thesis, Universidad de Buenos Aires, Buenos Aires
Quintana RD, Bolkovic ML (in press) Modelo de aptitud de hábitat: Carpincho. In: Zaccagnini ME, Rosatti V (eds) Serie: Modelos de aptitud de hábitat de biodiversidad de Argentina. INTA, Buenos Aires
Quintana RD, Rabinovich J (1993) Assessment of capybara (*Hydrochaeris hydrochaeris*) populations in the wetlands of Corrientes, Argentina. Wetlands Ecol Manage 2:223–230
Quintana RD, Bó RF, Merler JA, Minotti PG, Malvárez AI (1992) Situación y uso de la fauna silvestre en la región del Bajo Delta del río Paraná, Argentina. Iheringia, Sér Zool, Porto Alegre 73:13–33
Quintana RD, Bolkovic ML, Ramadori D, Rabinovich JE (2005) Perspectivas para el manejo de poblaciones de carpinchos en Argentina y la conservación de los humedales. In: Peteán J, Capatto J (compiladores) Humedales Fluviales en América del Sur. Hacia un manejo sustentable. Ediciones PROTEGER, Santa Fe, pp 521–534
Rabinovich JE, Ramadori D, Elisetch M, Bolkovic ML, Quintana RD, Malpartida A (2003) Proyecto Carpincho. Desarrollo de un plan de manejo sustentable de poblaciones silvestres de carpinchos (Hydrochaeris hydrochaeris). Secretaría de Ambiente y Desarrollo Sustentable de la Nación/Universidad de Belgrano, Buenos Aires
Santini M, Salceda S, de Santis L (2011) Análisis arqueofaunístico del sitio Sotelo I (Provincia de Chaco). In: Feuillet Terzaghi MR, Colasurdo B, Sartori JI, Escudero S (eds) Avances y perspectivas en la arqueología del nordeste. Municipalidad de Santo Tomé, Santo Tomé, pp 3–23

Sartori J, Colasurdo MB (2011) El análisis arqueofaunístico del sitio Playa Mansa: Nuevas perspectivas. In: Feuillet Terzaghi MR, Colasurdo B, Sartori JI, Escudero S (eds) Avances y perspectivas en la arqueología del nordeste. Municipalidad de Santo Tomé, Santo Tomé, pp 25–41

Schivo F (2009) Análisis a escala regional de la aptitud potencial de hábitat para el carpincho (*Hydrochoerus hydrochaeris*) en la provincia de Corrientes, Argentina. Graduate Dissertation, Facultad de Ciencias Exactas y Naturales, Universidad de Buenos Aires, Buenos Aires

Silva Neto PB (2005) Crianza racional de capibaras (*Hydrochoerus hydrochaeris*) en Brasil. In: Díaz NC, Vélez N (eds) Aprovechamiento del chigüiro en Latinoamérica. Grupo de Estudios Ecológicos OIKOS, Bogotá, pp 87–97

Stolk ME, Verweij PA, Stuip M, Baker CJ, Oosterberg W (2006) Valoración Socioeconómica de los Humedales en América Latina y el Caribe. Wetlands International, Wageningen

UNT (1943) Hacia allá y para acá (Una estada entre los indios Mocobíes; Crónicas del padre F. Pauke 1749–1767). Instituto Argentino, Universidad Nacional de Tucumán, San Miguel de Tucumán and Buenos Aires

Chapter 22
Counting Capybaras

Luciano M. Verdade, José Roberto Moreira, and Katia Maria P.M.B. Ferraz

22.1 Why Count Capybaras?

The decision-making process leading to management actions requires reliable estimates of the target population density or at least its abundance variation in space and time (Caughley 1977; Abercrombie and Verdade 1995). Capybaras apparently stand up well to anthropogenic pressures like environmental change (Verdade and Ferraz 2006) and hunting (Verdade 1996). For this reason, the species is not considered endangered in any country within its range. On the contrary, in some circumstances capybaras are considered a pest in agricultural and urban landscapes (Moreira et al. 2001; Cavalcanti 2003; Ferraz et al. 2003). In addition, they are hunted in many countries of South America (Ojasti 1991; Kaplan and Kopischke 1992). Therefore, there are two main reasons to count capybaras: sustainable use and damage control.

The species occurs in most ecosystems and biomes of the Neotropics, from the wetlands of the Brazilian Pantanal and Venezuelan Llanos, to the Amazon and the Atlantic rainforests, and even to humid microhabitats of drier biomes like the Cerrado and Caatinga. As part of these ecosystems, capybaras should also be

L.M. Verdade (✉)
Laboratório de Ecologia Isotópica, CENA, Universidade de São Paulo,
Caixa Postal 96, Piracicaba, São Paulo, Brazil
e-mail: lmverdade@usp.br

J.R. Moreira
Embrapa Recursos Genéticos e Biotecnologia, Parque Estação Biológica,
Final W5 Norte, 70770-900, Brasília, Distrito Federal, Brazil
e-mail: alencar_moreira@yahoo.com.br

K.M.P.M.B. Ferraz
Departamento de Ciências Florestais, Escola Superior de Agricultura "Luiz de Queiroz",
Universidade de São Paulo, Av. Pádua Dias 11, Piracicaba, São Paulo, Brazil
e-mail: katia.ferraz@usp.br

J.R. Moreira et al. (eds.), *Capybara: Biology, Use and Conservation of an Exceptional Neotropical Species*, DOI 10.1007/978-1-4614-4000-0_22,

monitored in the long-term biodiversity monitoring programs that have been established lately to watch species that can become endangered, pests, or economically valuable (e.g., Magnusson et al. 2005; Pezzini et al. 2012). In this chapter we describe the variables that affect counting capybaras and the steps to consider when planning the monitoring of capybara populations. We finally describe the methods that have been tested for counting capybaras or those that are currently available, and we recommend in which circumstances they should be used.

22.2 What You See

Counting capybaras, like virtually anything that moves, can be tricky because what you see is not necessarily what you have got. Therefore, some planning is required before going to the field. Two basic questions should guide this step: (a) What kind of information is needed? and (b) What kind of information is collectable in the field? Damage control, sustainable use, and long-term monitoring – the wildlife management actions into which capybaras usually fit – can vary in terms of the population information required to underpin the decision-making process (Caughley 1977; Davis and Winstead 1980; Lancia et al. 1996, 2005; Verdade 2004).

22.2.1 Uncertainties Associated with Counting Capybaras in the Wild

Unless the study site is very small and the focus species is really conspicuous, it is seldom possible to count the total number of individuals in an animal population (Coddington et al. 2009). Detecting capybaras in an open savanna is very different from detecting them in a forested habitat. In some situations capybaras can actively avoid human presence, which can significantly increase the observer's uncertainties in their population estimate. In addition, the time frame should be taken into consideration, because there is a difference in timescale among sampling, ecological, and evolutionary processes (Preston 1960). Therefore, the following sources of uncertainty (Schrader-Frechette 1995) should be considered when counting capybaras: time lapse of sampling and between samplings, individual detectability in space and time, and the consequences of these two factors in counting precision and accuracy (Fig. 22.1).

Surveys take time to carry out; some last days and others weeks to months (Sutherland 2006). Although capybaras exhibit some reproductive seasonality, the breeding season can span months (Ojasti 1973; Eisenberg and Redford 1999), and some females can have two litters in a single year (Lavorenti 1989). This pattern can result in significant monthly variation in capybara population sizes (Verdade and Ferraz 2006). It is, therefore, prudent to assume a capybara population as "open" in surveys that take more than a month to carry out, even within a single breeding

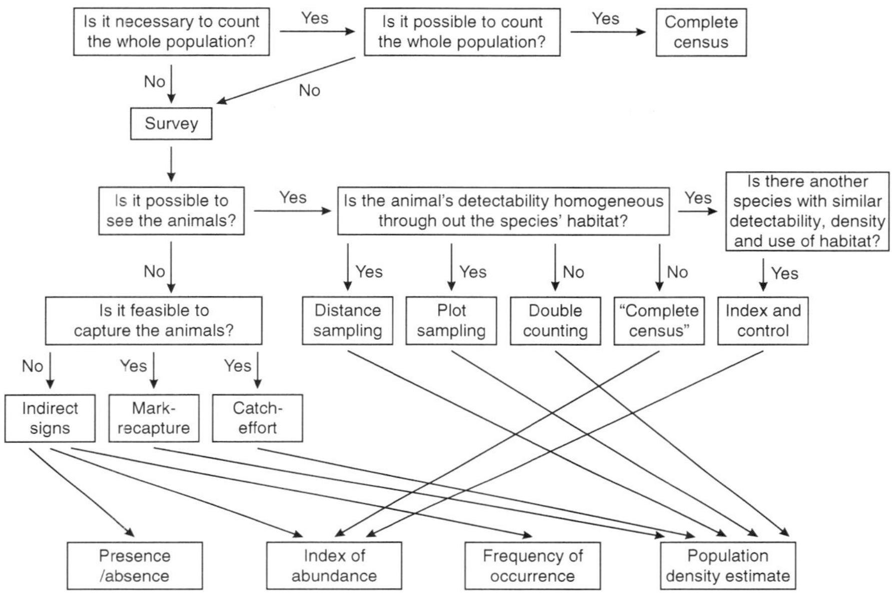

Fig. 22.1 Counting capybaras: field methods and the questions they can answer (From Lancia et al. (1996))

season (Mackenzie et al. 2006). Given these circumstances, population variation within a sampling period can be broader than its variation between separate sampling efforts, which may be detrimental to the decision-making process based on these data. The best way to reduce these uncertainties is to reduce the sampling period and to increase the number of sites sampled (Magnusson and Mourão 2003). In other words, when counting capybaras, it is often better to be fast, superficial, and broad than slow, in-depth, and restricted.

In order to estimate relative abundance of different species as well as the number of individuals from a single population, the detectability of all individuals is assumed to be equal (Nichols 1996), an assumption which is unlikely in practice. Individuals from any species have separate and independent chances of being detected by an observer. And to make things even worse, observers also vary in terms of fieldwork experience, familiarity with the species in question, and individual skills (Lehner 1996). These characteristics can increase uncertainty in surveys conducted by more than one observer. In counting capybaras, the distance to the animals can affect their detectability by observers, even on a relatively small spatial scale (i.e., <100 m) (Verdade and Ferraz 2006). In addition, environmental variables can affect capybara counts in different ways according to the environment (Ferraz et al. 2010). It is prudent to test the homogeneous detectability assumption as a hypothesis for the focused population. This procedure will permit some beneficial calibration of the methods used to count capybaras.

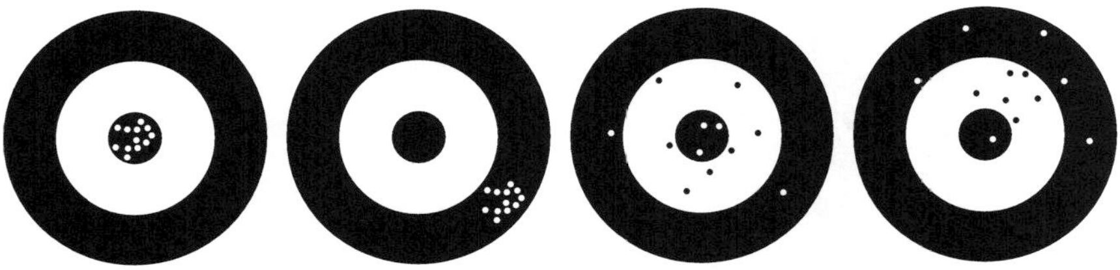

Fig. 22.2 Precision and accuracy illustrated by the target analogy, where the actual population size can be imagined as the bull's eye (From McCallum (2000))

Last but not least, some variation can occur in counting due to chance or sampling error and to observers' bias. Precision is inversely related to variance, whereas accuracy is inversely related to the observers' bias in repeated measurements (Fig. 22.2; McCallum 2000:23). Most wildlife surveys have low precision and unknown accuracy, which is possibly the most relevant uncertainty associated with the decision-making process in wildlife management, especially when real abundance estimates are needed.

22.2.2 Planning to Count Capybaras

The steps in planning to count capybaras should be based on the simple questions summarized in Fig. 22.1. It is seldom possible – or necessary – to determine the total population size. Therefore, a survey (i.e., capybara abundance estimate based on sampling) is usually more cost-effective than a census (i.e., counting the whole population) in the decision-making process for capybara management. The next step should consider the animals' visibility, as this can consistently vary among populations or even within the same population. Capybaras tend to be more conspicuous in open habitats with low hunting pressure than in forest habitats with high hunting pressure. Where animals are easily sighted the next question is: does individual detectability vary across the study site? If not, population estimates can be generated by aerial line transects (Buckland et al. 2011). If it does vary, counting visible animals can generate an index of abundance or at least a frequency of occurrence (Verdade and Ferraz 2006). Where animals are rarely seen the next question is: is it feasible to capture them? If not, signs like tracks and scats can also generate an index of abundance when they are abundant (therefore countable); when they are not, a simple presence/absence record can be created for each site (Pinto et al. 2006). Where capture is feasible, recapturing marked individuals can generate population density estimates (Seber 1973, 1982; Borchers et al. 2002; Williams et al. 2002; Lancia et al. 2005).

Budget and time constraints should be considered as limiting factors for any plan. These constraints determine two essential aspects of capybara management: what is currently possible and what we should improve in the future.

22.3 What You Get

Considering all uncertainties and limitations concerning capybara counts discussed above, the following methods have been tested for capybaras or are currently available: direct count, plot sampling, distance sampling, mark-recapture, indirect signs, index and control, and catch-effort method. Their results vary from presence/absence records to indices of abundance (Conroy 1996) to population density estimates (Table 22.1). These methods are briefly described below.

22.3.1 Direct Count Method

This method consists of counting visible animals on a preestablished trail or area, usually on the edges of water bodies, which can be done on foot, by boat, or in all-terrain vehicles. Two basic assumptions should be met: individual detectability should remain constant through time and there should be a strong and constant correlation between the number of animals seen and the actual population size. If these assumptions are largely fulfilled, then population fluctuation over time can be determined, which usually provides adequate information for the capybara management decision-making process.

This is the method that has so far been used most frequently to evaluate capybara population size, including Juhani Ojasti's (1973) classic work in Venezuela. Ojasti (1973) considered the possibility that he had not included some of the individuals hidden in grassy clumps in most of his counts. This method was used in various other works in different areas of capybara distribution (Bertelli et al. 2000; Ferraz et al. 2001; Vargas et al. 2007). Since the method does not provide a way to estimate the unseen animals it should not be used as a density estimate. It should be used only as an index of abundance.

Cordero and Ojasti (1981) used direct count to estimate capybara population size in a forested area. The standard error in the samples suggests that the detectability of the animals really affects the results of this method in areas where visibility is low due to dense vegetation. The detectability index of capybaras in forest habitats in the state of São Paulo, Brazil, was estimated to be 0.63 ± 0.32 (Pinto et al. 2006). The difficulty of seeing capybaras in forests detracts from the viability of direct count in such areas of vegetation.

22.3.2 Plot Sampling

This technique counts animals in sample units with previously fixed limits. Counting can be done along a preestablished transect where all the animals seen within the limit are recorded (Sinclair et al. 2006). Its basic assumption is that all animals present along a transect line will be detected and that are homogeneously distributed.

Table 22.1 Field methods of counting capybaras

Methods	Steps	Assumptions	Strengths	Weaknesses	Result	Examples
Direct count	Counting visible animals on a preestablished trail	Individual detectability constant through time Constant correlation between visible animals and actual population size	Low cost	Low precision Unknown accuracy, very rarely possible	Index of abundance	Ojasti (1973), Lord and Lord (1988), Bertelli et al. (2000), Ferraz et al. (2001), Verdade and Ferraz (2006), Vargas et al. (2007)
Plot sampling	Counting animals over a strip of land of known length covered at constant speed	All individuals present along transect lines should be detected	Easily applied	Effective only where capybaras are abundant, assumes random distribution of individuals	Population density estimate	Lord and Lord (1988)
Distance sampling	Counting animals over a strip of land covered at constant speed and recording their distance to the line and angle to the observer	Animals are more easily detected when closer to the observer	Low cost	Small sampling area, low precision, unknown accuracy, time consuming	Population density estimate	Schneider and Menegheti (1997)
Mark-recapture	Estimating population size based on the frequency of recapture of marked individuals	Homogeneous individual catchability, requires a large proportion of animals captured	Known accuracy for effective reproductive population	Expensive, time consuming, unknown precision, biased to dominant individuals	Population size	None

Double counting	Counting animals over a strip of land of preestablished length and width covered at constant speed and distance	Requires a large proportion of animals captured	Large sampling area, high precision	Observers should be screened from each other	Population density estimate	Mourão et al. (1994), Mourão and Campos (1995), Cunha et al. (2002), Pinha et al. (2003), Waga et al. (2003), Mourão and Magnusson (2004), Moreira et al. (2001)
Indirect signs	Counting field signs (i.e., scats and/tracks) on a preestablished trail or area	Constant correlation between signs abundance and actual population size	Low cost, fast data collection	Low precision, unknown accuracy	Presence/absence, index of abundance, frequency of occurrence	Pinto (2003), Pinto et al. (2006)
Index and control	Counting animals of the target species and individuals of another species of known density	Similar indices of abundance and detectability for both species	Low cost	Species' use of habitat, density, and detectability are rarely similar	Index of abundance	Lord and Lord (1988)
Catch-effort method	Calculating the index of abundance before and after a known number of individuals are included or removed from the population	Individual detectability constant before and after the manipulation; population closed through time	Low cost, high accuracy	Low precision	Population density estimate	Lord and Lord (1988)

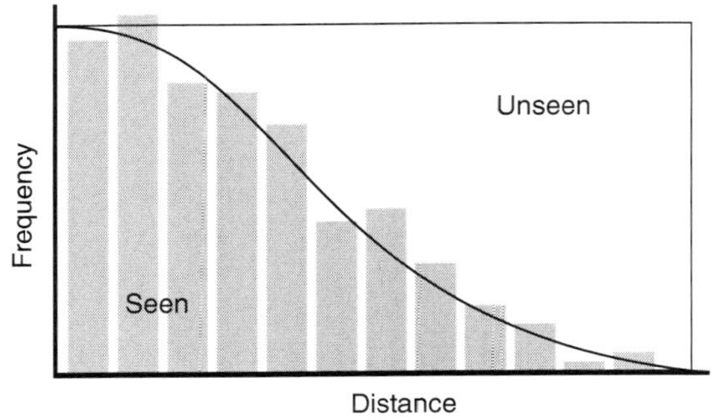

Fig. 22.3 Hypothetical detection function based on distance sampling (From Sinclair et al. (2006))

The total number of detected individuals over the area sampled gives an estimate of the total population density (Buckland et al. 2011).

This method was tested in Venezuela by Lord and Lord (1988), who counted all the capybaras present along the edges of roads driven in a vehicle. The limit was set at a distance of 100 m on either side of the road. The authors estimated the capybara population size for the whole ranch based on the number counted in the sampled area.

22.3.3 Distance Sampling

This is sample method in which the limits of the sampled area are not fixed (Sinclair et al. 2006). It developed from the premise that an animal closer to the observer has a greater probability of being seen than one farther away (Fig. 22.3; Krebs 1989). The method can be very useful for capybaras, but it demands some skill and time to calculate the angle of the animal in relation to the line covered and to estimate the distance of the animal from the observer. It is also very sensitive to errors in the estimates of distance, which means that these must be very precise. As capybaras usually inhabit wetlands and marshes, terrestrial linear transects are usually difficult to carry out. That is possibly the reason it is not commonly used for the species.

Larger groups of animals are seen at greater distances than small groups (Plumptre 2000), a bias which applies to capybaras and needs to be controlled for. The method was used in several types of vegetation in the Aracuri Ecological Station in Rio Grande do Sul state, Brazil (Schneider and Menegheti 1997). It was difficult to evaluate capybara group size because of the height of vegetation. The animals were often identified only by the noise they made when running away, or by alarm calls (Schneider and Menegheti 1997), which may have been a likely source of bias.

22.3.4 Mark-Recapture

This method consists of estimating population size based on the frequency of marked individuals' recapture. Its basic assumption is that all individuals have the same catchability. This method has not yet been tested on capybaras. It does not

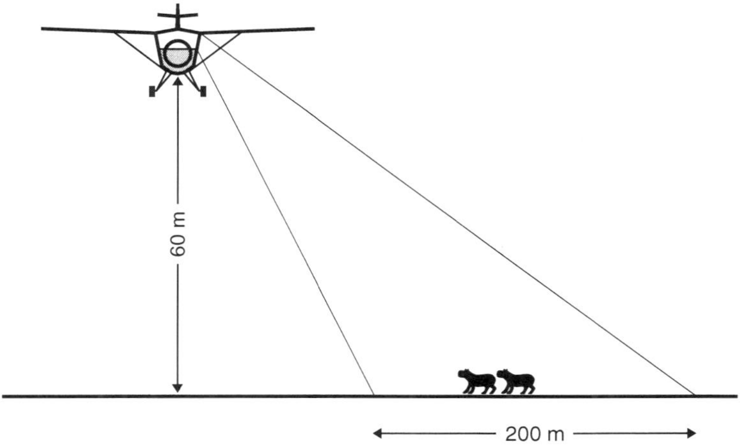

Fig. 22.4 Aerial survey of capybaras using double counting (From Mourão and Magnusson (2004))

work for extensive management systems where only one harvest takes place per year, but it can be very useful if many captures (as is sometimes the case for harvest) place in a year, as animals may be marked and released during these activities. It can also be useful if, immediately before capture (e.g., for slaughter), a large-scale capture for another management activity (e.g., for parasite control) takes place and the animals are marked during this time. Nevertheless, the assumption that all individuals have the same catchability is unlikely to be met with capybaras because of their complex social behavior (Ojasti 1973; Ferraz and Verdade 2001). Dominant males and females and their offspring tend to enter corrals before other individuals, actively preventing their access (Ferraz et al. 2001). Despite this bias, if information on effective reproductive population is pursued the method can have high precision and known accuracy. However, it tends to be relatively expensive and time-consuming.

22.3.4.1 Double Counting

This technique uses the same premise as mark-recapture – it is also a Petersen estimate. However, the two "captures" are done visually by two observers (Mourão and Campos 1995). The area needs to be divided (physically or temporally) for it to be possible to identify the 'recaptures' (those animals that both observers sighted).

Aerial surveys of capybaras (Fig. 22.4) and other wild species in the Brazilian Pantanal of Mato Grosso do Sul state have been carried out since the early 1990s, using the double counting method (Mourão et al. 1994; Mourão and Campos 1995). Capybaras are not counted individually but by groups ranked by size. This method has also been tested for capybaras in surveys on land and in water (Moreira et al. 2001; Cunha et al. 2002; Pinha et al. 2003; Waga et al. 2003). This is a very useful method for covering large areas, but does not work when less than seven capybara groups are counted.

22.3.5 Indirect Signs

This method is based on counting scats, tracks, and alarm calls (or any population attribute whose variation is related to that of the population) on preestablished trails or areas. Removal of these signs permits some quantification during repeated samplings at short intervals (e.g., daily), which tends to be relatively precise when abundance is intermediate, but it tends to become imprecise at low or high abundances. This pattern reduces the chance of meeting the basic assumption of the method, which is a constant correlation between sign abundance and actual population size. When quantification is not reliable, the method generates only presence/absence reports. When this kind of information is sufficient for the intended purpose, for example in simulation models of habitat use (e.g., Ferraz et al. 2009), the relatively low cost and fast data collection make this method cost-effective. When quantification is reliable, frequencies of occurrence or abundance indices are generated (Mills et al. 2005).

Counting piles of feces was tested as an index of species abundance in seven lagoons in the state of São Paulo, Brazil (Pinto 2003). Four of these areas were fenced and capybara population sizes were already known. A significant correlation was found between the index of feces abundance and the real size of the population.

22.3.6 Index and Control

The size of an animal population can be estimated by evaluating its relationship with a population of known size (Caughley 1977). This method was tested in Venezuela by Lord and Lord (1988), counting capybaras and cattle in the same area. The calculation of population size by the capybara X cattle ratio may be tendentious, given that cattle are more conspicuous than capybaras and use a different habitat. Caughley (1977) affirms that this method can only be used when the indices of abundance of the two species are equivalent. He also states that the two species should react in similar ways to the presence of an observer, possess similar detectability, and disperse themselves in a similar way. It is difficult to fulfill these conditions.

22.3.7 Catch-Effort Method

This technique compares species indices before and after removing part of the population (Caughley 1977). As the part removed is known, it is possible to estimate the total population using the difference between the two indices. This method assumes that: (a) no animals die apart from those removed; (b) the counting conditions of the index are the same before and after manipulation. These premises are not always true for capybaras. However, it is a method that may be very useful in extensive capybara production systems, in very open areas, where populations are not timid.

Lord and Lord (1988) tested this method adapted to the percentage of the number of animals counted before and after manipulation (commercial slaughter). Nevertheless, it was necessary to include natural death and birth rates for the period between the two counts to reduce the error in the population size estimate.

22.4 General Recommendations

Considering the needs, limitations, and available methods for capybara counting, the following general recommendations can be given to the would-be practitioner:

(a) For information on population fluctuation, an index of abundance is likely to be sufficient and one should not invest unnecessary time and funds estimating population density. In this case, direct counting visible animals (when animals are indeed visible) or counting indirect signs (when they are not) are probably the most cost-effective options.
(b) For a density estimate of a population from a large open habitat, aerial double counting is possibly the most cost-effective method as it covers large areas in relatively small periods of time.
(c) For a hunting program where hunters have easy (and safe) access to animals, the harvest itself (i.e., the total number of animals hunted) is an effective index of abundance achieved at low cost and with reasonable precision, if hunting pressure is constant. The catch-effort method could also be used in these situations.
(d) Distance sampling, index and control, mark-recapture, and camera-trapping (not described here), although extensively used with large terrestrial mammals, tend to be ineffective for counting capybaras, for the various reasons discussed above. They should be used with caution, or not used at all. In practice, an index of abundance based on direct counting visible animals, or even their signs, can be more cost-effective than these methods.
(e) Molecular biology is evolving fast. It is, therefore, reasonable to believe that specific molecular markers for capybaras will be available soon (Herrera et al. 2004). When this day comes, their use might be an effective way to estimate population size in areas where the other methods above do not work properly, like in agricultural landscapes where the species may eventually need to be managed and, therefore, counted.

Acknowledgments Fundação de Amparo à Pesquisa do Estado de São Paulo (FAPESP) and Conselho Nacional de Desenvolvimento Científico e Tecnológico (CNPq) partially funded this study. JRM is grateful for the support provided by Fundação de Apoio à Pesquisa Família Alencar (Brazil).

References

Abercrombie CL, Verdade LM (1995) Dinâmica populacional de crocodilianos: elaboração e uso de modelos. In: Larriera A, Verdade LM (eds) Conservación y Manejo de los Crocodylia de America Latina, vol 1. Fundación Banco Bica, Sao Tomé, pp 33–55

Bertelli PW, Rechenberg E, Conceição L, Walters P (2000) Levantamento preliminar quantitativo de *Hydrochaeris hydrochaeris* (Mammalia, Hydrochaeridae), na bacia do Rio Itajaí – SC. P.640. In: XXIII Congresso Brasileiro de Zoologia; programa e resumos. Sociedade Brasileira de Zoologia, Cuiabá, Resumo MA 230

Borchers DL, Buckland ST, Zucchini W (2002) Estimating animal abundance: closed populations. Springer, London

Buckland ST, Studeny AC, Magurran AE, Newson SE (2011) Biodiversity monitoring: the relevance of detectability. In: Magurran AE, McGill BJ (eds) Biological diversity: frontiers in measurement and assessment. Oxford University Press, Oxford, pp 25–36

Caughley G (1977) Analysis of vertebrate populations. Wiley, Chichester

Cavalcanti SMC (2003) Manejo e controle de danos causados por espécies da fauna. In: Cullen L Jr, Rudran R, Valladares-Padua C (eds) Métodos de Estudo em Biologia da Conservação and manejo da Vida Silvestre. Editora UFPR, Curitiba, pp 203–242

Coddington JA, Agnarsson I, Miller JA, Kuntner M, Hormiga G (2009) Undersampling bias: the null hypothesis for singleton species in tropical arthropod surveys. J Anim Ecol 78:573–584

Conroy MJ (1996) Abundance indices. In: Wilson DE, Cole FR, Nichols JD, Rudran R, Foster MS (eds) Measuring and monitoring biological diversity: standard methods for mammals. Smithsonian Institution Press, Washington, DC, pp 179–192

Cordero RGA, Ojasti J (1981) Comparison of the capybara populations of open and forested habitats. J Wildl Manage 45:267–271

Cunha HJ, Pinha PRS, Moreira JR, Carvalho JP, Hercos EAP (2002) Análise de poder para monitoramento de capivaras (*Hydrochoerus hydrochaeris*) no Lago Paranoá – Brasília. In: Simpósio Ecologia e Biodiversidade do Cerrado, vol 21, Perspectivas e desafios para o Século. Embrapa Sede, Brasília, p 43, Resumo 132

Davis DE, Winstead RL (1980) Estimating the number of wildlife populations. In: Schemnitz SD (ed) Wildlife management techniques manual, 4th edn. The Widlife Society, Washington, DC, pp 221–245

Eisenberg JF, Redford KH (1999) Mammals of the Neotropics. vol 3, Ecuador, Peru, Bolivia, Brazil. The University of Chicago Press. Chicago

Ferraz KPMB, Verdade LM (2001) Ecologia comportamental da capivara: bases biológicas para o manejo da espécie. In: Mattos WRS (ed) A Produção Animal na Visão dos Brasileiros. Sociedade Brasileira de Zootecnia, Piracicaba, pp 589–595

Ferraz KPMB, Santos-Filho RMF, Piffer TRO, Verdade LM (2001) Biologia e manejo da capivara: do controle de danos ao máximo rendimento sustentável. In: Mattos WRS (ed) A Produção Animal na Visão dos Brasileiros. Sociedade Brasileira de Zootecnia, Piracicaba, pp 580–588

Ferraz KPMB, Lechevalier M-A, Couto HTZ, Verdade LM (2003) Damage caused by capybara (*Hydrochaeris hydrochaeris*) on a cornfield in São Paulo, Brasil. Sci Agríc 60(1):191–194

Ferraz KMPMB, Peterson AT, Scachetti-Pereira R, Vettorazzi CA, Verdade LM (2009) Distribution of capybaras in an agroecosystem, Southeastern Brazil, based on ecological niche modeling. J Mammal 90(1):189–194

Ferraz KMPMB, Manly B, Verdade LM (2010) The influence of environmental variables on capybara (*Hydrochoerus hydrochaeris*: Rodentia, Hydrochoeridae) detectability in anthropogenic environments of southeastern Brazil. Popul Ecol 52:263–270

Herrera EA, Chemello ME, Lacey EA, Salas V, Sousa BF (2004) Characterization of microsatellite markers from capybaras, *Hydrochoerus hydrochaeris* (Rodentia: Hydrochoeridae). Mol Ecol Notes 4:541–543

Kaplan H, Kopischke K (1992) Resource use, traditional technology, and change among native peoples of lowland South America. In: Redford KH, Padoch C (eds) Conservation of neotropical

forests: working from traditional resource use. Collumbia University Press, New York, pp 83–107
Krebs CJ (1989) Ecological methodology. Harper Collins, Nova York
Lancia RA, Nichols JD, Pollock KH (1996) Estimating the number of animals in wildlife populations. In: Bookhout TA (ed) Research and management techniques for wildlife and habitats, 5th edn. The Wildlife Society, Bethesda, pp 215–253
Lancia RA, Kendall WL, Pollock KH, Nichols JD (2005) Estimating the number of animals in wildlife populations. In: Braun CE (ed) Techniques for wildlife investigations and management, vol 6. The Wildlife Society, Bethesda, pp 106–153
Lavorenti A (1989) Domestication and potential for genetic improvement of capybara. Rev Bras Genét 12(3):137–144
Lehner PN (1996) Handbook of ethological methods, 2nd edn. Cambridge University Press, Cambridge
Lord RD, Lord VR (1988) Cross checking censuses and a model of the annual cycle of mortality and reproduction in capybaras (*Hydrochaeris hydrochaeris*). Stud Neotrop Fauna Environ 23:213–224
Mackenzie DI, Nichols JD, Royle JA, Pollock KH, Bailey LL, Hines JE (2006) Occupancy estimation and modeling: inferring patterns and dynamics of species occurrence. Elsevier, Amsterdam
Magnusson WE, Mourão G (2003) Estatística sem Matemática. Editora Planta, Londrina
Magnusson WE, Lima AP, Luizão R, Luizão F, Costa FRC, Castilho CV, Kinupp VF (2005) RAPELD: a modification of the Gentry method for biodiversity surveys in long-term ecological research sites. Biota Neotrop 5(2):1–6
McCallum H (2000) Population parameters: estimation for ecological models. Blackwell, Oxford
Mills LS, Griffin PC, Hodges KE, McKelvey K, Ruggiero L, Ulizio T (2005) Pellet count indices compared to mark-recapture estimates for evaluating snowshoe hare density. J Wildl Manage 69:1053–1062
Moreira JR, Pinha PRS, Cunha HJ (2001) Capivaras do Lago Paranoá. In: Fonseca FO (ed) Olhares sobre o Lago Paranoá. Secretaria de Meio Ambiente e Recursos Hídricos, Brasília, pp 141–147
Mourão G, Campos Z (1995) Survey of the broad-snouted caiman *Caiman latirostris*, marsh deer *Blastocerus dichotomus* and capybara *Hydrochaeris hydrochaeris* in the area to be inundated by Porto Primavera dam, Brazil. Biol Conserv 73:27–31
Mourão GM, Magnusson WE (2004) Uso de levantamentos aéreos para o manejo de populações silvestres. Embrapa-CPAP, Corumbá, Brazil. Documentos 61:1–24
Mourão GM, Bayliss P, Coutinho ME, Abercrombie CL, Arruda A (1994) Test of an aerial survey for caiman and other wildlife in the Pantanal, Brazil. Wildl Soc Bull 22:50–56
Nichols JD (1996) Sources of variation in migratory movements of animal populations: statistical inference and a selective review of empirical results for birds. In: Rhodes OE, Chesser RK, Smith MH (eds) Population dynamics in ecological space and time. University of Chicago Press, Chicago, pp 147–197
Ojasti J (1973) Estudio Biologico del Chigüire o Capibara. FONAIAP, Caracas
Ojasti J (1991) Human exploitation of capybara. In: Robinson JG, Redford KH (eds) Neotropical wildlife use and conservation. The University of Chicago Press, Chicago, pp 236–252
Pezzini FFPHA et al (2012) The Brazilian program for diodiversity research (PPBio) information system. Biodivers Ecol 4: 10p
Pinha PS, Waga IC, Moreira JR (2003) Decréscimo populacional de capivaras (*Hydrochoerus hydrochaeris*) no lago Paranoá – Brasília. In: VI Congresso de Ecologia do Brasil: Anais de Trabalhos Completos. Universidade Federal do Ceará, Fortaleza, pp 564–565
Pinto GRM (2003) Contagem de fezes como índice de abundância de capivaras (*Hydrochaeris hydrochaeris*). M.Sc. Dissertation, Universidade de São Paulo, Piracicaba
Pinto GRM, Ferraz KMPMB, Couto HTZ, Verdade LM (2006) Detectability of capybaras (*Hydrochoerus hydrochaeris*) in forested habitats. Biota Neotrop 6(1):1–5
Plumptre AJ (2000) Monitoring mammal populations with line transect techniques in African forests. J Appl Ecol 37:356–368

Preston FW (1960) Time and space and the variation of species. Ecology 41:612–627
Schneider M, Menegheti JO (1997) Densidade e uso dos hábitats por capivaras, *Hydrochaeris hydrochaeris* (Linnaeus, 1766) (Mammalia; Rodentia) na Estação Ecológica de Aracuri, Muitos Capões, RS. Biociências 5:173–194
Schrader-Frechette K (1995) Methodological rules for four classes of scientific uncertainty. In: Lemons J (ed) Scientific uncertainty and environmental problem solving. Blackwell, Cambridge, MA, pp 12–39
Seber GAF (1973) The estimation of animal abundance and related parameters. Griffen, London
Seber GAF (1982) The estimation of animal abundance and related parameters, 2nd edn. McMillan, New York
Sinclair ARE, Fryxell JM, Caughley G (2006) Wildlife ecology, conservation, and management, 2nd edn. Blackwell, Malden, p 469
Sutherland WJ (2006) Ecological census techniques: a handbook, 2nd edn. Cambridge University Press, Cambridge, p 432
Vargas FC, Vargas SC, Moro MEG, Silva V, Carrer CRO (2007) Monitoramento populacional de capivaras (*Hydrochaeris hydrochaeris* Linnaeus, 1766) em Pirassununga, SP, Brasil. Ciên Rural 37:1104–1108
Verdade LM (1996) The influence of hunting pressure on the social behavior of vertebrates. Rev Bras Biol 56(1):1–13
Verdade LV (2004) A exploração da fauna silvestre no Brasil: jacarés, sistemas e recursos humanos. Biota Neotrop 4(2):1–12
Verdade LM, Ferraz KMPMB (2006) Capybaras (*Hydrochoerus hydrochaeris*) on an anthropogenic habitat in Southeastern Brazil. Braz J Biol 66(1b):371–378
Waga IC, Pinha PS, Moreira JR (2003) Efeito da pressão antrópica sobre o tamanho populacional de capivaras (*Hydrochoerus hydrochaeris*) em uma unidade de conservação em Brasília. VI Congresso de Ecologia do Brasil: Anais de Trabalhos Completos. Universidade Federal do Ceará, Fortaleza, pp 308–309
Williams BK, Nichols JD, Conroy MJ (2002) Analysis and management of animal poplations. Academic, San Diego

Chapter 23
Brazilian Spotted Fever: The Role of Capybaras

Marcelo B. Labruna

23.1 Introduction

Capybaras (*Hydrochoerus hydrochaeris*) are hosts for the tick species *Amblyomma dubitatum* and *Amblyomma cajennense*. The latter, popularly known in Brazil as "carrapato-estrela", is the main vector of the bacterium *Rickettsia rickettsii*, the etiological agent of Brazilian spotted fever, the most deadly rickettsiosis in the world. Current public opinion associates human cases of Brazilian spotted fever with capybaras and their ticks, and this has led to capybaras being blamed for the increasing occurrence of the disease over the last few decades in southeastern Brazil. In fact, the ecology of Brazilian spotted fever is more complex, involving many agents, including, but not restricted to, capybaras. This chapter discusses the role of capybaras in the occurrence of Brazilian spotted fever, especially in the state of São Paulo, southeastern Brazil, where this issue has been well studied. There may be a causal relationship between the rising capybara population and the re-emergence of the disease in the state of São Paulo, since both capybara populations and the number of Brazilian spotted fever cases have increased significantly in this state over the last three decades (Labruna 2009; Del Fiol et al. 2010). However, capybaras are not the sole vertebrate species associated with the reemergence of the disease. We present what is known of the current epidemiology of Brazilian spotted fever, in order to target control and prevention of the disease in areas where capybaras have been shown to play a primary role.

M.B. Labruna (✉)
Departamento de Medicina Veterinária Preventiva e Saúde Animal,
Faculdade de Medicina Veterinária e Zootecnia, Universidade de São Paulo,
Av. Prof. Orlando Marques de Paiva, 87, Cidade Universitária, 05508-270
São Paulo, Brazil
e-mail: labruna@usp.br

J.R. Moreira et al. (eds.), *Capybara: Biology, Use and Conservation of an Exceptional Neotropical Species*, DOI 10.1007/978-1-4614-4000-0_23,

23.2 Spotted Fever Group Rickettsiae

The genus *Rickettsia* encompasses bacteria of the alpha (α) subdivision of the class Proteobacteria, which are cocco-bacillar gram-negative organisms, with an obligate association with eukaryote cells. The *Rickettsia* species have been classified into three general groups based on antigenic, molecular, and ecological characteristics: the typhus group (TG), composed of the species *Rickettsia prowazekii* and *Rickettsia typhi,* primarily associated with lice and fleas, respectively; the spotted fever group (SFG), composed of more than 20 species, the majority primarily associated with ticks (the only exceptions within this group are *Rickettsia felis* and *Rickettsia akari,* which are associated with fleas and Gamasida mites, respectively); and finally, the ancestral group (AG) and other related basal groups, which include the tick-associated species *Rickettsia bellii* and *Rickettsia canadensis,* and other agents associated with annelids, and various insects (Roux et al. 1997; Kikuchi and Fukatsu 2005; Perlman et al. 2006).

Some *Rickettsia* species are transmitted by their invertebrate host (e.g., tick) to animals including humans, which may then become infected, sometimes severely ill, with a disease known, generically, as spotted fever. On the American continent, several SFG agents have been reported to cause disease in humans, including *R. rickettsii, R. parkeri, R. africae, R. massiliae, R. akari*, and *R. felis* (Parola et al. 2009). Various other SFG species are known to infect ticks, and sometimes cause infections in nonhuman animals, but their role as human pathogens is unknown (Blair et al. 2004; Labruna et al. 2004b, 2005a, 2007; Labruna 2009).

23.2.1 *Rickettsia rickettsii*

Rickettsia rickettsii is the etiological agent of the most severe form of rickettsiosis, referred to as Brazilian spotted fever in Brazil, or as Rocky Mountain spotted fever in the USA. The disease also occurs in Mexico, Costa Rica, Panama, Colombia, and Argentina. When infecting vertebrate hosts, *R. rickettsii* multiplies almost exclusively within endothelial cells. In ticks, the bacterium causes generalized infection, multiplying within the cells of the guts, ovaries, salivary glands, malpighian tubules, and in the hemolymph (Weiss and Moulder 1984). Known tick vectors of *R. rickettsii* to humans are *Dermacentor andersoni*, *Dermacentor variabilis*, and *Rhipicephalus sanguineus* in the USA (Demma et al. 2005); *R. sanguineus* and *Amblyomma cajennense* in Mexico (Bustamante and Varela 1947); *A. cajennense* in Panama (Rodaniche 1953), Colombia (Patino-Camargo 1941), and Argentina (Paddock et al. 2008); and *A. cajennense* and *Amblyomma aureolatum* in Brazil (Guedes et al. 2005; Pinter and Labruna 2006; Labruna 2009).

Transmission of *R. rickettsii* to a human through the parasitism of an infected tick results in classical symptoms that generally occur after an incubation period of 5–10 days: these include high fever, headache, and myalgia. A typical cutaneous rash also develops in most patients during this febrile period. Fatality rates of untreated cases can be as high as 80%. However, if treated with specific antibiotics (tetracyclines, chloramphenicol) started in the first days of fever, fatality rates are usually less than 10% (Galvão 1996).

Brazilian spotted fever has been known since the 1920s, but mostly restricted to the southeastern states of Brazil, namely, São Paulo, Minas Gerais, Rio de Janeiro, and Espírito Santo (Silva and Galvão 2004). There have been confirmed cases in the state of Bahia, in northeastern Brazil (Plank et al. 1979). More recently, during the last 10 years, the disease was first confirmed in southern Brazil (states of Paraná, Santa Catarina, and Rio Grande do Sul), as well as in central and northern Brazil (Del Fiol et al. 2010). However, the *Rickettsia* species responsible for the infection in these states has not been identified. It is suspected that at least some of these human cases were caused by a *Rickettsia* species other than *R. rickettsii,* since there were peculiar clinical and epidemiological characteristics observed that appeared to be distinct from the classical Brazilian spotted fever encountered in southeastern Brazil (Labruna 2009).

23.3 Ecology of Brazilian Spotted Fever

In the USA, *R. rickettsii* is maintained in nature between its tick vectors (*D. andersoni* and *D. variabilis*) and several small rodent species (*Microtus pennsylvanicus, Microtus pinetorum, Peromyscus leucopus* and *Sigmodon hispidus*; McDade and Newhouse 1986; Burgdorfer 1988). *R. rickettsii* is partially pathogenic to ticks so, although the bacterium undergoes transovarial transmission (hereditary transmission) between successive tick generations, the infection rate drops in the tick population with each tick generation because mortality rates are higher among infected than among uninfected ticks (Burgdorfer 1988; Niebylski et al. 1999; Labruna et al. 2011). Under these conditions, an amplifier host is required to maintain the bacterium active in nature. The amplifier hosts (in this case, rodents) maintain the bacterium in their bloodstream for some days or weeks, at sufficient levels to infect new tick cohorts, amplifying the rickettsial infection among the tick population (Burgdorfer 1988).

In general, a vertebrate host species has to fulfill the following requirements to function as an efficient amplifier host in this system (Labruna 2009): (1) It has to be abundant in the *R. rickettsii*-endemic area; (2) it has to be a major host for the tick vector; (3) it has to be susceptible to *R. rickettsii* infection; (4) once infected by *R. rickettsii,* the host has to develop a rickettsemia of sufficient length and degree to infect ticks that feed on this host; and (5) it has to be a prolific species, to continuously produce nonimmune animals into the host population.

In Brazil the situation differs from that in the USA, because *A. cajennense* (Fig. 23.1), the main tick vector of *R. rickettsii* in Brazil, does not feed on small rodents at any stage in its life cycle. The larvae and nymphs of *A. cajennense* feed preferentially on medium and large-sized mammals, while adults parasitize chiefly large mammals (Labruna et al. 2005b).

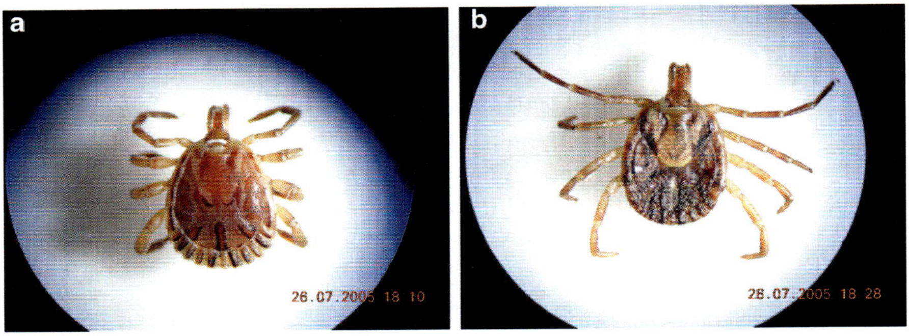

Fig. 23.1 The tick *Amblyomma cajennense*. (**a**) Male (**b**) Female

Box 23.1 Ecological Aspects of the Cycle of *Rickettsia rickettsii*

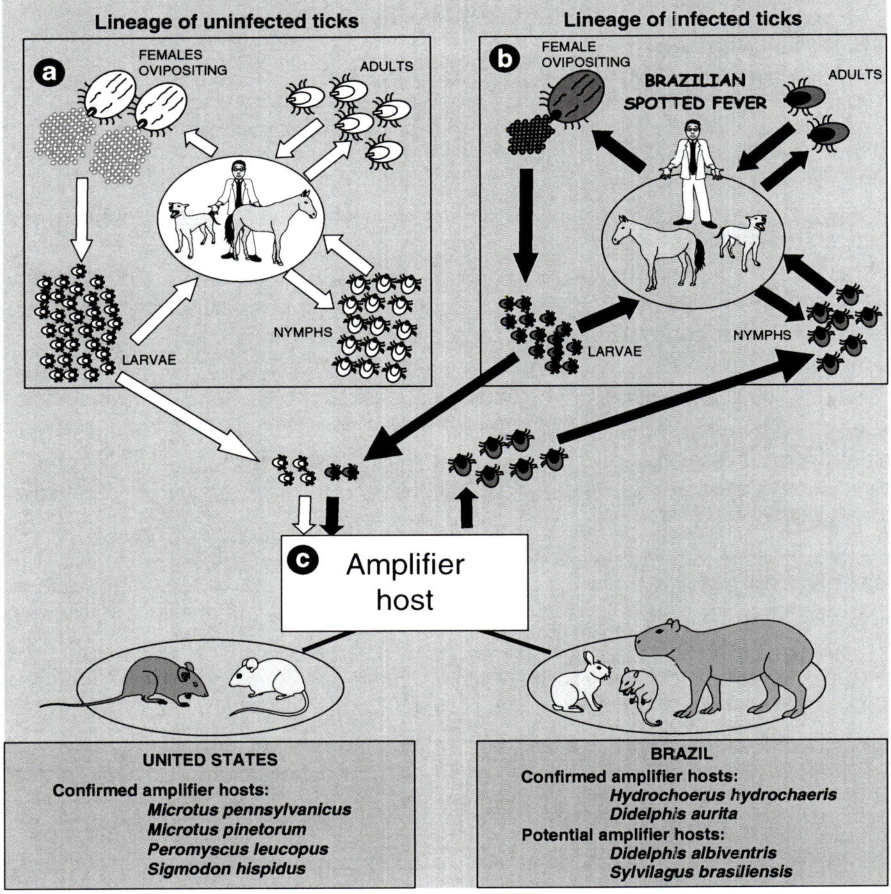

(continued)

Box 23.1 (continued)

(**a**) Larvae, nymphs, and adults of an uninfected tick lineage are maintained free of infection while feeding on uninfected hosts; hence, when they feed on humans, they do not transmit Brazilian spotted fever. (**b**) A lineage of *R. rickettsii*-infected ticks maintain the agent in the tick population, through transovarial and transstadial transmissions. Due to the lethal effect of *R. rickettsii* on some ticks, the infected lineage tend to have lower reproductive performance and survivorship throughout the generations. Thus, it is necessary the participation of amplifier hosts (**c**), which are vertebrate hosts that once primo-infected with *R. rickettsii* via an infected tick, they keep the bacterium circulating in their blood (rickettsemia) for a few days or weeks, when new uninfected ticks become infected while feeding on the host, starting new lineages of infected ticks.

Both transstadial (transmission between developmental stages) and transovarial transmission of *R. rickettsii* occurs in *A. cajennense* (Monteiro et al. 1932; Monteiro and Fonseca 1932; Brumpt 1933), and recent studies have demonstrated that both capybaras and opossums (*Didelphis aurita*) can act as amplifier hosts in Brazil (Horta et al. 2009; Souza et al. 2009). This latter finding corroborates earlier suspicions, dating back to the 1930s, when capybaras, opossums, and wild rabbits were incriminated as possible amplifier hosts (Box 23.1).

23.3.1 The Role of Capybaras in the Ecology of Spotted Fever

It is expected that capybaras play a major role in the maintenance and transmission of Brazilian spotted fever because this animal can act as the main host for the tick *A. cajennense*. On this basis, it could be predicted that the higher the population density of capybaras, the larger the tick population and, consequently, the higher the risk of human infection.

Capybaras fulfill very well the five requirements described above for an amplifier host for *R. rickettsii*. Indeed, capybaras are abundant in many Brazilian spotted fever-endemic areas, where they act as primary hosts for all parasitic stages of *A. cajennense*. Their susceptibility to *R. rickettsii* has been known since the 1940s (Travassos and Vallejo 1942; Souza et al. 2004), and it was recently shown that after being experimentally infected with *R. rickettsii*, capybaras maintained viable rickettsiae circulating in their blood (rickettsemia) for 1–2 weeks, when they infected 20–25% of the *A. cajennense* nymphs that fed on them. Capybaras are prolific, producing a mean of six pups per female per year (Ojasti 1973), generating a constant introduction of susceptible animals, not previously infected by *R. rickettsii*. For these reasons, capybaras have an important role in the ecology of Brazilian spotted fever in many endemic areas of Brazil. However, the capybara clearly cannot be the only vertebrate host involved because there are areas of Brazil where *R. rickettsii* is endemic but where capybaras are absent. Further, it is perhaps noteworthy that when experimentally infected with *R. rickettsii,* capybaras showed no clinical signs of infection (Souza et al. 2009).

23.3.2 Other Animals Involved or Possibly Involved in the Ecology of Brazilian Spotted Fever

Like capybaras, opossums (*Didelphis aurita*, and possibly *Didelphis albiventris*) also fulfill the five requirements for a suitable amplifier host of *R. rickettsii* for *A. cajennense* ticks. Opossums are very abundant in nearly all endemic areas, where they are frequently infested by the larvae and nymphs of *A. cajennense* (Horta 2006). The susceptibility of opossums to *R. rickettsii* infection has been documented since the 1930s (Moreira and Magalhães 1935; Horta 2006). A recent study in Brazil showed that *R. rickettsii*-experimentally infected opossums (*D. aurita*) developed rickettsemia lasting up to 3–4 weeks, when ≈5–20% of the *A. cajennnense* immature ticks that fed on them became infected by *R. rickettsii* (Horta et al. 2009). Opossums are also relatively prolific.

In the past, wild rabbits (*Sylvilagus* spp.) were considered an important amplifier host of *R. rickettsii* in the USA. Rabbits and their ticks (namely, the rabbit tick, *Haemaphysalis leporispalustris*) can be infected by *R. rickettsii*. Rabbits are abundant in *R. rickettsii*-endemic areas and they present rickettsemia of sufficient magnitude to infect ticks that feed on them (Parker et al. 1951; Shirai et al. 1961; Bozeman et al. 1967). However, in the USA, small rodents are much more frequently parasitized by tick vectors (*D. variabilis* and *D. andersoni*) than are rabbits, and they present rickettsemia of much higher magnitude, resulting in a larger proportion of infected ticks (Burgdorfer et al. 1980; Burgdorfer 1988). In Brazil, the wild rabbit *Sylvilagus brasiliensis* is abundant in many endemic areas, especially in the states of Minas Gerais and Rio de Janeiro. However, there is a lack of studies focusing on the role of these animals in the ecology of Brazilian spotted fever.

The role of the domestic dog (*Canis familiaris*) as amplifier host for *R. rickettsii* has been controversial, especially in the USA (Weiss and Moulder 1984; McDade and Newhouse 1986). Although Norment and Burdorfer (1984) reported that dogs, experimentally infected with *R. rickettsii*, did not develop rickettsemia of sufficient magnitude to infect more than 1% of *R. sanguineus* ticks that fed on them, earlier studies reported that rickettsemic dogs infected the majority of the *D. variabilis* ticks that fed on them (Price 1954; Keenan et al. 1977). More recently, human cases of Rocky Mountain spotted fever (*R. rickettsii*) vectored by *R. sanguineus* in Arizona, USA, resurrected the hypothesis that dogs might play a role as an amplifier host of *R. rickettsii*. All parasitic stages (larvae, nymphs, and adults) of *R. sanguineus* feed primarily on the domestic dog (Demma et al. 2005) and it is a primary host for the adult stage of *A. aureolatum,* one of the vectors of Brazilian spotted fever in the state of São Paulo (Pinter et al. 2004; Pinter and Labruna 2006).

The tick *R. sanguineus,* a confirmed vector of *R. rickettsii* in Mexico and USA, is the principal tick species that parasitizes dogs in Brazil, especially in urban areas (Labruna 2004). A recent study in Brazil showed that dogs could act as amplifier hosts for *R. sanguineus* ticks, after being experimentally infected with a Brazilian strain of *R. rickettsii* (Piranda et al. 2011). However, the role of *R. sanguineus* as a vector of *R. rickettsii* to humans in Brazil is unknown because, in Brazil, this tick very rarely bites humans.

Horses (*Equus caballus*), primary hosts for *A. cajennense,* are abundant in many Brazilian spotted fever-endemic areas in Brazil. Horta et al. (2004) demonstrated by serum cross-absorption techniques that horses became naturally infected by *R. rickettsii,* showing high antibody homologous titers. In some endemic areas of the states of São Paulo and Minas Gerais, where *A. cajennense* is a natural vector, 57–90% of the horses are serologically positive for *R. rickettsii* (Lemos et al. 1996; Horta et al. 2004). However, both the clinical symptoms (if any) induced by *R. rickettsii* in horses, or the role of horses as amplifier hosts, have yet to be investigated.

Tapirs (*Tapirus terrestris*) and possibly peccaries (*Tayassu pecari* and *Pecari tajacu*), despite being primary hosts for *A. cajennense*, are not present in any known Brazilian spotted fever-endemic area. Thus, there is no evidence that they play any direct role in the ecology of the disease.

It can be seen from the preceding discussion that any human interference that raises the reproduction rate of the amplifier host population would also increase the number of susceptible animals (that would develop rickettsemia) and, consequently, the *R. rickettsii*-infection rate in the tick population would be higher, increasing the risk of human disease.

23.4 Prevention of the Disease

As for any vector-borne disease, the occurrence of spotted fever is directly related to the size of the vector population (ticks). Human infestation by *Amblyomma* ticks is an accidental event, resulting from the large number of free-living ticks in the environment. Therefore, the most efficient method to minimize human infections, and prevent Brazilian spotted fever, is to reduce the size of the tick population.

Once a *R. rickettsii*-infected tick attaches itself to a host, it takes a minimum of 4–6 h to inoculate the bacterium into the host. Therefore, the faster a person removes a recently attached tick, the lower the chance of acquiring spotted fever. Since there are no commercial vaccines to prevent spotted fever in humans, the most efficient method to prevent the disease is to keep the tick population controlled at low levels and to reduce the chance of a tick coming into contact with a human, thus minimizing human infestations.

The establishment and growth of an *A. cajennense* population in a given area is dependent on the presence of at least one of its primary host species (horses, capybaras, tapirs, or peccaries) and the existence of suitable environmental conditions for its free-living developmental stages. The control of ticks, therefore, can be achieved directly by targeting either the parasitic population or the free-living population of ticks, or indirectly by reducing (or eliminating) the population of its primary host from the area. Educational programs for the human population inhabiting *R. rickettsii*-endemic areas are also important, and control efforts must be continuous to minimize the risk of transmission of spotted fever between nontreated and treated areas.

23.4.1 Control of *A. cajennense* by Targeting the Parasitic Population

The traditional method to control ticks is by treatment of parasitized animals with acaricides (pesticides that kill ticks and mites). A suitable method for continuous application of acaricides on free-living wild hosts, such as capybaras, however, is yet to be developed, so this approach is only suitable for domestic animals, such as horses. Details of treatment protocols for *A. cajennense* on horses can be found elsewhere (Leite et al. 1997; Labruna et al. 2004a; Cunha et al. 2007).

Any tick control program must be continuous, and results should only be expected in the medium or long-term, namely, after 1–3 years of control. There is a natural human desire for treatment to produce a rapid effect. However, immediate results, following a single acaricide application, will only be seen in a single infestation on a severely infested animal. The main aim of the control program is to reduce the environmental burden of ticks, through continuous treatments on animals. Occasional curative treatments have no effect on the free-living tick population; i.e., they do not control ticks.

23.4.2 Control of *A. cajennense* by Targeting the Free-Living Population

One study of 40 horse farms in the state of São Paulo showed that both the presence and the abundance of *A. cajennense* infestations on horses was associated with the presence of at least one overgrown mixed pasture on the farm (Fig. 23.2; Labruna et al. 2001). It is likely that the vegetation composition of a mixed pasture, composed of bushes and shrubs as well as grass, provides the ideal microclimate for the free-living developmental stages of the tick.

For isolated pastures, where there are no forests or preservation areas in the vicinity, the environmental tick burden can be reduced by destroying the microclimate conditions required for tick survival and development by mechanical mowing. The whole pasture should be mowed, close to the soil, at least once a year in the second half of the rainy season (January to March). After mowing, all cut grass must be immediately removed from the pasture and discarded. Annual mowing during

Fig. 23.2 Distinct situations with the presence of capybaras. (**a**) A pasture in good condition (without bushes and shrubs interspersed with grasses) inhabited by capybaras, but without *Amblyomma cajennense*, since that environment is not suitable for this tick species (Photo by J.R. Moreira). (**b**) A mixed overgrowth pasture inhabited by capybaras, with abundance of *A. cajennense*, since this denser vegetation favors the free-living stages of the tick (Photo by M.B. Labruna)

the rainy season prevents the formation of mixed overgrowth pasture, since it favors the regrowth of grasses over bushes and shrubs (which themselves are undesired invaders). Ideally, mowing should be carried out during the second half of the rainy season, when the vast majority of free-living ticks exist as eggs or unfed larvae in diapause prior to the larval infestation peak at the end of the rainy season and first half of the dry season (April to June; Labruna et al. 2003). Eggs and larvae are particularly sensitive to desiccation, so reducing the vegetation cover available should be extremely harmful to these tick stages.

For horses reared in forest or preservation areas, where mechanical destruction of the vegetation is not possible, the only feasible alternative is direct treatment of the horses with acaricides (above). However, restricting the horses' access to these forest areas will also give satisfactory results, since in the absence of primary hosts, most ticks will not be able to feed efficiently and will die of starvation. Nevertheless, the impacts of such an approach will not be seen for a year or two, because ticks can remain viable without feeding for up to a year or more in a suitable environment (Rohr 1909).

The worst situation arises when a forest area harbors high populations of *A. cajennense* that are sustained by wildlife (e.g., capybaras). In this case, neither direct treatment of horses nor mechanical removal of the vegetation will be effective. Local inhabitants can be advised not to enter forest areas harboring infected ticks, but educational activities (below) are crucial. Scientific research into methods to control populations of both ticks and capybaras in these areas is urgently needed.

23.4.3 Control of A. cajennense Through Exclusion of Primary Hosts

In places where there are no capybaras, and horses are the sole primary host for *A. cajennense*, the tick population can be successfully controlled by simply removing all horses from this area. However, it must be borne in mind that the free-living stages of *A. cajennense* (especially the adult stage) can remain viable without feeding in the environment for about 1 year (Rohr 1909). Therefore, results should only be expected 1 or 2 years after removing the primary host.

If capybaras are the primary host for *A. cajennense*, exclusion will only be efficient for tick control if *all* animals are removed, and if there are no means by which capybaras can recolonize the area. Partial exclusion of capybaras, without decreasing the food resources in a given area, will increase their reproductive rate (see Chap. 18), giving rise to a larger number of capybaras susceptible to infection by *R. rickettsii* (see requirement five for a species to be considered a good amplifier host). Although such a scenario has not been demonstrated under field conditions, it is quite feasible that an increased reproductive rate in capybaras could result in an increase in the *R. rickettsii*-infection rate in the tick population, since a greater number of capybaras would be acting as amplifier hosts. Consequently, the risk of humans contracting Brazilian spotted fever would also be likely to increase in the area. For this reason, partial removal of capybaras in a Brazilian spotted fever-endemic area, without decreasing food resources, is banned by law.

23.4.4 Educating the Population to Decrease Risks of Spotted Fever

In *R. rickettsii* endemic areas, only a small portion of the tick population is actually infected with *R. rickettsii*. For *A. cajennense,* this portion is usually below 1% (Sangioni et al. 2005). Thus, in these endemic areas, the more ticks on a person, the higher the chances are of that person harboring an infected tick. For this reason, it is mandatory to inform the human population of appropriate methods of tick control as described above.

In parallel with appropriate tick control, public health services need to inform local human populations about the risks of acquiring spotted fever. In areas where the disease is known to occur, people should search for and remove ticks from their bodies every 2–3 h while in the tick-infested area. When entering such areas, clothes that provide physical barriers against ticks should be worn (arms and legs should be covered and long boots should be worn with trousers tucked inside them). Light-colored clothes are best, because ticks can be more easily seen on them, before they can reach the skin.

Finally, it is important to remember that Brazilian spotted fever is a bacterial disease easily treated with antibiotics (tetracyclins or chloramphenicol), provided these drugs are prescribed in the very first days of the febrile period. For this reason, it is mandatory to inform local medical services of the risks, symptoms and treatment. They should therefore be aware that any case of acute febrile disease may be a clinical sign of spotted fever, and to ensure that an appropriate antibiotic therapy is given as early as possible in all suspected or probable cases.

References

Blair PJ et al (2004) Characterization of spotted fever group rickettsiae in flea and tick specimens from northern Peru. J Clin Microbiol 42:4961–4967

Bozeman FM, Shirai A, Humphries JW, Fuller HS (1967) Ecology of Rocky Mountain spotted fever II. Natural infection of wild mammals and birds in Virginia and Maryland. Am J Trop Med Hyg 16:48–59

Brumpt E (1933) Transmission de la fiébre pourprée des Montagnes rocheuses par la tique américaine *Amblyomma cajennense*. Comptes Rendues des Seances de la Societé de Biologie. Compt Rend Soc Biol 144:416–419

Burgdorfer W (1988) Ecological and epidemiological considerations of Rocky Mountain spotted fever and scrub typhus. In: Walker DH (ed) Biology of Rickettsial diseases, vol 1. CRC Inc, Boca Raton, pp 33–50

Burgdorfer W, Cooney JC, Mavros AJ, Jellison WL, Maser C (1980) The role of cottontail rabbits (*Sylvilagus* spp.) in the ecology of *Rickettsia rickettsii* in the United States. Am J Trop Med Hyg 29:686–690

Bustamante ME, Varela G (1947) Distribuicion de las rickettsiasis en Mexico. Rev Inst Salubr Enferm Trop 8:3–14

Cunha AP, Bello ACP, Leite RC, Bastianetto E, Ribeiro ACCL, Freitas CMV, Oliveira PR (2007) Controle estratégico de Amblyomma cajennense (Fabricius,1787) (Acari:Ixodidae) em equinos em Minas Gerais, Brasil – Parte I. Rev Bras Parasitol Vet 16:212–219

Del Fiol FS et al (2010) Febre maculosa no Brasil. Rev Panam Salud Publica 27:461–466

Demma LJ et al (2005) Rocky Mountain spotted fever from an unexpected tick vector in Arizona. N Engl J Med 353:587–594

Galvão MAM (1996) Febre maculosa em Minas Gerais: um estudo sobre a distribuição da doença no estado e seu comportamento em área de foco periurbano. Ph.D. thesis, Universidade Federal de Minas Gerais, Belo Horizonte

Guedes E, Leite RC, Prata MCA, Pacheco RC, Walker DH, Labruna MB (2005) Detection of *Rickettsia rickettsii* in the tick *Amblyomma cajennense* in a new Brazilian spotted fever–endemic area in the state of Minas Gerais. Memórias do Instituto Oswaldo Cruz. Compt Rend Soc Biol 100:841–848

Horta MC (2006) Estudo Epidemiológico de *Rickettsia felis* em Áreas Endêmicas e Não-endêmicas para Febre Maculosa Brasileira do Estado de São Paulo. Ph.D. thesis, Universidade de São Paulo, São Paulo

Horta MC et al (2004) Prevalence of antibodies to spotted fever group rickettsiae in humans and domestic animals in a Brazilian spotted fever endemic area in the state of São Paulo, Brazil: serological evidence for infection by *Rickettsia rickettsii* and another spotted fever group rickettsia. Am J Trop Med Hyg 71:93–97

Horta MC, Moraes-Filho J, Casagrande RA, Saito TB, Rosa SC, Ogrzewalska M, Matushima ER, Labruna MB (2009) Experimental infection of opossums *Didelphis aurita* by *Rickettsia rickettsii* and evaluation of the transmission of the infection to ticks *Amblyomma cajennense*. Vector Borne Zoonotic Dis 9:109–117

Keenan K et al (1977) Pathogenesis of infection with *Rickettsia rickettsii* in the dog: a disease model for Rocky Mountain spotted fever. J Infect Dis 135:911–917

Kikuchi Y, Fukatsu T (2005) *Rickettsia* infection in natural leech populations. Microb Ecol 49:265–271

Labruna MB (2004) Biologia-ecologia de *Rhipicephalus sanguineus* (Acari: Ixodidae). Revista Brasileira de Parasitologia Veterinária 13(suppl.):123–124

Labruna MB (2009) Ecology of *Rickettsia* in South America. Ann N Y Acad Sci 1166:156–166

Labruna MB, Kerber CE, Ferreira F, Faccini JLH, De Waal DT, Gennari SM (2001) Risk factors to tick infestations and their occurrence on horses in the State of São Paulo, Brazil. Vet Parasitol 97:1–14

Labruna MB, Amaku M, Metzner JA, Pinter A, Ferreira F (2003) Larval behavioral diapause regulates life cycle of *Amblyomma cajennense* (Acari: Ixodidae) in southeast Brazil. J Med Entomol 40:171–178

Labruna MB, Leite RC, Gobesso A, Gennari SM, Kasai N (2004a) Controle estratégico do carrapato *Amblyomma cajennense* em eqüinos. Ciência Rural Santa Maria. Compt Rend Soc Biol 34:195–200

Labruna MB et al (2004b) *Rickettsia bellii* and *Rickettsia amblyommii* in *Amblyomma* ticks from the state of Rondonia, Western Amazon, Brazil. J Med Entomol 41:1073–1081

Labruna MB, Camargo LMA, Camargo EP, Walker DH (2005a) Detection of a spotted fever group *Rickettsia* in the tick *Haemaphysalis juxtakochi* in Rondonia, Brazil. Vet Parasitol 127:169–174

Labruna MB, Camargo LMA, Terrassini FA, Ferreira F, Schumaker TTS, Camargo EP (2005b) Ticks (Acari: Ixodidae) from the State of Rondônia, western Amazon, Brazil. Sys App Acarol 10:17–32

Labruna MB et al (2007) Prevalence of *Rickettsia* infection in dogs from the urban and rural areas of Monte Negro Municipality, western Amazon, Brazil. Vector Borne Zoonotic Dis 7:249–255

Labruna MB et al (2011) Experimental infection of *Amblyomma aureolatum* ticks with *Rickettsia rickettsii*. Emerg Infect Dis 17:829–834

Leite RC, Oliveira PR, Lopes CML, Freitas CMV (1997) Alguns aspectos epidemiológicos das infestações por *Amblyomma cajennense*: uma proposta de controle estratégico. In: Veríssimo CJ, Augusto C (eds) II Simpósio sobre Controle de Parasitos: Controle de parasitos de eqüinos. Instituto de Zootecnia, Nova Odessa, pp 9–14

Lemos ER, Machado RD, Coura JR, Guimarães MA, Chagas N (1996) Epidemiological aspects of the Brazilian spotted fever: serological survey of dogs and horses in an endemic area in the State of Sao Paulo, Brazil. Rev Inst Med Trop Sao Paulo 38:427–430

Mcdade JE, Newhouse VF (1986) Natural history of *Rickettsia rickettsii*. Annu Rev Microbiol 40:287–309

Monteiro JL, Fonseca F (1932) Typho exanthematico de S. Paulo – Novas experiências sobre transmissão experimental por carrapatos (*Boophilus microplus* e *Amblyomma cajennense*). Memórias do Instituto Butantã. Compt Rend Soc Biol 7:33–40

Monteiro JL, Fonseca F, Prado A (1932) Typho endêmico de São Paulo – VI. Pesquisas sobre a possibilidade da transmissão experimental do vírus por Ixodidae. Compt Rend Soc Biol 46:49–52

Moreira JA, Magalhães O (1935) Thypho exanthematico em Minas Gerais. Bras Med 44:465–470

Niebylski ML, Peacock MG, Schwan TG (1999) Lethal effect of *Rickettsia rickettsii* on its tick vector (*Dermacentor andersoni*). Appl Environ Microbiol 65:773–778

Norment BR, Burgdorfer W (1984) Susceptibility and reservoir potential of the dog to spotted fever-group rickettsiae. Am J Vet Res 45:1706–1710

Ojasti J (1973) Estudio Biologico del Chigüire o Capibara. FONAIAP, Caracas

Paddock CD, Fernandez S, Echenique GA, Sumner JW, Reeves WK, Zaki SR, Remondegui CE (2008) Rocky Mountain spotted fever in Argentina. Am J Trop Med Hyg 78:687–692

Parker RR, Pickens EG, Lackman DB, Bell EJ, Thraikill FB (1951) Isolation and characterization of Rocky Mountain spotted fever Rickettsiae from the rabbit tick *Haemaphysalis leporispalustris* Packard. Publ Health Rep 66:455–463

Parola P, Labruna MB, Raoult D (2009) Tick-borne Rickettsioses in America: unanswered questions and emerging diseases. Curr Infect Dis Rep 11:40–50

Patino-Camargo L (1941) Nuevas observaciones sobre un tercer foco de fiebre petequial (maculosa) en el hemisferio americano. Boletin de la Oficina Sanitaria Panamericana. Compt Rend Soc Biol 20:1112–1124

Perlman SJ, Hunter MS, Zchori-Fein E (2006) The emerging diversity of *Rickettsia*. Proceedings. Biological sciences 273:2097–2106

Pinter A, Labruna MB (2006) Isolation of *Rickettsia rickettsii* and *Rickettsia bellii* in cell culture from the tick *Amblyomma aureolatum* in Brazil. Ann N Y Acad Sci 1078:523–529

Pinter A, Dias RA, Gennari SM, Labruna MB (2004) Study of the seasonal dynamics, life cycle, and host specificity of *Amblyomma aureolatum* (Acari: Ixodidae). J Med Entomol 41:324–332

Piranda EM, Faccini JL, Pinter A, Pacheco RC, Cançado PH, Labruna MB (2011) Experimental infection of *Rhipicephalus sanguineus* ticks with the bacterium *Rickettsia rickettsii*, using experimentally infected dogs. Vector Borne Zoonotic Dis 11:29–36

Plank SJ, Teixeira RS, Milanesi ML (1979) Febre maculosa em Salvador: descrição de um caso. Rev Med Bahia 25:330–334

Price WH (1954) The epidemiology of rocky mountain spotted fever. II. Studies on the biological survival mechanism of Rickettsia rickettsii. Am J Trop Med 19:103–108

Rodaniche EC (1953) Natural infection of the tick *Amblyomma cajennense* with *Rickettsia rickettsii* in Panama. Am J Trop Med Hyg 2:696–699

Rohr C (1909) Estudo sobre Ixódidas do Brasil. Inst. Oswaldo Cruz, Rio de Janeiro, 220 pp

Roux V, Rydkina E, Eremeeva M, Rault D (1997) Citrate synthase gene comparison, a new tool for phylogenetic analysis and its application for the Rickettsiae. Int J Syst Evol Microbiol 47:252–261

Sangioni LA et al (2005) Rickettsial infection in animals and Brazilian spotted fever endemicity. Emerg Infect Dis 11:265–270

Shirai A, Bozeman M, Perri S, Humphries JW, Fuller HS (1961) Ecology of Rocky Mountain spotted fever. I. *Rickettsia rickettsii* recovered from a cottontail rabbit from Virginia. Proc Soc Exp Biol Med 107:211–214

Silva LJ, Galvão MAM (2004) Epidemiologia das rickettsioses do gênero *Rickettsia* no Brasil. Revista Brasileira de Parasitologia Veterinária 13(Suppl.):197–198

Souza CE et al (2004) O papel da capivara *Hydrochaeris hydrochaeris* na cadeia epidemiológica da febre maculosa brasileira. Revista Brasileira de Parasitologia Veterinária 13(Suppl):203–205

Souza CE, Moraes-Filho J, Ogrzewalska M, Uchoa FC, Horta MC, Souza SS, Borba RC, Labruna MB (2009) Experimental infection of capybaras *Hydrochoerus hydrochaeris* by *Rickettsia rickettsii* and evaluation of the transmission of the infection to ticks *Amblyomma cajennense*. Vet Parasitol 161:116–121

Travassos J, Vallejo A (1942) Comportamento de alguns cavídeos (*Cavia aperea* e *Hydrochoerus capybara*) às inoculações experimentais do vírus da febre maculosa. Possibilidade desses cavídeos representarem o papel de depositários transitórios do vírus na natureza. Compt Rend Soc Biol 15:73–86

Weiss E, Moulder JW (1984) Bergey's manual of systematic bacteriology. In: Kreig NR, Holt JG (eds) Bergey's manual of systematic bacteriology, vol 1. Williams & Wilkins, Baltimore, pp 687–739

Chapter 24
The Capybara Paradigm: From Sociality to Sustainability

David W. Macdonald, Emilio A. Herrera, Katia Maria P.M.B. Ferraz, and José Roberto Moreira

24.1 Introduction

Capybaras are not merely unusual, they are extraordinary. Of course, it is widely known that they are the largest living rodent (*c.* 50 kg). It is less well known that their biology is exceptional in many other regards, as documented in the foregoing chapters of this book, *Capybara: Biology, Use and Conservation of an Exceptional Neotropical Species.*

A snapshot of capybara extremes begins with their digestive system: they are not only the largest rodents, but also the largest species to both engage in cecal fermentation and be a cecotrophous mammal (Herrera 1985, 2012a; Borges et al. 1996; Hirakawa 2001, 2002; Mendes and Nogueira-Filho 2012). Furthermore, the occlusal morphology of their cheek teeth is so peculiar that it was necessary to develop a specific nomenclature for them (Vucetich et al. 2012) – capybara have unpaired

D.W. Macdonald (✉)
Wildlife Conservation Research Unit Zoology Department, University of Oxford
The Recanati-Kaplan Centre Oxford, Tubney House, Abingdon Road,
Tubney, Abingdon OX13 5QL, UK
e-mail: david.macdonald@zoo.ox.ac.uk

E.A. Herrera
Departamento de Estudios Ambientales, Universidad Simón Bolívar, Apdo. 89000,
Caracas, 1080-A, Venezuela
e-mail: eherre@usb.ve

K.M.P.M.B. Ferraz
Departamento de Ciências Florestais, Escola Superior de Agricultura "Luiz de Queiroz",
Universidade de São Paulo, Av. Pádua Dias 11, Piracicaba, São Paulo 13418-900, Brazil
e-mail: katia.ferraz@usp.br

J.R. Moreira
Embrapa Recursos Genéticos e Biotecnologia, Parque Estação Biológica, Final W5 Norte,
70770-900 Brasília, Distrito Federal, Brazil
e-mail: alencar_moreira@yahoo.com.br

J.R. Moreira et al. (eds.), *Capybara: Biology, Use and Conservation of an Exceptional Neotropical Species*, DOI 10.1007/978-1-4614-4000-0_24,

ever-growing cheek teeth with a very intricate occlusal surface design that becomes more complex throughout life (Vucetich et al. 2005). As Vucetich et al. (2012) point out, this growth pattern is unique amongst rodents, and was originally misinterpreted (see Vucetich et al. 2005): small fossil specimens with simpler morphology were originally thought to be primitive, whereas larger specimens with more intricate morphology were interpreted as more derived ones. Now, with new insights into the ontogeny of their dental morphology, some of these small "species" are recognized to be juveniles of larger ones. Functionally, these molars enable capybaras to reduce plant food to particle sizes of 0.001–0.3 mm, maceration comparable to that achieved by similarly sized ruminants (Ojasti 1973; Fritz et al. 2009; Barreto and Quintana 2012; Herrera 2012a) with the result that, overall, the digestive efficiency of capybaras is comparable to that of ruminants (González-Jiménez 1977). In South America, the capybara can be considered an ecological equivalent of the medium-sized ungulates of Africa (Ojasti 1983; Macdonald et al. 2007). Nonetheless, capybara group ranges (as documented by Herrera and Macdonald 1987) are small compared with other tropical herbivores of similar size – perhaps due to the pattern of primary productivity in their habitats and little competition.

Turning to their reproduction, capybaras are again unusual and have the longest gestation known amongst rodents at 150.6 ± 2.8 days (López-Barbella 1987; Miglino et al. 2012). While there is a negative correlation between litter size and body mass among South American Hystricognaths (Kleiman et al. 1979), capybaras have both the largest litter size (1–8, mean 4) and body mass (*c.* 50 kg) of the infra-order. These unexpectedly productive reproductive characteristics are thought to be adaptations to the extremes of drought and flooding of neotropical savannas, which may cause high infant mortality. Nonetheless, capybara litter sizes are still small compared with those of other (much smaller in body size) rodents like the Myomorphs (rats, mice, and the like).

In terms of reproductive anatomy, capybaras are unusual in that, despite being the largest rodent, adult males have one of the lowest gonadosomatic indices (i.e., the percentage of body weight allocated to the testicle) amongst rodents (0.12%; Kenagy and Trombulak 1986; Moreira et al. 1997b; Paula 1999). This suggests that there is low sperm competition among capybara males. The testes are not only small (32 g), but remarkably uniform in size (Paula 1999; Paula and Walker 2012). Furthermore, nearly a third of the adult testicle is comprised of Leydig cells (the part of the testicle that produces testosterone; the other main component is seminiferous tubules, which produce sperm) in capybaras (Paula et al. 2007). This is the highest volumetric proportion of Leydig cells so far described for any mammal (Moreira et al. 1997a; França and Russell 1998; Paula 1999; Costa and Paula 2006; Costa et al. 2006; Paula et al. 2007). The approximate count of 126.4 million of these cells per gram of capybara testicle (Costa et al. 2006) contrasts dramatically with only six million per gram for guinea pigs (Mori et al. 1980; Zirkin and Ewing 1987). Nonetheless, given the much higher gonadosomatic index (4.26%) of rats, they have twice the quantity, proportional to body weight, of Leydig cells to that found in capybaras. Capybaras appear to invest more in the production of testosterone than in sperm-producing tissue (Moreira et al. 1997b), perhaps an adaptation to the maintenance of a strict dominance hierarchy year round (Herrera and Macdonald 1993). Continuing the pattern of extremes, the number of round spermatid per Sertoli cell is low (5.6) relative to other rodents (Wing and Christensen 1982; Parreira 1990; Patil and Saidapur 1991; Rocha et al. 1999), whereas the duration of one seminiferous epithelium cycle in capybaras (11.9 days), is among the highest for rodents. A very short period of sexual receptivity (8 h; López-Barbella 1982) appears to help dominant males to have exclusive access to females; it is very unlikely that two females will be receptive at the same time, allowing the dominant male to prevent other males from mating with a receptive female. This contributes to the presumed rare occurrence of sperm competition. Still, subordinate males seem to compensate for their smaller testes (compared to dominants) by having a greater proportion of sperm-producing tissue in their testes (López et al. 2008).

Reproduction involves communication, and capybaras are unique among caviomorph rodents in having a nasal gland (the morrillo) and uniquely sexually dimorphic structures in their anal glands (Macdonald and Herrera 2012). The simplest (though not infallible) way to distinguish male capybaras from female capybaras is by the size of the morrillo, which is larger in males (Macdonald et al. 1984; Macdonald 1985), and also larger in dominant males than in subordinate ones (Herrera and Macdonald 1994).

The list of capybara unorthodoxies seems unusually long. Although they are relatively long-lived, capybaras have a generation length almost as short as that of small rodents (Moreira and Macdonald 1996; Eisenberg and Redford 1999). Another snippet from the list of peculiarities is that they are the only rodents to have subcutaneous sweat glands (Pereira et al. 1980). There may be other unusual aspects of their biology, but these have been little studied. One such is the immune system. Capybaras appear to resist numerous parasitic infections with little effect on their general condition or their reproductive success (Cueto 2012). Admittedly, many so-called parasites are actually commensal or even mutualistic, but the immune system is an aspect of capybara biology which needs more attention. Still, despite their unusual features, we will now argue that capybaras have the potential to serve as paradigms for two spheres of study for which they provide revealing models. The first is intraspecific variation in social behavior and the second is sustainable use.

24.2 Capybara Societies: Group Size and the Ecology of Social Behavior

The multi-male, multi-female groups of capybaras are described in Herrera (2012b) and in Herrera et al. (2011). While elements of their social system reflect those of other mammals, they are unusual amongst rodents (see Macdonald et al. 2007), and some aspects of their social behavior are unique. As an example, take the extraordinary observations – yet to be followed through to document final outcomes for reproductive success – of a subordinate male seemingly developing its own social groups by "kidnapping" the immature members of a crèche and dispersing with

Fig. 24.1 Map of South America with the locations of the places referred to in the text

them; this phenomenon is unparalleled amongst mammals (Herrera and Macdonald 1987; Herrera 1992). It is also relevant to understanding the formation of new groups, the patterns of genetic relatedness therein, and the ontogeny of the dominance hierarchy, another extreme – its unparalleled rigidity and linearity – requiring further study.

The first detailed study of capybara behavioral ecology was conducted by Macdonald (1981), who reported that they were seen singly or in groups of up to 67 animals in the Llanos of Venezuela (Fig. 24.1), where the mean of 345 groups (of two or more) was 10.9. He showed that some social units aggregated in the dry season, that group size was directly linked to the geometry of available water and critical resources, and that possible sociological benefits of grouping included protection from predators, access to mates, parental care, and kin selection. Interestingly, as suspected by Ojasti (1973), groups coalescing in the dry season did not lose their social integrity and went back to their original formation when rains returned (Herrera and Macdonald 1987). There were significant differences in mean group size among habitats (differing in type and extent of surface water at the height of the dry season), and groups were largest on rivers (surrounded by parched savanna). They were also large in dry savanna areas consisting of a large lagoon surrounded by dry land and were smallest in wet savanna, which was marshy and dotted with small temporary pools or lagoons, separated from each other by 100 m or less. Subsequently, working in the same study area, Herrera and Macdonald (1989) explored in greater depth the determinants of group and territory size, presenting data on stable groups of 4–40 individuals (between 4 and 16 adults). They found that territory size was positively correlated with group size, and varied between 5 and 16 ha (mean = 10.4). All home ranges (with one exception) included a section of each of four major habitat types, of which two (bushy scrub and low-lying grassy patches) correlated significantly with group size. Further, the area of bushy scrub in each home range correlated with female reproductive success. They concluded that grassy patches are essential for survival in the dry season, while bushy scrub is vital in the wet season (providing food and harborage, especially for females during parturition and their newborn pups). Capybaras depend on access to permanent surface water, and where this is available, it appears that territories are configured to encompass sufficient resources to ensure survival under widely different seasonal conditions. Activity centers were always located close to water, and distances from the main grazing patches to the nearest pond were never greater than 300 m (Herrera and Macdonald 1989). In conclusion, the waterhole is the key resource that makes a territory viable and defendable for year-round survival; adult group size is limited by the area of bushy scrub.

Herrera and Macdonald (1989) explain these variations in capybara group sizes by invoking the Resource Dispersion Hypothesis (RDH), which posits that territory size and group size might be determined independently, and respectively, by the dispersion and abundance ("richness") of available resources (see Macdonald 1983; Carr and Macdonald 1986). They observed a correlation between territory size and group size, and suggested several hypotheses that might explain this. The first hypothesis was that, within minimum territories, the relationship between group size and territory size could be explained solely by a correlation between patch dispersion and patch richness; for example, if territories are large enough to encompass widely spaced ponds and/or low-lying grassy patches in the dry season automatically encompassed additional bushy scrub for the wet season. In this case, the dispersion of one (or two) critical resources in one season would be correlated with the richness of the third critical resource (in this case, the bottleneck) in the other season. The observation that larger territories are apparently configured to embrace two ponds, and thereby acquire additional intervening bushy scrub (and low-lying grassy patches), is compatible with this suggestion. Alternatively, a second hypothesis proposes that if the territories are larger than the minimum needed to sustain the basic social unit, the capybaras would meet the criteria that define expansionists (sensu Kruuk and Macdonald 1985). Expansionism is a successful strategy when the advantages of larger group size outweigh the costs of maintaining the increased area of territory necessary to support the additional group members (interaction of such costs and benefits are discussed in Macdonald and Carr 1989). Several lines of evidence suggest that even if it pays capybaras to expand up to group sizes of around 10, the individual marginal benefits thereafter may reduce: Yáber and Herrera (1994) found no benefits in terms of shared vigilance to members of groups larger than 10, and Salas (1999) observed no increase in reproductive success to members of groups of more than 15 adults.

There are, however, intermediate explanations which take into account the annual variation in the severity of the alternating floods and droughts and the costs of adjusting territory size to such fluctuating resource availability, which are likely to be high (as emphasized by von Schantz 1984; see also Carr and Macdonald 1986). For example, territorial configuration of capybaras could be adapted to the worst conditions likely to occur over a number of years. In the mild, intervening years, the capybaras resist the pressure to contract their borders because of the costs of such a short-term strategy (e.g., territorial clashes in the subsequent expansion in the next severe year). Furthermore, precipitous readjustment of territorial borders would forestall any opportunity of benefiting from the larger group sizes that might develop on the additional resources (e.g., slightly lower banco that is not submerged and slightly shallower ponds that are not evaporated to dryness). Only in exceptionally wet or dry years, when the resource bottleneck patches are at their "narrowest," would the group size drop to its minimum level, thereby possibly destroying the correlation that would otherwise generally hold between group size and home range size.

Two main inferences can be drawn from the observations above. First, the discussion opened by Herrera and Macdonald (1989) raises various possible relationships between group size and various ecological resources, and these could illustrate principles affecting species far beyond capybaras. Second, the variations in group sizes associated with different parts of his study area led Macdonald (1981) to emphasize the intra-specific variation in capybara group size and composition under different circumstances (as summarized in Table 24.1, below). In several chapters of this book, and in other publications, this variation is further documented. For example, sex ratio in social groups varies from 1:1.7 (male:female; Herrera and Macdonald

Table 24.1 Mean group size of capybaras occupying habitats that differed in the type and extent of surface water at the height of the dry season (From Macdonald (1981))

Habitat	N	Group size (SD)
River 1	16	21.8 (15.2)
River 2	10	27.0 (16.7)
Reservoir 1	20	7.8 (5.4)
Reservoir 2	13	6.7 (4.7)
Dry savanna 1	6	15.0 (6.0)
Dry savanna 2	7	18.6 (16.2)
Wet savanna	6	7.2 (3.9)

1989) to 1:3 (Alho and Rondon 1987). In comparison to Herrera and Macdonald's (1989) observations at Hato El Frio (in the Venezuelan Llanos), larger group sizes (8–28) were reported on a different ranch (El Cedral, 50 km away) but in the same general ecosystem (Salas 1999). These differences are probably attributable to resources: El Cedral had a more homogeneously distributed, more abundant, and less seasonally variable resource base than El Frio. Similarly, Salas (1999) found larger groups in a location with both greater habitat homogeneity and more abundant (and less seasonal) resources, whereas she found rather small groups where resources are homogeneously dispersed and/or patch richness is low. Salas (1999) found that 40% of all animals at El Cedral were unaffiliated to any particular group, while Alho and Rondon (1987) in the Brazilian Pantanal (Fig. 24.1) reported that 8% of capybaras were "satellite" to a group. Studies in the Pantanal revealed larger home ranges (21–200 ha) than those in the Llanos (Schaller and Crawshaw 1981; Alho et al. 1987). Herrera et al. (2011) document some of this variation in Table 24.2, below.

The substantial point we make here is that capybara group sizes (and doubtless their social dynamics) vary very significantly. This variation has been sufficiently explained in terms of resource dispersion to make it clear that this species provides an exceptional model for shedding light on the ecological basis of societies. There are plenty of anecdotes confirming that this intraspecific variation exists throughout the species' range – for example, Moreira et al. (2012a) point out that on the River Pilcomayo in the Chaco Seco of Paraguay (Fig. 24.1), they are often spotted as individuals and small groups. An overlooked study by Soini (1992) in the Peruvian Amazon rainforest (Fig. 24.1) reported capybaras living in pairs or, rarely, trios (one male, two females) with offspring from one to two litters. Home ranges for these small groups were similar (17–22 ha; Soini and Soini 1992) to those found for groups in savanna, which suggests a much lower carrying capacity of this habitat. This extreme variability from the most common polygynous groups of the savannas to the possibly monogamous pairs of the forest is rare: very few mammals have shown such wide intraspecific variation in their mating system (e.g., Randall et al. 2005; see also Waterman 2007).

In short, we propose that the capybara is an excellent model for the study of variation in mammalian social systems and its adaptive significance. Nevertheless, more research should be done on capybara social behavior across its vast range. Very little is known beyond the Venezuelan Llanos.

24.3 The Sustainable Use of Capybaras

The group-living, reproductively productive life-styles of capybaras are adaptations to the habitat in which they have evolved. The biological traits of capybaras (large, grazers, relatively docile, group-living in open habitat and diurnal) and their good protein yield make them accessible for harvest by people, with the consequence that they have financial value and are traded. Furthermore, capybaras can thrive under

Table 24.2 Variation in capybara demography. M = male, F = female (From Herrera et al. (2011))

Locality	Density (ind/km²)	Adult group size (SD)	Adult sex ratio (M:F)	Group home range size (ha)	Source
Pantanal, Brazil	9.5	10.6 (3.8)	1:2.9	12	Alho and Rondon (1987)
Pantanal, Brazil	12.5	5.9 (5.8)	1:2.5	12–200	Schaller and Crawshaw (1981)
Llanos, Venezuela	*c.* 150	9.6 (3.8)	1:1.7	10	Herrera and Macdonald (1987)
Llanos, Venezuela	172.9	11.25	1:1.2	–	E.R. Congdon personal communication
Llanos, Venezuela	200–400	16 (IQR = 9.5)	1:2.0	14	Salas (1999)
Esteros del Iberá, Argentina	Probably high	15.2 (3.4)	1:2.8	–	M.J. Corriale personal communication
Llanos, Colombia		–	–	10–56	Perea and Ruiz (1977)

conditions that are difficult for other livestock species, due to their heat tolerance and resistance to local diseases and parasites, and even on poor nutritional quality food, and thus they may offer particular opportunities for economic development and the integration of wildlife and agriculture (Nogueira-Filho and Nogueira 2004; Nogueira-Filho et al. 2012). Innovative possibilities include integrated systems, for example, combining capybara and fish farming. Indeed, Herrera and Barreto (2012) highlight the low costs associated with sustainable management of capybaras, which requires little habitat management except perhaps for flood control and drought prevention – both of which are implemented for the conventional land use of extensive cattle ranching. Capybara farming also has minimal environmental impact and motivates protection of the species due to its economic value. So, sustainable capybara management, together with the species' potential for eco-tourism, can trigger a cascade of conservation benefits.

Wildlife trade is a highly topical issue, with direct relevance to the wider issue of sustainable use of natural resources, and encompassing such diverse, and sometimes controversial, issues as biodiversity conservation, environmental economics, development and poverty alleviation, human health, and animal welfare. Because capybaras are harvested widely, and their biology lends itself to potentially sustainable harvest (as illustrated by Moreira et al.'s (2012b) simulations and Herrera and Barreto's (2012) analysis of capybara management in Venezuela), we argue that just as they are an excellent exemplar for the study of intraspecific variation in social behavior (above), so too, and for similar reasons, are they a model species for considering the pros and cons, from biological to ethical, of the sustainable use of wildlife.

Before evaluating issues arising from the various uses of capybaras, and the sustainability of these uses, we first examine the main components of trade in this species, by tackling the following questions: (a) how and where capybaras are used, (b) what issues are associated with their use, (c) where the demand for capybaras lies, and where the sources of supply are, (d) to what extent trade in capybaras is regulated, (e) what the drivers of trade in capybaras are, and (f) what external levers might influence the trade.

24.3.1 Uses

Capybaras have been used traditionally for hide (especially in Argentina; Fig. 24.1) and meat (especially in Venezuela, Colombia, and Brazil; Fig. 24.1) (Nogueira-Neto 1973; Ojasti 1973, 1991; Aldana-Domínguez et al. 2012; Herrera and Barreto 2012; Moreira and Pinheiro 2012), and their oil/fat is reputedly used in traditional medicine for the treatment of asthma, rheumatism and allergies (Moreira and Macdonald 1996), and wounds (Pinheiro and Moreira 2012); some indigenous tribes attribute medicinal qualities to capybara bones (Pinheiro and Moreira 2012), but none of these claims have been medically tested. In Venezuela, eating capybara involves a traditional (religious) element insofar as it is sanctioned for consumption by Roman Catholics during Lent. In Brazil, while some Amazonian people eat capybaras acquired from illegal trade, capybara meat (and other wildlife meat) produced in legal captivity is sold predominantly in state capitals and large towns, mainly to wealthy consumers (Pinheiro and Moreira 2012). In passing, we note that regulated public health inspection of capybara meat is an issue that has not yet emerged widely, but may do so soon. In Argentina, in contrast, it is mainly a subsistence resource for the poor and minority ethnic groups (Quintana et al. 1992; Barbarán 2000; González 2001; Arenas 2003; Quintana and Bolkovic 2012). Similarly, in Bolivia (Fig. 24.1), farmed capybara meat is consumed largely by the indigenous poor. Capybara leather is increasingly marketed as a "traditional" product, and in some places in Brazil their bones and teeth are incorporated into traditional handicrafts sold to tourists (Pinheiro and Moreira 2012). The main use of capybaras in Argentina is as a source of high-quality leather for footwear, clothing, luggage, and handicrafts. Other monetisable uses of capybaras include eco-tourism, sport hunting (e.g., capybara safaris in Argentina – http://www.scirecordbook.org/capybara/), and sometimes for zoo collections and pets. Capybara droppings can be used as manure, and they have been proposed as organ donors (xenotransplantation) for humans (Chaline 1994; Pinheiro and Moreira 2012).

Overall, then, categories of use for capybaras include being hunted for subsistence food and illegally traded locally by traditional populations; being hunted commercially for meat or hide, or for sport; being managed on ranches, or farmed for luxury meat. Their body parts have value in traditional medicine and tourist memorabilia, although these are only local and probably minor uses. A further motive for killing capybaras in some countries is seeking to limit their perceived role as pests. In this context, a potential emergent issue is conflict between different groups over the pest status and lethal control of capybaras. For example, in affluent south and southeast Brazil (Fig. 24.1), capybaras are thriving and becoming perceived as pests by some stakeholders who would have them killed. Others, with a different ethical perspective, oppose this. Increasingly, such divisions of opinion may put capybaras at the center of conflict.

24.3.2 Legal Status of Capybaras

In Colombia, there is a legal Lenten harvest for export to Venezuela, with a total harvest quota of 70,000 individuals in the Orinoquia Region (Fig. 24.1), but otherwise trade in Colombia is illegal and its prohibition is poorly enforced (López et al. 2002; Aldana-Domínguez et al. 2012). Venezuela has a state-regulated management program through which ranches obtain permits to cull 20% of their censused population, provided a previously defined minimum population size is reached. The program is based on Ojasti's (1973) sustainable yield design and is relatively, albeit fragmentarily, well enforced and successful (Herrera and Barreto 2012). In Argentina, trade in capybaras is regulated provincially and varies between total prohibition, subsistence hunting only, seasonal commercial and sport hunting quotas, and hunting allowed in response to damage to crops (Quintana and Bolkovic 2012). Brazil opted for a total hunting ban in 1967 (Moreira and Pinheiro 2012) because government lacked an adequate system to monitor or regulate the sustainable use of wildlife. Although laws now exist to regulate the commercial use of wildlife, control of wildlife farms and traders is still minimal; on the other hand, harvest for captive breeding is allowed in Brazil, as is "pest control." Meanwhile, concern for wildlife conservation and animal welfare has developed, especially in the urban areas of the southeast and south of Brazil, and this culminated in the embargo, in 2008, on sport hunting in the State of Rio Grande do Sul (Fig. 24.1; the only state in Brazil where sport hunting had been legal).

The trafficking of capybara meat is difficult to regulate, given the large areas over which the species is distributed and the limited capacity of the authorities to control intricate trade routes (Aldana-Domínguez et al. 2012; Moreira and Pinheiro 2012; Quintana and Bolkovic 2012). Trade in by-products is even more complex due to the wide range of items involved (see Fig. 21.2, Chap. 21). In general, local markets are poorly regulated, although export may be easier to control.

24.3.3 Capybara Trade

Most trade appears to be domestic, with two exceptions. Colombia's occasional exports of capybara meat to Venezuela depend on whether supply covers demand in Venezuela, and much of this trade can be illegal (Aldana-Domínguez et al. 2012). Argentina exports hides, mainly to the European Union, but worryingly, it has been difficult to certify whether the hides have legal provenance (Quintana and Bolkovic 2012). Despite trials in several South American countries, Brazil is the only nation where capybara farming – as opposed to ranching, which is successful and sustainable in Venezuela – is practiced (Alho 1986; Nogueira-Filho 1996; Pinheiro 2007). Nevertheless, its economic feasibility is questionable and the result is that only a few farms, which use a ranching approach, are productive (Moreira 2004; Moreira and Pinheiro 2012).

There is extensive capybara hunting in Colombia both for commercial sale and for consumption by indigenous people, and hunting is especially prevalent where illegal trade in capybaras is stimulated by proximity to the Venezuelan border. In Venezuela capybara meat is not important for indigenous communities, but they are harvested for meat largely for middle-class consumption (Herrera and Barreto 2012). They are hunted for meat in the Argentinean Chaco (Fig. 24.1), and their skins are exported, mainly to the European Union (Quintana and Bolkovic 2012). In Brazil, where commercial hunting of wild capybaras is illegal, capybara meat consumption varies nationwide. In the poor and dry northeast (Fig. 24.1) capybaras are rare and here, as in the poor north, consumption of illegal capybara meat is restricted to the less privileged. In the wealthy midwest, southeast, and south the meat of captive-bred capybaras is sold (only domestically), but consumption appears to have gone down remarkably in recent years. This is mainly because of the high price, but partially because of scaremongering about Brazilian spotted fever contamination (Labruna 2012; Moreira and Pinheiro 2012). Even in these wealthier regions, capybaras are still illegally hunted and eaten by traditional peoples. In Bolivia, capybaras are eaten only by indigenous people, and the same situation may apply in other countries for which there is minimal information, such as Uruguay, Panama (the lesser capybara, *Hydrochoerus isthmius*), the Guyanas, Peru, and Paraguay (Fig. 24.1). One consequence, with wider environmental relevance, of the value of capybaras is that some Colombian cattle ranchers are reported to provide wells to increase water availability for capybaras, suggesting that there are substantial populations (Caro et al. 2005; Aldana-Domínguez and Ángel-Escobar 2007). Similarly, in Venezuela, large reservoirs built for cattle have improved habitat for capybaras as well as for many other water-loving wild species such as wading birds, ducks, caiman, and turtles.

In a literature review of animal welfare in the context of wildlife trade, Baker et al. (submitted) identified several drivers of trade, most of which have at least some relevance to the capybara trade. These include luxury goods and food and culture, and together these two are probably the principal drivers of the capybara trade, with large quantities of capybara meat being eaten by urban dwellers over Lent (permitted by religious decree) in Venezuela and growing demand by the middle classes for "wild meat." Similarly, sales of capybara leather products in Argentinean markets increased by 76.5% between 2003 and 2010 due to demand by affluent tourists (Quintana and Bolkovic 2012). Third, subsistence food also acts as a driver of trade in capybara meat, especially in the Amazon due to the vastness of the region, but probably any meat traded is sold within 200 km of where it was killed, and for illegally hunted meat in wealthy regions the figure may be closer to 20 km. However, while capybaras can be an important source of food for indigenous and traditional peoples, they are not necessarily so (e.g., in Venezuela only 0.07% of the wildlife biomass consumed by indigenous people consists of capybara meat; compared to the amounts produced and sold commercially, consumption by indigenous peoples is negligible; Herrera and Barreto 2012). Traditional medicine creates demand for capybara oil and fat which are purported to have hitherto untested medicinal properties, but the volume of this trade is minimal. Finally, as pets/entertainment: capybaras are kept only occasionally as pets, but have great potential for eco-tourism. The volume of this trade is negligible.

A common justification of wildlife trade is the use of individuals killed during problem animal control (Dutton et al. in press). Thus in Brazil, in an area where capybaras damaged a quarter of a field of crops, Ferraz et al. (2003) suggest that a rational by-product of reducing the numbers of pest capybaras would be to hunt them sustainably, perhaps in conjunction with other mitigations, such as planting sacrificial crops (Matschke et al. 1984), and exploring financial mechanisms to encourage coexistence (e.g., Wagner et al. 1997; Dickman et al. 2011).

24.3.4 Conservation Impact of Trade

The interface between trade, as a form of use (sustainable or otherwise) of wildlife, and conservation is reviewed by Dutton et al. (in press). The cost-benefit evaluation of any particular trade involves diverse currencies, many of which are incommensurable, while others are extremely difficult to measure. There are two categories of problem potentially associated with the various uses of capybaras: we will discuss in order sustainability and animal welfare.

First, there is the question of whether a use is sustainable, in terms of the long-term viability of the population. The potential for an uninformed or miscalculated harvest to be unsustainable is illustrated by the threat caused by illegal hunting of capybaras in Colombia, where it has caused local extinctions (Hernández et al. 1983; Aldana-Domínguez et al. 2012), and there is concern for populations in northeastern Brazil and the Chaco Seco region of Argentina where hunting used to be intense and the habitat conditions adverse. Also, in Venezuela, despite a general enforcement of the program, many ranches that do not apply for licenses are left unmonitored and some have allowed their capybara populations to reach very low numbers by not controlling poaching. An unhappy cautionary note might be drawn from Asia, with the situation of the pangolin, *Manis* spp., which has been overhunted in China and is now sourced from nearby countries progressively further to the south as each, successively, is overharvested. Aldana-Domínguez et al. (2012) also document the very heavy harvest of lesser capybara in Colombia in the 1980s, and their decline there in the Caribbean region (Fig. 24.1) where hunting has exacerbated the damage done by habitat loss (Ballesteros 2001). A further problem with the lesser capybara in Colombia is the risk of hybridization with the more common "greater" capybara, since the latter is known to have been introduced into the distribution of the former (Aldana-Dominguez et al. 2012). However, while overharvest is a risk, Herrera and Barreto (2012) use the case study of Venezuela to illustrate that legal, regulated harvests of wild capybaras can be sustainable, and Moreira et al. (2012b) present a model of how the sex of harvested individuals affects sustainability. As is generally the case for mammalian population dynamics, the issues are not merely the numbers of individuals culled, but characteristics such as their age (Cordero and Ojasti 1981; Herrera 1992) and sex which may affect the subsequent demography of the population: for example, younger females may have lower reproductive success (Ojasti 1973) and smaller litters (Moreira 1995; Moreira et al. 2012c), with consequences for population growth. Tuyttens and Macdonald (2000) draw attention to behavioral perturbation effects of culls on the behavior of survivors, and Moreira et al. (2012b) raise the question of whether such effects impact capybaras.

The viability of a harvest is inextricably linked to knowledge of population sizes and characteristics, and thus to the quality of monitoring, and similarly linked to the quality of regulation. Herrera and Barreto (2012) point out that in Venezuela, a lack of official supervision or monitoring of annual population censuses on some ranches has led to population overestimates and the granting of licenses above the maximum sustainable yield. There are parallels in the consequences of legal but miscalculated

– and thus unsustainable – harvests of wildlife around the world, for example, lions for trophy hunting (Loveridge et al. 2007). Clearly, the financial costs of proper monitoring of capybara numbers, and effective regulation of a harvest, all have implications for the economic viability of harvesting them.

24.3.5 Impact of Trade on Animal Welfare

Capybara hunting involves two methods, both of which raise animal welfare concerns. First, they may be shot, with the inevitable risk of wounding (and the practical disadvantage of damaging the pelt); this is the only practical option in forests. Welfare issues associated with shooting have been documented in other cases, for example, red foxes in Britain (Fox et al. 2005). The British Deer Society recognizes the potential for wounding deer during shooting and is currently researching the causes of inaccurate shooting, or the wounding of animals, as well as how best to recover and dispatch wounded deer (http://www.bds.org.uk/accuracy_and_recovery.html). Second, in more open country, capybaras are herded by horsemen into a tight huddle, whereupon individuals selected by the slaughter man on foot are then killed with a sharp blow to the head using a cudgel (commonly a thick piece of metal tubing). This method is predominant in the open floodplain savannahs of Colombia (Aldana-Domínguez et al. 2012) and Venezuela (Herrera and Barreto 2012) and involves two elements where welfare is a concern. First, while the horsemen are herding the capybaras (in the same way they might gather domestic stock) this may be stressful in itself, especially if individuals breaking from the herd are chased back to it. Second, although the slaughter man may be experienced and skilled, and the capybaras may often die instantly when struck, there is the possibility of an inaccurate or otherwise unsuccessful blow, particularly where animals are jostling each other. Another issue is that of stress possibly experienced by those not selected, as a result of their close proximity to dying conspecifics. For this reason, domestic stock slaughtered in UK slaughterhouses is legally required to be killed out of the sight of other animals. However, Herrera and Barreto (2012) note that there is no observable panic or sound amongst capybaras that are not selected for culling.

A second arena in which welfare concerns are relevant is in the farming of capybaras. Nogueira-Filho and Nogueira (2012) and Moreira and Pinheiro (2012) point out that early husbandry practices were poorly aligned with the species' behavior, but that lessons learned have led to many improvements, including the replacement of confined systems by semi-confined systems (Nogueira-Filho et al. 2012). Nonetheless, even in more modern, refined systems, capybaras are kept at higher densities than they would be in the wild (as, of course, is typical of farmed stock), with the attendant risks of social and disease problems. For example, Ferraz et al. (2012) mention the need to attend to concerns over social stresses associated with feeding from communal troughs and the artificial assembly of social groupings; indeed, Allekotte (2003) reports up to 80% mortality due to inappropriate grouping of wild-caught females. They also cite the need to isolate females (and provide them with shelter) during parturition in inappropriately formed groups and point out the diverse social factors that increase the risk of high infant mortality rates and infanticide (Nogueira-Filho et al. 2012). Other aspects of husbandry to which authors in this book draw attention are the control of predation by black vultures, the standards of sanitary and medical procedures, over-feeding and the impact of endoparasites, especially on neonates (Nogueira-Filho and Nogueira 2012). Of course, none of these issues is unexpected in the context of farming any wild species, or indeed domestic species (although they, by definition, have been selected for farmed conditions), but the crucial questions are whether effective means of mitigating these problems exist, and whether these mitigations are effectively enforced to maximize welfare. In general, handling should be minimized (for a general example of the consequences of handling rodents see Gelling et al. 2009). Best practice in a semi-confined system is documented in Nogueira-Filho et al. (2012), and key elements include provision of dispersed food to minimize agonistic interactions (Ferraz et al. 2012), provision of vitamin C (in fresh grass or as a supplement, which capybaras cannot synthesize but require to prevent scurvy and for successful gestation; Nogueira-Filho and Nogueira 2012; Cueto 2012), provision of adequate water sources (for "escape," health, and social behavior), and control of scabies mites (González-Jiménez 1995; Cueto 2012). Finally, for the success of captive rearing, all the females in the group should be close relatives or familiar to one another from at most 60 days of age to prevent conflict and infanticide (Nogueira-Filho et al. 2012). This has been the most important reason for the failure of capybara captive farms. When capybara welfare impacts are minimized, the ultimate question is whether the benefits of capybara use justify its impacts on their welfare.

In categorizing welfare issues relevant to wildlife trade, Baker et al. (submitted) used five domains of welfare impact originally developed by Mellor and Reid (1994) for assessing the welfare impact of experimental procedures on animals, and based on the Farm Animal Welfare Council's Five Freedoms. These are listed in Table 24.3 and checked for each of the categories of trade identified for capybaras.

Baker et al. (submitted) concluded that the animal welfare implications of trade involving capture of live wild animals were likely to be greater than those involving killing animals in situ. Hunting and killing capybara in situ (including pest control) may have implications for welfare domains 3 and 5, depending on how efficiently the animals are killed. A quick clean kill should not be regarded as a welfare issue, although it might be considered an ethical one. The extent of welfare impacts for capybaras taken into captivity (including ranching, depending on how this is done) potentially involve all of the domains, and the duration of suffering could be substantially longer, for example, for farmed animals, than suffering incurred by animals killed or injured in situ. One area that requires illumination is the suffering typified by best practice when gathering and slaughtering capybaras on ranchland, and whether best practice is judged to be good enough, and indeed whether it can be widely achieved and regulated.

Table 24.3 Welfare issues relevant to capybara trade. Following Baker et al. (submitted), the welfare issues identified here are based on five domains of welfare impact originally developed by Mellor and Reid (1994) for assessing the welfare impact of experimental procedures on animals, and based on the Farm Animal Welfare Council's Five Freedoms

	Water/food deprivation/ malnutrition	Environmental challenge	Disease/injury/functional impairment	Behavioral/interactive restriction	Anxiety, fear, pain, distress
Subsistence hunting			X[a]		X[b]
Commercial hunting			X		X
Sport hunting			X		X
Management/ranching					
Farming	X[c]	X[c]	X[c]	X[c]	X[c]
Harvest from the wild (for farming stock)	X	X	X	X	X[d]
Pest control			X[e]		X

[a]Assuming some animals are injured when clubbed
[b]Assuming that animals breaking from the herd and being chased back by horsemen are stressed
[c]Improved husbandry practices mitigate all these domains
[d]Similar to farming, but killing the animal in the wild has the potential for serious stress
[e]Depending on methods used for killing

24.3.6 Levers That Might Influence the Trade

Complementing the drivers that encourage trade, there will be a set of levers that influence its acceptability and may be used to reduce trade. In their review, Baker et al. (submitted) identified conservation, economics, human health, and animal welfare as levers in wildlife trade. In the case of capybaras, there is evidence throughout the chapters of this book of the damage that unregulated and excessive hunting of capybaras can do to the conservation of their populations, but equally there is frequent mention of the feasibility of sustainable harvests where they are abundant and the cull is tightly regulated and evidence based. Baker et al. (submitted) identified human health risks as a potentially important lever in wildlife trade in developing countries, the risk of zoonotic diseases being a disbenefit to the hunting and consumption of bush meat. However this does not seem to be a high risk factor in the case of capybaras, although people pursuing them are likely to be exposed to diseases associated with their habitats. Workers may be exposed to tick-borne diseases or others transmitted by insects, in the absence of capybaras, simply by working near livestock in a tropical rural setting. However, exposure to *Brucella* by contact with infected capybara blood is a potential and unknown danger for workers (Cueto 2012). Baker et al. (submitted) suggested that animal welfare might be a powerful lever among consumers in developed countries – this could be relevant to the demand for capybara skins in the EU.

Regarding the economic implications of capybara trade, the question arises as to whether a biologically sustainable capybara harvest (from the wild or in farms), carefully regulated and adhering to the highest welfare standards, is economically sustainable. Concerns for animal welfare, either directly or as a result of reputational consequences, seriously affect the balance sheet of costs and benefits to all wildlife trade (see Dutton et al. in press; Baker et al. submitted). In the context of capybaras, the negative impacts on welfare associated with the herding, killing, or live capture of wild capybaras and those associated with the husbandry of farmed ones, will heavily influence opinion as to the acceptability of these enterprises.

24.4 Concluding Remarks

In some cases, wildlife can be used sustainably, but sustainability is rarely considered as a requirement before trade begins. This is because much of the trade in wildlife has its roots in traditional or subsistence living and, as is the case for capybaras, it has ballooned to meet demand from sometimes international, and particularly luxury, markets. When trade burgeons in this way, there are two characteristic risks from the standpoint of conservation, including animal welfare: first, practices that may have been unsustainable, and even cruel, on a small scale may become embedded on a larger scale; second, under the pressure of commercialization, margins may be squeezed to the detriment of conservation and welfare concerns. Furthermore, in terms of monitoring, regulation, and enforcement, details of the quantity and location of trade are often lacking, particularly in the case of illegal trade, and this complicates the already difficult task of predicting the likely sustainability of harvest and the welfare impacts on the animals concerned.

Potential sources of welfare concern in capybara trade are: a) how the animals are killed during hunting and ranching, b) how they are handled and otherwise treated during ranching (including effects on animals not selected for culling), and c) the husbandry of farmed capybaras. Although some information is available on aspects of these, as with most other types of wildlife trade, the information is patchy and lacks detail and, most troublingly, has not benefited from scrutiny and quantification under the impartial lens of animal welfare science. In their review of the welfare impact of the wildlife trade, Baker et al. (submitted) concluded that animal welfare is rarely considered in the literature on wildlife trade and that a more structured approach is required in assessing and reporting this trade's welfare impacts; they affirm that recommendations with direct benefit for animal welfare need to be made. This is a technically difficult remit, and there are few good models from which to seek guidance. In the context of hunting, two reports from the UK attempted to synthesize the data on scientific measurements of the stress experienced by hunted mammals; one of these dealt with foxes (Macdonald et al. 2000), the other concerned deer (Bateson and Bradshaw 1997), both hunted by hounds. Perhaps the most thorough attempt to evaluate, and then improve, the welfare of harvested wildlife was the mix of behavioral, physiological, and biochemical measures made of another South American mammal, the vicuna, *Vicuna vicuna*, which is shorn for its wool (Bonacic and Macdonald 2003; Bonacic et al. 2003; Gordon 2009). Subsequently, more refined immunological techniques for measuring stress, and thereby reducing it, have been developed (e.g., Gelling et al. 2010). These techniques would be helpful in evaluating the pros and cons of capybara harvests and farming, and are thus relevant to the interwoven aims of conservation and animal welfare (McLaren et al. 2006), and to the even more challenging aim of further integrating with them the wider goals of development. Certainly, research is needed to determine the sustainability and welfare impact of capybara use. In particular, detailed information is required on the impacts incurred in the five welfare domains described in this chapter. Standard Operating Procedures and Best Practice Guidelines could be developed for hunting, ranching, and farming capybaras and legal quotas determined to conduct these at sustainable levels and ensuring humane treatment of capybaras. Furthermore, the biology and circumstances of capybaras may make them an unusually insightful and compelling general model for the systematic scientific study of sustainable wildlife trade.

Acknowledgments We are very grateful to Dr. Lauren Harrington for her help in assembling material for this chapter, and to Drs. Sandra Baker and Adam Dutton for insightful comments on an earlier draft.

References

Aldana-Domínguez J, Ángel-Escobar DC (2007) Evaluación del tamaño y densidad de poblaciones silvestres de chigüiros en el departamento del Casanare. In: Aldana-Domínguez J, Vieira-Muñoz MI, Ángel-Escobar DC (eds) Estudios sobre la ecología del chigüiro (*Hydrochoerus hydrochaeris*), enfocados a su manejo y uso sostenible en Colombia. Instituto Alexander von Humboldt, Bogotá, pp 33–48

Aldana-Domínguez J, Vieira-Muñoz MI, Bejarano P (2012) Conservation and use of the capybara and the lesser capybara in Colombia. In: Moreira JR, Ferraz KMPMB, Herrera EA, Macdonald DW (eds) Capybara: biology, use and conservation of an exceptional neotropical species. Springer, New York, pp 321–332

Alho CJR (1986) Criação e manejo de capivaras em pequenas propriedades rurais. Empresa Brasileira de Pesquisa Agropecuária EMBRAPA-DDT Série Documentos, 13, Brasília

Alho CJR, Rondon NL (1987) Habitats, population densities, and social structure of capybaras (*Hydrochaeris hydrochaeris* Rodentia) in the Pantanal, Brazil. Revista Brasileira de Zoologia 4:139–149

Alho CJR, Campos ZMS, Gonçalves HC (1987) Ecologia da capivara, *Hydrochaeris hydrochaeris*, Rodentia) do Pantanal: I. Habitats, densidades e tamanho de grupo. Rev Bras Biol 47:87–97

Allekotte R (2003) La cria del carpincho. Ediciones INTA, Buenos Aires

Arenas P (2003) Etnografía y alimentación entre los Toba-Nachilamoleek y Wichí-Lhuku'tas del Chaco Central (Argentina). Latin Gráfica, Buenos Aires

Baker SE, D'Cruze NC, Cain R, van Kesteren F, Zommers Z, Macdonald DW (submitted) A systematic review of the animal welfare impact of global wildlife trade

Ballesteros J (2001) Estado de conservación del chigüiro o ponche (*Hydrochaeris hydrochaeris isthmius*) en el departamento de Córdoba, Colombia. M.Sc. Dissertation, Universidad Pontificia Javeriana, Bogotá

Barbarán FR (2000) Recursos alimenticios derivados de la caza, pesca y recolección de los Wichi del Río Pilcomayo (Provincia de Salta, Argentina). In: Cabrera E, Mercolli C, Resquin R (eds) Manejo de Fauna Silvestre en Amazonia y Latinoamérica. CITES Paraguay – Fundación Moisés Bertoni – University of Florida, Asunción, pp 507–527

Barreto GR, Quintana RD (2012) Foraging strategies and feeding habits of capybaras. In: Moreira JR, Ferraz KMPMB, Herrera EA, Macdonald DW (eds) Capybara: biology, use and conservation of an exceptional neotropical species. Springer, New York, pp 83–96

Bateson P, Bradshaw EL (1997) Physiological effects of hunting red deer (*Cervus cervus*). Proc Roy Soc Lond Ser B 264:1707–1714

Bonacic C, Macdonald DW (2003) The physiological impact of wool-harvesting procedures in vicunas (*Vicugna vicugna*). Anim Welf 12:387–402

Bonacic C, Macdonald DW, Villouta G (2003) Adrenocorticotrophin-induced stress response in captive vicunas (*Vicugna vicugna*) in the Andes of Chile. Anim Welf 12:369–385

Borges PA, Domínguez-Bello MG, Herrera EA (1996) Digestive physiology of wild capybara. J Comp Physiol B 166:55–60

Caro C, López HF, Sánchez P, Mesa E, Atuesta N, Lenis AR, Camargo A, Hernández Y (2005) Fortalecimiento al control y seguimiento a las poblaciones naturales de chigüiro para la implementación de proyectos de uso sostenible en Casanare- Municipios de Paz de Ariporo y Hato Corozal. Fase III. Informe final. UNILLANOS-Ministerio de Ambiente, Vivienda y Desarrollo Territorial, Villavicencio

Carr GM, Macdonald DW (1986) The sociality of solitary foragers: a model based on resource dispersion. Anim Behav 34:1540–1549

Chaline J (1994) Xenotransplants: why not the capybara? Lancet 343:126

Cordero RGA, Ojasti J (1981) Comparison of the capybara populations of open and forested habitats. J Wildl Manag 45:267–271

Costa DS, Paula TAR (2006) Testosterone level, nasal gland volume and Leydig cell morphometry in capybaras (*Hydrochoerus hydrochaeris*). Arq. Bras. Med. Vet. Zootec. 58(6):1086–1091

Costa DS, Paula TAR, Matta SLP (2006) The intertubular compartment morphometry in capybaras (*Hydrochoerus hydrochaeris*) testis. Anim Reprod Sci 91:173–179
Cueto GR (2012) Diseases of capybara. In: Moreira JR, Ferraz KMPMB, Herrera EA, Macdonald DW (eds) Capybara: biology, use and conservation of an exceptional neotropical species. Springer, New York, pp 169–184
Dickman AJ, Macdonald EA, Macdonald DW (2011) A review of financial instruments to pay for predator conservation and encourage human-carnivore coexistence. Proc Nat Acad Sci USA 108(34):13937–13944
Dutton AJ, Gratwicke B, Hepburn C, Herrera EA, Macdonald DW (in press) Tackling unsustainable wildlife trade. In: Macdonald DW, Willis K (eds), Key topics in conservation 2, Blackwell, Oxford
Eisenberg JF, Redford KH (1999) Mammals of the Neotropics: the Central Neotropics, vol 3, Ecuador, Peru, Bolivia, Brazil. The University of Chicago, Chicago, pp 609
Ferraz KPMB, Lechevalier M-A, Couto HTZ, Verdade LM (2003) Damage caused by capybara (*Hydrochaeris hydrochaeris*) on a cornfield in São Paulo, Brasil. Sci. Agric. 60(1):191–194
Ferraz KMPMB, Izar P, Sato T, Nishida SM (2012) Social and spatial relationships of capybaras in a semi-confined production system. In: Moreira JR, Ferraz KMPMB, Herrera EA, Macdonald DW (eds) Capybara: biology, use and conservation of an exceptional neotropical species. Springer, New York, pp 243–260
Fox NC, Blay N, Greenwood AG, Wise D, Potapov E (2005) Wounding rates in shooting foxes (*Vulpes vulpes*). Anim Welf 14(2):93–102
França LR, Russell LD (1998) The testis of domestic animals. In: Regadera J, Martinez-Garcia F (eds) Male reproduction. A multidisciplinary overview. Churchill Livingstone, Madrid, pp 197–219
Fritz J, Hummel J, Kienzle E, Arnold C, Nunn C, Clauss M (2009) Comparative chewing efficiency in mammalian herbivores. Oikos 118:1623–1632
Gelling M, McLaren GW, Mathews F, Mian R, Macdonald DW (2009) Impact of trapping and handling on leukocyte coping capacity in bank voles (*Clethrionomys glareolus*) and wood mice (*Apodemus sylvaticus*). Anim Welf 18:1
Gelling M, Montes I, Moorhouse TP, Macdonald DW (2010) Captive housing during water vole (*Arvicola terrestris*) reintroduction: does short-term social stress impact on animal welfare? PLoS One 5(3):e9791
González G (2001) Conociendo lo nuestro. La cultura de los grupos guaraníticos del río Paraná y su construcción del sistema alimentario actual. BS thesis, Universidad de Belgrano, Buenos Aires
González-Jiménez E (1977) Digestive physiology and feeding of the capybara. In: Rechcigl M (ed) Handbook series in nutrition and food, section G, vol 1. CRC Press, Cleveland, pp 167–177
González-Jiménez E (1995) El capybara (*Hydrochoerus hydrochaeris*) – Estado actual de su producción. FAO, (Série Estudio FAO). Producion y Sanidad Animal, Rome, p 122
Gordon I (ed) (2009) The Vicuña: the theory and practice of community-based wildlife management. Springer, New York
Hernández J, Rodríguez J, Pachón J (1983) Evaluación de poblaciones de chigüiro (*Hydrochaeris hydrochaeris*) en los hatos "Brasilia", "Guamito", "La Aurora", "La Borra", "El Danubio", "La Veremos" y "Mapurisa", municipio de Hato Corozal. INDERENA, Bogotá
Herrera EA (1985) Coprophagy in the capybara, *Hydrochoerus hydrochaeris*. J Zool (London) 207A:616–619
Herrera EA (1992) The effect of harvesting on the age structure and body size of a capybara population. Ecotropicos 5:20–25
Herrera EA (2012a) Capybara digestive adaptations. In: Moreira JR, Ferraz KMPMB, Herrera EA, Macdonald DW (eds) Capybara: biology, use and conservation of an exceptional neotropical species. Springer, New York, pp 97–106
Herrera (2012b) Capybara Social Behavior and Use of Space: Patterns and Processes. In: Moreira JR, Ferraz KMPMB, Herrera EA, Macdonald DW (eds) Capybara: biology, use and conservation of an exceptional neotropical species. Springer, New York, pp 195–207
Herrera EA, Barreto GR (2012) Capybaras as a source of protein: utilization and management in Venezuela. In: Moreira JR, Ferraz KMPMB, Herrera EA, Macdonald DW (eds) Capybara:

biology, use and conservation of an exceptional neotropical species. Springer, New York, pp 305–319
Herrera EA, Macdonald DW (1987) Group stability and the structure of a capybara population. Sympo Zool Soc Lond 58:115–130
Herrera EA, Macdonald DW (1987) Social significance of scent marking in capybaras. J Mammal 75:410–415
Herrera EA, Macdonald DW (1989) Resource utilization and territoriality in group-living capybaras. J Anim Ecol 58:667–679
Herrera EA, Macdonald DW (1993) Aggression, dominance and mating success among capybara males. Behav Ecol 4:114–119
Herrera EA, Macdonald DW (1994) Social significance of scent marking in capybaras. J Mammal 75:410–415
Herrera EA, Salas V, Congdon ERC, Corriale MJ, Tang-Martínez Z (2011) Capybara social structure and dispersal patterns: variations on a theme. J Mammal 92:12–20
Hirakawa H (2001) Coprophagy in leporids and other mammalian herbivores. Mammal Rev 31:61–80
Hirakawa H (2002) Supplement: coprophagy in leporids and other mammalian herbivores. Mammal Rev 32:150–152
Kenagy GJ, Trombulak SC (1986) Size and function of mammalian testes in relation to body size. J Mammal 67(1):1–22
Kleiman DG, Eisenberg JF, Maliniak E (1979) Reproductive parameters and productivity of caviomorph rodents. In: Ecology V, Eisenberg JF (eds) Vertebrate ecology in the northern neotropics. Smithsonian Institution Press, Washington, DC, pp 173–183
Kruuk H, Macdonald DW (1985) Group territories of carnivores: empire and enclaves. In: Ecology B, Sibly R, Smith R (eds) Behavioural ecology. Blackwell, Oxford, pp 521–536
Labruna MB (2012) Brazilian spotted fever: the role of capybaras. In: Moreira JR, Ferraz KMPMB, Herrera EA, Macdonald DW (eds) Capybara: biology, use and conservation of an exceptional neotropical species. Springer, New York, pp 371–383
López HF, Mesa E, Sánchez JH (2002) Implementación de la fase inicial del programa de manejo de chigüiro (*Hydrochaeris hydrochaeris*) en semicautiverio, como estrategia de conservación de la biodiversidad en la Orinoquia colombiana y acopio de información complementaria. Informe final. Universidad Nacional de Colombia – Ministerio de Ambiente, Vivienda y Desarrollo Territorio, Bogotá
López M, Muñoz MG, Herrera EA (2008) Reproductive morphology of capybaras, *Hydrochoerus hydrochaeris* (Rodentia: Hystricognathi): no evidence for sperm competition? Mammal Biol 73:241–244
López-Barbella S (1982) Determinación del ciclo estral en chigüires, *Hydrochoerus hydrochaeris*. Acta Cient Venez 33:497–501
López-Barbella S (1987) Consideraciones generales sobre la gestacion del chigüire (*Hydrochoerus hydrochaeris*). Acta Cient Venez 38:84–89
Loveridge AJ, Searle AW, Murindagomo F, Macdonald DW (2007) The impact of sport-hunting on the population dynamics of an African lion population in a protected area. Biol Conserv 134:548–558
Macdonald DW (1981) Dwindling resources and the social behavior of capybara, (*Hydrochoerus hydrochaeris*) (Mammalia). J Zool 194:371–391
Macdonald DW (1983) The ecology of carnivore social behaviour. Nature 301:379–384
Macdonald DW (1985) The rodents IV: suborder Hystricomorpha. In: Brown RE, Macdonald DW (eds) Social odours in mammals. Clarendon, Oxford, pp 480–506
Macdonald DW, Carr GM (1989) Food security and the rewards of tolerance. In: Standen V, Folley R (eds) Sociobiology of animals and man. Blackwell, Oxford, pp 75–97
Macdonald DW, Herrera EA (2012) Capybara scent glands and scent-marking behavior. In: Moreira JR, Ferraz KMPMB, Herrera EA, Macdonald DW (eds) Capybara: biology, use and conservation of an exceptional neotropical species. Springer, New York, pp 185–193

Macdonald DW, Krantz K, Aplin RT (1984) Behavioral, anatomical and chemical aspects of scent marking amongst capybaras (*Hydrochoerus hydrochaeris*) (Rodentia, Caviomorpha). J Zool 202:341–360

Macdonald DW, Tattersall FH, Johnson PJ, Carbone C, Reynolds J, Langbein J, Rushton SP, Shirley M (2000) Managing British mammals: case studies from the hunting debate. WildCRU, Oxford

Macdonald DW, Herrera EA, Taber AB, Moreira JR (2007) Social organization and resource use in capybaras and maras. In: Wolff JO, Sherman PW (eds) Rodent societies, an ecological and evolutionary perspective. University of Chicago Press, Chicago, pp 393–402

Matschke GH, de Calesta DS, Harder JD (1984) Crop damage and control. In: Halls LK (ed) White-tailed deer: ecology and management. Stackpole Book, Harrisburg, pp 647–654

McLaren G, Bonacic C, Rowan A (2006) In: Macdonald DW, Service K (eds) Key topics in conservation biology. Blackwell, Oxford, p 120

Mellor DJ, Reid CSW (1994) Concepts of animal well-being and predicting the impact of procedures on experimental animals. In: Baker R, Jenkins G, Mellor DJ (eds) Improving the well-being of animals in the research environment. Australian and New Zealand Council for the Care of Animals in Research and Teaching, Wellington, pp 3–18

Mendes A, Nogueira-Filho SLG (2012) Feeds and nutrition of farmed capybaras. In: Moreira JR, Ferraz KMPMB, Herrera EA, Macdonald DW, Macdonald DW (eds) Capybara: biology, use and conservation of an exceptional neotropical species. Springer, New York, pp 261–274

Miglino MA, dos Santos TC, Kanashiro C, Ferraz RHS (2012) Morphology and reproductive physiology of female capybaras. In: Moreira JR, Ferraz KMPMB, Herrera EA, Macdonald DW (eds) Capybara: biology, use and conservation of an exceptional neotropical species. Springer, New York, pp 131–146

Moreira JR (1995) The reproduction, demography and management of capybaras (*Hydrochaeris hydrochaeris*) on Marajó Island- Brazil. D.Phil. thesis, University of Oxford, Oxford

Moreira JR (2004) Avaliação do atual estado de uso e da conservação da capivara no Brasil. In: XLI Reunião Anual da Sociedade Brasileira de Zootecnia. Embrapa Gado de Corte, Campo Grande, pp 517–527

Moreira JR, Macdonald DW (1996) Capybara use and conservation in South America. In: Taylor VJ, Dunstone N (eds) The exploitation of mammal populations. Chapman & Hall, London, pp 88–101

Moreira JR, Pinheiro MS (2012) Capybara production in Brazil: captive breeding or sustainable management? In: Moreira JR, Ferraz KMPMB, Herrera EA, Macdonald DW (eds) Capybara: biology, use and conservation of an exceptional neotropical species. Springer, New York, pp 333–344

Moreira JR, Clarke JR, Macdonald DW (1997a) The testis of capybaras (*Hydrochaeris hydrochaeris* - Rodentia). J Mammal 78:1096–1100

Moreira JR, Macdonald DW, Clarke JR (1997b) Correlates of testis mass in capybaras (*Hydrochoerus hydrochaeris*): dominance assurance or sperm production? J Zool (Lond) 241:457–463

Moreira JR, Alvarez MR, Tarifa T, Pacheco V, Taber A, Tirira DG, Herrera EA, Ferraz KMPMB, Aldana-Domínguez J, Macdonald DW (2012a) Taxonomy, natural history and distribution of the capybara. In: Moreira JR, Ferraz KMPMB, Herrera EA, Macdonald DW (eds) Capybara: biology, use and conservation of an exceptional neotropical species. Springer, New York, pp 3–37

Moreira JR, Verdade LM, Ferraz KMPMB, Macdonald DW (2012b) The sustainable management of capybaras. In: Moreira JR, Ferraz KMPMB, Herrera EA, Macdonald DW (eds) Capybara: biology, use and conservation of an exceptional neotropical species. Springer, New York, pp 283–302

Moreira JR, Wiederhecker H, Ferraz KMPMB, Aldana-Domínguez J, Verdade LM, Macdonald DW (2012c) Capybara demographic traits. In: Moreira JR, Ferraz KMPMB, Herrera EA, Macdonald DW (eds) Capybara: biology, use and conservation of an exceptional neotropical species. Springer, New York, pp 147–167

Mori H, Shimizu D, Takeda A, Takioka Y, Fikinishi R (1980) Stereologic analysis of Leydig cells in normal guinea pig testis. J Electron Microsc 29:8–21

Nogueira-Filho SLG (1996) Criação de Capivaras. In: Centro de Produções Técnicas (CPT) (ed). Viçosa

Nogueira-Filho SLG, Nogueira SSC (2004) Captive breeding program as an alternative for wildlife conservation in Brazil. In: Kirsten S, Fragoso JMV, Bodmer R (eds) People in nature: wildlife manegement and conservation in Latin America. Columbia University Press, New York, pp 171–190

Nogueira-Filho SLG, Nogueira SSC (2012) The impact of management practices on female capybara reproductive parameters in captivity. In: Moreira JR, Ferraz KMPMB, Herrera EA, Macdonald DW (eds) Capybara: biology, use and conservation of an exceptional neotropical species. Springer, New York, pp 275–282

Nogueira-Filho SLG, Pinheiro MS, Nogueira SSC (2012) Confined and semi-confined production systems for capybaras. In: Moreira JR, Ferraz KMPMB, Herrera EA, Macdonald DW (eds) Capybara: biology, use and conservation of an exceptional neotropical species. Springer, New York, pp 229–241

Nogueira-Neto P (1973) Ordem Rodentia. In: A criação de animais indígenas vertebrados: peixes, anfíbios, répteis, aves e mamíferos. Edições Tecnapis, São Paulo, pp 257–273

Ojasti J (1973) Estudio del chigüire o capibara. Fondo Nacional de Investigaciones Agropecuarias, Caracas

Ojasti J (1983) Ungulates and large rodents of South America. In: Bourlière F (ed) Tropical savannas. Elsevier, Amsterdam, pp 427–439

Ojasti J (1991) Human explotation of capybara. In: Robinson JG, Redford KH (eds) Neotropical wild life use and conservation. The University of Chicago Press, Chicago, pp 236–252

Parreira GG (1990) Morfologia e variação sazonal da atividade dos testículos e órgãos genitais acessórios de *Bolomys lasiurus* Lund, 1841 (*Rodentia, Cricetidae*). M.Sc. Dissertation, Universidade Federal de Minas Gerais, Belo Horizonte

Patil SB, Saidapur SK (1991) Kinetics of spermatogenesis in the wild squirrel *Funambulus palmarum* (Linnaeus). Acta Anat (Basel) 141(4):352–363

Paula TAR (1999) Avaliação histológica e funcional do testículo de capivaras adultas (*Hydrochoerus hydrochaeris*). Ph.D. thesis, Universidade Federal de Minas Gerais. Belo Horizonte

Paula TAR, Walker NJ (2012) Reproductive morphology and physiology of the male capybara. In: Moreira JR, Ferraz KMPMB, Herrera EA, Macdonald DW (eds) Capybara: biology, use and conservation of an exceptional neotropical species. Springer, New York, pp 107–129

Paula TAR, Matta SLP, Silva VA Jr, Costa DS, Fonseca CC, Neves MTD (2007) Intertubular space characterization in adult capybara (*Hydrochoerus hydrochaeris*) testis. Braz Arch Biol Technol 50(2):289–297

Perea J, S Ruíz (1977) Organización social y hábitos territoriales del cacó. Graduate Dissertation, Biología. Universidad Nacional, Santa Fé de Bogotá. 109 pp

Pereira JN, McEvan D, Finley E (1980) The structure of the skin of the capybara. Acta Cient Venez 31:361–364

Pinheiro MS (2007) Criação de Capivara em Sistema Intensivo. Empresa Brasileira de Pesquisa Agropecuária, Pelotas, p 43

Pinheiro MS, Moreira JR (2012) Products and uses of capybaras. In: Moreira JR, Ferraz KMPMB, Herrera EA, Macdonald DW (eds) Capybara: biology, use and conservation of an exceptional neotropical species. Springer, New York, pp 211–227

Quintana RD, Bolkovic ML (2012) Use of capybaras in Argentina. In: Moreira JR, Ferraz KMPMB, Herrera EA, Macdonald DW (eds) Capybara: biology, use and conservation of an exceptional neotropical species. Springer, New York, pp 345–356

Quintana RD, Bó RF, Merler JA, Minotti PG, Malvárez AI (1992) Situación y uso de la fauna silvestre en la región del Bajo Delta del río Paraná, Argentina. Iheringia, Sér. Zool. Porto Alegre 73:13–33

Randall J, Rogovin K, Parker PG, Eimes JA (2005) Flexible social structure of a desert rodent, *Rhombomys opimus*: philopatry, kinship, and ecological constraints. Behav Ecol 16:961–973

Rocha DC, Debeljuk ML, França LR (1999) Exposure to constant light during testis development increases daily sperm production in adult Wistar rats. Tissue Cell 31(3):128–136

Salas V (1999) Social organisation of capybaras in the Venezuelan Llanos. Ph.D. thesis, University of Cambridge, Cambridge, UK

Schaller GB, Crawshaw PG (1981) Social organization in a capybara population. Säugetierkundliche Mitteilungen 29:3–16

Soini P (1992) Un estudio de la dinámica poblacional del ronsoco o capibara (*Hydrochoerus hydrochaeris*) en el río Pacaya, Perú. Folia Amazonica 5:139–156

Soini P, Soini M (1992) Ecologia del ronsoco o capibara (*Hydrochoerus hydrochaeris*) en la reserva nacional Pacaya-Samiria, Peru. Folia Amazonica 4:135–150

Tuyttens FAM, Macdonald DW (2000) Consequences of social perturbation for wildlife management and conservation. In: Gosling LM, William JS (eds) Behaviour and conservation. Cambridge University Press, Cambridge, pp 315–329

von Schantz T (1984) 'Non-breeders' in the red fox, *Vulpes vulpes*: a case of resource surplus. Oikos 42:59–65

Vucetich MG, Deschamps CM, Olivares AI, Dozo MT (2005) Capybaras, size, shape, and time: a model kit. Acta Palaeontol Pol 50:259–272

Vucetich MG, Deschamps CM, Pérez ME (2012) Paleontology, evolution and systematics of capybara. In: Moreira JR, Ferraz KMPMB, Herrera EA, Macdonald DW (eds) Capybara: biology, use and conservation of an exceptional neotropical species. Springer, New York, pp 39–59

Wagner KK, Schimidt RH, Conover MR (1997) Compensation programs for wildlife damage in North America. Wildl Soc Bull 25:312–319

Waterman J (2007) Male mating systems in rodents. In: Wolff JO, Sherman PW (eds) Rodent societies. Chicago University Press, Chicago, pp 27–41

Wing TY, Christensen AK (1982) Morphometric studies on rat seminiferous tubule. Am J Anat 165:13–25

Yáber MC, Herrera EA (1994) Vigilance, group size and social status in capybaras. Anim Behav 48:1301–1307

Zirkin BR, Ewing LL (1987) Leydig cell differentiation during maturation of the rat testis: a stereological study of cell number and ultrastructure. Anat Rec 219:157–163

Appendix: List of Reviewers

Ludmilla Moura de Souza Aguiar Universidade de Brasília, Departamento de Zoologia, 70910-900, Brasília, Distrito Federal, Brazil
ludmillaaguiar@unb.br

Natália Inagaki de Albuquerque Embrapa Amazônia Oriental, Travessa Enéas Pinheiro s/n, Marco 66095-100, Belém, Pará, Brazil
natalia@cpatu.embrapa.br

Cleber J.R. Alho Universidade de Brasília, Departamento de Ecologia, SHIS QL 6, conj. 3, casa 1371620-035, Brasília, Distrito Federal, Brazil
alho@unb.br

Paulo Cesar Machado Andrade Projeto Manejo Sustentável de Quelônios - Pé-de-pincha, Praça Dom Bosco, 176, Centro 69010-190, Manaus, Amazonas, Brazil
pandrade@ufam.edu.br

Arrilton Araújo Universidade Federal do Rio Grande do Norte, Departamento de Fisiologia, 59072-970, Natal, Rio Grande do Norte, Brazil
arrilton@cb.ufrn.br

Brigitte Luis Guillermo Baptiste Instituto de Investigación de Recursos Biológicos Alexander von Humboldt, Calle28A#15-09 Bogotá, D.C., Colombia
brigittebaptiste@humboldt.org.co

Renato Campanarut Barnabe Universidade de São Paulo, Faculdade de Medicina Veterinária e Zootecnia, 05508-000, São Paulo, São Paulo, Brazil
rbarnabe@usp.br

Roberto F. Bó Universidad de Buenos Aires, Facultad de Ciencias Exactas y Naturales, Intendente Güiraldes 2620, Ciudad Universitaria, Pabellón II, 4to Piso 1428EHA, Buenos Aires, Argentina
rober@ege.fcen.uba.ar

J.R. Moreira et al. (eds.), *Capybara: Biology, Use and Conservation of an Exceptional Neotropical Species*, DOI 10.1007/978-1-4614-4000-0,

Maria Cristina Bressan Instituto Nacional de Recursos Biológicos, Unidade de Recursos Genéticos, 2005-048, Vale de Santarém, Portugal
mcbressan1@gmail.com

Zilca Campos Embrapa Pantanal, Rua 21 de Setembro, 1880, Bairro Nossa Senhora de Fátima, 79320-900, Corumbá, Mato Grosso do Sul, Brazil
zilca@cpap.embrapa.br

Juan M. Campos-Krauer Centro Chaqueño para la Conservación e Investigación, Proyecto Taguá, Primer Presidente casi Julio Correa 2988, Asunción, Paraguay
jmcampos@cccipy.org

Paula Courtalon Universidad de Buenos Aires, Facultad de Ciencias Exactas y Naturales, Intendente Güiraldes 2620, Ciudad Universitaria, Pabellón II, 4to Piso 1428EHA, Buenos Aires, Argentina
pcourt@ege.fcen.uba.ar

Wilfried W. de Jong Radboud University, Nijmegen Center for Molecular Life Sciences, 6500 HB Nijmegen, The Netherlands
w.dejong@ncmls.ru.nl

Marta Elisetch Pueyrredon 1548 piso 8, 1118, Buenos Aires, Argentina
marta.elisetch@gmail.com

César Roberto Esper Universidade Estadual Paulista Júlio de Mesquita Filho, Faculdade de Ciências Agrárias e Veterinárias de Jaboticabal, Via de Acesso Prof. Paulo Donato Castellane, s/n, Rural 14884-900, Jaboticabal, São Paulo, Brazil
cresper@fcav.unesp.br

Paula Federico The Ohio State University, Mathematical Biosciences Institute, 1735 Neil Avenue, 379 Jennings Hall, Columbus, Ohio 43210, USA
pfederic@utk.edu

Jimena Forero-Montaña University of Puerto Rico-Rio Piedras, Institute for Tropical Ecosystem Studies, San Juan, Puerto Rico, 00931-3341, USA
jimefore@yahoo.com

Diva Anelie de Araújo Guimarães Universidade Federal do Pará, Instituto de Ciências Biológicas, Rua Augusto Correa S/N, Guama, 66075-900, Belém, Pará, Brazil
diva@ufpa.br

Fábio Morais Hosken Zoo Way - Assessoria para criação de animais silvestres, Rua Bolivia 509/701, 30330-360, Belo Horizonte, Minas Gerais, Brazil
planetarural@terra.com.br

Gilson E. Iack-Ximenes Universidade Estadual do Sudoeste da Bahia, Departamento de Ciências Naturais, Estrada do Bem Querer km 04, 45083-900, Vitória da Conquista, Bahia, Brazil
giliackx@yahoo.com

Devra G. Kleiman Late of Smithsonian National Zoological Park, Division of Conservation and Science (In Memoriam)

Romário Cerqueira Leite Rua Cordilheiras 86, Serrano 30882-040, Belo Horizonte, Minas Gerais, Brazil
romarioleite@gmail.com

Théa Mirian Medeiros Machado Universidade Federal de Viçosa, Departamento de Zootecnia, Av. PH Rolfs s/n, Campus Universitário 36571-000, Viçosa, Minas Gerais, Brazil
thea@ufv.br

Rodiney de Arruda Mauro Embrapa Gado de Corte, Rodovia BR 262, Km 4 79002-970, Campo Grande, Mato Grosso do Sul, Brazil
rodiney@cnpgc.embrapa.br

Pedro G. Mayor-Aparicio Universitat Autònoma de Barcelona, Department of Animal Health and Anatomy, 08193 Bellaterra, Barcelona, Spain
mayorpedro@hotmail.com

Alvaro Mones Franzensbadstr. 7 b D-86199, Augsburg, Germany
a-mones@t-online.de

João Alves de Oliveira Universidade Federal do Rio de Janeiro, Museu Nacional, Departamento de Vertebrados., Quinta da Boa Vista, s/n., São Cristóvão, 20940-040, Rio de Janeiro, Rio de Janeiro, Brazil
jaoliv@mn.ufrj.br

Adriano Pereira Paglia Universidade Federal de Minas Gerais, Departamento de Biologia Geral, Av. Antônio Carlos 6627, Pampulha 31270-901, Belo Horizonte, Minas Gerais, Brazil
apaglia@icb.ufmg.br

Gustavo Porini Secretaría de Ambiente y Desarrollo Sustentable de la Nación, Dirección de Fauna Silvestre, San Martín 451, 2do Piso 1004AAI, Ciudad Autónoma de Buenos Aires, Argentina
gporini@ambiente.gob.ar

Roberto de Oliveira Roça Universidade Estadual Paulista Júlio de Mesquita Filho, Departamento de Gestão e Tecnologia Agroindustrial, Rua José Barbosa de Barros, 1780 18610-307, Botucatu, São Paulo, Brazil
robertoroca@fca.unesp.br

James Joseph Roper Universidade de Vila Velha, Caixa Postal 19034 83309-279, Curitiba, Paraná, Brazil
jjroper@gmail.com

Juan Pablo Roux Universidad Nacional del Nordeste, Facultad de Ciencias Veterinárias, Sargento Cabral 2139 3400, Corrientes, Argentina
jproux@vet.unne.edu.ar

Cristina Sainz-Borgo Universidad Simón Bolívar, Departamento de Biología de Organismos, Caracas 1080, Venezuela
cristinazainzb@gmail.com

Viviana Salas Asociación Civil Bioparques, Avenida Este, Edificio Imperial, Oficina 7-B, La Candelaria, Caracas, Venezuela
vivianasalas@gmail.com

Sandra Aparecida Santos Embrapa Pantanal, Rua 21 de Setembro, 1880, Bairro Nossa Senhora de Fátima, 79320-900, Corumbá, Mato Grosso do Sul, Brazil
sasantos@cpap.embrapa.br

Daniel de Brito Candido da Silva Universidade Federal de Goiás, Departamento de Ecologia, 74001-970, Goiânia, Goiás, Brazil
brito.dan@gmail.com

Pablo R. Stevenson Universidad de los Andes, Departamento de Ciências Biológicas, Carrera 1 No 18A-10, CO-4976 Bogotá, Colombia
pstevens@uniandes.edu.co

Matias Pablo Juan Szabó Universidade Federal de Uberlândia, Faculdade de Medicina Veterinária., Avenida Pará, 1720, Umuarama 38400-902, Uberlândia, Minas Gerais, Brazil
szabo@famev.ufu.br

Fritz Trillmich University of Bielefeld, Behavioural Biology, 10 01 31 33501, Bielefeld, Germany
ftrillmich@uni-bielefeld.de

Index

A

Amblyomma cajennense. *See also* Spotted fever
- abundance, 378
- amplifier host, 376
- free-living population, 378–379
- growth, 377
- horses, 377
- mammals, 373
- parasitic population, 378
- primary hosts, 375, 379–380
- tapirs, 377
- tick vector, 373, 374

Antilocapra americana, 74
Asteromys, 45

B

Birth frequency and fertility rate calculation, 151, 152
Brachyodont teeth, 43
Brazilian spotted fever. *See* Spotted fever
Brucellosis, 176–177

C

Captive breeding/sustainable management, Brazil
- captive rearing, 335–337
- commercial captive exploitation, 335
- confined production system, 335
- grill-house restaurants, 336
- human conflicts, 337–338
- leather, 334, 335
- meat and oil, 334
- population, 337
- production, 339–341
  - breeding farms, 339–340
  - contamination, 340
  - farmers, 340
  - management defense, 341
  - road traffic dangerous, 340
  - sanitary inspection, 341, 342
- semi-confined production system, 336
- vaquinha, 335
- wildlife exploitation, 336
- wildlife use, 338–339

Capy-bara Brasiliensibus, 6
Carcass yield, 212
Cardiatherium chasicoense, 47, 49, 51
Cardiatherium orientalis, 51–52
Cardiatherium paranense, 50–52
Cardiatherium patagonicum, 47, 49, 50
Cardiomys, 45
Caviodon, 45
Cavioidea, 44, 46
- capybaras, 69–70
- classification scheme, 66, 67
- genetic investigations, 72–75
- molecular phylogeny, 68
- sociality and mating system, 70–72
- subfamilies, 67

Caviomorph rodents, 387
- body size and basal metabolic rate, 62
- classification schemes, 66, 67
- hypsodont herbivores, 61
- morphological and diversity, 62
- phylogenetics
  - hystricognath rodents, 63–65
  - hystricomorpha, 64
  - nuclear and mitochondrial sequences, 63

Cecotrophe production, 264, 265
Cecotrophy, 103, 104

J.R. Moreira et al. (eds.), *Capybara: Biology, Use and Conservation of an Exceptional Neotropical Species*, DOI 10.1007/978-1-4614-4000-0,

Cheek teeth
dental structure, 48, 49
enamel microstructure, 47
morphology, 385–386
Chemical communication, 202
Coccidia, 174–175
Colloncuran, 46
Confined production system
cost, 237–239
feeds and nutrition
concentrate supplements, 268–270
corn silage, 269
cut grass, 266
elephant grass, 266, 268
feedstuff, 266
gestation period, 269
protein/energy supplements, 269, 271
fence characteristics husbandry structures, 234–235
improved, 231–233
prophylactic management practices, 236–238
reproductive husbandry regime, 230
sex determination, growth and identification system, 236
traditional, 230–231
Corporaciones Autónomas Regionales (CARs), 323, 324

D

Death
age structure, 156
life time calculation, 153–154
sex ratio, 156–157
static life table, 153–156
survivorship curves, 155
Demographic traits
agriculture, 148
birth, 148
conservation, 148
death
age structure, 156
life time calculation, 153–154
sex ratio, 156–157
static life table, 153–156
survivorship curves, 155
growth, 149–150
human health, 148
maturity, 150
mortality rates, 148, 155
population dynamics
birth season, 160–162
environmental resources, 157
growth rate, 162–163
in Marajó Island, 158, 159
in Pirassununga, 158
in Venezuelan Llanos, 158, 159
reproduction, 150–152
Dental structure
Cardiatherium orientalis, 51–52
Cardiatherium paranense, 50–52
Cardiatherium patagonicum, 50
cheek teeth, 48, 49
early Pliocene Capybaras, 52
Hydrochoeridae, 53
Kerodon, 52–53
ontogenetic tooth development, 51
Digestive adaptations
cecotrophy, 103, 104
dry season, 104–105
feeding and reingestion, 98
flexibility, 104
foregut fermentation, 97
hindgut fermentation, 97–98
morphology
dental formula, 98
diastema, 98, 99
digestive tract, 100, 101
incisors, 98, 99
stomach, 100
physiology, 101–103
symbionts fermentation, 97
Direction of Wild Fauna and Flora (DFyFS), 352
Diseases
brucellosis, 176–177
captive breeding, sanitation, 170
causes, 169–170
coccidia, 174–175
control, 169
filariae, 173–174
foot-and-mouth disease, 176
helminths, 172–173
intensive breeding systems
sanitary management, 178
scurvy, 179–180
stress, 178–179
leptospirosis, 177
rabies, 176
scabies, 171
ticks, 171–172
transmission, 170
trypanosoma, 175
Dispersal, 204
Dolichotis, 45
Dominance hierarchy, 200–202

Dominance structure
agonistic interactions, 250–252
females, 253, 254
food resources, 253
males, 254
trough feeding, 251–252

E
Ecological and socioeconomy, Argentina
anthropological records, 348
Buenos Aires province, 346
current legislation and trade issues, 349–350
del Plata Basin, 346
Entre Ríos province, 347, 348
high-quality leather, 348
hunting pressure, 348
Iberá Natural Reserve, 348, 349
land use changes, 346
oil extraction, 347
potential habitat suitability, 346, 347
production system, 350–351
trade network, 347, 348
tropical and subtropical species, 346
typical regional products, 345
wild populations, 352–353
Eocardia, 45
Euhypsodont teeth, 43

F
Feeds and nutrition
confined and semi-confined system
concentrate supplements, 268–270
corn silage, 269
cut grass, 266
elephant grass, 266, 268
feedstuff, 266
gestation period, 269
protein/energy supplements, 269, 271
digestible protein, 265
digestive system
cecotrophe production, 264, 265
energy and protein metabolisms, 262–264
grass-only diet, 266, 267
growth and reproduction, 266
microbial fermentation, 265
minerals and vitamins, 266
Female reproductive parameters
birth intervals, 278–279
birth weights, 277–278
breeding age, 279–280
breeding seasons, 276
captive breeding data, 275
litter size, 277–278
newborn and young mortality, 277, 280
sex ratio, 277–278
Filariae, 173–174
Foot-and-mouth disease, 176
Foraging strategy
cattle, 91–94
cecum, 83
diet composition
aquatic herb, 91
beard algae, 91
Cyperaceae consumption, 88, 90, 91
grasses species, 88, 90
insular and non-insular areas, 90
microhistological techniques, 88, 89
digestive efficiency, 83
effective feeding time in dry and wet season, 87
feeding habits, 84
food quality, spatial and temporal variation, 88
habitat and space use, 85–87
herbivore foraging patterns, 84
resource selection, 84
water hyacinth consumption, 88
Fossil specimens, 386

G
Galea musteloides, 74
Geographic distributions of capybara, 11, 12, 47
Argentina
Buenos Aires Province, 27–28
ecological and socioeconomy (*see* Ecological and socioeconomy, Argentina) North, Central and South, 25–26
Province of Mendoza, 25
Provinces of Santiago del Estero and Córdoba, 26–27
Bolivia, 21, 22
Brazil, 17–20
Colombia, 14–16
Ecuador, 20
French Guiana, 15, 16
Gran Chaco Region, 22–24
Guyana, 15, 16
Panama, 13
Paraguay, 24, 25
Peru, 20–21
Suriname, 15, 16
Uruguay, 24, 25
Venezuela, 14–16

H
Helminths, 172–173
Historia Naturalis Brasiliae, 5
Hunting
Brazil (*see* Captive breeding/sustainable management, Brazil)Venezuela
animal welfare, 310
breeding stock, 309
carcasses, 310–312
chronology, 311–313
culled animals, 310
dense aggregation, 309, 310
dried meat, 307
ecological convergence factors, 311
Finnish biologist Juhani Ojasti, 308
harvest, 313–314
indicating overexploitation, 314
management program, 316–318
meat uses, 311
ranches analysis, 315
salted meat, 307, 310–311
slaughter populations impact, 315–316
subsistence hunting, 306–307
Trypanosoma evansi, 318
Wildlife Service, 309
yields range, 315
Hunting techniques, 286
Hydrochoerus gaylordi, 47
Hydrochoerus hydrochaeris and isthmius Colombia
bibliographical survey, 323
biogeographic regions, 321, 322
breeding approach, 328
CARs, 323, 324
cattle ranches, 330
conservation and management, 322
distribution and abundance, 329
hunting, 326–327
INDERENA, 327
intensive breeding scheme, 327
legally hunted animals, 328
rainforests, 322, 330
scarce species, 323, 324
semi-extensive breeding scheme, 327
sustainable use, 329
visual encounter survey, 328
wild population density, 325–326
Hystricognath rodents, 41–44

I
Intensive breeding system
sanitary management, 178
scurvy, 179–180
stress, 178–179
International Commission on Zoological Nomenclature (ICZN), 6

K
Kerodon, 52–53

L
Leptospirosis, 177
Leydig cells, 386
Luantus, 45

M
Marmota marmota, 74
Meat
blood and bones, 219
capybara ceremony, 222, 224
carcass yield, 212
fat and oil, 219
flavor and aroma, 215–218
handicrafts, 220, 222–223
head, 219
leather, 219–221
leisur, 221–222
manure, 224
meat cuts, 212–214
medicinal use, 223
neotropical species, 224
organ donor, 221
processing form, 218–219
quality, 215
viscera, 219
Minimum Spanning Trees, 244–246, 257
Mus porcellus, 6

N
National Institute of Renewable Natural Resources and Environment (INDERENA), 327
Natural history
bacculum, 10
bipartite uterus, 10
body mass, geographical distribution, 7
cranium and lower jaw, 9
ecological and life characteristics, 10–11
height, 8
herbivorous diet, 9
nasal gland, 10
sweat glands, 8
weight, 7

O
Odocoileus virginianus, 74
Ovis canadensis, 74

P
Paleobiology, 53–54
Penis, 109
caudal trajectory, 111
copulation, 112
dorsal view, 114
external cranial-caudal manipulation, 111, 112
felines, 111
histological section, 114
invagination, 111
longitudinal section, 112, 113
penile bone, 112, 113
retractor muscles, 113
ventral view, 111–113
Phugatherium, 47, 49, 52
Physiographic units, 86
Placenta
amnion, 141
characteristics, 138
chorioallantoic/principal type, 139–141
conceptuses, 138
decidua, 141–142
interlobe region, 140
junction zone, 140
location and shape, 138
median sagittal section, 139
prominent region, 142
trophoblast, 139
vascularization, 142
vitelline type, 139–141
Pleistocene-Holocene period, 47
Population density method
beneficial calibration, 359
budget and time constraints, 360
catch-effort method, 366–367
decision-making process, 357, 358
direct count method, 361
distance sampling, 364
field methods, 358, 359, 361–363
hunting pressure, 360
index and control, 366
indirect signs, 366
mark-recapture, 364–365
plot sampling, 361, 364
precision and accuracy, 360
recommendations, 367
sampling error, 360
spatial scale, 359
surveys, 358
Population dynamics
birth season, 160–162
environmental resources, 157
growth rate, 162–163
monthly variations in Pirassununga, 158
monthly variations in Venezuelan Llanos, 158, 159
seasonal variations in Marajó Island, 158, 159
Protohypsodont teeth, 43

R
Rabies, 176
Reproductive anatomy, 386
Reproductive morphology and physiology
female capybara
embryonic development, 142–144
estrus cycle, 136–137
gestation, 131
hystricomorph female genital organ structure, 132
mating and gestation, 137–138
ovaries, 132–134
placenta (*see* Placenta) uterine tubes, 134
uterus, 132, 134–135
vagina and vulva, 132, 135–136
male capybara
anal sacs, 111
foreskin, 111
glandular scent marking, 108
gonadosomatic index, 108–110
intertubular compartment, 118–121
penis (*see* Penis) progeny production, 107
prostate gland, 114–115
reproductive investment, 107–108
seminiferous tubules, 121–122
spermatic production quantification, 125
spermatogenic process, 122–125
sperm production, 108
testicle structure, 115–118
testicular parenchyma, 108, 117–118
testicular positioning, 109–110
testosterone, 108
Resource Dispersion Hypothesis (RDH), 389
Rickettsia rickettsii. *See* Spotted fever

S

Scabies, 171
Scent glands
  chemical composition, 189
  morphology and histology
    anal gland, 187–188
    nasal gland, 186–187
Scent-marking behavior
  anal gland rubbing, 189, 190
  courtship, 189, 191
  functions of, 191
  in mammals, 185, 192
  nasal gland rubbing, 189, 190
  social interaction, 191
Scurvy, 179–180
Secretary of Environment and Sustainable Development (SAyDS), 352
Semi-confined production system
  cost, 237–239
  feeds and nutrition
    concentrate supplements, 268–270
    corn silage, 269
    cut grass, 266
    elephant grass, 266, 268
    feedstuff, 266
    gestation period, 269
    protein/energy supplements, 269, 271
  fence characteristics husbandry structures, 234–235
  grazing and trampling, 233
  life cycle, 233
  lowland flood plains or areas, 233
  prophylactic management practices, 236–238
  protein/energy supplements, 233–234
  sex determination, growth and identification system, 236
  social and spatial relationships
    affiliative interactions, 254–256
    agonistic interactions, 246–249
    captivity management, 259
    dominance structure (*see* Dominance structure)
    graph theory methods, 244–246
  social proximity (*see* Social proximity)
  stocking density, 233
Sexual receptivity, 387
Social behavior
  bushy scrub, 389
  demography variation, 391, 392
  dry land, 389
  ecology, 388
  floods and droughts, 390
  habitats, 390–391
  kidnapping, 387
  RDH, 389
  size and ecology resources, 390
  territory, 389–390
Social proximity
  daily activity, 257
  males and females, 257–259
  mammal species, 256
  Minimum Spanning Trees, 244–246, 257, 258
Social structure
  ecological factors, 205
  group living advantage, 200
  group size and composition, 198–200
South American Land-Mammal Age (SALMA), 40
Space and time, 196–198
Spermatogenic process, 122–125
Sperm-producing tissue, 386
Spotted fever
  ecology
    A. cajennense, 373, 374
    amplifier hosts, 373–375
    domestic dogs, 376
    horses, 377
    opossums, 376
    parasitic stages, 375
    population density, 375
    R. sanguineus, 376–377
    tapirs, 377
    tick vectors, 373
    wild rabbits, 376
  prevention of disease
    A. cajennense (*see Amblyomma cajennense*) acute febrile disease, 380
    bacterial disease, 380
    tick control, 380
    vector-borne disease, 377
  Rickettsiae, 372–373
Stratigraphic chart, 40
Subsistence hunting, 306–307
Suriname, 5, 15, 16
Sustainable harvest
  age class, 291–293
  biannual exploitation, 294
  deterministic seasonal age-structured Leslie model, 285
  exploitation system, 285
  frequency and seasonality effects, 286
  Hunting techniques, 286
  inseminate females, 289
  male-biased harvest, 288
  management modeling, 300

neotropics productivity, 299
populations
confounding survival mortality, 295–296
habitat change, 299
hunting pressure, 295
incomplete data sets, 297–298
size, 289, 290
wildlife management, 294
reproductive and population data, 285
sensitivity analysis, 289, 291
sex ratio, 287–289
social behavior effects, 293–294
Vortex 8.0 software, 285
yield, 288

T
Tamias amoenus, 74
Taxonomic history, 6–7
Ticks, 171–172
Trypanosoma, 175
Trypanosoma evansi, 318

U
Unorthodoxies, 387

V
Visual communication, 202–203

W
Wildlife trade
animal welfare impact, 398–400
Argentina exports, 395–396
biological traits, 391
Colombia's occasional export, 395–396
conservation trade impact, 397–398
legal status, 394–395
levers, 401
natural resources, 393
planting sacrificial crops, 396
traditional medicine, 396
uses, 393–394

Printed by Publishers' Graphics LLC